Networks for Computer Scientists and Engineers

Youlu Zheng

Shakil Akhtar

New York • Oxford
OXFORD UNIVERSITY PRESS
2002

Oxford University Press

Oxford New York
Athens Auckland Bangkok Bogotá Buenos Aires Cape Town
Chennai Dar es Salaam Delhi Florence Hong Kong Istanbul Karachi
Kolkata Kuala Lumpur Madrid Melbourne Mexico City Mumbai Nairobi
Paris São Paulo Shanghai Singapore Taipei Tokyo Toronto Warsaw

and associated companies in
Berlin Ibadan

Published by Oxford University Press, Inc.
198 Madison Avenue, New York, New York, 10016
http://www.oup-usa.org

Oxford is a registered trademark of Oxford University Press

Library of Congress Cataloging-in-Publication Data

Zheng, Youlu.
Networks for computer scientists and engineers / Youlu Zheng, Shakil Akhtar.
p. cm.
Includes bibliographical references.
ISBN 0-19-511398-5
1. Computer networks. I. Akhtar, Shakil. II. Title.

TK5105.5 .Z46 2001
004.6--dc21 2001050004

Printing number: 9 8 7 6 5 4 3 2 1

Printed in the United States of America
on acid-free paper

To
My parents, Fatlai and Yi Tong Cheng
My wife, Yan, and daughters, Nan and Gayle
—Youlu Zheng

To
My parents, Imad and Jamila
My wife, Kishwar, and sons, Osama, Omair, and Saad
—Shakil Akhtar

CONTENTS

PREFACE

This is a data communications and networks textbook covering both fundamental theory and new technologies. It is written for computer science and software engineering students at the junior and senior undergraduate level. However, it may be used for first year graduate students and advanced students in specialized community colleges and professional schools. In addition, the book may be used by IT professionals for performing an independent study to gain knowledge on networking fundamentals and practices. Prerequisite/corequisite courses include: data structures, programming languages, operating systems, and computer architecture. The book is unique in its coverage as it includes several network projects for a variety of platforms including Linux and Microsoft Windows/NT.

Features:

- Clear explanations and illustrations of the fundamental theory of computer networks by simple language with numerous examples, figures, and sample programs. A progressive and cumulative approach with each chapter building on previous chapters gives students confidence and a better understanding of complex concepts.
- Intensive coverage of practical applications through the use of software projects in a lab setup provides hands-on experience with network construction and programming skills.
- Particular attention to current innovative technologies enhances understanding in learned theory and allows students to grasp industry trends.

The book project was motivated by the result of an email survey that showed a strong need for a new network textbook for undergraduate students in computer science and engineering. Later, the book and curriculum development project was reviewed and unanimously recommended by a committee of five computer science professors and led by the National Science Foundation. This book is a result of numerous years of experience in academia and industry. Dr. Youlu Zheng was a tenured Professor at University of Montana from 1987 to 1996 before joining KPMG Consulting and Bellcore (now Telcordia) as Network Architect and Principal Consultant. He was involved in teaching Computer Networks, Operating Systems, and Computer Graphics courses using the material in the book, conducting network and communications related researches, and supervising graduate students at the Department of Computer Science, University of Montana. Dr. Zheng is now the General Partner of RockyTec Partnership involving investment activities in high-tech industry. Dr. Shakil Akhtar worked for Central Michigan University from 1988 to 2000 before joining Lucent Technologies where he is currently involved with the research in switching systems performance. He has taught several undergraduate and graduate level courses in Computer Networks for a number of years using the material in the book.

The book chapters are organized as follows:

Chapter 1 deals with the evolution of data and communication networks over the years. Recent changes as well as future trends in the field of computer networks have been discussed. It is shown how the telephony and telegraphy are the roots of current communication system. In addition, a fundamental perspective of the network is provided in light of protocols, standardization, and distributed system architecture.

The theory of data communications is important in understanding the computer networks. Chapter 2 covers the fundamentals of data communications including the modulation and modem technologies. In addition, the theory of wireless communication media is compared and contrasted with the several wired medias.

Chapter 3 is about the seven-layer OSI model of computer communication. The model is presented in detail and its academic importance is illustrated. In addition, the subject of network performance is introduced.

Chapter 4 introduces the Local Area Networks. LAN protocols, hardware, services, and operating systems are discussed in this chapter. The chapter uses mainly an engineering approach to explain the LAN devices and connectivity. However, it provides careful relevance to the subject matter covered in the previous chapters.

Chapter 5 deals with TCP/IP and Internet. After presenting the seven-layer model in Chapter 3, a contrast is presented here in form of a practical four-layer model used in the Internet. Internet architecture is discussed in detail. In addition routing protocols and IP are discussed in addition to many Internet services such as FTP, SMTP and TELNET. Also the transport protocols, UDP and TCP, the DNS, and the WWW application are examined.

Chapter 6 introduces the several high-speed networking technologies such as ISDN, cable modem, frame relay, FDDI, and SONET. Understanding of modern wide area networks would not be complete without an understanding of these high-speed technologies. The chapter illustrates how these emerging technologies are currently being integrated into the Internet that may give rise to better videoconferencing, voice-over IP and multimedia networking standards.

Today's switching systems are discussed in Chapter 7. It has been shown that using the switches and routers how the LANs have already exceeded the boundary of a building or a campus. Starting with the hub and switching technology, virtual LAN emulation is shown for the ATM and non-ATM networks.

Understanding of network performance and evaluation has been emphasized in Chapter 8. Several analytical as well as simulation modeling techniques are explained with many examples. In addition, the chapter examines the subjects of network monitoring and queuing theory as applied to computer networks.

Chapter 9 is about network management. Building on the previous chapter, it is shown why and how network management is important. The management protocols based on SNMP, RMON, and TMN have been discussed and explained.

Network security is addressed in Chapter 10. Several cryptography techniques are examined. Firewalls, digital certificate, and various virtual private network (VPN) structures are discussed and explained.

Chapter 11 is about network programming. Several types of network programming are addressed here, namely serial/parallel port, NetBios, socket, winsock, and RPC programming. This chapter is included at the end to allow its easier integration with the previous chapters.

Depending upon a desired coverage the text could be used in many ways that cover two semesters of content. Chapters 1–4 provide the course fundamentals. For students with a strong programming background, additional chapters may include Chapter 5, 6, 10 and 11 or 5, 7, 10 and 11. For students with a strong engineering background, the additional chapters may include Chapters 6–8 followed by either Chapter 9 or 10. For students with both programming and engineering backgrounds (such as CSE or EECS), additional chapters may include Chapters 5–8 followed by one of Chapters 9, 10 or 11. However, if the students have some knowledge and background in computer networks (i.e., in a first year graduate level), then the course may be taught using Chapters 4 to 11.

The book includes examples, summaries, review questions, a glossary, and references. An Instructor's Manual that includes a lab manual and solutions to the text problems will be available to adopting professors. The lab manual provides software and guidance for the development of software projects. In addition, an exclusive online access will be provided for instructors using the text with a set of PowerPoint presentation slides, link to web resources, errata, and solutions to textbook problems and software projects.

The authors have been supported by colleagues, friends, and numerous students in the preparation of the book. Graduate students in the Department of Computer Science at the University of Montana—Yan Zhu, Saxon Holbrook, Sean Xiaoan Hou—have contributed their course project work to the book. Professor Alden Wright has been particularly encouraging to the development of the materials for the book. Mr. Curt Taylor and John Hunt helped in the early development of the book. Special thanks also to several undergraduate and graduate students at Central Michigan University who took part in developing special projects included in the book and lab manual. They are Piyush Bhatt, Travis May, Julie Hill, Bhushan Balani, Yanjie Cheng, Wei Wang, Veeraraghavan Naranammalpuram, Chee Ming Hee, Ling Chen, Mei Li, Niroshena Jayawardena, Qingshan Xie, and several other students. Shakil Akhtar would like to specially acknowledge his colleague and friend Professor Gongzhu Hu, who is the chair of Department of Computer Science at Central Michigan University, for providing encouragement to undertake this project. Also, we are thankful to the reviewers for providing valuable guidance in the development of manuscript. We would also like to acknowledge the support and help provided by Central Michigan University and Lucent Technologies, Inc.

Special thanks go to the National Science Foundation for the sponsorship of the curriculum development project, which makes this book project possible.

1

INTRODUCTION

In 1965 a young scientist named Gordon Moore predicted that the power of the silicon chip would double every 18 to 24 months, with proportionate decreases in cost. This statement became known as *Moore's law*. In 1980 Robert Metcalfe, a coinventor of Ethernet and the founder of the U.S. network company 3COM, presented *Metcalfe's law*:

The value of a network grows as the square of its number of users.

Both claims have proven to be true. The result has been tremendous growth in both use and size of computer networks. For many of those who use networks regularly, the world has indeed become a "global village." This combination of televisions, telephones, computers, and networks continues to grow in size and function, creating the public access "information superhighway" and changing the way we live. Efficient access to information is critical to economic development in many parts of the world. The introduction of efficient, high-speed communication systems has changed the world and will continue to do so for years to come.

1.1 EVOLUTION OF DATA COMMUNICATIONS AND NETWORKS

One of the earliest telecommunications systems recorded by human history can be traced back to the beacon fires used to transmit border alarm messages between the towers on the Great Wall of China more than 3000 years ago. The use of carrier pigeons to transport important messages also has a long history. But there is quite a difference between these crude methods of sending short messages and the high-speed, data-intensive methods of today. Such progress has not come easily.

In twenty-first century technology, data communications, and networks technology in particular, play an increasingly vital role in reshaping the way we work, entertain, and educate ourselves and our children. With telecommunication and network technology tools becoming more affordable and ubiquitous, distance and time will be less important factors in our life. Progress in these areas also has not come easily.

1.1.1 Changes in Telecommunications in the Late Twentieth Century

In January 1984 a court ruling broke up AT&T and the Bell Telephone System. Seven regional Bell operating companies (RBOCs) were barred from providing long distance telephone services between court-approved geographic areas known as local access and transport areas, or LATAs. The Bell operating companies were further barred from providing information services and from the design, development, and manufacturing of telecommunications equipment. Congress also enacted legislation that prohibits local telephone companies, including the Bell companies, from competing with local cable television systems by providing video programming to subscribers in their respective telephone service areas.

In the summer of 1991 the Internet, the world's largest electronic network, lifted its decade-old ban on business use. Data communications resources that had been available only to government and educational institutions suddenly were made available to the business world. As a result, Internet use skyrocketed. Tens of millions of new users joined the Internet. The relatively open, interactive nature of the Internet made it easy for the new users to find new uses for this powerful electronic infrastructure. At the same time, personal computer technology was reaching maturity. As an ever-increasing number of PC users found it easier to use the growing power of their machines, they naturally demanded more from any data communications resources available.

In early February 1996 Congress overwhelmingly passed the controversial Telecommunications Bill, which represented an attempt to open local telephone services to free market competition. Long distance telephone companies and cable television operators would be free to offer local calling services, in direct competition with the RBOCs. The bill called for a sweeping revolution in the way information is distributed to consumers and proposed the most far-reaching and profound changes in domestic communications in the U.S. history. Among the bill's provisions are the following.

- The complete deregulation of cable television rates, allowing each company to set its own rates in exchange for its own menu of services.
- A provision for telephone companies to deliver video to homes.
- The deregulation of local phone service, opening the market to cable, utility, and long distance phone companies.
- An increase in television bandwidth, providing for digitally based programming and hundreds of new channels.
- A provision for the installation of the so-called V-chip in all new television sets, which allows parents to block the transmission of violent and sexually explicit programming
- A controversial provision that outlaws transmission over the Internet of indecent material to minors.
- Additional regulations included pricing discounts for educational institutions, rural areas, and hospitals, and new ownership laws for companies that provide multiple media services like radio and television.

The most important provisions are those that try to eliminate all the legal and regulatory barriers that prevented local telephone companies, long distance carriers, cable TV operators, and other producers of information products and services from competing in each other's markets. This spirit of deregulation has induced many other countries to consider similar divestiture policies.

What will be the consequences of this legislation? There is much speculation. The weakest link in the web of global communications is the "local loop." It is relatively easy and cost-effective to upgrade international communications by launching more satellites or by laying fiber optic cables, but it is another matter to rewire the thousands of homes connected to a local service provider. The Federal Communications Committee (FCC) wishing to make it easier and less expensive for long distance providers, cable companies, and others to compete with RBOCs, attempted to accelerate local network innovation, stimulate the introduction of more sophisticated services, and encourage telecom firms to lower prices. To support this policy, the FCC did the following.

- Let competitors buy the RBOCs' services and network capacity at 17–25% below retail prices. This measure was intended to lower competing carriers' cost of entry into the local services market.
- Allowed competitors to connect to and buy individual elements of an RBOC network so they could create sophisticated local solutions for corporate customers.
- Pegged network segment prices to current and future costs, replacing a higher price structure based on old technology and accounting methods.

1.1.2 The Progress and Future of Telecommunications

People expected that the increasing competition for local exchange company (LEC) services and carrier access services would allow long distance companies, competitive local exchange companies (CLEC), and even cable TV companies to offer local telephone service. However, competition in residential local services remains a matter of words on paper, and the general consensus is that the local telephone market was still tightly controlled by RBOCs at the turn of the century because of the RBOCs' furious resistance to competition.

However, increasing productivity along with decreasing costs, including more use of fiber optics and digital switches, has led to significantly lower maintenance requirements and significantly greater capacity. Technological progress provides many innovative ways for access to information at speeds much higher than those formerly available to the public. Cable companies are now offering Internet access over cable companies' existing networks through cable modem (Chapter 4, LAN Technologies) in addition to television services. They also plan to offer telephone services and video-on-demand. In the meantime, local phone carriers tempt consumers with a wide variety of Internet-related services, from online bill payment to online banking and shopping. On the other hand, long distance phone companies are also starting to offer data connections, including Internet access, in addition to voice.

The competition for local telephone customers is increasing as cellular telephone use and personal communication service (PCS) become less expensive, and more callers use their mobile telephones to make an increasing proportion of their calls, including long distance. According to a 1995 U.S. government report (*Economic Impact of Deregulating U.S. Communication Industries*), the cost per wireless minute of use is falling and is soon expected to be less than wireline service. Moreover, wireless has the potential to provide more than just narrowband services. Direct broadcast satellite (DBS) is already using digital communications to compete with cable television. Many technologists have proposed a low earth orbit satellite system to provide digital multimedia communications within a decade. Terrestrial, cellular-type, video/multimedia wireless networks are a likely alternative to wireline telephone networks, as well as cable TV networks.

Recent technological trends are pushing the capacity of wireless systems to expand more rapidly than the demand for mainstream telephone services, intensifying competition among wireless carriers. Thus, the price of wireless services will fall, resulting in an even greater increase in usage. A commensurate drop in wireline communication usage will occur.

Voice services are the traditional telecommunications services. The whole vast network of twisted-pair lines that makes up a large portion of the international telephone network is based on analog voice communication. Many Internet users still use analog modems and voice-grade lines for digital purposes. This is satisfactory for text and fax, but insufficient for imaging, video, and multimedia applications. A two-way video call that provides VCR quality through data compression requires a 1.5 Mbps (megabits per second) (T1) data transmission rate. This is far in excess of the capability of traditional twisted-pair telephone lines unless new electronic gears, such as high-bit-rate digital subscriber line (HDSL) are deployed. CD-ROM or high-definition TV (HDTV) quality video requires an even greater data transmission rate of 10 Mbps. These rates are simply not possible when analog methods are used.

Technological progress in digital-based optoelectronics has increased the capacity of fiber optic systems dramatically. For example, a 2.5 Gbps commercial system, commonly used today, can carry roughly 37,500 simultaneous voice calls on a single pair of fiber optic cables. The capacity of commercial fiber optic systems has been doubling about every two years by means of dense wave division multiplexing (DWDM) technology. Most experts believe that telecommunications could accomplish the same rate of improvement (50% per year), in accordance with Moore's law. Others have suggested that progress in telecommunications will increase even faster as advances in microelectronics continue. Here too, data transmission capacity may far exceed the demands of traditional telecommunications services. New services will be offered to take advantage of the increased capacity.

Traditional transmission and switching equipment has been designed to handle narrowband, circuit-switched voice services. On the other hand, data communications equipment has been developed to handle bursty packet switching. The new generation of asynchronous transfer mode (ATM) and synchronous optical network (SONET) equipment is optimized to handle all types of service, including voice, data, image, video, and multimedia traffic. With these new technologies, interactive information systems employing full-motion video and high-quality audio will be incorporated into networked applications as easily as text is today. The attractiveness of such services may well stimulate demand in ways that will dwarf current levels of interest in the Internet. With high-speed transmission and ATM technology, the wide-area network (WAN) becomes an extension of the computer, freeing the processing, storage, and retrieval functions from dependence on location. Large-scale software and programming packages could be distributed through the high-speed networks instead of CD-ROMs. Education, training, medical, marketing, entertainment, and other multimedia applications could be delivered by tens of thousands of program providers.

The potential of a global communication system based on digital data is tremendous. Speculation about its eventual makeup and implications is endless.

1.1.3 Present Solutions

Data communications and voice communications are converging. Video is no longer a one-way broadcast service. The newer and streamlined frame relay and ATM services, fast Ethernet, gigabit Ethernet, and IP-based streaming media technologies are being widely deployed. Broadband versus narrowband, video versus telephony, one-way versus interactive: each of these creates complex choices between today's demands and those of the future. In some ways, the network and communication services provider is expected to build for a future, which the market itself has not yet defined. To some extent, the next century is easier to predict than next year.

Someday, maybe, there will be an all-digital global network. But today an integrated solution transporting both analog and digital data is most cost-effective. Traditional copper wire is giving way to fiber optic cables, but the two systems must coexist for years. This constantly changing industry is particularly demanding for network architects, who must make cost-effective decisions. The industry is moving, and the one standing still is the loser even if the finish line is, as yet, undetermined.

A winning approach in the network industry is the one that gives an end-to-end, future-proof solution that ensures maximum utilization of bandwidth and the existing infrastructure, while providing for virtually unlimited expansion to accommodate future communication needs by either upgrading the existing plant or expanding into new platforms. The greatest challenge to network designers is not just to provide service, but to do it cost-effectively: to provide maximum service without overequipping or overspending, to reduce the costs of provisioning and maintenance, and to have exactly the right component for every job, be it large or small.

Meeting that challenge in the real world may be next to impossible, but the goal remains a worthy one.

1.2 TELEPHONE SYSTEMS AND COMPUTER TELEPHONY

The telephone has been and will be, for a long time to come, the most widely used telecommunication device. Telephone systems also play an important role in data communications because of their ubiquity and low cost in comparison to leased lines. No meaningful discussion of data communications and networks is possible without detailed knowledge of today's telephone systems.

For years, many people thought about telecommunications solely in terms of telephone systems. Many computer scientists and engineers, on the other hand, thought that the technologies of computing and telecommunications were merging and that the computer business was like the telephone business.

They have been shocked and disappointed!

Historically, no computer company has been successful in telecommunications, including Big Blue, IBM, which tried years ago but failed to deliver profitable PBX (private branch exchange) products. The telephone industry has scored even worse. AT&T, which owned Bell Labs and the powerful Unix operating system, bought NCR, but had to spin off the computer company with a big financial loss.

Because telecommunication always involves many parties far beyond the categories of technological designers and manufacturers, telephone companies face extremely heavy regulations, much government involvement, and other difficulties that firms in the computer industry may avoid. Service providers, government regulators, and international organizations all must agree in numerous matters before any telecommunications can actually happen, inevitably retarding both business and technical progress. There are also many differences, particularly cultural differences, between the computer and telephone businesses. As computer telephony integration begins making inroads into our life, some comparisons between the two are interesting:

Price: Computer hardware and software prices keep dropping about 20–30% each year with increased computing power. The fee for local telephone services, on the other hand, has remained flat, even slight increasing for years, without much improvement in services. Although long distance prices have dropped dramatically as a result of competition between long distance phone carriers, the reduction in price can never match that for computers.

Competition: Computer businesses have always faced cutthroat competition from both domestic and international rivals, whereas the telephone industry, local telephone companies in particular, historically had a captive customer base. Moreover, it was necessary to please only one customer—the local regulatory agency.

Technical innovation: Since the first computer was constructed in the 1940s, both hardware and software have experienced exploding innovations. A desktop PC today is much more powerful than a mainframe computer just 10 years ago, at about one-hundredth the cost. However, the telephone industry has been selling the same equipment and service—an analog phone, twisted wire, and 3 kHz bandwidth—for each phone line for 120 years. It is a common practice for computer companies to hire executives from outside industries. Until recently, however, almost no high-ranking officers in the telephone industry have come from outside the traditional circle.

Training and education: Computer books can be found everywhere, and many computer-related training courses, seminars, and certificate programs are available to the public in very low cost. The telephone industry, on the other hand, has kept its end users in the dark, probably not by accident.

Such apparent differences, at least partly, led to the deregulation of the telecommunications industry, which is generating far-reaching effects.

1.2.1 Telephone Functioning

An analog telephone, a *teleset* in telecommunications jargon, consists of two major components: a handset, or receiver (Figure 1.1) and a dial pad. The handset contains both a receiver (Rx) for listening and a transmitter (Tx)—a microphone. The ringer sounds when there is an incoming call; the hook switch is open when the telephone is on the hook and closes when the handset is lifted (off-hook). A volume control switch and/or a mute button may also present on the handset. The handset may be hardwired to the dial pad, or combined with the dial pad and connected through radio signals to the telephone plug.

The dial pad, is also called the keypad, or touch-tone button pad. The technical term is DTMF (dual-tone multifrequency) pad.

The old-style rotary dial pad uses the loop current to send different number of pulses to represent the 10 digits. This type of dial pad has been mostly replaced by the newer

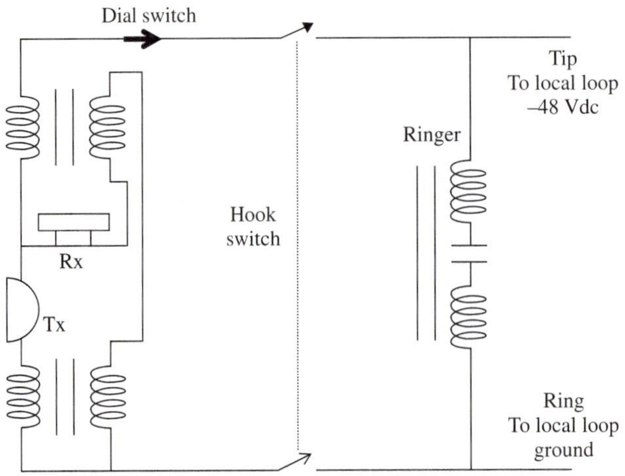

Figure 1.1 Simplified schematic of a telephone handset.

DTMF buttons, which use sounds to represent 0–9, #, and *, all of which are called digits. Each digit is assigned a unique pair of frequencies, hence the name DTMF digits. Pressing one of the buttons generate a pair of tones—a low-frequency tone for each row and a high-frequency tone for each column. (Figure 1.2). Since there are four rows and three columns in a standard dialing pad, the seven (3 + 4) frequencies can represent twelve (3 × 4) frequency combinations. For example, pressing button 5 generates 770 Hz lower tone and a 1336 Hz higher tone. (Figure 1.3) The frequencies and the dial pad layout have been internationally standardized, although slight variations in frequency may be tolerated.

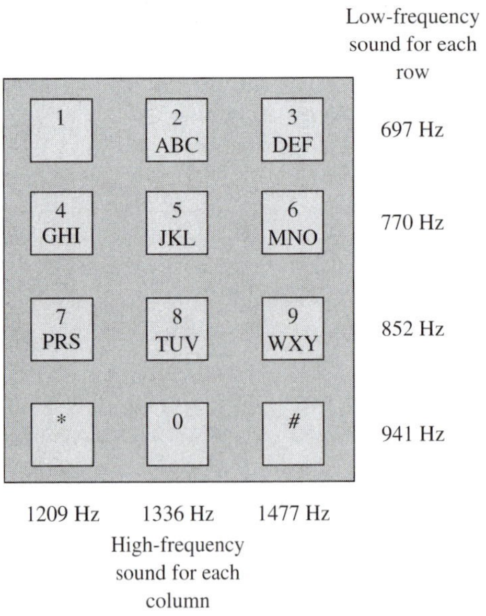

Figure 1.2 A standard DTMF touch-tone pad consists of 12 buttons that can generate 12 combinations of low- and high-frequency tones.

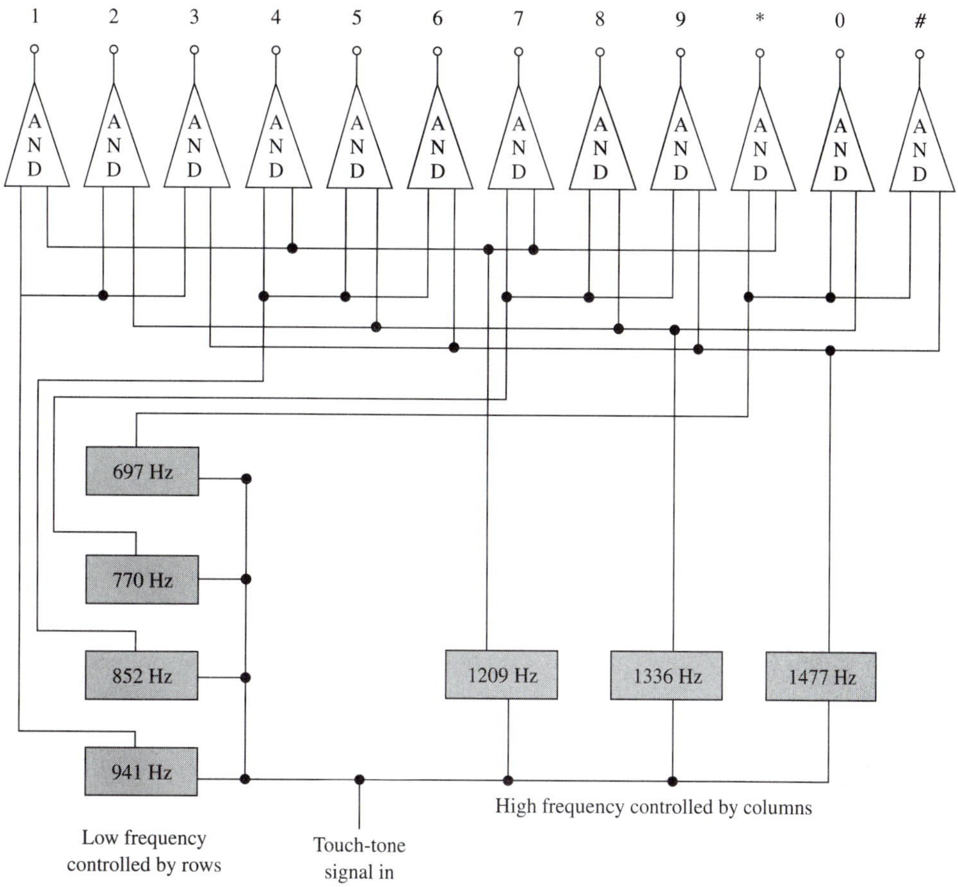

Figure 1.3 A DTMF decoder circuit has 12 AND gates to produce 12 dual tones representing the 12 bottoms on a telephone dial pad.

The tone dialing structure is much faster than the pulse tone structure, and the touch-tone signals can be easily converted into digital form and transmitted efficiently by most switches. The DTMF pad is used not only to dial telephone numbers, but also to interact with interactive voice response systems and voice mail. The pulse dialing signals, on the other hand, are difficult for voice processing equipment to recognize.

When a phone call is made, even the least expensive telephone will handle the following eight functions:

1. When the handset is picked up, or the "talk" button is pushed, a signal is sent to the telephone local switch to request the use of a telephone line.
2. A dial tone is received to indicate that the system is normal and ready for the call.
3. The telephone number is sent, by either touch tone or pulse tone.
4. A ringing or busy signal is sent back from the local switch that directly connects to the called phone.

5. The telephone rings to indicate an incoming call.
6. The speech of the person answering one call is first transformed into electrical signals for local loop transmission. A channel bank then converts the analog signal into digital signals to be transmitted through a digital trunk that may include fiber optics. And finally, somewhere near the called phone, the digital signals are converted back to analog and received by the called telephone.
7. When the telephone is hung up, a signal is sent to both the local switch and instrument of the called party.
8. The telephone is powered by the telephone line, and it can adjust for power fluctuation automatically.

1.2.2 Telephone Networks

Telephones are connected to the telephone company's nearest exchange, called the central office (CO), or end office, by means of a cable containing two conducting wires, called ring and tip (Figure 1.4). The CO also provides power to the local loop and to the telephone itself. The local loop sometimes is also called subscriber line. When the handset is taken off-hook, the circuit is completed, loop current flows, and the CO responds by generating a dial tone. When an incoming call occurs, the CO switch applies ac voltage to the line, causing the phone to ring.

Electronic switches in the telephone network cross-connect the physical circuits to other physical circuits so that complete connection is made throughout the network. This is called circuit switching. In a CO, channel banks will likely convert the analog signals to digital and assemble many telephone channels (most likely 24) into sets of T1 channels. The signals are then transferred through a toll connecting trunk to a toll office. Thirty-two T1 channels may be combined into a T3 intertoll trunk by means of fiber optics. The intertoll trunk connects with another toll office, with intermediate switching offices in between if necessary. Then the digital signals are converted back into analog speech signals (Figure 1.5). Terms like channel bank, circuit switching, T1, and T3 are discussed in more detail in Chapters 2 and 3.

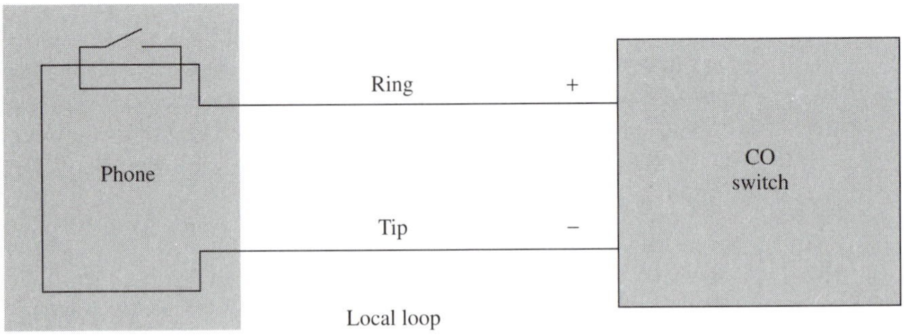

Figure 1.4 A telephone is connected to the local telephone company's CO by a ring wire and a tip wire.

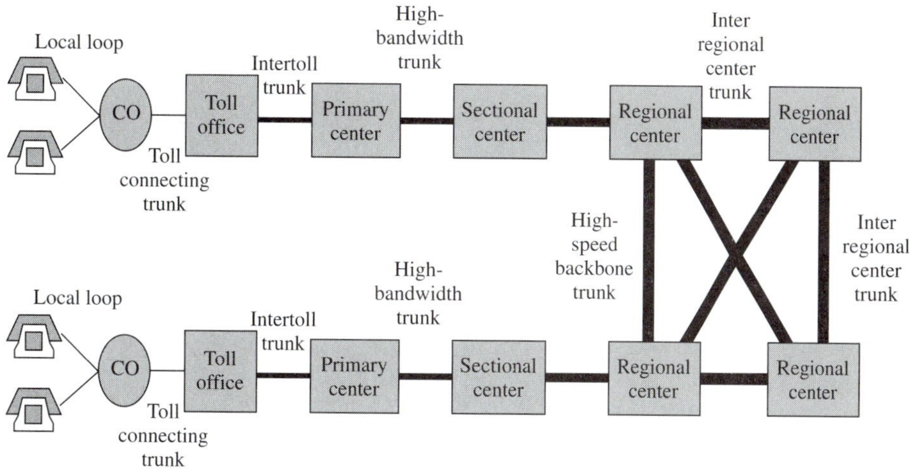

Figure 1.5 A telephone line comprises a local loop, a toll connecting trunk, an intertoll trunk, and many switches.

In the United States there is a five-level hierarchy of switches. At the bottom are the CO switch, or the end office, which directly connects to the subscriber lines (the local loop), and a toll center (or toll office), which may connect many COs together and, in turn, connects to a higher level intermediate switching office. Further levels are primary center, sectional center, and regional centers.

The regional center is the highest level switch center. There are 10 regional centers, which are completely connected; that is, each regional center is connected to each of the other nine by a high-speed, high-volume trunk, making a total of 45 links. A telephone is connected to only one CO through a local loop. Each of the four lower levels is connected in a tree or star topology. Generally speaking, the CO is connected to only one toll office, and so on, and a switch in the middle level is connected to only one next higher level switch. However, a tandem switch that connects many switches in the same level may be introduced to reduce the number of interoffice trunks required.

When a call is made, a path through the telephone network is established at the lowest level possible. If both telephones are attached to the same CO, a direct connection will be established through that CO. If they are attached to different COs but both COs are directly connected to a common toll office, the call will be routed through that toll office, and so on.

Before 1984, the Bell System provided both local and long distance services covering most of the United States. When the Bell System broke up in 1983, the seven RBOCs then known as Nynex, Bell Atlantic, BellSouth, Ameritech, Southwestern Bell, US West, and Pacific Telesis, along with a few independent telephone companies, notably GTE and Centel, became responsible for local telephone services. The twenty-two BOCs are also referred to as Incumbent Local Exchange Carriers (ILECs). AT&T, along with MCI, Sprint and a few smaller companies—all called **i**nter**ex**change **c**arriers (IXC), remained responsible only for long distance services.

Under RBOCs, there are about 160 local access and transport areas, which enable the RBOCs to offer long distance services, or intraLATA calls within a LATA through the local exchange carrier (LEC). A LATA covers an area about the size covered by an area code. interLATA calls are handled by IXCs through a switching office called a point of presence (POP) within LATAs. The LEC is required to connect each IXC to every CO, for a fee. Since LATAs are monopolies, it is not uncommon for intraLATA calls to cost consumers more than some interLATA long distance calls.

On the other hand, a Competitive Local Exchange Carrier (CLEC) may compete with the already established LECs by providing its own network and switching. Many CLECs, however, complain that the main hurdle to expanding consumer choice is that RBOCs, who own the local lines to homes (also called "the last mile"), charge competitors too much to access these lines. Consequently, MCI Worldcom, AT&T, and Sprint have already exited from the residential local business. AT&T acquired TCI in 1999 and started AT&T Broadband and Internet Services offering high speed Internet and cable services. Sprint, Verizon and many other smaller companies started to offer Internet access via wireless and the battle for the "last mile" is still ongoing as of this writing.

1.2.3 Telephone Numbering

A complete telephone number includes country code, area code, and local number. The area code identifies the region. In North America and many places elsewhere in the world, the area code is three digit long, and there used to be some rules for the area code and the seven-digit local numbers, notably:

- The area code could have only 0 and 1 as the second digit.
- The seven-digit local number would never have 0 or 1 as a second digit.

This scheme, called numbering plan area (NPA), allowed a switch to distinguish an area code from a local number easily by verifying the second digit. This resulted in shorter call completion because the switch could start routing local calls, which, of course, occur much more frequently than long distance calls, right after the seventh digit has been received, without waiting for more digits. Under the NPA, however, there were fewer than 200 possible area codes, and the available three-digit prefixes for local numbers were significantly reduced. Due to the increase in demand for phone numbers for fax service, computer modems, pagers, and cellular phones, these limits were modified in mid-1990s. Now both area codes and the three-digit prefixes of local numbers are represented as NXX:

$$N = \text{any digit from 2 to 9}$$

$$X = \text{any digit from 0 to 9}$$

Therefore, in some regions, the full 10 digits must be entered for nonlocal calls within the same area code.

Certain area codes are reserved for special services. For example, area code 700 may be defined by the long-distance companies for special purposes. The codes 800, 888, and 877 are used for toll-free numbers. Premium rate or pay-per-call numbers begin with 900.

Formerly, users had to register a separate toll-free number for each country. But since early 1997, it has been possible to apply for a single, international toll-free number that can be dialed from anywhere in the world. The International Telecommunications Union (ITU) has established a standard for Universal International Free-phone Numbers (UIFN)—E.169. The new standard uses eight digits, instead of the current seven, after the number 800. Under standard E.169, companies based in the United States can choose global toll-free numbers that are similar to their existing toll-free numbers. A company can choose a number *X* (any digit 0–9) and affix it to the beginning or end of the seven-digit portion of its North American 800 or 888 number. For example, the company *Network* may pick 8 and place it at the beginning of its toll-free number, making its UIFN number (800) 8-NETWORK. But another company may use (800) NETWORK-8 for its own UIFN toll-free number, introducing potential confusion. When a foreign caller dials a country code and 800 or 888 followed by eight digits, the global switched public network will route the call to the correct U.S. carrier, which will deliver it just like a domestic toll-free call.

In the United States, the local number is always a seven-digit number. The first three-digit prefix in *NXX* code, with x00, x11, and 555 excluded, specifies the CO switch the line is connected to. A single switch may have more than one *NXX* code assigned. That is why you can keep the same phone number if you move within an area served by the same switch. The last four digits comprise the subscriber line ID. The local number 555-1212, with any area code including 800 and 888, is for directory assistance.

The demand for phone numbers continues to mount since more users are installing second phones and businesses are growing. As local phone markets open to competition, demand grows even more quickly for two additional reasons:

First, the blocks of numbers each carrier reserves are nontransferable. If a carrier does not use all the numbers in a block, they sit idle.

Second, most business would not consider changing local service providers if they had to change phone numbers. Under local competition rules, users who switch to a different carrier must be allowed to keep their old phone numbers. While it appears that the number follows the subscriber to the new company, the underlying method is actually a form of *call forwarding*—calls to the old numbers are forwarded to a second number held by the new carrier. That second number is invisible to the user but nonetheless cannot be used for any other purpose. The additional switching can delay the call by up to 1.2 seconds and makes it impossible for the new provider to offer some features.

Call forwarding is a temporary solution for letting subscribers keep familiar phone numbers. The industry is working on permanent solutions that will directly route calls to new service providers' switches, thereby eliminating the restrictions of the remote call-forwarding method. The one of the new methods is based on the use of either Intelligent Network (IN) or Advanced Intelligent Network (AIN) 0.1, in combination with a local number portability database for the routing information necessary to terminate calls. With this method, each provider is assigned a unique three-digit carrier portability code (CPC) for every NPA (numbering plan area, or area code), where service is provided. The CPC is stored with the subscriber's directory number in the local number portability database and replaces the NPA during call routing. The CPC solution requires minimal switching software modification and offers transparent end-user feature operation.

In many countries, area codes vary in length: in United Kingdom, area codes are two digits in London, but three digits in Sheffield. Singapore and Panama do not use area codes. The length of local numbers may also vary from country to country, mostly depending on the number of telephone lines in a particular region.

With the passage of the 1996 Telecommunications Bill, the shape of the telephone system in the United States is changing. Eventually all U.S. consumers will be able to choose their local telephone company the same way they choose their long distance carrier. Cable companies will start the one-stop shopping digital services that provide digital telephone with complete voice, data, and video-on-demand service. Other innovations in telephone services include personal communications services (PCS) or personal communications network (PCN) (covered in Chapter 2), and follow-me phone numbers, such as the personal 500 numbers offered by some long distance carriers.

A follow-me number enables a call to be automatically routed to a variety of physical locations and communications devices such as traditional or cellular phones, fax machines, pagers, laptops, PDAs, or notebooks. Two technological approaches are currently used for these services: new dialing plans, such as 500 numbers, or software enhancements that add follow-me features to telephone numbers. Users can be reached via a single number, no matter where they are. Follow-me numbers also allow receivers to reroute calls to other people or locations.

With 500 number services, the system can sequentially search different phone numbers or locations to find users in a specified order. With this requested service, subscribers will be able to preprogram the telephone network to route calls automatically to different locations at certain times of day. Currently, 500 number services are handed by different long distance carriers, cellular service providers, and local phone companies, who must program their own networks with information on where a call should be terminated. The numbers are assigned in 10,000-number blocks. Each block begins with an NXX three-digit code. For local carriers to know which 500 provider should receive a given call, each NXX is assigned to a specific long distance carrier or other provider. For example, AT\&T has been assigned the 263 code. Any call that begins with 500 263 is hence handed off to AT\&T for termination.

1.3 COMPUTER TELEPHONY

Telephones are ubiquitous. For decades, they have been prevalent in homes, offices, public building, and everywhere people go. Computers and networks are revolutionizing the way the world communicates. However, even though a telephone and a computer often sit side by side on a desk, they historically have remained separate technologies.

Computer telephony is a new wave of technology that connects the world with the best of the two technologies, which allow people to exchange information more conveniently and efficiently. Computer telephony can store, retrieve, and manipulate computer-based information and control many devices over a telephone network. The power that drives the computer telephony industry is universal telephone access to computer information by convenient and easy-to-use terminal devices, including telephones, pagers, fax machines, and personal computers via wire or wireless telephone lines and the Internet.

1.3.1 Overview of Computer Telephony

Since early 1980, the computer telephony industry has grown to encompass many diverse and advanced multiuser computer telephony applications, technologies, and services, as well as development tools. Computer telephony systems can range from simple voice mail to multi-media gateways. The equipment used in these systems includes voice response units (VRUs), fax servers, speech recognition and voice recognition hardware. Computer telephony applications, services, and development tools generally fall into the following four categories.

1. Information access and processing applications, including audiotex, fax-on-demand, interactive voice response, interactive fax response, and simultaneous voice and data.

- **Voice processing** is the fundamental technology at the core of most computer telephony systems. It may incorporate both the processing and the manipulation of audio signals in a computer telephony system, including digitizing, filtering, analyzing, compressing, expanding, recording, storing, and replaying signals.
- **Audiotex** and **interactive voice response (IVR)** provide prerecorded or synthetic information to callers and provide callers with a single message or a choice of messages through touch-tone or automatic speech recognition (ASR).
- **Fax server and fax-on-demand** service allows users to send documents to a large number of people in a very short period of time. Documents that reside on a fax server can be retrieved on demand.
- **Automatic speech recognition (ASR)** or **voice recognition** technology recognizes certain human speech, such as discrete or continuous strings of numbers and short commands. Speaker-independent ASR can recognize and identify a limited group of words from any caller for a password-controlled systems and hands-free work environments.
- **Simultaneous voice and data** system can process voice and data simultaneously in an application, which greatly enhances the quality of communication for such collaborative applications as technical support, training, and group reviews.

2. Messaging applications, including voice mail, paging, unified messaging, and electronic mail readers.

- **Voice mail** allows telephone users to record, store, and manipulate spoken messages. Organizations can easily distribute recorded information to a large audience at any predetermined time.
- **Unified messaging** lets users access messages of many kinds from a single point. At the computer, one can point and click on the screen to check for electronic mail, fax, and voice messages. Conference calls are easily set up by clicking on screen icons and choosing from a list of names.
- **Text-to-speech (TTS)** technology generates synthetic speech from text stored in computer files. TTS provides an audible interface to frequently updated information and information stored in computer databases.

- **Electronic mail readers** reside on a media server that uses TTS technology. E-mail readers translate the ASCII text of an e-mail message into voice to be retrieved by callers through a telephone.

3. Connectivity applications, including call center and help desk automation, international callback services, operator services, conferencing, automatic telemarketing, Internet telephony, and predictive/auto dialing.

- **Automatic connection systems** can handle person-to-person calls that used to require a human operator. This type of computer telephony system can route a call from a main number to an extension and verify that a called party will accept collect call charges. Computer telephony interface boards with related software can be integrated with a PC-based computer telephony system that works with all the popular telephone networks around the globe. A computer telephony system with switching and conferencing technology can handle the routing, transfer, and connection of more than two parties in a call, thus providing an inexpensive alternative to *private branch exchange* (*PBX*). The switching-based conference call system typically includes a conference bridge that allows people to call into a central number, then switch over to a party line for their particular conversation.
- **Predictive/auto dialing** stores phone numbers in a central database, automatically dials numbers, and transfers each call to an available agent when the call is connected. Some sophisticated systems can determine when to place the next call based on the volume and average length of calls.

4. Central office (CO)/advanced intelligent network (AIN) applications leverage the capabilities of public and private telephone network providers. These capabilities include sophisticated call routing, switching, access to database systems, and automatic call billing services. CO/AIN applications offer subscribers a wide variety of services, including central office voice mail, voice-activated services, and enhanced services for wireline, wireless, and cellular networks.

The term private branch exchange (PBX) applies to larger business phone systems that provide internal telephone services function like CO switches. Typical PBX functions include in-house dialing, dialing outside lines, and most of the additional common telephone services such as call forwarding, call transferring, and conference calls. A PC-based voice response unit (VRU) is a simple computer telephony system that can be integrated into many different telecommunications environments and perform PBX functions. In addition, it can provide voice mail, voice processing, and responding services. Such a computer telephony system also can be easily integrated into a LAN.

A simple PC-based VRU can be built around a voice board, or speech card, attached to a personal computer bus (AT, EISA, Micro Channel, VESA/PCI local bus, etc. for PCs). The voice board includes the following components:

- A telephone line interface, which may be RJ-11 or RJ-14 jacks, a digital T1/E1 interface, or a bus interface (this may require an additional board connecting with a phone line).

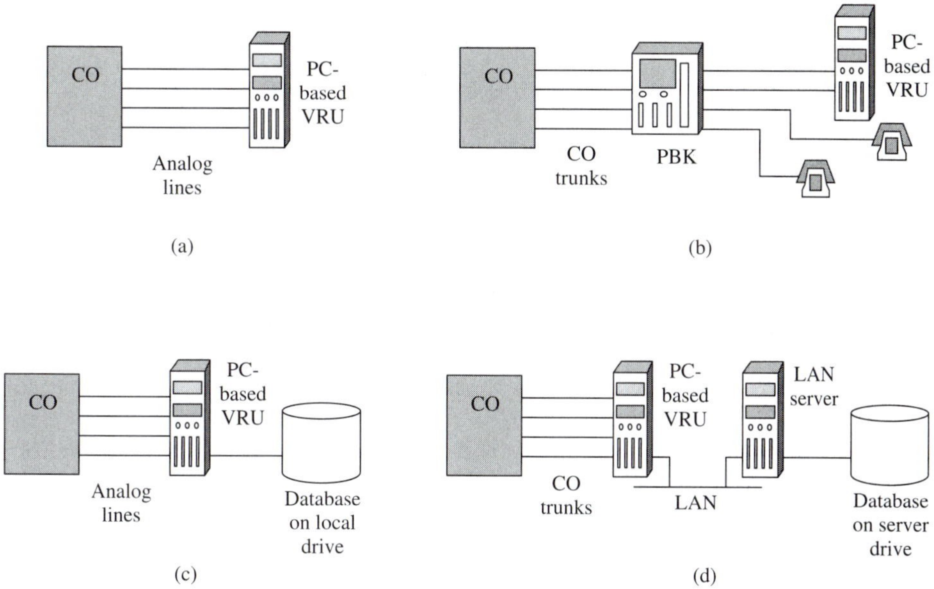

Figure 1.6 The commonly used PC-based computer telephony systems. (a) The simplest standalone VRU configuration. (b) A PC-based VRU is connected as one or more extensions behind a company PBX. (c) A simple IVR system consisting of a VRU plus a database on a local hard disk. (d) The IVR system with database on a LAN server drive.

- A voice bus, independent of the computer I/O bus, through which audio and controlling signals can be passed among several voice boards and different voice processing components.
- Device drivers, which provide an interface between the voice board and the operating system. Usually the software device drivers come from the voice board vendor.

Figure 1.6 shows the configurations of some commonly used PC-based VRU and IVR systems.

Until recent years, computer telephony applications required intensive expertise about telephone interconnection processes as well as networking and database programming to perform even the most rudimentary tasks. These two skill sets are rarely available to an individual who is trained as either as a computer scientist or an electrical engineer. With personal computers becoming more powerful and less expensive, more and more advanced computer telephony systems can be integrated into an inexpensive PC, providing endless applications and services in daily life. Many standard application program interface (API) tools lower the cost and complexity of computer telephony applications. Some actual computer telephony applications include the following.

- Bank customers use telephones to access personal account information stored in the bank's database. And a stock brokerage firm can provide stock information for investors and take stock trading orders from them.
- Business customers can use the telephone to receive information about a product automatically, through a fax machine.
- A person in a foreign country who initiates a call to the United States may be able to activate the international callback service to take advantage of the lower U.S. rate.
- A customer using an application can click on a help button that automatically dials the application's call center hotline with a simultaneous voice and data system. While the customer is describing the problem to the support engineer, the application can simultaneously transmit a copy of the customer's screen over the same phone line to the engineer. In response, the engineer can send a new configuration file to the customer while simultaneously explaining the solution to the problem.
- Employees can access computer-managed voice, fax, and even data (text messages and other information) through telephones, computers, or both, to effectively connect off-site workers to the office and to expand relationships with the external community.
- Many household tasks in a high-tech family residence may be managed automatically through a computer telephony system.

1.3.2 Internet Telephony

The TCP/IP protocol suite and Internet (Chapter 5) technology provide a very cost-effective way of handling telephone calls. The Internet Protocol (IP) telephone system, or Voice-over IP (VoIP), provides inexpensive long distance service without the need of computers for the caller and receiver. The caller dials a local number using a regular analog phone and is connected to the Internet telephone service provider's computer. The voice signals are digitized, converted into datagram frames and packets (discussed in Chapter 3), and transmitted through the Internet to the server of another Internet telephone service provider near the receiver. The frames are reassembled in sequence and converted back into voice signals to be sent to the receiver's phone via a regular telephone local loop (Figure 1.7).

The anticipated convergence of voice, data, and video networks will establish a new world order in the network and telecommunication industry. IP may be the vehicle that finally delivers the voice/data convergence. Many telecommunication and data communication companies are currently building large IP telephony networks; cable companies are also developing IP telephony standards. The sound quality of VoIP network services almost equals that of the public switched telephone networks. The sound delays on international private IP networks average less than 0.1 second, which is well below the threshold of perception. By using the public Internet, international private data networks, or leased transmission lines, IP calls avoid the huge termination fees (as much as a dollar per minute) charged by foreign carriers. The great ease and reduced cost of providing multimedia services over IP networks combined with cost savings may eventually tip the scales in voice-over IP's favor against the expensive conventional telephone switched net-

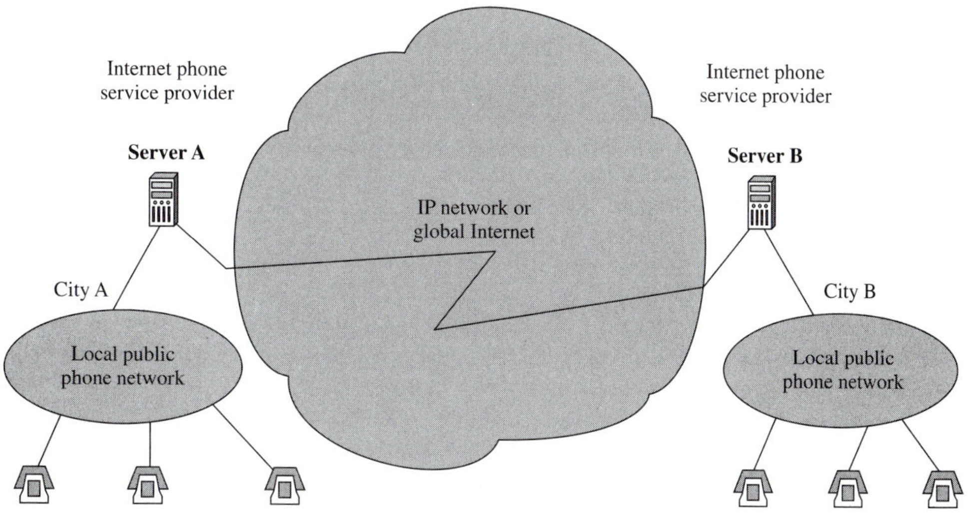

Figure 1.7 Internet telephone service providers take advantage of free access to the long haul global Internet.

works. How can IP telephony increase efficiency and reduce communications costs? The reasons are multifold.

- Mixing different kinds of traffic is the natural way to provide multimedia services. Traditional networks struggle to provide multimedia services, and the expensive equipment makes upgrading slow and expensive. Substituting a general IP network for many specialized networks should lower costs. IP telephony's greatest appeal may be that it allows a single IP network to carry all voice, fax, video, data, and Internet communications.
- Because IP telephony uses packet switching instead of circuit switching (Chapter 3), which requires full circuit occupancy, it lowers transmission costs by reducing the capacity required for a high-quality phone call from 64 kilobits per second (Kbps) to as little as 12 Kbps. Thus the same number of calls can be carried at one-fifth the cost.
- IP networks offer big savings in switching costs. An intelligent open architecture IP switch may cost considerably less than a telephone switch such as 5ESS.
- Network costs constitute a major barrier to entry of telecommunication business. VoIP greatly reduces the network costs for smaller carriers that cannot achieve the economies of scale enjoyed by large carriers. Thus by reducing start-up costs, price competition is encouraged.
- IP networks are replacing expensive nonintelligent and proprietary hardware solutions with inexpensive intelligent software solutions, thus making it easy to add new code and upgrade the entire system.

There are obstacles to be overcome, of course, before Internet telephony achieves universal availability. For example, reliability presents a problem for Internet phone systems and is also a big marketing issue. An ordinary telephone will connect with the number being dialed unless the line is busy or damaged. The Internet, however, is a "best-efforts technology" (see Chapter 5), and depending on the amount of network traffic, it might or might not be able to complete a call to 911. Such failures would be rare, but the possibility of their occurrence highlights the reliability issue.

1.3.3 Unified Messaging

For most traveling professionals, staying in touch is a chore that requires them to carry communications gadgets ranging from cell phones and pagers to palm-sized computers and personal digital assistants. But increasingly, companies are striving to simplify the task of connection. The so-called unified messaging systems allow people to send and receive electronic mail, faxes, and phone messages whether they are using a computer or a telephone (Figure 1.8). Unified messaging is the latest effort in a field that is slowly picking up steam, and is being joined by telephone equipment companies, start-up communications companies, software makers, and voice mail providers.

Phone users can dial up a unified mailbox to retrieve voice messages as well as e-mails. They can also respond to e-mail by sending an e-mail or fax message, or by recording a voice message to be played on a recipient's PC via audio. Likewise, a PC user can open up a mailbox to access not just electronic mail, but faxes and voice messages.

Combined with electronic mail technology, voice processing technology, and powerful server computers, a unified system stores and manages the unified mailboxes, translates text into voice, and provides voice mail to its clients everywhere. Unified messaging systems will be the first to address the large-scale needs of telephone and Internet companies that typically handle millions of calls and messages each day.

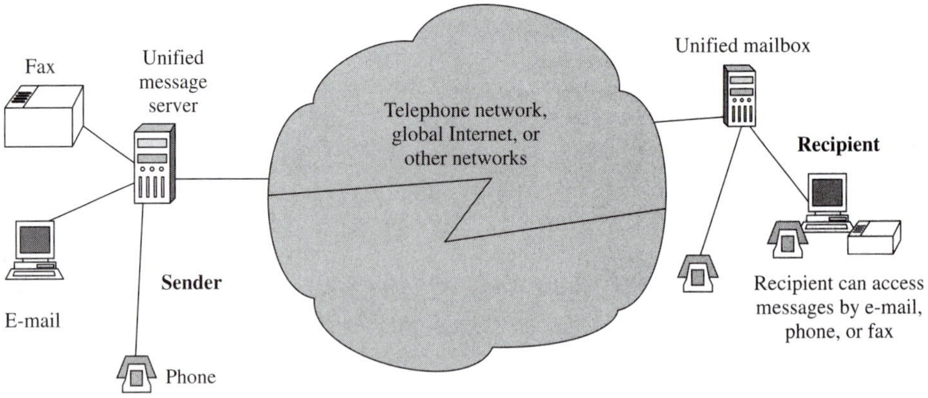

Figure 1.8 Unified messaging systems allow people to use a phone or a computer to send and receive e-mail, voice mail, and faxes through a unified message server and unified mailbox, which can translate text into voice.

1.3.4 Network Structures and Network Topologies

A computer network is an interconnection of computers, peripherals, and data transfer devices with related software that performs certain data communications functions.

- The types of computer associated with a network range from microcomputers through mainframes and supercomputers. Generally, a host computer in a network is referred as a *node*. The computer peripherals include storage devices such as an array of discs, CD-ROM and tape drives, printers, plotters, fax machines, and telephony devices.
- The data transfer devices are responsible for digital signal transmission, amplification, data switching, and routing functions. They include repeaters, bridges, routers, switches, modems, and various data transmission devices such as transponders on a satellite.
- Network software can be part of the host computer operating system, special network operating, controlling, management, and security software (firmware) residing in a host computer or in the data transfer devices.

Typical functions of computer networks include the following:

- Communications among two or more computers or network devices
- Sharing computer peripherals and other equipment
- Sharing information including software

A computer network may have a point-to-point connection or broadcasting/multicasting connections (Figure 1.9). A point-to-point connection may be implemented through a bridge or a high-speed modem. Or, the connection can be established by means of a series of connections between individual pairs of computers, hub, routing, or switching devices. An individual transmission is sent to the routing or switching device wherein the switch or router will determine where to forward the data. A broadcasting/multicasting network, in contrast, has a single communication channel that is shared by all the network devices in the network. Any message transmitted by a device will be received by all (or multiple) devices in the network. A switch can send traffic to all parties in a switched network, but this may not be economical.

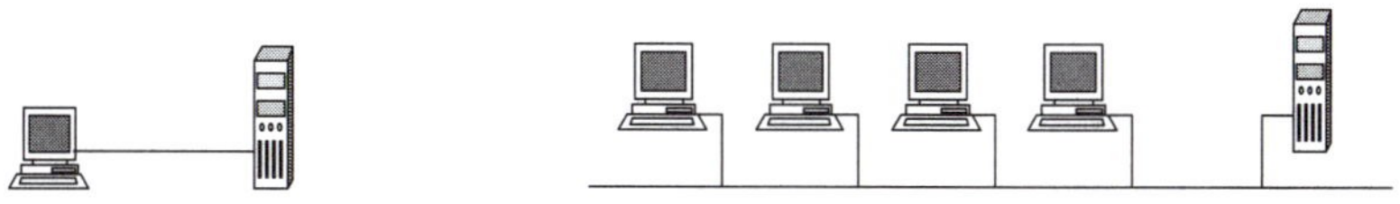

(a) (b)

Figure 1.9 Direct links: (a) point-to-point connection and (b) broadcasting or multicasting connection.

Networks are classified by the size and distance between the network devices as follows.

- A local-area network (LAN), which may span a few hundred meters distance and is confined to the same or near-by building. LANs almost always have a broadcasting or multicasting connection. A LAN offers extremely high-speed communications among network devices ranging from 10 Mbps to 1 Gbps.
- A metropolitan-area network (MAN) may cover the whole campus of an organization with many buildings, or an entire metropolitan area. A MAN often has a mixture of multiple point-to-point and broadcasting/multicasting connections.
- A wide-area network (WAN) usually connects remotely located devices as far as a hundred kilometers to nationwide or even worldwide, mostly through point-to-point connections. A WAN often interconnects multiple LANs. A WAN's speed is decided by its uses, ranging from 56–64 Kbps for a leased line connection to multiple gigabits per second for a network backbone connection through fiber optics. Cable, radio, microwave, and satellite communications are all often involved with WANs.

A network with three or more nodes connected together is called multipoint network. The physical layout of a multipoint network is called the network's topology. Network devices can be connected in different topological forms. Figure 1.10 depicts the most frequently used network topologies: star, ring, mesh, bus, and tree.

The bus and ring topologies are most often seen in a LAN environment, while star, tree, and mesh topologies are used more often in MANs and WANs. In a 10BaseT or to-

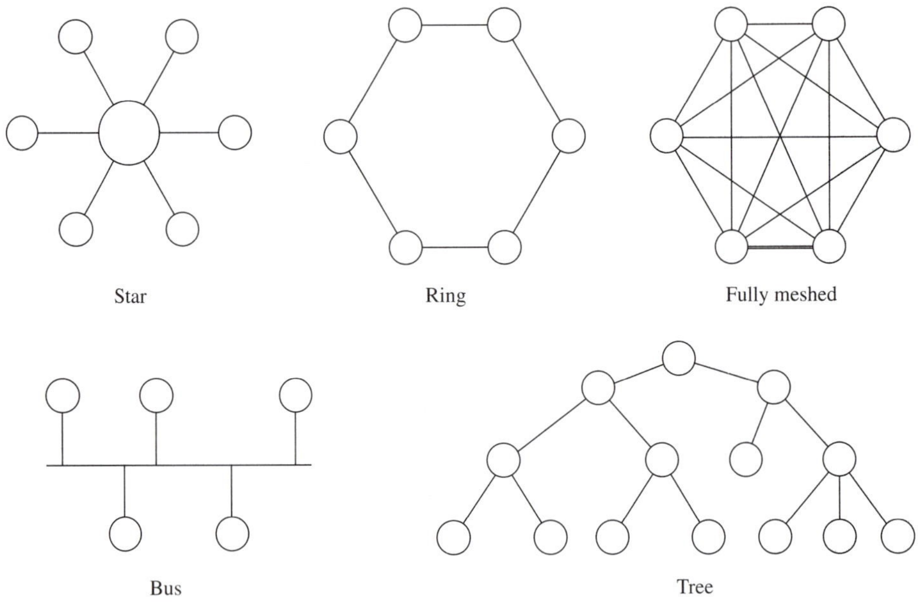

Figure 1.10 The common network topologies: star, ring, mesh, bus, and tree.

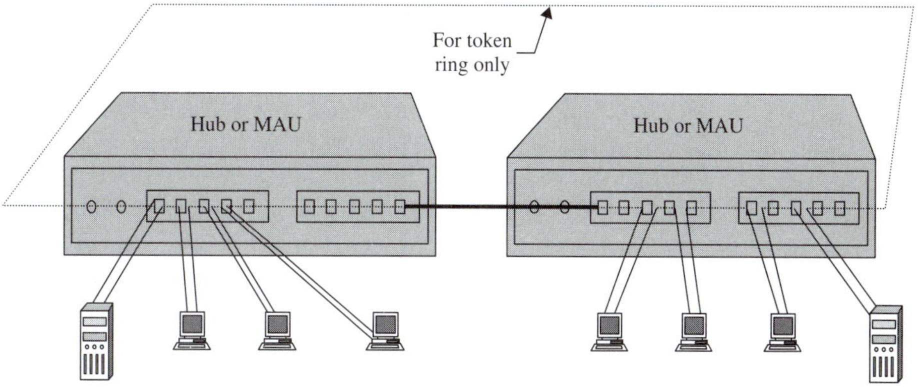

Figure 1.11 Actual 10BaseT and token ring connections.

ken ring LAN (discussed in Chapter 4), many network nodes (computers, etc.) are connected by cables to a central device called a hub for 10BaseT or a MAU (multiple-access unit) for a token ring. These configurations have a starlike topology and facilitate network debugging and management. However, from Figure 1.11 we note that a 10BaseT LAN actually has a bus topology and the token ring LAN has a ring topology if the cable is stretched from the hub or the MAU.

A large network may contain many smaller networks. These networks are internetworked, commonly with a mixture of topologies. The Internet is a huge network that includes numerous small heterogeneous networks with various architectures and topologies. (Figure 1.12b). However, from the user's point of view, each computer appears to attach to a single large network represented as a cloud—the term used among computer scientists (Figure 1.12a).

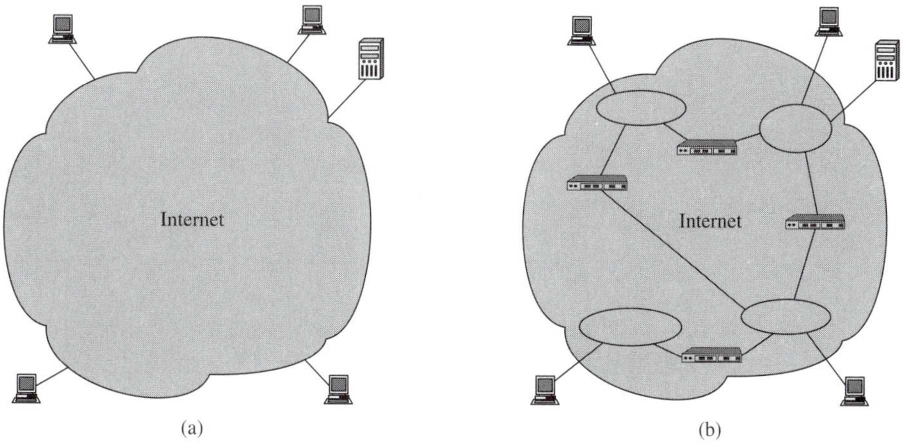

Figure 1.12 The user's view of the Internet. (b) The actual Internet structure.

Finally networks may be classified as public or private. A "public" network provides network services to the public but is often owned by a private enterprise. The owner of the public network, often called the carrier, also rents the network to the public. AT&T, MCI, and Sprint, separately, are well-known public telephone carriers in the United States. The only true public network, probably, is the Internet. In contrast, a private network is owned, used, and managed by an organization for its own purposes, with access to outsiders strictly limited.

1.4 DISTRIBUTED SYSTEMS AND CLIENT–SERVER MODELS

In the past two decades, rapid development and progress of personal computer and network technologies caused the computer industry to experience an important change from centralized computing to distributed computing. The decentralized approach of distributed systems makes use of inexpensive but powerful networked PCs and workstations to replace expensive mainframe computers. Distributed computing is also a direct result of the increasing demand for a user-friendlier graphical user interface (GUI) environment. With a centralized system, massive amounts of graphical data must be transported from the central computer to the terminals for display on individual monitors.

1.4.1 Centralized Computing vs Distributed and Client–Server Systems

LAN technology, internetworking technology, and client–server applications are the core of distributed computing systems. Network technology is the foundation of client–server computing, and client–server is the standard model for network applications.

In the client–server computing model, the server is a program that provides certain services to the client programs. The connection between client and server is normally by means of a message passing mechanism, often over a network, and a specific protocol is used to encode the client's requests and the server's responses (Figure 1.13). For exam-

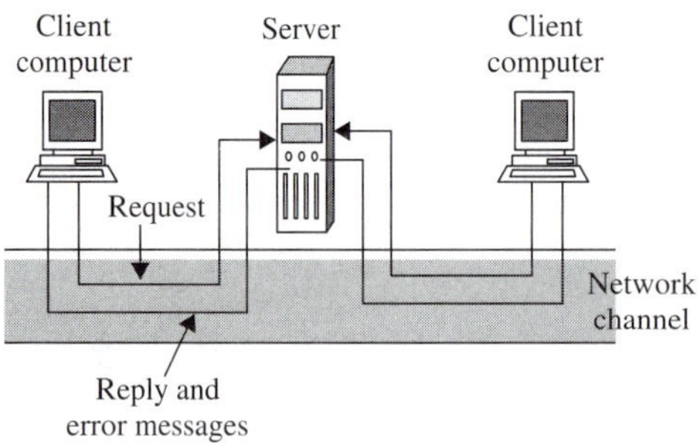

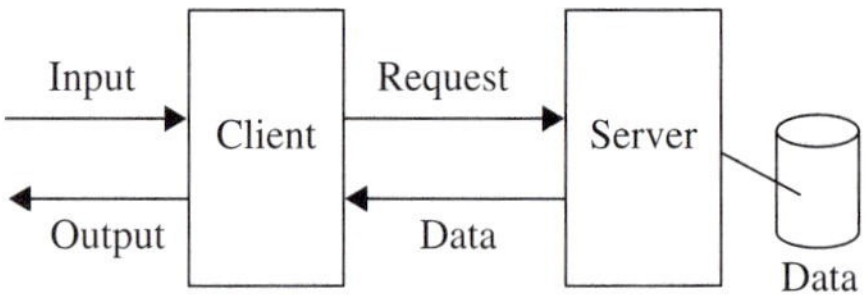

Figure 1.13 The data flow (a) for the client–server model (b).

ple, one or more computers (the file servers) may be dedicated to storing files of information for all the other computers. Various machines in the network can ask the file servers to deliver copies of files on demand. Another set of computers may be dedicated to providing printing services, ranging from draft-quality work from a high-speed matrix printer to high-quality laser printing.

In many cases, the client program residing in a local computer provides a GUI, takes a request, and sends the request through the network to the server in a remote computer. The server does the actual job, such as access the database information, then sends the information back through the network to be displayed on the client machine, graphically in many cases. This way the client software and the server software can be implemented separately and independently, even from completely different hardware platforms, providing strong productivity, portability, and interoperability. Typical client–server applications include the following:

- Distributed database applications and other similar systems, where the users enter requests from a front computer through the client program. The requests are sent through a network connection to a server running in a remote machine that accesses the database and sends the data back to the clients local to the users. One server can serve many clients, of course.
- Network file systems (NFS) on Unix and Microsoft NT, and Novell's NetWare system, in which a workstation requesting the contents of a file from a file server is a client of the file server. The client can read and write files on the file server's disk, as if on the client's local disk.
- Hardware sharing for printers, scanners, fax machines, and so on.
- X Windows system.
- Internet World Wide Web (WWW) services and HTTP protocols (Chapter 5).

Note that the client and server programs can be in the same machine or in different machines. Also the roles of server and client may be reversed in terms of location. For example, the client program may be running in a remote machine while the server is on the local computer, as with the X Windows system protocols.

X Windows has a special form of client–server system in which the user's computer terminal, including display, keyboard, and mouse, is controlled by the local X server software, which is apparently hardware dependent, while the application programs, connected via TCP/IP, are the clients. An application may run on a remote host or on the same machine with the server. Applications make calls to the server when they need input, output, or the user's attention, that is, the server software provides input (from keyboard and mouse) and output (display) services to the client program. Under this arrangement, the client program is entirely independent of the display without worrying about many different graphical attributes—colors, resolutions, fonts, frequency, and so on. The best thing is that by multitasking, the X server is able to work with more than one client application, be it local or remote (Figure 1.14).

An important feature of client–server systems is that the services provided by a remote server or the requests from a remote client through networks are completely trans-

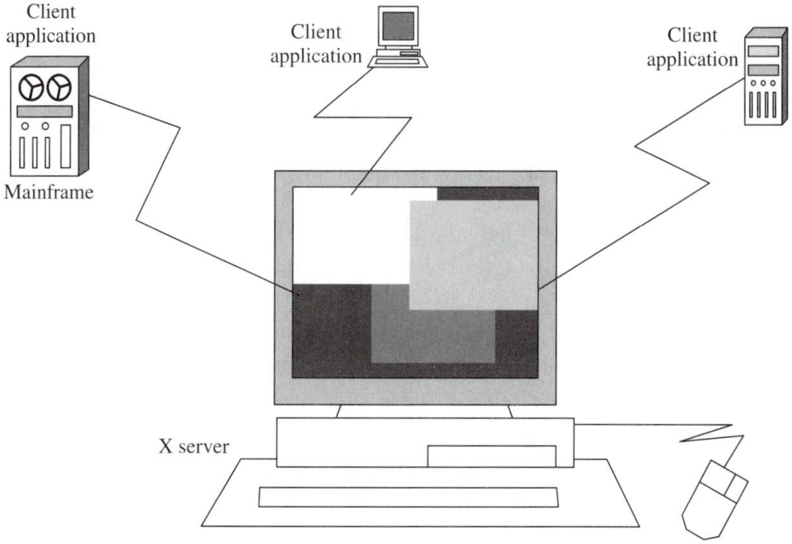

Figure 1.14 In the X Windows system, the server is a local program and the client applications may be remote, running on entirely different computers. The X server can work with many remote client applications.

parent to the user. From the point of view of the user, the service is provided the same way as on a local machine.

Under the Unix operating system environment, a server program often runs continuously as a daemon, waiting for requests to arrive, or it may be invoked by a higher level daemon that controls a number of specific servers. Besides NFS, mentioned earlier, there are many other servers associated with the Internet: Network Information Service (NIS), the domain name system (DNS), FTP, news, finger, Network Time Protocol, and others. On Unix, a long list can be found in /etc/services or in the NIS database "services". Chapters 5 and 11 cover the Internet and related client–server programming issues in detail.

1.4.2 Three-Tier Client–Server Systems

In the typical two-tier client–server application environment, the client software runs on a front-end machine interacting with the user while sending requests to and receiving data from the back-end server. Clients usually have all the application logic and user interface code, the server is responsible for the actual job—providing the services, typically by means of access to a shared database system. This simple arrangement of two-tier client–server applications may be adequate for relatively small applications. However, compared with what distributed systems are expected to deliver, these two-tier structures have proved to be primitive.

Because of the increasing demand for access to large-scale distributed database applications and Internet WWW services, more advanced and complex three-tier client–server architectures are emerging as a better solution (Figure 1.15).

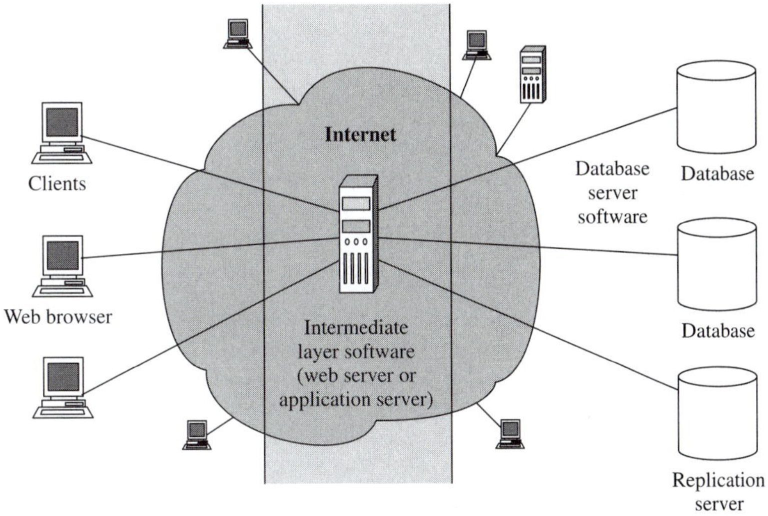

Figure 1.15 The three-tiered client–server model.

Generally speaking, three-tier applications supply a tier of distribution separate from the front-end presentation and the back-end database access. Under the three-tier client–server approach, the application is split into three pieces: user interface, program and business logic, and data storage. Three-tier client–server systems promise to increase applications performance, flexibility, and robustness, which are the most important factors in developing mission-critical applications. With the three-tier client–server system, the user interface code runs on the client, while program logic is split between an application server and a database server with transaction processing resources. The three-tier structure adds an intermediate layer of servers that support application logic and distributed computing services. Among other things, this approach separates database access, which is a very complicated task on its own, from the intermediate layer, and makes it easier to change any one part of an application without affecting the entire program. Software developers can port the three-tier client–server system to separate networked machines and move parts of an application to network locations that provide the best speed, security, stability, and so on. Three-tier architectures could be a stepping-stone to building applications composed of true distributed objects, in which all application resources interact as monolithic components and peers.

Applications based on Internet technologies will also take advantage of the new architecture to solve scalability problems. With the three-tier structure, ideally, web browsers will simply become a new front-end option for presenting the application. The existing database and transaction processing resources remain in place, while the web servers become the middle tier between the other two to distribute queries and requests from web users to the database, and to deliver the data from the database and transaction processing resources to users through the browsers. The emergence of complex web applications has validated the multitier client–server architectures, and the entire application infrastructure will accommodate the web technologies seamlessly (Figure 1.15).

Because of technological complications, however, three-tier systems may be more difficult for software developers and users to build and learn to use than the traditional two-tier structure. For example, to update a two-tier application that houses application logic solely on the client, software engineers must often work in more than one programming language and multiple APIs. Nor does the situation improve for those working with the Open Software Foundation's (OSF) distributed computing environment (DCE), which relies on distributed, yet static, stub files representing application components. Once an application has been set up, a developer may not be able to move a component from its home computer to another machine without breaking other sections of the application that depended on the location of that component's stub file. It is also a challenge for web developers to tie in the existing legacy applications and data stores under the three-tier structure.

Many special software packages, especially middleware tools, have been emerging for development of three-tier client–server applications, and these are allowing software engineers to move workgroup applications to three-tier systems without imposing unexpected performance woes or unanticipated rigidity. The tools also can dynamically control the distribution of application logic at run time, providing built-in middleware to handle the detailed communications coding necessary to direct application requests around the underlying network.

1.4.3 Web-Based Application Servers and Electronic Commerce

Thanks to the Internet and the rapidly developing World Wide Web technology, a new software technology makes electronic commerce realistic. This revolutionary software technology, which could rival the multibillion-dollar database software market, produces application servers to help connect users of web browsers to a company's information resources via intranet/extranet, or the global Internet. These servers take much of the complexity out of building electronic commerce sites and other large web-based applications. With the emergence of electronic commerce, companies representing all aspects of information technology are vying for a piece of the application server pie. Numerous software and hardware vendors and start-ups emerge and pitch their own technology as the new platform upon which to build programs in the Internet age. Potential customers range from financial institutions to airlines and web-based retailers.

The interest in application servers is fueled by a technology shift beyond the emergence of electronic commerce. Many corporate applications over the past years have been built to run partly on personal computers and partly on powerful server computers running database applications. Maintaining those programs on a huge number of PCs throughout a corporation has become an increasingly costly and unruly chore. As traditional corporate computer systems have grown in complexity, computer scientists and engineers have looked for better ways to design application software systems.

The software technology of application servers is based on the fledgling three-tier client/server technology, which makes it easy to build complex systems by tying software components together in one integrated application. The idea is similar to a car manufacturing assembly line. The components, once assembled, act as the engine infrastructure that pipes data between corporate databases and the front-end user interfaces, linking product catalogs and shipment tracking systems, membership lists, and order-taking programs.

Like an assembled car engine, the application server remains hidden from users, who only get a view of the finished product with increased sophistication—a web site for everything from complicated tax return forms to online stock transactions. For example, an online stock trading system using application servers on a web site can support as many as 25,000 simultaneous users on the Internet throughout the world. Online transactions over the web require complex, behind-the-scenes computing resources. Transactions in a web-based online stock trading system may go through the following steps:

- The user connects to the Internet through a PC, workstation, network computer, or WebTV set, then locates the broker's web site and logs in.
- The application server connects the web browser with at least two databases, which may reside on different computers—the user account database to verify the user and stock quote database for real-time quotes. As many as three small windows may appear on the same browser screen for user account information, stock quote, and trading entry, respectively.
- The user reviews his or her stock portfolio in the user account information window and requests a quote for a particular stock through the quote information window. The application server responds with a real-time quote from the stock quote database.
- The user decides to buy 1000 shares of stock XYZ and enters the order by the way of trading entry window. The trading server verifies the order and responds that the user does not have enough cash for the transaction.
- The user finds orders entered earlier by clicking a tab in the account information window that connects to account database. Previous orders may then be modified or canceled.
- The user enters the XYZ order again. The application server verifies the order, approves and confirms it, and then sends it to the appropriate stock exchange.
- The order is filled. The application server updates the user account database and automatically calls the user or sends him or her e-mail to confirm the transaction.
- The user can retrieve account activities through the user account information window, which is connected by the application server to the user account database.

Many application servers find use in the popular corporate intranet environment. Almost every organization has software that gives employees access to their benefits packages—a suite of separate human resources programs allowing workers to manage everything from sick leave to retirement plans and other benefits. Each program used to be independently maintained on each PC. Now, however, not only are all the separate programs running on the server, the server takes care of piping data among the users' web browsers, the application programs, and the databases through the corporation intranet.

In fact, the technology might become the next platform for writing software, a virtual operating system that controls how corporate applications are developed for the Internet age. Most application servers have rallied around the Java programming language and related technologies, which are suited for environments with a mix of systems including Windows/NT, Unix, and mainframes, while Microsoft touts its integrated Windows-based technology. The standards war is similar to the battles over Java/ActiveX and browsers from Netscape and Microsoft. The two camps are likely to coexist for the foreseeable future. Corporate planners are becoming increasingly aware that the selection of the right application server is as important strategically as the choice of a database or operating system.

1.5 PROTOCOLS AND STANDARDIZATION

Data communications and networks always involve at least two parties. The formal specifications and conventions that govern and control the communications and data exchanges among the communication parties are called protocols. All parties must observe the communication and network protocols for a particular platform, be it hardware or software. Without such agreement, communications are simply impossible. Protocols are the standards for communications, networking, and internetworking.

1.5.1 Protocol Syntax and Semantics

Each of the various forms of information has syntax, or rules for presentation. Information also has semantics, by which the meaning of the information is interpreted. Generally speaking, low-level protocols define the protocol syntax that specifies electrical and physical aspects of data exchange, including the following.

- Signal levels, polarity, voltage, simplex vs duplex communications, and the number of stops in asynchronous communications (Chapter 3)
- Network composition, identifying the different media and the physical connecting components (Chapter 4).
- Character set (7- and 8-bit character sets)

Protocol semantics specifies high-level properties related to control information for coordination and error handling, including the following.

- Error detection and correction of the bit stream. There are various ways of error handling: simple parity checking, arithmetic check sum, and CRC (cyclic redundant code), to name a few (Chapter 3). To be able to communicate, all communications parties must use the same protocol.
- Bit and byte ordering of the signals to be transferred. For example, the lowest order bit of a character is usually transmitted first. But this protocol can be big-endian, meaning that the higher order byte of a number transmits first (TCP/IP protocol suite) or little-endian, for the opposite case (e.g., Intel, DEC).
- Data and message formatting, sequencing, and fragmentation. These protocols specify how a data frame or packet is assembled and disassembled and what kind of header a frame or packet may have.

- Signal timing and synchronization mechanisms, to specify the properties of data transmission and flow control (Chapter 3)
- The terminal-to-computer dialog, or end-of-file, to specify how a piece of message is to be terminated.

Almost all networks have a layered structure that seeks to reduce network design complexity by partitioning the network functions into different layers. A list of protocols used by a network system, one protocol for each layer of the network, is called a protocol stack. More details about the layered network structure are discussed in Chapter 3.

1.5.2 National and International Standards

In the data communications and network world, there are numerous standards from which to choose. A standard may exist in fact, or it may be imposed by law—*de facto* or *de jure* in Latin. For example, the TCP/IP is the de facto standard for the Internet, while OSI is the interconnection standard de jure. Standards are critical to the industry because they provide guidance for those trying to ensure compatibility, internetworking, and interoperability. On the other hand, standards often become a source of confusion, division, and duplication of effort because of the enormous number of competing documents.

The International Organization for Standardization (ISO) is an international body, founded in 1946, that proposes, drafts, discusses, specifies, and approves standards. ISO is responsible for creating international standards in many areas, including computers, communications, and networks. It is ISO that produced the well-known OSI seven-layer model for network architecture. ISO members are the national standards organizations of 89 countries, including the American National Standards Institute (ANSI), and the British Standards Institution (BSI). ISO has numerous technical committees (TCs), currently about 200, and the number is increasing. TCs are responsible for specific subjects, and each TC has subcommittees as working groups. ISO is often one of the first organizations to formally define a common way to connect computers and networks.

Besides ANSI, which is responsible for approving U.S. standards in many areas, including computers and communications, the National Institute of Standards and Technology (NIST), formerly the National Bureau of Standards, is another governmental body that provides assistance in developing U.S. standards.

1.5.3 Telecommunication Standards and Organizations

The telecommunications world is full of various organizations that try to coordinate and provide guidance in protocol development and standards specifications. In the United States, the telecommunications industry has become quite open and diverse. In other countries, however, despite a trend toward deregulation and liberalization, the telecommunications industry remains highly regulated and is dominated by the respective national governments through the organization commonly known as the Postal, Telegraph, and Telephone Administration (PTT).

The International Telecommunications Union (ITU), which is affiliated with the United Nations, is an international treaty organization, while ISO is a voluntary, nontreaty organization. ITU is the standards organization which is more closely related to the communications industries. ITU has three main sectors: ITU-D (Development Sector), ITU-R (Radiocommunications Sector), and ITU-T (Telecommunications Standardization Sector). Before March 1, 1993, ITU-T was known as the Consultative Committee for International Telephony and Telegraphy (CCITT). ITU sectors hold plenary sessions regularly to introduce new standards, as well as improve existing ones.

ITU-T sets standards and makes technical recommendations about telephone and data (including fax) communications systems for PTTs and suppliers. ITU-T works closely with all standards organizations to form an international uniform standard system for communication. For example, Study Group XVII is responsible for recommending standards for data communications over telephone networks. It publishes the V.XX standards for serial communications and modem standards. ITU-T also defined the standards for X.25 packet switching network protocols (Chapter 3). ETSI, the European Telecommunications Standards Institute, is a European version of the ITU-T.

In the United States, the Federal Communications Commission (FCC) is the government arm that manages the regulation of interstate radio, TV, cable, and telephone communications and communications policy related to other countries. All proposed interstate services must be reviewed and approved by the FCC.

1.5.4 Networks Standards and Organizations

There are hundreds of network protocols for different media, platforms, and environments. Many protocols are defined by Open Systems Interconnect (OSI), the Institute of Electrical and Electronic Engineers (IEEE), and request for comments (RFCs).

OSI is the umbrella name for numerous open, well-recognized, and formal standard protocols and specifications comprising the OSI reference model, including Abstract Syntax Notation 1 (ASN.1), Basic Encoding Rules (BER), Common Management Information Protocol and Services (CMIP and CMIS), X.400 (message handling system, or MHS), X.500 (Directory Service), Z39.50 (search and retrieval protocol used by WAIS), and many others.

The International Electrotechnical Commission (IEC) consists of committees with members drawn from many technologically advanced countries. IEC and ISO cooperate in defining and specifying standards for electrical and electronic technologies; typically, IEC covers lower layer standards and ISO deals with more general aspects. The IEEE often acts as the representative of the United States on ISO and IEC committees. IEEE specifies, among others, many computer- and network-related (standards including the widely implemented LAN protocol set 820.1–802.11, discussed in Chapter 4.

Since the days of Arpanet, the original Internet, developed by the U.S. Defense Advanced Research Projects Agency (DARPA) in 1969, hundreds of Internet informational documents and standards have been publicized in the form of requests for comments (RFC). The RFCs are submitted by technical experts acting on their own initiative and are reviewed by the digital community at large, rather than being formally presented through an institution. For this reason, the proposals and documents are called RFCs even when followed by the Internet and Unix communities and become adopted as standards. In the TCP/IP world—the source of the dominant internetworking protocols for Internet (discussed in Chapter 5)—all standards are recorded in RFC form even though far from all RFCs are standards.

The RFC series is numbered sequentially in chronological order as RFCs are proposed. When an RFC is placed on the standards track, it must progress through three states: proposed standard, draft standard, and Internet standard. At each stage, the RFC is reviewed, along with implementation and deployment experience. After each step, two years is needed to demonstrate feasibility, implementability, and usefulness. For a proposed standard to become a draft standard, there must be two independent implementations with significant implementation experience. Similarly, to progress to full Internet standard status, there must be several independent implementations along with extensive deployment and considerable interoperability experience.

During the course of each review, changes may be made to the documents. Depending on the severity of the changes, the documents is either reissued in its current state or is reduced to proposed standard status, whereupon another two-year demonstration cycle begins.

The de facto Internet management standard, Simple Network Management Protocol (SNMP), is a good example of an RFC that passed through this rigorous procedure.

Implementation, deployment, and interoperability are all important criteria in Internet standard creation. Further, to foster understanding and availability, an openly available reference implementation is required. In addition to assigning a standardization state, to each RFC, a protocol status is assigned. The protocol status indicates the level of applicability for the technology documented in the RFC:

Required: A system must implement this protocol.

Recommended: A system should implement this protocol.

Elective: A system may or may not implement this protocol. Usually a given technology area has multiple elective protocols from which to choose.

Limited-use and **not-recommended:** The protocol is not recommended, or a system may implement it only in limited circumstances because of its specialized nature and limited functionality.

When an RFC is superseded, it is termed historic.

The Internet Architecture Board (IAB), the technical body that oversees the development of the Internet suite of protocols, has two task forces: the Internet Engineering Task Force (IETF) and the Internet Research Task Force (IRTF).

The Internet Engineering Task Force is a large, open international community of network designers, operators, vendors, and researchers whose purpose is to coordinate the operation, management, and evolution of the Internet and to resolve short- and mid-range protocol and architectural issues. It is a major source of proposals for protocol standards, which are submitted to the IAB for final approval. The IETF meets three times a year, and extensive minutes are included in its Proceedings. The IRTF is responsible for longer term Internet-related research. IETF and IRTF each has a steering group—the Internet Engineering Steering Group (IESG) and the Internet Research Steering Group (IRSG), respectively—to manage its affairs.

The Internet Network Information Center (INTERNIC), sponsored by the National Science Foundation (NSF) in the United States, is responsible for maintaining and distributing information about the global Internet and the closely related TCP/IP network protocols. To aid INTERNIC and to make document retrieval easier, numerous Internet sites scattered in many countries keep the archives of all RFCs in various formats. Here are just a few of them:

ds.internic.net

nisc.jvnc.net

ftp.isi.edu

ftp.ncren.net

ftp.sesqui.net

wuarchive.wustl.edu

Numerous other standards involving the data communications and networks industry are introduced in later chapters when appropriate.

1.6 CHAPTER SUMMARY

This chapter is an overview of current computer communications technologies and how these technologies evolved over the last few decades. We discussed the developments in the fields of computer and telecommunications that are likely to affect lifestyles in the future. Advances in PC technology and competition due to open architecture have caused a considerable drop in the prices of hardware and software alike. That coupled with Internet growth has resulted in new modes of communications, which continue to be developed in the twenty-first century.

As soon as homes and small businesses obtain network access at reasonably fast speeds (of the order of a few megabits per second) and at affordable prices, a variety of web-related applications will change the way we communicate, do business, and perform our jobs and many other day-to-day activities. Videoconferencing, Web TV, and e-commerce are some of those applications. The effects of the web are already being noticed as bigger corporations are introducing intranet/extranet setups to support such applications.

The relatively low data rate of e-mail has already allowed that application to prove its worth. A big percentage of the computer-using population prefers sending e-mail to writing letters or notes. With the availability of cable modem, ISDN, and other higher speed technologies at affordable prices, other applications requiring the use of computer networks will certainly appear. Later chapters deal with these topics in detail.

In addition to discussing current networking technologies, we also talk about the standardizing process and bodies. Standards organizations such as ISO, IEEE, ITU-T, and IETF help develop and govern the administration of standards and protocols to support computer communications. Members of these organizations, including leading equipment vendors, telecommunication companies, government agencies, scientists, and engineers, help to develop the standards that may be used as guidelines for building products. The reference section includes the URLs of some of the leading standards organizations. These online resources may be very helpful in learning more about the operation and role of the standards organizations.

1.7 REFERENCES

BOOKS

Baker, R. H., *Networking the Enterprise: How to Build Client/Server Systems That Work*. New York: McGraw-Hill, 1994.

Bellamy, J. C., *Digital Telephony*, 2nd ed. New York: Wiley, 1991.

Berson, A., *Client–Server Architecture*. New York: McGraw-Hill, 1992.

Gibilisco, S., *Handbook of Radio and Wireless Technology*. New York: McGraw-Hill, 1998.

Green, J. H., *The Irwin Handbook of Telecommunications*, 3rd ed. New York: McGraw-Hill, 1997.

Laino, J., *The Telephony Book: Understanding Telephone Systems and Services*, 3rd ed. San Francisco: Miller Freeman Books, 1999.

Linthicum, D., *David Linthicum's Guide to Client/Server and Intranet Development*. New York: Wiley, 1997.

MacKie-Mason, J. K., and C. Lee, *Telecommunications Guide to the Internet*. Rockville, MD: ABS Group.

Meyer, R., *Old Time Telephones! Technology, Restoration and Repair*. New York: McGraw-Hill, 1994.

Renaud, P. E., *Introduction to Client/Server Systems*, 2nd ed. New York: Wiley, 1996.

Robert, D. H., *Client/Server Programming with OS/2 2.0*. New York: Van Nostrand Reinhold, 1992.

Stallings, W., *SNMP, SNMPv2, and CMIP: The Practical Guide to Network Management Standards*, 2nd ed. Reading, MA: Addison-Wesley, 1996.

Stallings, W., *SNMP, SNMPv2, SNMPv3, and RMON 1 and 2*, 3rd ed. Reading, MA: Addison-Wesley,1 999.

Smith, P., *Client/Server Computing*. Indianapolis: SAMS Publishing, 1992.

Travis, D. T., *Client/Server Computing*. New York: McGraw-Hill, 1993.

Walters, R., *Computer Telephony Integration*, 2nd ed. Norwood, MA: Artech House, 1998.

ARTICLES

Adler, R. M., "Distributed Coordination Models for Client/Server Computing," *Computer* 28, no. 4 (April 1995), 14–22.

Dickman, A., "Two-Tier Versus Three-Tier Applications," *Informationweek* 553 (November 13, 1995), 74–80.

Edelstein, H., "Unraveling Client/Server Architecture," *DBMS* 34, no. 7 (May 1994).

Gallaugher, J., and S. Ramanathan, "Choosing a Client/Server Architecture. A Comparison of Two-Tier and Three-Tier Systems," *Information Systems Management Magazine* 13, no. 2 (Spring 1996), 7–13.

RFCS

RFCs may be obtained via file transfer protocol from one of the following primary repositories:
ftp://ftp.nis.nsf.net

ftp://ftp.nisc.jvnc.net
ftp://ftp.isi.edu
ftp://ftp.wuarchive.wustl.edu
ftp://ftp.src.docic.ac.uk
ftp://ftp.ncren.net
ftp://ftp.sesqui.net
ftp://ftp.nic.it
ftp://ftp.imag.fr
http://www.normos.org
http://www.fags.org/rfcs
These are also available at ftp://ftp.cis.ohio-state.edu/pub/rfc
RFC-1157, A Simple Network Management Protocol (SNMP), 1990.
RFC-1310, The Internet Standards Process, 1992.
RFC-1441, Introduction to Version 2 of the Internet-Standard Network Management Framework, 1993

WORLD WIDE WEB SITES

Telecommunications News/Events

European Telecommunications Standards Institute
http://www.etsi.org/
Byte Magazine
http://www.byte.com/
Infoworld Magazine
http://www.infoworld.com
Network World Fusion Magazine
http://www.nwfusion.com/
Telstra (Telecom Australia) communications information sources page (contains links to many telecommunication companies worldwide)
http://www.telstra.com.au/info/communications.html

Associations and Standardization Bodies

Association for Computing Machinery (ACM)
http://info.acm.org/
ATM Forum
http://www.atmforum.com/
Computer and communication standards and organizations
http://www.cmpcmm.com/cc/standards.html
European Computer-Industry Research Center (ECRC)
http://www.ecrc.de/
IEEE Computer Society
http://www.computer.org/
Institute of Electrical and Electronics Engineers (IEEE)
http://www.ieee.org/
ISO (International Standards Organization)
http://www.iso.ch/welcome.html
ITU (International Telecommunication Union)
http://www.itu.int/

Telecommunication Scciety of Australia
http://www.tsa.org.au/
TERENA
http:/www.terena.nl/

Telecommunications-related corporate sites

AT&T Bell Labs
http://www.research.att.com/
Bell Atlantic
http://www.ba.com//ba.html
Bellcore ISDN
http://info.bellcore.com/ISDN/ISDN.html
Broadcom
http://www.broadcom.ie/
Ericsson
http://www.ericsson.nl/
EuroKom
http://www.eurokom.ie/
FOKUS
http://www.fokus.gmd.de/
Hewlett-Packard
http://www.hp.com/
Lucent Technologies, Inc.
http://www.lucent.com
Motorola
http://www.mot.com/
Nokia
http://www.nokia.com/
Northern Telecom
http://www.nt.com/

2

DATA COMMUNICATIONS

We are in the midst of an information technology revolution, and computer networks are playing an important role in it. In this information age, the trend is toward more digital data as opposed to data in the traditional analog form. In this sense, the terms *data communications* and *digital communications* are closely related and are interchangeable in many situations. But strictly speaking, "data communications" is the more general term because it may include both digital and analog parts. For example, modems are often used to modulate and digitize computer data into analog signals, which can be sent over analog telephone lines. At the receiver side, the data are demodulated and restored back to digital form to be used by another computer. On the other hand, digital communications may be included in a nondigital communications. A typical example is the long distance telephone call that is transmitted, partly in digital form, to take advantages of the many benefits a digital system offers over an analog one.

2.1 ADVANTAGES OF DIGITAL COMMUNICATIONS

There are many ways information can be carried and exchanged. For instance, voice, picture, light, and body language all convey information. However, the most useful and convenient forms supporting today's data communications are electrical and optical signals. An analog signal like an electrical wave needs to be converted into digital form before being transmitted through the digital communications channel (Figure 2.1), and similarly a modem must be used to convert a digital signal to analog form before the signal can be transmitted via an analog channel (a phone line).

Analog is the original and natural expression of information. So it is important to ask why we prefer digital over analog for computer communications. The simple answer to

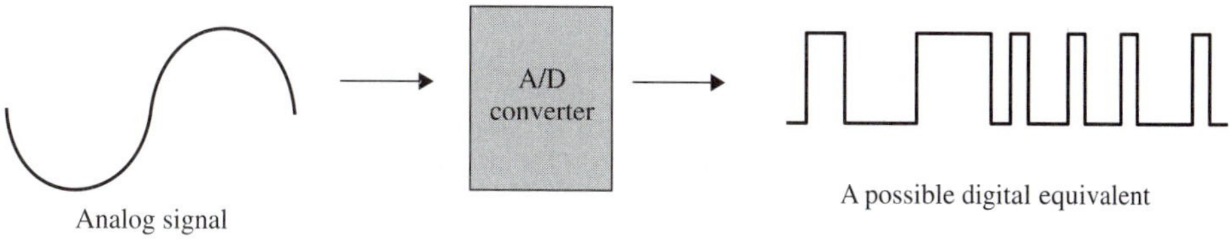

Figure 2.1 An analog signal and a possible form of its digital equivalent.

the question may be that digital signals are more error free and easier to manipulate than analog. More detailed explanations can be summarized as follows.

• Any signal, be it digital or analog, electrical or light, always undergoes nonuniform weakening and distortion in the transmission procedure. With analog signals, the nonuniform declination (or attenuation) and distortion due to distance, interference, and noise are impossible to remove completely at the receiver side. An amplifier in the middle of the transmission will also increase the level of interference, noise, and distortion, thus offering little improvement. When the distortion becomes too serious because of the cumulative effect, the original information may be completely changed or lost. In contrast, digital signals can always be regenerated at the receiver side or by a repeater in the middle of the transmission, since the only thing that needs to be detected is the absence (binary 0) or presence (binary 1) of a pulse. The regenerated digital signals are as good as the original. Consequently, digital transmission is not subject to cumulative distortion and can stand much higher noise-to-signal ratios than analog transmission.

• Because of the increasing power of digital circuits and microprocessors, digital signals are substantially easier than analog to manipulate and maintain. Switching, routing, multiplexing (i.e., putting many communications channels together to be transmitted), compression, and encryption are all easy and efficient for today's extremely high speed digital circuits (ICs) or microprocessor-controlled devices.

• Voice, video, and any other analog information can be easily converted into digital formats and transmitted by high-speed digital communications facilities, including the backbone telephone networks, the Internet, and private networks, using coaxial cables, optical fiber, radio, microwave, and satellites. Asynchronous transfer mode (ATM) technology especially makes it possible for a single communications channel to carry data, voice, and multimedia information (see Chapter 6).

• Digital transmission is inherently more secure than analog transmission. The older analog scramblers are not effective. Digital encryption is much more difficult to decipher and costs less.

A data communications system comprises five main parts.

1. The transmission medium that carries the traffic between communication parties
2. The devices that modulate the signals, (i.e. convert them from analog to digital) to be transmitted
3. The devices that multiplex and switch data paths to allow channel sharing
4. Terminal devices
5. Communications protocols and software that control the data communications systems

Transmission media, modulation/demodulation and digitization techniques, and terminal devices are introduced in this chapter. Multiplexing and switching techniques, and communication protocols and related software, are discussed in subsequent chapters.

We turn first to a discussion of how communications system components work together to fulfill the communications tasks. Comparisons between different techniques used in the industry are useful for this purpose.

2.2 TRANSMISSION MEDIA

The transmission medium is the physical foundation for all data communications. Such media include communications channels, paths, links, trunks, and circuits. At the bottom of the most frequently used telephone system for data communications, the local loop still uses two copper wires to connect residential customers' phones to the telephone company's central office. The distance and diameter of the wire pair limit its transmission capacity. The American Wire Gauge (AWG) is the U.S. standard for measuring the size of wires. Smaller gauge numbers represent larger wire sizes with greater transmission capacity. Larger wire diameter means smaller resistance, which causes signal strength reduction. This property is particularly critical for high-frequency signals, which tend to travel on the outer surface of a conductor. The higher the frequency of the signal, the higher its tendency to travel on the outer surface of the cable. Additionally, higher frequency means higher bandwidth (or capacity, as we explain shortly).

The local loop of a telephone system, normally uses 22–26 gauge copper wire, often openly suspended on poles or bundled and hidden in an underground pipe. The primary communication trunks for handling larger volume typically uses 19 gauge or higher grade wires.

A leased line is a set of normal telephone facilities using the same wiring. A leased line is continuously available because its trunks are hardwired together to form a circuit and the switching equipment is bypassed. Bypassing the switching equipment minimizes impulse noise due to part of the circuit passing through an older electromechanical central office switch. A leased line is often conditioned to exhibit lower echo interference by having all the taps in the path removed. A conditional line is optimized for the frequency response and delay characteristics of the data transmission. With a much higher price than shared telephone lines, a leased line offers consistent quality, which is guaranteed from end to end.

While cheap and available everywhere, the wire pair connection is weak owing to its limited bandwidth. Therefore this connection, used mostly for subscriber loops, is often considered to be the last mile of the information superhighway to be conquered.

2.2.1 Shielded and Unshielded Twisted-Pair (STP and UTP) Wiring

Unshielded copper wires, each clad in plastic, are often paired and twisted together to reduce outside electromagnetic interference and cross talk between wires. Two pairs (four wires) or more are usually encased together in a single sheath. The first token ring LANs, installed in the mid-1980s, used twisted but shielded pair of wires. The most popular Ethernet LAN standard—10BaseT—uses UTP wiring for the connections. Sometimes, UTP technology permits the use of Ethernet over existing telephone wiring without adding new cables. The highest grade UTP cable, which is very tightly twisted, thus can sustain greater data rates because it is more immune to interference, can be used for fast Ethernet LANs that run at 100 Mbps.

In shielded twisted-pair wiring, a layer of conductor braid surrounds the twisted pairs of copper wires to further reduce interference. The resulting STP wiring is used for the new copper distributed data interface (CDDI) standard that runs at 100 Mbps. CDDI is the copper counterpart of the fiber distributed data interface (FDDI) standard. More detailed information about STP and UTP wiring is provided in Chapter 4, which describes current LAN technology.

Twisted-pair cable is usually balanced because the two wires have the same electrical references to ground potential. The signals in the wire pair have a phase difference of 180°, which tends to cancel out, enhancing resistance to outside interference. At 100 Mbps, twisted-pair copper wiring is running close to its maximum potential speed. For LANs that must cover larger distances, coaxial cables or optical fiber are required.

2.2.2 Coaxial Cable

Telecommunications for voice began with the simple wire pair and UTP wiring. A pair of wires was needed to support two-way communications, and the twisted pair can reduce cross talk by canceling out interference. The increasing demand for higher capacity and quality of data communications spawned a variety of new cabling types to meet the more rigorous requirements.

The first Ethernet LANs in the early 1980s used a heavy-duty coaxial cable called thick cable, which had four layers to minimize electrical interference. Thick cable is heavy, expensive, and provides only the capacity of 10 Mbps over a distance of 500 m. Technological progress makes it possible to produce much lighter and less expensive coaxial cable called *thin cable* for Ethernet use. As illustrated in Figure 2.2, thin cable has a central copper conductor surrounded by a layer of insulator, a copper braid layer, and finally a protective outer plastic sheath. Repeaters can be used to extend the distance covered by coaxial cables.

A similarly structured coaxial cable used for analog transmission is called broadband coaxial cable. A single video signal is about 2000 times bigger than that needed for a telephone call, so it must be carried over high-capacity coaxial cables rather than twisted-pair wires, with their limited bandwidth. Broadband cables, which can carry many amplified high-frequency analog channels at 300–500 MHz for nearly 100 km, are widely used in the cable industry.

In contrast to twisted-pair cable, a coaxial cable is an unbalanced circuit: the center conductor carries the current and the braided metal shield acts as another conductor operating at ground potential. The shield prevents energy from radiating and gives the signals better protection from outside interference.

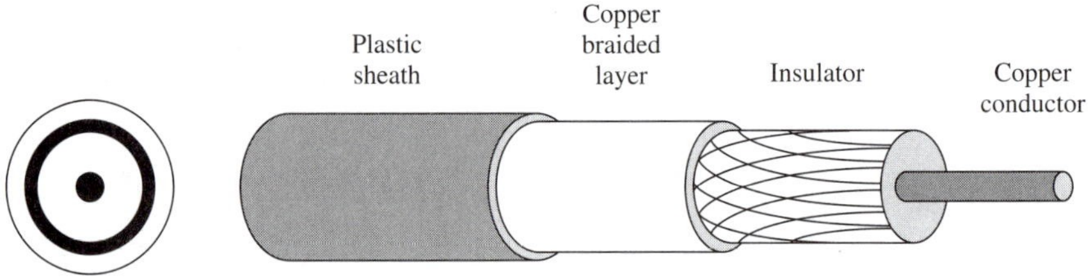

Figure 2.2 Coaxial cable structure.

2.2.3 Optical Fiber

The switch from analog to digital communications, the tremendous increase in the number of personal computers, and the popularity of Internet access have created a strong demand for communications media capable of carrying massive amounts of information. Wired pairs are adequate for transmitting telephone signals. For a network backbone that requires covering large areas and long distances at extremely high speeds (>45 Mbps), even coaxial cable cannot perform well.

An optical fiber is a thin glass or plastic core with a diameter of 2–125 μm surrounded by a cladding layer. The boundary between the core and the cladding is not visible because both components are made of transparent glass or plastic. The cladding has a lower index of refraction, meaning that light travels faster in the cladding than in the core. When light reaches the boundary between the core and the cladding, the change in velocity causes it to completely reflect or bend back toward the core (total internal reflection). This structure forms a flexible medium capable of carrying an optical ray over long distances, just as metallic waveguides carry microwave signals from a source to a destination with little loss (Figure 2.3). Optical fibers are so thin that hundreds of them can be bundled into a single cable. The outermost layer, surrounding one or a bundle of cladded fibers, is the jacket cover composed of plastic or other layered protective material.

The light source for fiber optic communications is either a laser diode or a light-emitting diode (LED). The light source at 8500, 13,000, or 15,000 Å (all near-infrared wavelengths), representing a digital bit stream, is switched on and off by an electronic digital pulse oscillator. A laser diode that emits a purer laser beam with a wavelength of 15,000 Å provides the highest bandwidth and covers the longest distance, while most local applications use an 8500 Å LED light source, at lower cost. A silicon photodiode senses light pulses from the optical fiber at the receiving end and recovers the original electrical signals. A simple intensity modulation method is used to transmit binary signals through optical fiber—a short pulse of light represents 1 and the absence of light stands for 0.

There are actually two basic types of optical fiber—multimode and single-mode fibers. The core of multimode fiber has a larger diameter, which causes light rays to be reflected with different angles inside the core. With multimode transmission, the presence of mul-

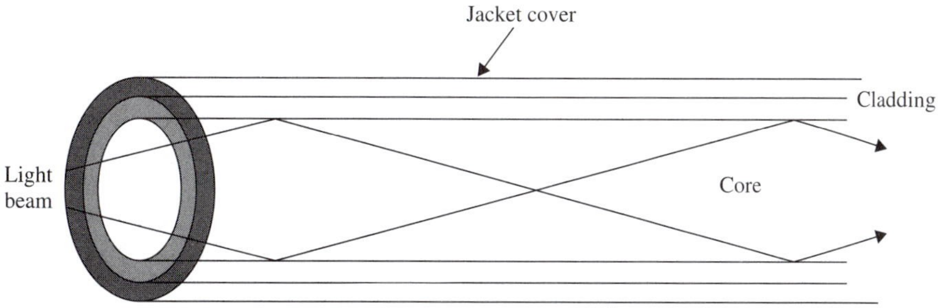

Figure 2.3 Optical fiber is made of a core surrounded by a reflective cladding layer with a jacket cover. Light beams traverse the core and are reflected at the boundary between the core and the cladding layer.

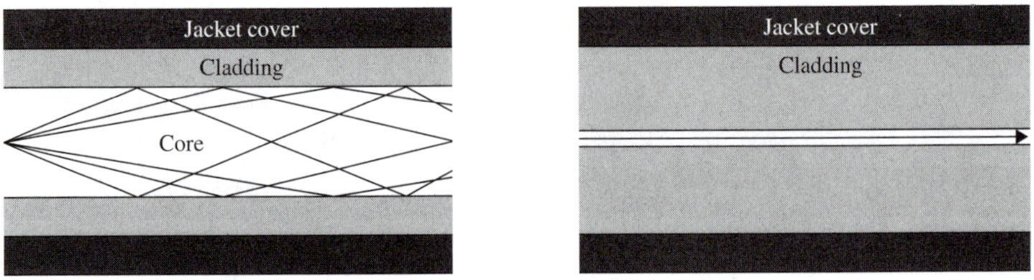

(a) (b)

Figure 2.4 (a) Multimode fiber creates multiple reflections, which limits the transmission rate. (b) Single-mode fiber allows only one transmission path, hence produces clean, sharp signals at the destination.

tiple propagation paths (modes) means different path lengths and different times of traversing fiber, and these discrepancies greatly limit the transmission rate. With a single-mode fiber, the diameter of the core is greatly reduced (to 1.5–5 μm, which is comparable to the carrying wavelength of light). This design, which allows only one transmission path for the light beam, produces a clean, sharp stream of signals at the receiving end (Figure 2.4). A very expensive, ultrapure single-mode glass fiber, which is thinner than a human hair yet stronger, inch for inch, than steel, can keep signal losses extremely low. Plastic fibers are much less costly, but these mainly multimode media still provide good performance, especially for shorter distance.

Optical fibers have tremendous advantages over their copper wire counterparts, not only in capacity but in distance covered, reliability, durability, security, weight and size, and cost.

• Because of its high frequency and low attenuation, the highest grade single-mode optical fiber may have a bandwidth/distance parameter as high as 1000 GHz/km, meaning that it can carry signals at 1000 GHz frequency to a distance of 1 km. Thus 50,000 telephone calls can be carried simultaneously on one optical fiber. An experimental system has reached data rates of 5 Gbps over a distance of 100 km without repeaters. In comparison, the bandwidth/distance parameter for wire pair and coaxial cable is about 1 and 20 MHz/km, respectively. (The actual data rate/distance parameters should be higher than bandwidth/distance parameters because of better modulation and coding techniques, as explained later.)

• Light signals are immune to interference by electromagnetic and radiofrequency waves and from noise generated by lightning, switching devices, nearby motors, and many other sources. The error rate with fiber optics can be as low as 10^{-10}. It is estimated that on average, there are about one to two failures per thousand sheath kilometers of optical fiber, with more than 80% of the failures attributable to cable cuts.

• Fiber optics are more durable than copper cables in adverse environments.

• Light signals do not radiate energy, thus they cause little interference and provide a high degree of security from eavesdropping. Optical fiber is also inherently difficult to tap into. (Although this is also a disadvantage for multipoint optical fiber lines, which are correspondingly difficult to split.)

- Optical fiber cables are understandably very small and very lightweight—about 1/6 the weight of coaxial cable for an equivalent span, with much higher capacity.

- Copper is becoming more and more expensive, while the source of silicone, or sand, is endless.

Light waves are capable of transmitting data at very high bandwidth. To date, only a fraction of the capacity of optical fiber has been exploited. All modern network backbones exclusively rely on fiber optics. Indeed, for several years the operational benefits of optical fiber have been reducing operating expenses and delivering reliable, enhanced signal quality to cable TV subscribers. Thus fiber deployment has been a key component of the cable TV industry's network modernization strategy, and cable TV has been the fastest growing fiber market segment in North America. Only two decades ago, optical fiber was a scientist's dream. Today, this extraordinary technology is a reality, transforming the way the world sends and receives information.

2.2.4 Radio and Microwave

It has been known for more than a hundred years that electromagnetic waves can carry information. Both radio waves and microwaves are electromagnetic waves, which are used for various communication purposes. Electromagnetic waves can be generated by inducing a high-frequency current in an antenna that has roughly the same dimensions as the wavelength. Because wavelength and frequency have the following relation:

$$c = f\lambda \tag{2.1}$$

where c is a constant, the speed of light in a vacuum (300,000 kim/s), and f and λ are frequency and wavelength, respectively. A higher frequency means a shorter wavelength, hence a shorter antenna. For example, a 1 MHz wave is 300 m long.

Radio frequency ranges from about 100 kHz to under 1 GHz (3000 m–30 cm wavelength), while microwave transmission is usually above 1 GHz. The higher the frequency, the more bandwidth and capacity can be provided. Lower frequency electromagnetic waves are widely used for radio broadcasting (100 kHz–50 MHz). TV stations use from 70 to about 300 MHz. The frequency for pagers, cellular phones, and PCS (personal communications service) ranges from 800 MHz to above 5 GHz. When the frequency is higher than 100 MHz, the signal can and must be focused in certain directions for effective transmission. Above 1 GHz, microwave is a directed line-of-sight transmission medium. Therefore, a microwave system always uses pairs of parabolic (dish) antennas, accurately aligned with each other, to concentrate all the energy into a small beam. Because of the limited curvature of the earth's surface, the maximum distance for microwave transmission is about 50 miles when antenna towers 100 m high are used for transmission. To cover larger distance, a series of relay stations must be constructed every 50 miles (Figure 2.5).

A receiving microwave dish antenna provides signal gain and a transmitting dish focuses the signal energy. The larger the dish size, the better the gain and focus. Microwave

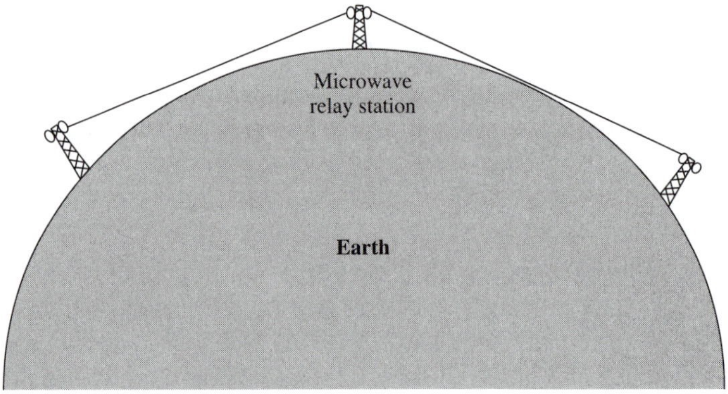

Figure 2.5 Because of the line-of-sight limit imposed by the curvature of the earth, the long distance transmission of microwave signals requires a series of relay stations 50 miles apart.

transmission is impractical at frequencies above, say, 10 GHz because at very short wavelengths, the signals are absorbed by rain.

Microwave has a high communications capacity and is relatively inexpensive. It used to be the dominant medium for long distance telephone transmission and is still widely used for wireless communications. Because of the limited electromagnetic wave spectrum, however, most of the usable regions have been allocated for different purposes, and a severe shortage of usable radio and microwave spectrum has developed. One way to solve the problem of lack of usable spectra is to prevent conflicts and interference by using less powerful radio and microwave signals in a tiny area. This is the fundamental concept for cellular communications and personal communications systems.

2.2.5 Satellite Communications

Satellite systems have been used for communications for many years. The world's first communications satellite, *Early Bird*, was launched in 1965. Since then, hundreds of communications satellites have been orbited for various purposes. In the 1970s geosynchronous satellites were successfully launched in Canada, which has greatly benefited from satellite communications because of its large area and the low population density—an environment seemingly made for this medium of communications.

The early satellites were passive devices that only reflected the uplink signals received from the ground back to the earth. Later, thanks to the solid state electronics techniques, which provide powerful transmitters and long-lasting solar batteries, active communications satellites became possible. Modern satellite communication systems are essentially two microwave transmission systems with three components (Figure 2.6):

- One or more uplink ground stations that transmit signals to the communication satellite
- The satellite, which receives signals from the uplink ground station, amplifies the signals, and retransmits the signals to the ground
- The ground receivers

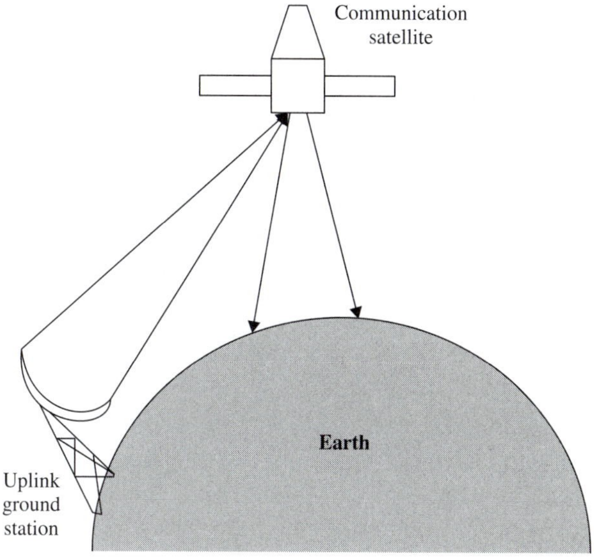

Figure 2.6 An uplink ground station transmits signals to a communication satellite, which in turn retransmits the signals to receivers on earth.

According to Kepler's law, the orbital period of a satellite is proportional to the radius to the 3/2 power. When a satellite is sent into orbit about 36,000 km (23,400 miles) above equator, its orbital period becomes about the same as that described by the revolution of the earth. Thus from the point of view from the earth the satellite appears to be static. This geosynchronous arrangement allows the use of a simple parabolic (dish) antenna to focus on the satellite for transmission and receiving. The orbiting station is called a geosynchronous satellite. Theoretically, three properly located geosynchronous satellites can cover the earth's surface (Figure 2.7), making possible communications between any two spots on the globe. Geosynchronous satellites have some unique features, including the following.

- Cover large areas of the earth's surface
- Very high frequency (2–40 GHz) and very high transmission bandwidth (as high as 100 Mbps)
- Long propagation delays (250–300 ms typical) due to the long transmission distances
- Limited number of geosynchronous satellite slots and heavy competition for the available wave spectrum

The earlier communications satellites operate in the 6 and 4 GHz bands, called the *C band*. A large parabolic antenna (3 m diameter dish) must be used for the ground station because of the longer wavelength and relatively weaker downlink signals. Increasingly, satellite are using higher frequencies in the 11/12/14 GHz (*Ku band*) and 20/30 GHz (*K band*) range, which use much smaller antenna [45–90 cm (18–36 in.) dishes]. Besides higher bandwidth and smaller size antenna, the *Ku* and *K* band satellites have another critical advantage—narrower and better focus. Even though there is plenty of space in the sky, the *C* band satellites, 36,000 km above the earth, operating at 6/4 GHz must be 2900 km

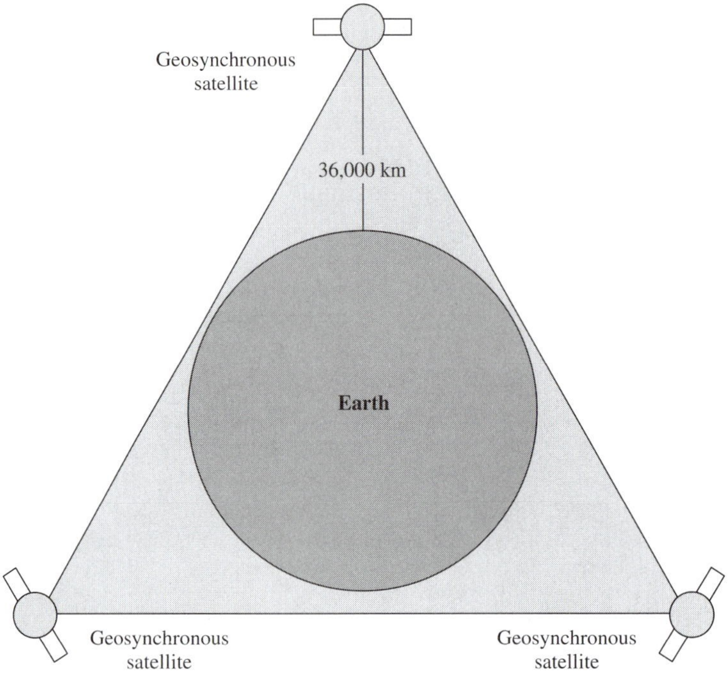

Figure 2.7 Theoretically, three geosynchronous satellites, each at a height of 36,000 km, can cover the entire surface of the earth.

(1800 miles) away from each other to prevent radio interference from adjacent satellites. That is equivalent to 4° of separation, meaning that at most 90 such satellites can be deployed above the equator. With the wavelength reduced and transmission more focused, *K* band satellites can be placed about 1° apart. This opens up many new slots in the crowded sky for increasing communications needs.

The newest addition to the family of communications satellites is the Direct Broadcasting Satellite (DBS). Since 1993, such devices have been launched to provide direct TV programs to homes and businesses equipped with digital satellite system (DSS) home receiving units. These high-powered satellites transmit signals at *K* bands via as many as sixteen 120 W transponders. A DSS system consists of an 18 in. (45 cm) dish, a digital set-top decoder box, and remote control. Programming is delivered to the uplink facility via satellite, optical fiber, and through the use of digital video tape. All programming is digitally compressed and encrypted, then transmitted to the DBS satellites.

The little DSS dishes are so easy to set up, and locking onto the signal is so easily done that many drivers are installing DSS on the cab of their trucks. Setup time can be less than a minute with a truck-mounted unit. The only problem with ultrahigh-frequency satellite communications is susceptibility to rain, particularly when the receiver is located far from the central receiving area. In such cases, the signals must travel through much thicker atmosphere at a larger angle, and as noted earlier, the short signal wavelengths under these conditions put signals at risk for absorption by raindrops.

Another relatively new application of communications satellites is the very small aperture terminal (VSAT). Ideal for wide-area distribution and corporate networking, VSAT satellite systems are becoming a major growth area in connecting remote locations with sparse populations where other means of communication are prohibitively expensive. A VSAT system has one or more powerful ground stations equipped with a mainframe computer as a hub for data processing and exchange. Smaller subscriber stations use small antennas (1 m) and small power ($\sim$1 W) to exchange messages at the *Ku* band with other subscribers through the hub. The primary use of VSAT is for data communications, but telephone services are often piggybacked in remote areas in developing countries that have no existing telephone network infrastructure.

Finally, satellites play critical role in mobile communications, cellular telephony, and in particular, in the Global Positioning System (GPS) and in personal communications service, introduced later in this chapter.

2.3 FUNDAMENTALS OF DIGITAL TRANSMISSION

Different media differ in their capacity for carrying information. Two types of transmission medium used for data communications are guided (wires, coaxial cables, and fiber optics) and unguided (radio and microwave) media. With the same medium, the transmission data rate is dependent on the frequency used. This section gives a summary on the frequency spectrum for various media, examines the factors that decide the maximum communications capacity, and discusses the kinds of relation that exist among frequency, bandwidth, and transmission data rate, as well as related theories.

2.3.1 Frequency Spectrum, Bandwidth, and Data Rate

For a periodic signal, its period T is the amount of time for one cycle duration. There is a well-known and simple relationship between T and f, the frequency in hertz of the periodic signal:

$$f = \frac{1}{T} \tag{2.2}$$

A higher frequency signal has a shorter period. For example, a signal of 1 MHz has a period of 1 microsecond (μs), for a signal of 1 GHz, $T = 1$ ns (nanosecond).

On the other hand, a periodic signal's frequency f is inversely proportional to its wavelength because the speed of the wave transmission is considered to be constant, usually the speed of light—about 300,000 km/s (or 187,500 miles/s) as described in Equation (2.1). That is, higher frequency means shorter wavelength. Figure 2.8 gives the frequency ranges of the frequently used transmission media and electromagnetic waves.

The simplest digital signal can be represented by a sequence of square waves in electrical form. However, the ideal square wave is extremely difficult to obtain because of the limits of many physical properties—(existence of resistance, stray capacitance, inductance, etc.). The resulting waves may look like the one shown in Figure 2.9.

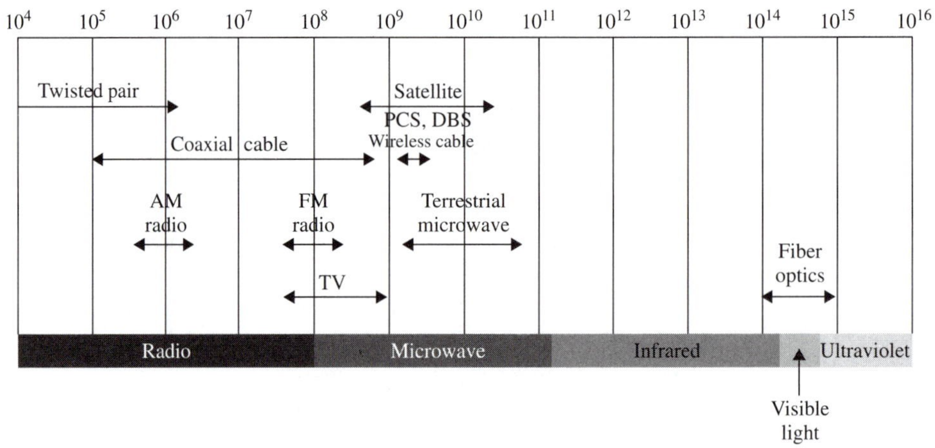

Figure 2.8 Frequency ranges of the most useful transmission media and electromagnetic waves.

Most signals carried by communication channels are modulated forms of sine waves. A sine wave is described mathematically by the expression

$$s(t) = A \sin(2\pi f t + \phi) \tag{2.3}$$

where A, f, and ϕ are the *amplitude*, *frequency*, and *phase* of the sine wave, respectively. (Note that, if the ϕ equals $n\pi/2$, where n is an integer, then $s(t)$ is actually a cosine wave.) The sine wave has a single frequency f.

In fact, according to the well-known *Fourier analysis* theory:

Any signal of time may be represented as the sum of a set of sine waves of different frequencies and phases.

Mathematically, we write

$$s(t) = A_1\sin(2\pi f_1 t + \phi_1) + A_2\sin(2\pi f_2 t + \phi_2) + A_3\sin(2\pi f_3 t + \phi_3) + \cdots \tag{2.4}$$

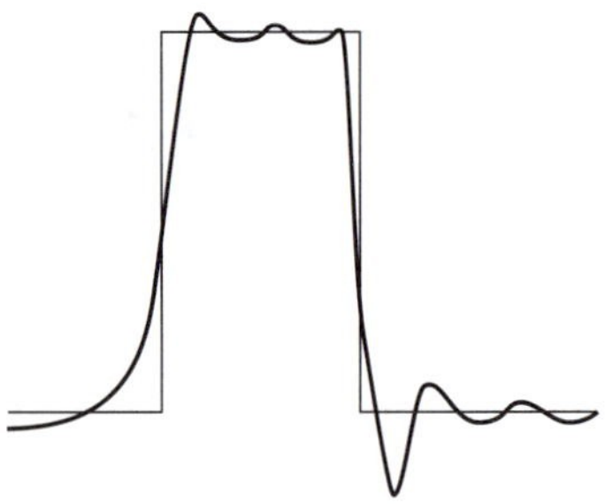

Figure 2.9 An ideal square wave and the actual wave, distorted by many factors.

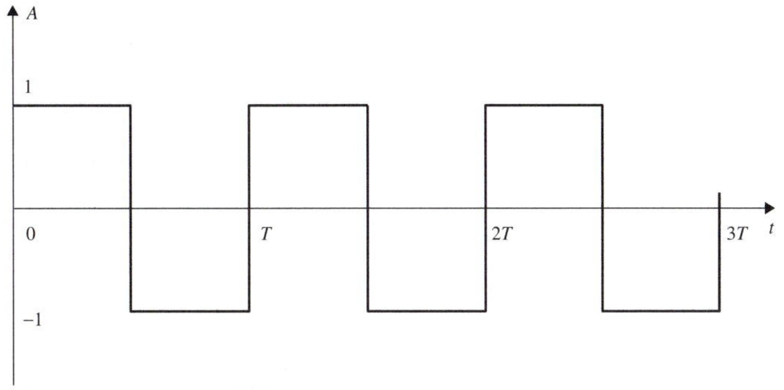

Figure 2.10 A square wave that can be decomposed into an infinite number of harmonic sine waves.

Generally speaking, any periodic wave with period T can be decomposed into one *fundamental* sine or cosine wave with frequency $f = 1/T$ and a series of *harmonic* sine and cosine waves, which have fractions of period T, or an integer multiples of the fundamental frequency f. It is possible that there are an infinite number of harmonic waves for a specific wave. All the harmonics may have different amplitudes, which are the functions of the frequency. The sum of all the root-mean-square (rms) amplitudes of the component waves equals the square root of the original signal amplitude. It can be easily proved that to compose an ideal square wave, a sine wave with infinite frequency must be generated. That is why it is impossible in practice to produce an ideal square wave.

As an example, consider the square wave (Figure 2.10):

$$\begin{aligned} s(t) &= 1;\ 0 < t < \pi,\ 2\pi < t < 3\pi,\ \cdots \\ &= -1;\ \pi < t < 2\pi,\ 3\pi < t < 4\pi,\ \cdots \end{aligned} \tag{2.5}$$

The corresponding Fourier series to the square wave is

$$s(t) = \sin(t) + (1/3)\sin(3t) + (1/5)\sin(5t) + \cdots \tag{2.6}$$

In other words, this square wave can be represented by an infinite number of sine waves with increasing frequencies and decreasing amplitudes. Or, adding all the sine wave signals in Figure 2.11 together will approximately recover the original square wave of Figure 2.10. The corresponding graph of the spectrum has a line at the odd harmonic frequencies, 1, 3, 5, . . . , whose respective amplitudes decay as 1, 1/3, 1/5, . . . (Figure 2.12).

The description of a signal in terms of its constituent frequencies, or the frequency range that a signal contains or a communications line can carry, is called the *frequency spectrum* of the signal or the communications line. The *bandwidth* is defined as the width of the spectrum. Therefore if a signal consists of a group of sinusoid waves ranging from a lower frequency f_1 to a higher frequency f_2, this signal's bandwidth is $f_2 - f_1$. The effective bandwidth is largely determined by a limited number of the dominant components' frequencies. Signals whose spectra consist of isolated lines are periodic; that is, they repeat themselves indefinitely. The lines in this spectrum are infinitely thin; they have zero bandwidth. Shannon's law as described later, tells us that the maximum information rate

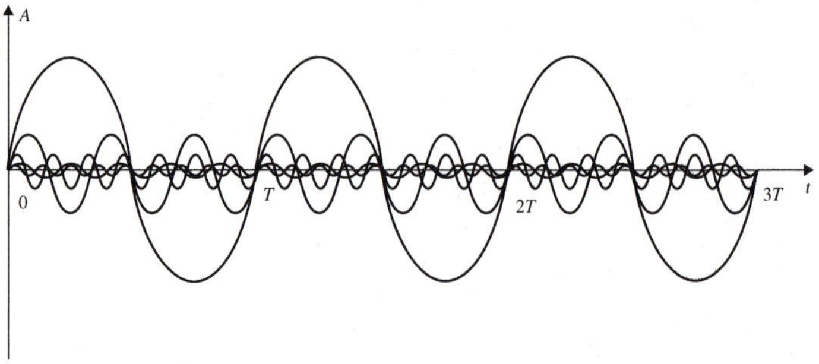

Figure 2.11 Adding all the sine waves described by Equation (2.6) results in a square wave.

of a zero-bandwidth channel is zero. Thus, zero-bandwidth signals carry no information. If a signal contains a zero-frequency component, that component is a direct current (dc) component, or constant component. The dc component of a signal can be easily filtered out by a capacitor.

All transmission media and communication lines are limited in bandwidth, which in turn limits the signals' high-frequency components. This bandwidth limit distorts the original signals and hence limits the maximum data transmission rate. In general, the lower the bandwidth, the more distortion is caused in the original signal, and the more probable transmission errors are. The higher the bandwidth, the greater the transmission quality. Figure 2.13 shows the effects of different bandwidths on a binary digital signal.

To transmit digital signals on an analog network, each interface must contain electronics to convert the outgoing bit stream to an analog signal and the incoming analog signal to a bit stream. Depending on the type of electronic device used, 1 bit per second (bps) may occupy roughly 1 Hz of bandwidth for wired transmission media. At higher frequencies, advanced modulation and coding techniques make possible multiple bits per

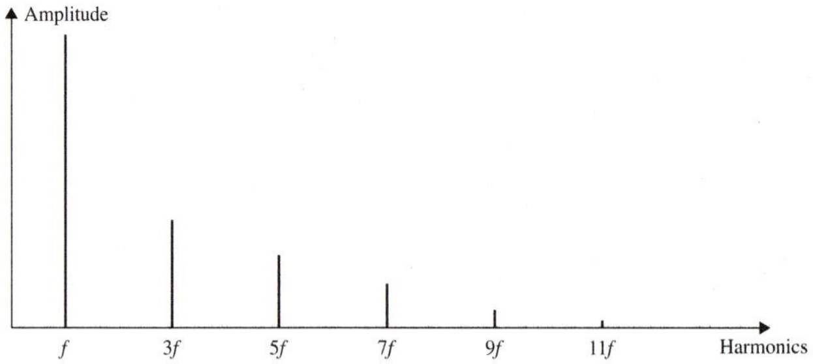

Figure 2.12 The frequency spectrum of the square wave of Figure 2.10.

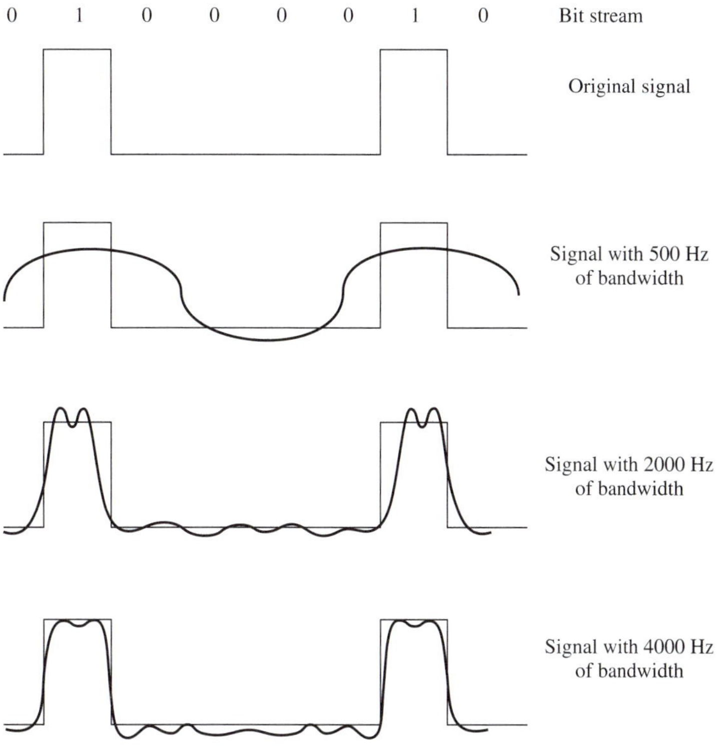

Figure 2.13 Comparison of the effects of different bandwidths on a digital signal.

second per hertz. Shannon's law gives a more quantitative relation between the data rate and bandwidth (along with other factors).

2.3.2 Shannon's Law

The transmission capacity, measured in bits per second, is an important measure for digital communications system. It provides us with some idea of transmission quality and speed. For example, we can figure out that an image file of size 2 Mbyte needs about 5 minutes to download from an Internet site by a 56 Kbps modem. The same image will take about 3 seconds to download via ADSL asynchronous digital subscriber line (ADSL, see Chapter 6). The voice grade telephone line has a very limited bandwidth (about 3000 Hz), and this is the dominant factor limiting the data transmission rate at which communications can be carried.

Besides bandwidth, signal power and noise on the communications channel limit the data rate. Increasing signal strength apparently helps to increase the maximum data transmission rate and extends the distance of signal propagation. But excessive power increases the heat generated by the transmission and switching devices. For wireless networks, higher power increases interference with other communications. More power, of course, means high cost, too. In any event, according to Shannon's law, discussed shortly, compared with bandwidth, increases in transmission power play a relatively insignificant role in increasing communications capacity.

Noise is generated from various sources: mechanical and electrical switches, motors, lightning, and so on. The most important kind of noise actually comes from the random and constant movement of electrons and molecules, which always exists in all objects that have a temperature above absolute zero ($-273°C$). This inherent noise is called thermal noise, Gaussian noise, white noise, or background noise. Noise also limits transmission capacity because when it is strong enough relative to the original signal, noise may change the signal from 1 to 0 or from 0 to 1. When bandwidth or data rate increases, the interval between bits shrinks. The same noise pattern will affect more signal bits and cause a higher error rate. So usually, increasing the bandwidth also increases the level of noise and transmission error rate at the same noise level. The extent to which perfect transmission is approached is measured by the bit error rate (BER), which is measured over a certain time interval and calculated by the following formula:

$$\mathrm{BER} = \frac{B_e}{RT} \tag{2.7}$$

where R is the channel speed (bps), B_e is the number of bits in error, and T is the measurement period, in seconds. Typically acceptable BERs are of the order 10^{-6} to 10^{-7}. An important measurement of the communications channel quality is the ratio of signal power *(S)* to noise power *(N)*. Fundamental communications theory—*Shannon's law*—describes the relation among the maximum channel capacity, bandwidth, and the *S/N* ratio as follows:

$$C = B \log_2(1 + S/N) \tag{2.8}$$

where C is the maximum capacity (bps), B is the channel bandwidth (Hz), and *S/N* is the ratio of signal power to noise power. The ratio between S and N is usually measured in decibels (dB).

$$N_{\mathrm{dB}} = 10 \log_{10}(P_1/P_2) \tag{2.9}$$

where N_{dB} is the number of decibels, and P_1 and P_2 are the relative power values to be measured. The decibel is the most suitable unit for indicating the gain or loss of signal levels, including electrical, voice, light, and so on. A 100-fold power gain is represented by 20 dB as $\log_{10} 100 = 2$. Consider a voice grade telephone line operating at frequency range of 300–3400 Hz. It has a bandwidth of 3100 Hz and an *S/N* level of about 30 dB—equivalent to an *S/N* ratio of 1000:1 (only thermal noise is considered). By Shannon's law, the maximum data rate will be:

$$C = 3100 \log_2 (1 + 1000) = 30{,}898 \text{ (bps)}$$

This is the theoretical maximum data rate achievable by a regular "analog" modem on an analog voice grade telephone line. The newer dial-up devices, such as the ISDN and HDSL adapters, discussed in Chapter 6, are another story. To achieve higher rates, a close to perfect line with a much higher signal-to-noise ratio is required; alternatively, compression techniques can be applied. From Figure 2.14, we can see that bandwidth plays a more important role in increasing the data transmission rate than increasing signal level because

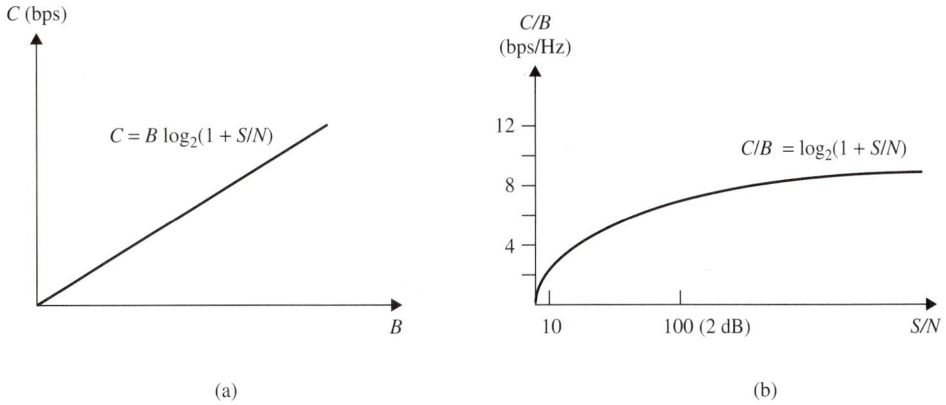

Figure 2.14 Based on Shannon's law (a) C and B have a linear relationship when S/N is a constant. (b) C/B increase slowly when S increases.

of the logarithmic relationship between channel capacity and signal level. Because the bandwidth and channel capacity are so closely related, sometimes the two terms are used interchangeably. Note that Shannon's law gives only the theoretical upper limit; it does not indicate how the channel capacity is achieved. Providing high channel efficiency is the goal of modulation and coding techniques.

2.3.3 The Sampling (Nyquist) Theorem

To allow signals to carry information, nonperiodic change of the signals must be present. The consequence of a nonperiodic change is to introduce a spread of frequencies into the signals. For a digital computer to process analog information or send analog signals over a digital communication system, analog signals must be converted into digital ones. An analog-to-digital (A/D) converter performs this process. The analog signal is sampled (i.e., measured at periodic intervals), and then quantified (i.e., converted to discrete numeric values). The greater the number of quantification levels, the lesser the quantification error. A digital-to-analog (D/A) converter performs the converse operation. It was proved that *a signal of bandwidth B may change in a nonperiodic fashion at a maximum rate of 2B.* In other words, any analog signal can be reconstructed from its samples if sufficient samples in a unit time period are taken at a rate at least twice the signal's bandwidth:

$$f_s \geq 2B \tag{2.10}$$

This is the sampling theorem, proved mathematically by H. Nyquist in 1924 at AT&T's Bell Telephone Laboratories. This sampling rate is known as the Nyquist rate in honor of its discoverer. The sampling theorem makes no observation concerning the number of the sampling levels. Usually, the sample is assigned a level based on the signal amplitude at the time of the sampling. The more levels a sample can take, the higher the information rate required. Thus, if each time the signal changes it can take one of n levels, the information rate is increased to

$$R = 2B\log_2(n) \quad \text{(bps)} \tag{2.11}$$

This means that the rate at which an A/D converter generates bits depends on how many bits the converter uses for each sampling:

$$R = 2SB \tag{2.12}$$

For example, a speech signal has an approximate bandwidth of 4000 Hz. If this is sampled at the Nyquist sampling rate by an 8-bit A/D converter ($n = 256$ sampling levels), the bit rate R is:

$$R = 2 \times 8 \text{ bits} \times 4000 \text{ Hz} = 64{,}000 \text{ bps}$$

Voice grade telephone signals actually occupy a 4 kHz bandwidth with a 500 Hz guard band in each side of the band, so a T1 carrier system uses a rate of 8000 samples each second, and a T1 channel contains 24 digitized conversations, each at a data rate of 64,000 bps.

Equation (2.11) states that as n tends to infinity, so the information rate changes logarithmically. The noise level sets the actual limit. If we continue to subdivide the magnitude of the sample levels into ever decreasing intervals, we reach the point of being unable to distinguish the individual levels because of the presence of noise. Noise therefore places a limit on the maximum rate at which we can transfer information. Obviously, this limit is expressed and governed by Shannon's law.

2.4 DIGITIZATION AND MULTILEVEL TRANSMISSION

The process of converting an analog signal into a binary code is called digitizing. Digitizing an analog signal consists of three steps: sampling, quantization, and encoding. We discuss various digitizing techniques here. The details of coding techniques are covered in Chapter 3.

As we discussed in the last section, to first convert an analog signal into binary data and then reconstruct the analog signal completely from the binary data, the signal must be sampled at regular intervals of time. The sampling rate must be higher than twice the highest significant signal frequency for the samples to contain all the information of the original signal, which can be restored by means of a low-pass filter using the sample.

2.4.1 Pulse Code Modulation (PCM)

Pulse code modulation (PCM) is the most frequently used pulse modulation method in digital communication systems. Even though PCM contains the word "modulation," it is more a digitizing technique than another type of modulation that is intensively used by modem—the device that converts digital signals into analog to be transmitted through analog channel, in particular by the legacy analog telephone line.

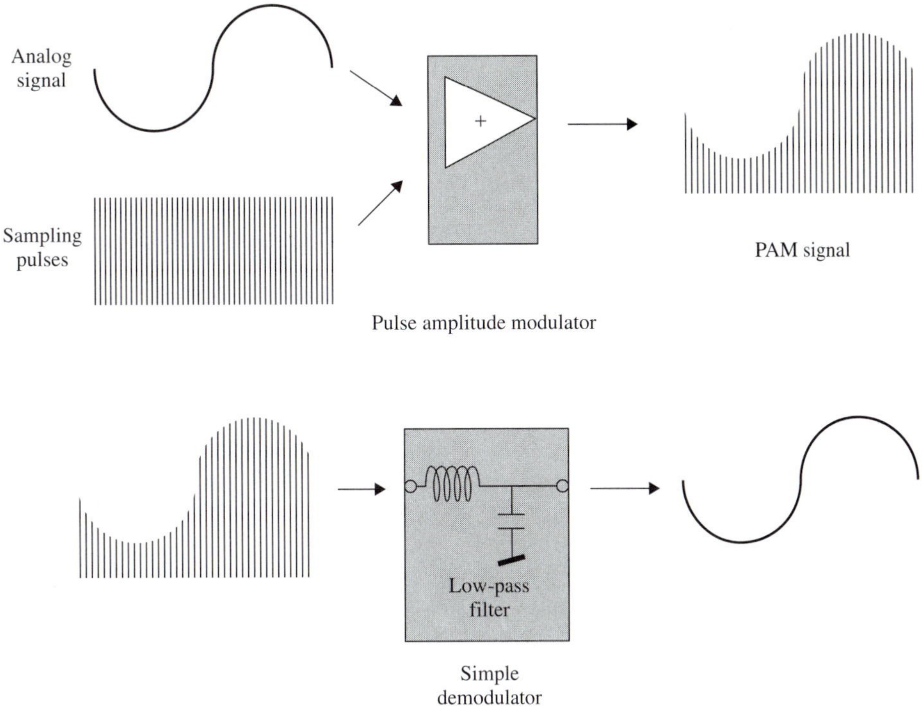

Figure 2.15 A pulse amplitude modulator adds the analog signal to the sample pulses, producing a PAM signal. The demodulator could be simplified as a low-pass filter that allows only low-frequency signal to pass through.

PCM converts the samples of an analog signal into a series of 1's and 0's to allow signal to be transmitted over a digital communication medium. The first step of PCM is to do an analog modulation called pulse amplitude modulation (PAM), in which the original analog signal with bandwidth B is sampled at a rate $2B$, or with a period of $1/2B$ second. These samples are taken as narrow pulses, whose amplitude is proportional to the value of the original signal (Figure 2.15).

Note that without further processing, the output of PAM is still analog, or continuous numbers, which are useful in their own right. The main disadvantages of PAM are those inherent in any other analog system:

- PAM requires high-bandwidth analog amplifiers to be transmitted any distance.
- PAM signal is very susceptible to noise. Once a PAM signal has been contaminated by noise, the noise is difficult to remove.

To produce binary output, or digital PCM data, the PAM samples are quantized by an A/D converter, in which the amplitude of each PAM pulse is approximated by an n-bit integer. In Figure 2.16 $n = 3$, so 8 (2^3) levels are used to represent the PAM pulses. The levels that can be represented by the binary number are called quantizing levels. The

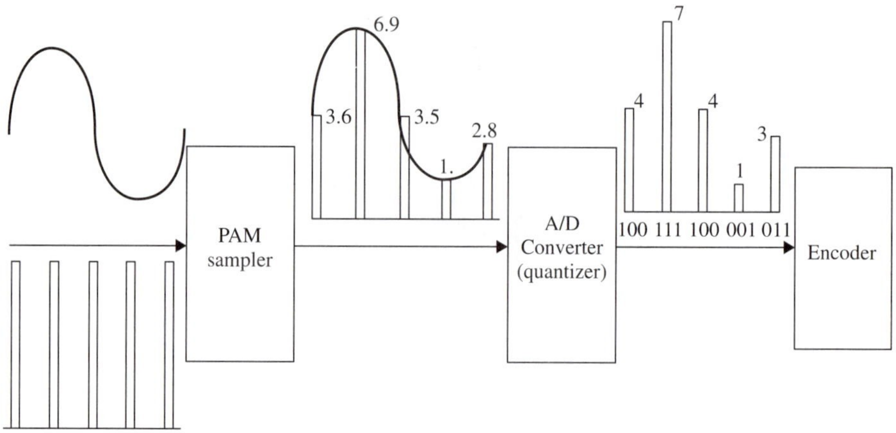

Figure 2.16 Pulse code modulation (PCM).

quantized data still need more processing to be encoded prior to transmission over a digital communication device.

To restore the original analog signal, the digital data must be demodulated. The demodulator circuit can be a simple low-pass filter (Figure 2.15), in which the capacitor charges with each input pulse and slowly discharges between the pulses. The slow discharge smoothes out the spaces between the pulses and produces an output signal closely resembling the original analog wave form. Figure 2.17 gives a block diagram of complete

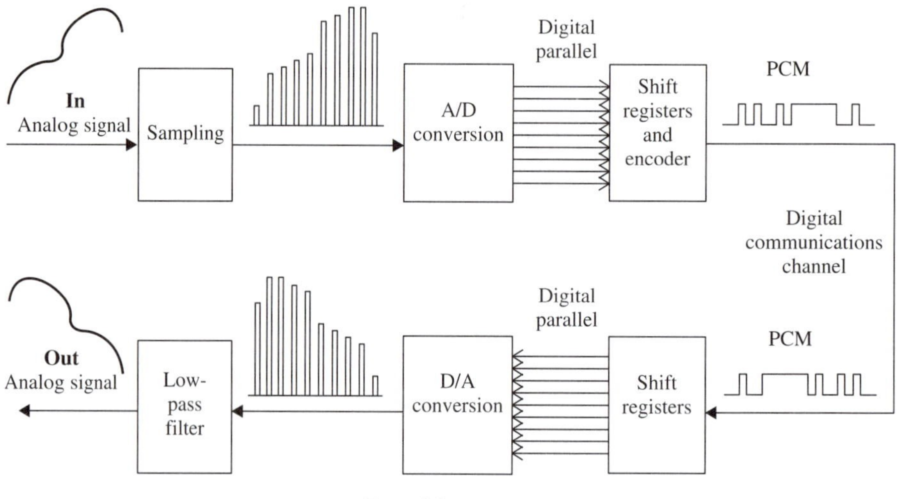

Figure 2.17 Block diagram of a PCM modulator and demodulator.

PCM modulation and demodulation procedures. A coder/decoder can be included in a single integrated circuit (IC) that performs all the functions shown in the diagram and can convert an analog signal to PCM and restore analog signal from the PCM code.

2.4.2 Advanced Digitization Techniques: CPCM, DPCM, and ADPCM

The voice grade telephone line has an actual bandwidth 4000 Hz, which is sampled 8000 times per second, with 8 bits representing each sample, thus resulting in a 64 Kbps voice digitization rate (VDR). The quantizing steps are noncontinuous, thus introducing discrete errors to analog signals. The resulting inaccuracy is called quantizing noise. Low-level signals are particularly susceptible to quantizing noise when linear quantization technologies are used. To improve quality and reduce channel capacity requirement, a variety of quantization schemes have been presented.

The companded (compress–expand) *PCM* uses nonlinear quantizing levels to step up lower amplitude signals, effectively boosting the weaker signals and attenuating the larger amplitudes (Figure 2.18). At the receiver side, the compress–expand procedure is reversed to restore the original signals. Because human ear sensitivity is measured logarithmically, and the ear is more sensitive to low sound levels, the adaptive enhancement of CPCM is particularly significant. AT&T's D2 channel band system uses this technique and achieves superb quality in voice reproduction.

Considering that the signal changes relatively slowly in comparison to the sampling frequency, the variation of PCM method called differential pulse code modulation (DPCM) digitizes not the amplitude itself but the difference between the current value of amplitude and the preceding one. Thus 64 levels (using 6 bits), instead 256 levels, might be adequate to describe the amplitude changes. Thus a channel capacity of only 48 Kbps is sufficient to transmit a conversation, since speech transmission is less sensitive to occasional errors.

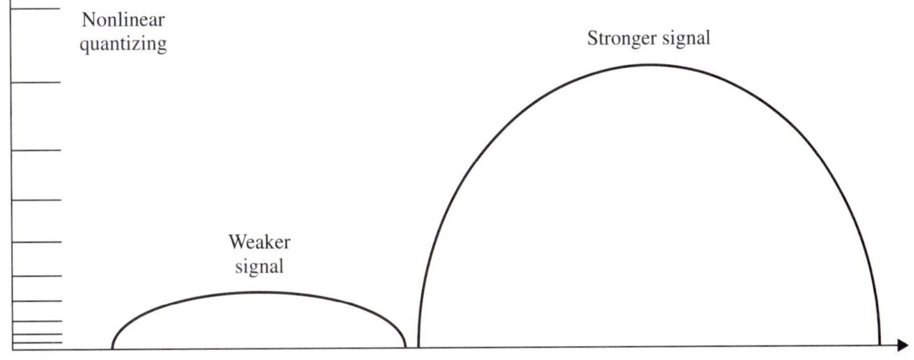

Figure 2.18 Companded PCM uses nonlinear quantizing steps to reduce quantizing error for low-level signals.

Combining the ideas of CPCM and DPCM results in a more complex and more advanced version of PCM. When signal amplitude is greater, the possible changes of amplitude tend to be larger, and this more sophisticated version of DPCM dynamically alters the quantizing levels based on the input signal's amplitude. This approach is called adaptive differential pulse code modulation (ADPCM), which can further reduce the requirement for channel capacity to 32 bps or even lower. This allows a regular voice grade telephone line to provide two simultaneous conversations of acceptable quality. ADPCM is a practical encoding and data compression technique that has wide applications, including use in digital video disc (DVD) devices.

Other variations of PCM exist, and they are attractive from the standpoint of reducing the voice digitizing rate.

2.5 MODULATION TECHNIQUES AND MODEMS

The direct transmission of most natural signals (sound, light, etc.) in their original form, is very inefficient, even after conversion into electrical signals, and thus some form of modulation is required. Originally used by radio systems, modulation "adds" the original low-frequency signal to a higher frequency electromagnetic wave to be transmitted into the air through an antenna. In such applications the higher frequency electromagnetic wave is called the carrier wave. For digital communications:

Modulation is the process of encoding the source data in digital form onto a carrier signal to be transmitted through an analog channel.

A modem (*mod*ulator/*dem*odulator) is a device that converts digital signals into analog to be transmitted through an analog channel, in particular a telephone line. Its operation is based on Equation (2.3):

$$s(t) = A \sin(2 \pi f t + \phi)$$

Here, altering either A, f, or ϕ (Figure 2.19) in the carrier sine wave function comprises one of the three types of basic modulation techniques: amplitude modulation (AM), frequency modulation (FM), and phase modulation (PM). These three modulation methods, sometimes called amplitude shift keying (ASK), frequency shift keying (FSK), and phase shift keying (PSK), all use sine waves as carrier waves, and the resulting signals have a bandwidth centered on the carrier wave frequency. All these methods are conceptually and practically simple and are used in practice. A modem device may use one method or a combination of types to achieve better performance and bandwidth utilization.

2.5.1 Amplitude Modulation

The AM method "modulates" the carrier wave's *amplitude* in accordance with the modulating digital bit stream (Figure 2.20). The amplitude of the carrier wave is raised or low-

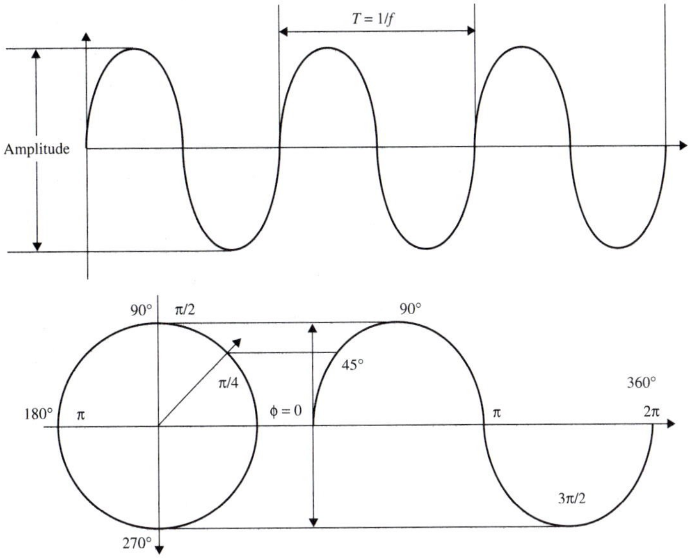

Figure 2.19 The amplitude, frequency, and phase of a carrier sine wave.

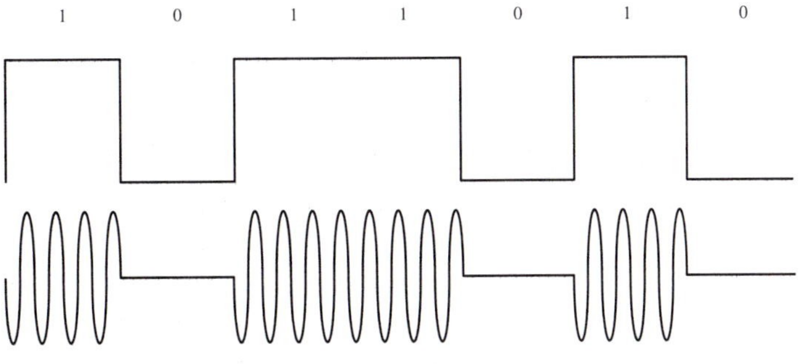

Figure 2.20 Amplitude modulation alters carrier wave's amplitudes, while keeping frequency and phase fixed.

ered to represent a 1 or a 0. Mathematically, under AM, the carrier wave can be expressed as follows:

$$s(t) = \begin{cases} 0 & \text{binary } 0 \\ A_1 \sin(2\pi f t + \phi) & \text{binary } 1 \end{cases}$$

AM translates the spectrum of the modulating signal to the carrier frequency, although the bandwidth of the signal remains unchanged. This allows AM to simply shift the signal spectrum to a more suitable value to be transmitted and then restored by the receiver. In this simplest form, AM is inefficient and sensitive to distortion, so it is rarely used alone in a modem faster than 1200 bps. However, combined with phase modulation, AM adds another dimension to the modulation providing a performance superior to either AM, FM, or PM alone.

2.5.2 Frequency Modulation

Frequency modulation was originally developed to provide high-fidelity sound for use in FM radio and the TV sound channels (Figure 2.21). With digital FM, this method alters the frequency of the carrier wave in accordance with the digital signal, while the amplitude and phase of the carrier wave are kept constant. In its plain form, a binary 1 (the mark) is represented by a certain frequency and a binary 0 (the space) by another (Figure 2.22). The carrier wave under FM can be mathematically expressed as follows:

$$s(t) = \begin{cases} A \sin(2\pi f_0 t + \phi) & \text{binary } 0 \\ A \sin(2\pi f_1 t + \phi) & \text{binary } 1 \end{cases}$$

FSK is classified as wideband if the difference between the two carrier frequencies is larger than the bandwidth of the spectrums of the modulating signal or narrowband when the difference between carrier wave frequencies is smaller than the bandwidth of the original signal. FSK was the first type of modulation to be widely used for data communications because it is less susceptible to noise interference than AM modems. Today,

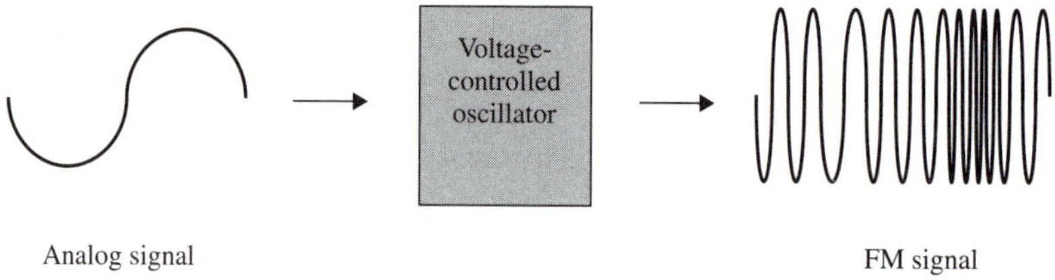

Figure 2.21 A frequency modulator consists of a voltage-controlled oscillator integrated circuit (IC).

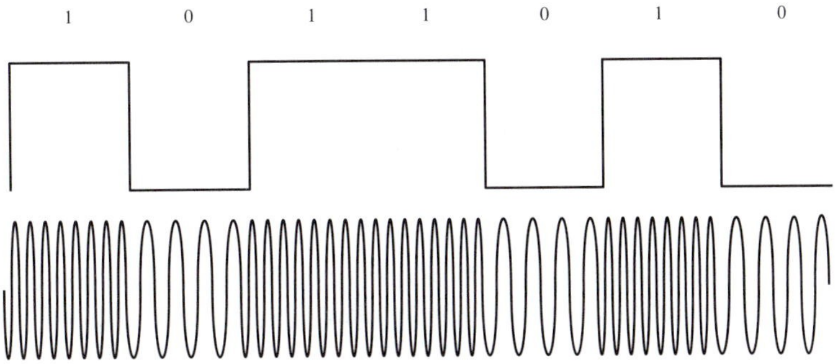

Figure 2.22 Frequency shift keying modulation uses different frequencies to represent 0 and 1.

however, it is used only in lower speed modems (≤1200 bps). For practical purpose, besides FM radio, FSK is also used in LANs at higher frequencies that use coaxial cable.

2.5.3 Phase Modulation

Phase modulation is the modulation technique that alters the phase of the carrier sine wave. Mathematically:

$$s(t) = \begin{cases} A\sin(2\pi f t + \phi_0) & \text{binary } 0 \\ A\sin(2\pi f t + \phi_1) & \text{binary } 1 \end{cases}$$

Figure 2.23 shows the simplest phase modulation method, binary phase shift keying (BPSK), where the signal is changed from 0 to 1 by shifting the phase angle 180° (or π radians). The practical phase modulation techniques are much more sophisticated than this simple approach.

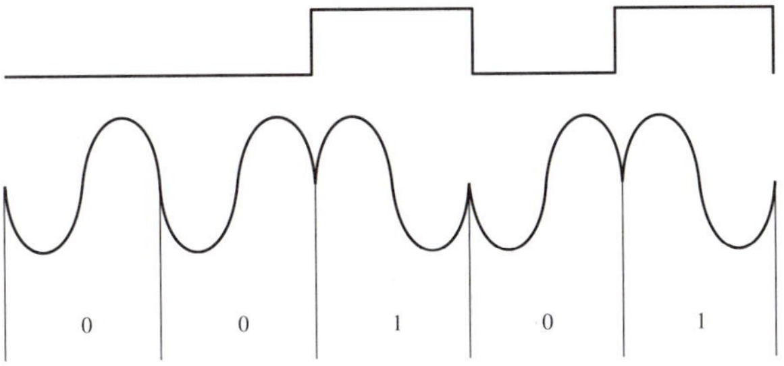

Figure 2.23 Binary phase shift keying, the simplest phase modulation scheme.

The number of times the signal parameter (amplitude, frequency, phase) is changed per second is called the modulation rate or signaling rate. It is measured in baud (1 baud = 1 change per second). Equation (2.13) describe the relation among data rate R, baud rate B, and the number of different signal elements (levels) L:

$$B = \frac{R}{\log_2 L} \tag{2.13}$$

With binary modulations such as ASK, FSK, and BPSK, the baud rate equals the bit rate. With quadrature phase shift keying (QPSK), the bit rate doubles the baud rate. A simple way of increasing the number of signal elements (levels) in the transmission without increasing the bandwidth is by introducing smaller phase shifts. QPSK modulation uses four different phases. For a given bit rate, QPSK requires half the bandwidth of PSK and is widely used for this reason.

A combination of PSK and ASK modulation methods called quadrature amplitude modulation (16QAM) uses 12 different phases, among which four have two amplitude values (Figure 2.24), thus yielding 16 signal elements ($l = 4$, $L = 16$). At the modulation rate $B = 2400$ baud, the modem can transmit 4 bits at once, providing an effective data rate of 9600 bps. This is the ITU v.32 modem standard. Table 2.1 is the truth table for this modem.

2.5.4 Modem Technology

Modems of the new generation often contain very powerful digital signal processors (DSPs) and general-purpose microprocessors like the other sophisticated network devices (switches, routers, etc.). These devices enable the modems to perform routine control functions such as dialing telephone numbers and interpreting commands from a computer, as well as the tasks of data compression and error detection and correction (Figure 2.25).

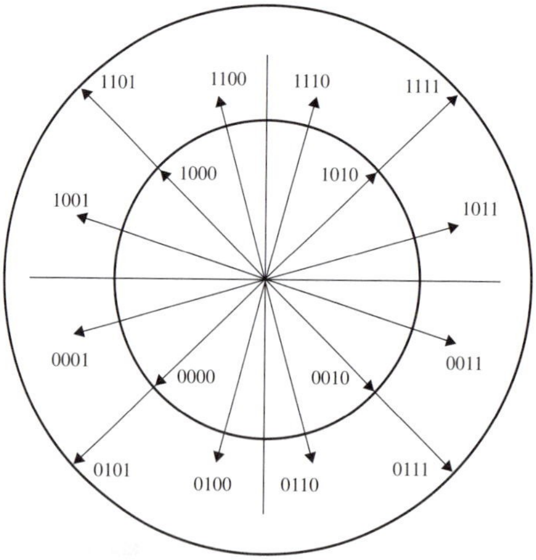

Figure 2.24 The 16QAM constellation diagram shows the 16 combinations of phase angles and amplitudes that can transmit 4 bits per time unit (baud).

TABLE 2.1
Truth Table for the v.32 Modem

Binary Code	Amplitude (V)	Phase (°)
0000	0.311	−135
0001	0.850	−165
0010	0.311	−45
0011	0.850	−15
0100	0.850	−105
0101	1.161	−135
0110	0.850	−75
0111	1.161	−45
1000	0.311	135
1001	0.850	175
1010	0.311	45
1011	0.850	15
1100	0.850	105
1101	1.161	135
1110	0.850	75
1111	1.161	45

Microprocessor and DSPs execute programs that are "burned" into the read-only memory (ROM) or stored in flash memory—a newer type of writable nonvolatile memory. In the latter case, the modem can be upgraded by downloading programs with new features from the manufacturer's web site. Another type of modem, called the softmodem, or host signal processor (HSP) modem, relies heavily on software, cheaper application-specific integrated circuit (ASIC) chips, and the processing power of the host computer. HSP modems can also be upgraded via a software download.

The newer V.90 modems provide speeds of up to 56 Kbps. The asymmetrical modem has a speed of 56 Kbps in one direction and 14,400 bps in the other, which is a perfect fit for World Wide Web Internet access, where downloading files and images de-

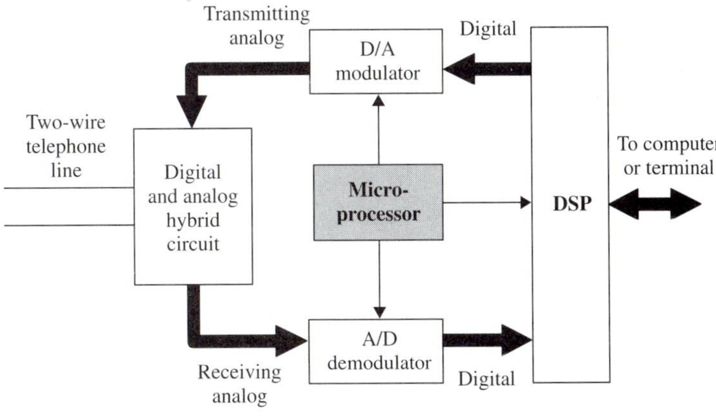

Figure 2.25 Newer modems use a powerful microprocessor and a digital signal processor (DSP) to control operations and convert signals.

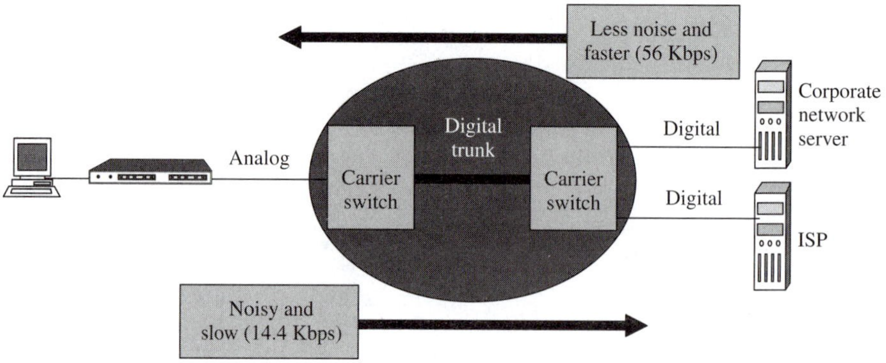

Figure 2.26 An asymmetrical modem is faster in the downstream direction than in the upstream direction.

mand high speed, while upstream transmission is not very demanding. This kind of modem takes advantage of two facts:

1. The subscriber loop between an end user and the carrier's central office is the only part that is analog over most of its length to the Internet service provider (ISP). The connection between ISP/corporate network and the carrier switch usually is digital (T1, T3, ISDN, etc.).
2. The downstream traffic is free of the quantizing noise introduced when analog signals are digitized and fed into the carrier's digital network.

Therefore for an Internet connection, the signals are converted from analog to digital at only one place, where the analog line meets the central office, and the quantizing noise introduced at this point reduces speed. In the other direction, from digital to analog, the downstream speed can be much faster (Figure 2.26).

While broadband access methods such as DSL and cable modems are making inroads against dial-up modems, analog modems may still dominate remote access for few more years. The V.90 56K bps modems have been popular and were touted as the end-all in analog modems. However, a new standard V.92 came at the end of 2000. Many modem vendors now offer V.90 modem owners software upgrade to the V.92 standard. The V.92 modem provides

1. Shorter setup times to as low as 10 seconds instead of 15–30 seconds
2. Faster and adjustable uplink speed at 48 kbps rather than 33.6 kbps to favor either the downstream or upstream transmission or to balance the two.
3. The call-waiting or on-hold feature that can put the data connection on hold while taking a voice incoming call without breaking the data connection.

The redesign of the upstream modulation to accommodate the PCM codec required the greatest technological effort. Current modems assume that each connection is from a new location. In reality, the same phone lines connect repeatedly. With V.92, the training handshake process has been designed to be intelligent and flexible. Quick connect feature shortens a modem's time to learn a phone line's characteristics by reusing some information previously learned. The on-hold feature lets modems gracefully break a connection and stand by while another call is taken. Used in conjunction with quick connection, the on-hold feature can make resumption of the data call faster and more seamless. Furthermore, by enabling servers to pause rather than drop long-idled users, the on-hold feature can greatly enhance quality of service and potentially increase worker productivity.

A fast modem automatically falls back to a slower speed when it encounters a lower speed modem at the other end of the communication. Most modems that use the v.42bis standard also have MNP-5 capability, and they will fall back to MNP-5 if the modem at the other end can use only MNP-5 compression. With the help of data compression, a 28,800 bps modem may reach the peak speed of 115,200 bps over a very clear channel.

Modems using the digital simultaneous voice and data (DSVD) protocol can establish virtual multichannel computer-to-computer connections that will let users talk and exchange digital data, images, and video files, but will not support live videoconferencing. DSVD-compliant modems intercept the analog voice signal and digitize it into a format that can be sent via a digital channel along with other data. At the receiver's end, the process is reversed. DSVD is a handshake and data handling protocol that resides on top of the normal v.32 modem protocol and adds a voice handling procedure to the modem protocol set.

When a home subscriber places a local telephone call to an Internet service provider, in many cases, the only analog portion of the total connection is the short distance from the user's home to the local telephone company's central office. To conquer this "last mile" of the information superhighway, pure digital technology must be employed. ISDN is the first step that raises the data rate to 128 Kbps in both directions. More advanced digital technologies such as high-data-rate digital subscriber line (HDSL) and asynchronous digital subscriber line (ADSL) can boost the channel capacity of telephone wires to much higher levels. We will study HDSL and ADSL in Chapter 6.

2.6 TERMINAL DEVICES

A modem is a data communications equipment (DCE) device, which is responsible for connecting the data terminal equipment (DTE) device that is, in turn, typically attached to a computer or terminal, in the context of a point-to-point communications system. Each end of the communications system consists of a DTE and a DCE. Data are transmitted between the two ends of each communication through some appropriate medium. DCE is the interface between the digital DTE and the communications medium, such as the local loop of an analog telephone circuit.

2.6.1 Dumb and Intelligent Terminals

Many types of terminals communicate with networks through DTE and DCE devices. A personal computer that connects to the Internet, an automatic teller machine that communicates with the mainframe computer of a bank, or an X-terminal that receives graphical data from client programs on a remote computer are all examples of terminals.

A terminal is essentially an input/output device for entering and displaying data. Typically, a terminal consists of a keyboard and screen connecting to a local or remote computer. The older generation of terminals is mainly used for text entry and display. They are "dumb" because they have little memory storage and very limited processing and graphical capability.

A terminal may interface with a host computer through an asynchronous teletypewriter (TTY) interface, which is basically an RS-232-type serial communications interface. In the mainframe category, the most widely used terminal devices are the IBM 3270 family of terminals, which connect to other IBM equipment through the IBM 327x interface. Because the IBM interface requires coaxial cable and relies on voltages and signals significantly different from those of TTY interfaces, it is often difficult to connect IBM equipment with the outside world.

Smart or intelligent terminals usually have a powerful microprocessor and a significant amount of random access memory (RAM). A personal computer equipped with powerful software makes an excellent intelligent terminal that can fulfill many functions locally, thus shifting the computation load from the remote computers. An X terminal running X server software takes data and commands from the remote client program, then processes these data and displays them on the screen based on local hardware configurations (available color pallets, resolutions, etc.). In the meantime, an X terminal also responds to external events, such as mouse movements, and to possible error conditions, besides taking input from the keyboard.

The newest generation of terminal devices is the so-called network computer (NC) specially designed for the Internet World Wide Web. An NC has a powerful microprocessor and the necessary internal RAM. To reduce costs, however, the NC may rely heavily on the resources provided by the network, including hard disk storage and even software. Furthermore, an NC uses a TV screen as its monitor to save money. In fact, some NCs are simply advanced TV sets with network connections built in. It is perhaps disputable whether NCs will ever be practical for Internet access, but the distinction between these devices and PCs may disappear when the price of PCs drops even lower.

2.6.2 Data Terminal Equipment and Data Communications Equipment

A terminal, or a computer, simultaneously sends data to and receives data from DTE in parallel 8 bits (1 byte) or more, depending on the bus structure in use, while data communications use serial transmission (i.e., one bit at a time) almost exclusively. That is probably why *bit* per second, instead of *byte* per second, is used as the data rate measurement for digital communications. Because a computer is much faster than most communication lines, it often can handle chores in addition to communicating with the networks. Therefore, it is more efficient for a computer to deal with the DTE in *interrupt* mode rather than busy/waiting, which ties the computer to a communications channel.

The DTE will convert the parallel data from the computer (terminal) into serial data and will exchange information with the DCE. Typical DTEs for point-to-point serial com-

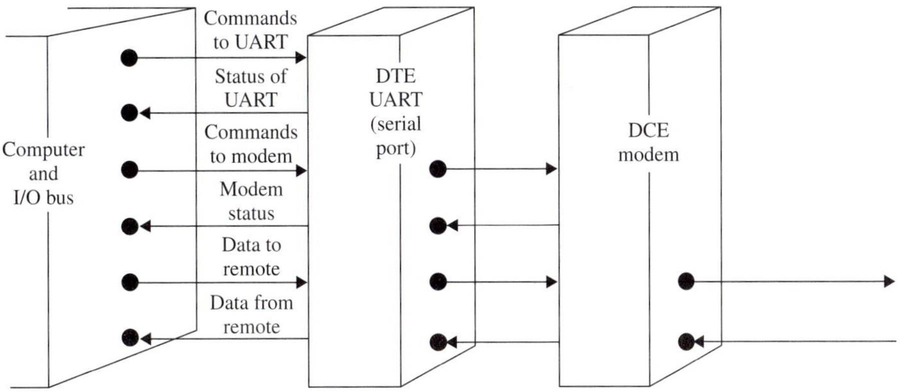

Figure 2.27 The information flow among a computer, a UART, and a modem for serial communications through a telephone line.

munications are the universal asynchronous receiver/transmitter (UART) and the universal synchronous receiver/transmitter (USRT). A UART is usually a special large-scale integrated circuit. When transmitting, a computer that has a 32-bit I/O bus sends to the UART one or as many as 4 bytes at once. The UART then converts the data into serial format and sends them to the DCE device one by one. When all the bits have been sent out, an interrupt request (IRQ) is generated by the UART. If no other higher priority interrupts are waiting, the computer responds to the IRQ by sending another byte or bytes to the DCE when an instruction is completed, until the job is done.

When receiving, the UART assembles data from the DCE in serial form into parallel form and interrupts the computer every time it receives one or more whole bytes. The computer responds to each interrupt by leaving any program running and storing the byte(s) in a buffer (memory). Then the computer returns to the program until it receives another interrupt request. The computer also sends control signals to and receives status reports from the UART and a DCE device. Figure 2.27 illustrates the flows of data, commands (control signals), and status reports among a computer (or terminal), the DTE (the UART), and the DCE (the modem) devices.

For point-to-point serial communications, the Electronics Industries Association (EIA) specified its Recommended Standard 232 (RS-232) as the standard interface between a DTE device and a DCE device. The original RS-232 interface uses the DB-25 connector, which has 25 pins that are assigned to more than 20 signals for data, controls, and states. A recent version, EIA 232-D, added three signals for testing purposes: local loopback, remote loopback, and test mode. The local loopback signal permits a test on the loop between the DTE and the local DCE. The remote loopback allows a test on the loop from the local DTE, via the local DCE, over the transmission lines to the remote DCE, and then back. Test mode indicates whether the local DCE is in a test condition. The most essential lines in the EIA 232-D interface are connected to pins 2, 3, and 7. Because the newer UART and the fast modems are "smart," all the signals are sent and received through the same lines (TD and RD), most other lines are not actually used. In a personal computer, the DB-25 connector is often replaced by the smaller 9-pin DB-9 connector, and

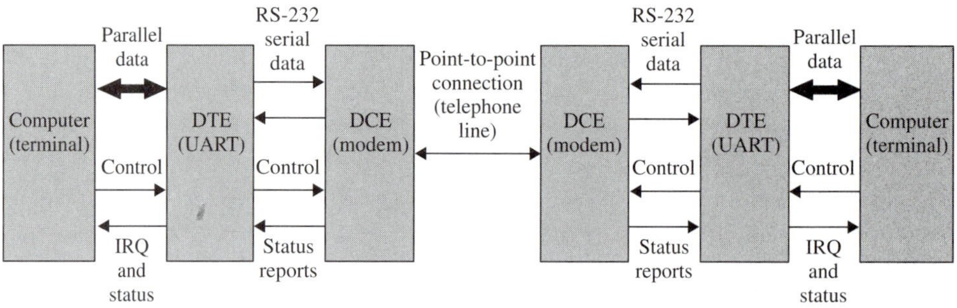

Figure 2.28 Computers (terminals) make point-to-point serial communications through DTE and DCE devices.

many unused signals are dropped. Figure 2.28 shows the block diagram of the serial communications between two computers through a telephone line.

Table 2.2 gives the signal names, sources, and corresponding pin numbers for the EIA 232-D standard.

To connect two terminal devices (computers) directly through their serial ports without going through a modem and telephone line, there must be a null modem between the two DTE device slots. Figure 2.29 shows a fully implemented null modem. However, for asynchronous serial communications (Chapter 3), a simple cable of three wires that connect SG, TD, and RD will be adequate for the practical information exchange purposes.

TABLE 2.2
List of Pins for EIA Standard 232-D

Pin	EIA name	Mnemonic	Source	Description
1	AA	PG		Protective ground
2	BA	TD	DTE	Transmit data
3	BB	RD	DCE	Receive data
4	CA	RTS	DTE	Request to send
5	CB	CTS	DCE	Clear to send
6	CC	DSR	DCE	Data set ready
7	AB	SG		Signal ground
8	CF	DCD	DCE	Data carrier detect (received line signal detector)
12	SCF	SDCD	DCE	Secondary received line signal detector
13	SCB	SCTS	DCE	Secondary clear to send
14	SBA	STD	DTE	Secondary transmitted data
15	DB	TC	DCE	Transmit clock
16	SBB	SRD	DCE	Secondary received data
17	DD	RC	DCE	Receive clock
18	LL	LL		Local loop
19	SCA	SRTS	DTE	Secondary request to send
20	CD	DTR	DTE	Data terminal ready
21	RL	RL	DCE	Remote loopback
22	CE	RI	DCE	Ring indicator
23	CH or CI		DTE or DCE	Data signal rate selector
24	DA	TC	DTE	Transmit clock
25	TM	TM		Test mode

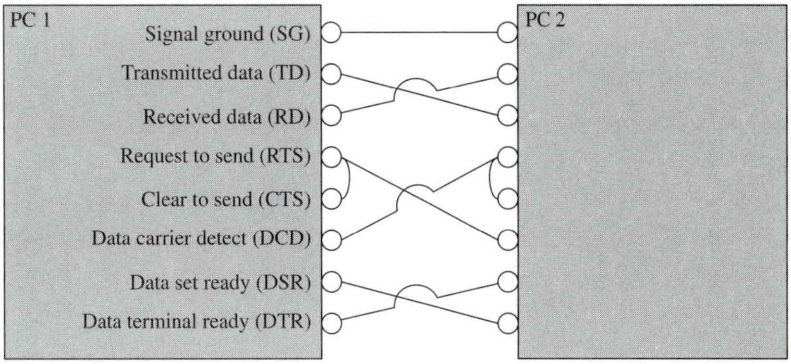

Figure 2.29 A null modem connects the signal ground and swaps the transmission and receiving lines.

Although EIA 232 (RS-232) continues to be the most popular serial interface standard, its 20 Kbps limit on data rate and 15 m (50 ft) limit on distance make it unusable. Responding to the need for an interface that can carry data at higher speed over longer distances, the newer RS-449 standard family, which includes RS-422 and RS-423 electrical interfaces, can satisfy current demands better. RS-422 is the electrical specification that allows communications speeds of up to 10 Mbps at distances up to 12 m (90 ft), or communications at speeds of 100 Kbps at distances up to 1222 m (4000 ft). These high speeds are possible because RS-422 uses balanced circuits, which have a separate return path for each signal, while the unbalanced circuits in RS-232 reference all signals to a common ground.

The RS-449 standard was slow to be deployed by the industry because of its unnecessary complexity. Most serial communications ports still use some variation of the RS-232. Furthermore, the new Universal Serial Interfaces standards are gaining popularity and are replacing the old standards by virtue of superior speed and flexibility.

2.7 WIRELESS COMMUNICATION

Wireless communications have many advantages over their wired counterparts. Before early 1980, citizens band (CB) radio was the major vehicle for mobile communications. In the United States a total of only forty 10 kHz channels were available at a frequency of about 27 MHz because limited bandwidth assignments had led to the saturation of CB radio. As a result, CB finds limited applications today. Another type of wireless communication system is specialized mobile radio (SMR), which is a regional service with less roaming support. SMR is often associated with police and taxi dispatch.

Modern mobile communications systems work by limiting transmitter powers to avoid interference and conflict. This restricts the range of communication to a small region called a cell, with a diameter of 5–10 miles. Within a cell, all the users communicate with a single transmitter in the center of the cell. Outside this region, other transmitters can oper-

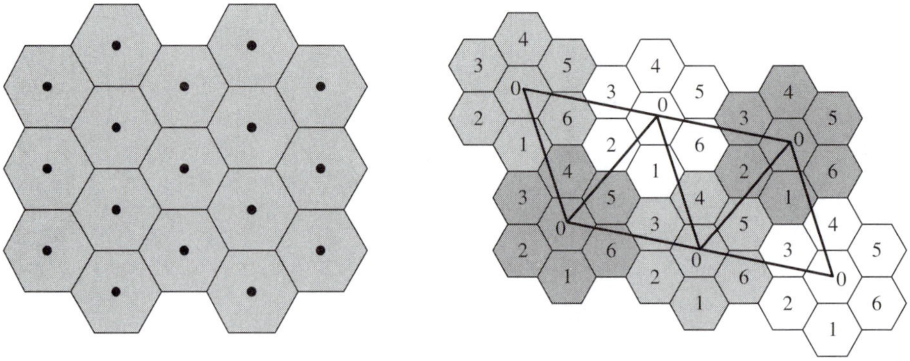

Figure 2.30 Adjacent hexagonal cells cover the entire designated communication area (numbers indicate different groups of channels). When cells become smaller, more users can be included in the same area, whereupon cellular phones and the transmitter may, and must, use lower transmitting power.

ate independently to allow more users. These cells are hexagonal but often their exact configuration is determined by local propagation characteristics. Together the cells completely cover the entire designated area (Figure 2.30). Within a single cell, a small number of channels separated by different frequencies are used. When a mobile phone user initiates a call, it is allocated to an idle channel within the current cell by the mobile services switching center (MSC). As the mobile device approaches a cell boundary, the signal strength fades, the user is passed on to a transmitter from the next cell, and a new idle channel is assigned within that new cell.

Mobile communications usually are allocated in the 50 MHz to 1 GHz band. The American advanced mobile phone services (AMPS) is a first-generation cellular phone system that uses frequency division multiplexing (FDM), an analog multiplexing technique to be discussed in Chapter 3. AMPS uses a 40 MHz band in the 800–900 MHz spectrum. This band is split into a 20 MHz transmitting subband and a 20 MHz receiving subband. Each 20 MHz band is partitioned into 666 two-way channels, each having a bandwidth of 30 kHz. These 666 kHz channels are subdivided into seven groups with three sets in each group. In turn, each set contains 31 channels, each of which uses a bandwidth of 30 kHz (Figure 2.31). A module of seven adjacent hexagonal cell patterns is assigned to the seven groups of channels—a central one and its six nearest neighbors. Each cell is assigned a different group that contains 3×31 channels. The 666 channels are reused by another module of seven cells with the same structure. In this arrangement, any two cells that use the same channels are physically two cells apart (Figure 2.30).

The advantage of the cellular telephone network is the ease of scaling. As demand rises, the cell size can be reduced, then the same number of channels is available in a smaller area. The same sets of frequencies and channels are reused in another group of cells, as indicated in Figure 2.30, increasing the total number of channels per unit area. AMPS is just one of six incompatible first-generation systems that exist around the world. Currently, second-generation systems using digital technologies are being deployed.

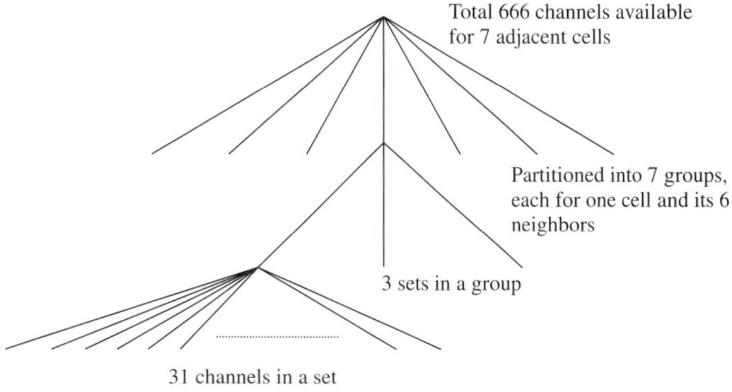
Total 666 channels available
for 7 adjacent cells
Partitioned into 7 groups,
each for one cell and its 6
neighbors
3 sets in a group
31 channels in a set

Figure 2.31 In an AMPS cell phone system, 666 channels are partitioned into seven groups with three sets for each group. Each set contains 31 channels. A cell is assigned a group of three sets of 31 channels. The group of channels assigned to a given cell should be different from the respective groups of its six adjacent neighbors.

2.7.1 Digital Cellular Telephone and Personal Communication Services (PCS)

One might mistakenly think that because the frequencies used are so high, channel capacity (or data rate) available for wireless communication channels would be almost unlimited. In reality, because of the very low *S/N* ratio, actual channel capacity is quite limited. Shannon's law can explain this. We have:

$$
\begin{aligned}
C &= B \log_2(1 + S/N) \\
&= \frac{B \ln(1 + S/N)}{\ln 2} \\
&= 1.44B \ln(1 + S/N)
\end{aligned}
$$

Using logarithmic expansion, we write

$$\ln(1 + x) = \frac{x - x^2}{2} + \frac{x^3}{3} - \frac{x^5}{5} + \cdots$$

When x is small ($x << 1$), which is the case for most radio transmissions, all terms in the expansion except the first are negligible. So we have:

$$C = 1.44B(S/N)$$

It is easy to see, for a 20 MHz bandwidth channel with the *S/N* level at about –30 dB (or $S/N = 1/1000$) that the maximum transmitting rate is about 30 Kbps.

Digital cellular phones have many advantages over analog technology. In particular, they provide the following:

- Universal access to national and international telephone system

- Secure communications
- Data-independent transmission
- An infinitely extendible numbering system

Another goal of the digital mobile systems was to develop one global standard, allowing use of the same mobile phone anywhere in the world. Since, however, there are currently at least three digital standards in use, this seems unlikely. Two rival multiplexing technologies—time division multiple access (TDMA) and code division multiple access (CDMA) (a North American digital standard IS-95)—are fiercely competing for market share. Today's analog cellular service assigns one user per 9.6 Kbps channel. The pure digital systems allow better bandwidth utilization and higher bandwidth for each channel than analog systems. However, it is not easy to choose between a TDMA and a CDMA system. CDMA is a spread-spectrum technique that allows all transmissions to share the same channel by marking each conversation with a unique ID—the code; calls are sorted out at the receiver. It is claimed that between 10 and 20 users can be accommodated per channel to 14.4 Kbps.

TDMA, better known as D-AMPS, for digital AMPS in the wireless communications industry, was introduced in 1992 as a way of upgrading existing AMPS networks to digital. Digital AMPS divides a large chunk of bandwidth into small time slots allocated to different channels. It can accommodate up to three users per channel at 9.6 Kbps per channel. CDMA is a newer and more innovative technology, while TDMA is a more mature and proved technology that delivers good voice quality and is in wide use.

The pan-European standard known as GSM (global systems for mobile communications), available in the United Kingdom, provides similar services to those for ISDN such as call forwarding and charge advice. GSM is now used in 80 countries at 900 MHz and 1.8 GHz. GSM has an array of 200 kHz physical channels each offering a data rate of 270 Kbps. Currently, one physical channel is split between 8 users by TDMA, each of which uses 13 Kbps. The remaining 166 Kbps is used for channel overhead.

3G is a short term for third-generation wireless, used for mobile communications. The third generation follows the first generation (1G) and second generation (2G) in wireless communications. The 1G period began in the late 1970s and lasted through the 1980s. These systems featured the first true mobile phone systems based on analog voice signaling, such as AMPS (explained earlier). The 2G cell phone features digital voice encoding. Examples include CDMA and TDMA based on Telecommunications Industry Association's Interim Standards 95 and 136 (IS-95 and IS-136), and GSM (Global System for Mobile Communication). Cdma-One is the original IS-95 standard and has the potential to support traffic up to 115 Kbps. 3G is an evolution of Cdma-One, which is also referred to as Cdma2000.

3G includes capabilities and features to support data rates up to 2 Mbps for indoor traffic, 384 Kbps for pedestrian traffic, and 144 kbps for high-mobility traffic. With these high data rates over wireless links, mobile applications to support enhanced multimedia (voice, data, video, and remote control) and all popular modes such as cellular telephone, e-mail, paging, fax, videoconferencing, and Web browsing are supported. In addition, the system will operate at higher data rates under stationary conditions. 3G features bandwidths of 1.25 MHz (1x environment) and also 3.75 MHz (3x environment), with a potential of going up to 15 MHz (12x environment) enabling much higher data rates.

Another development in the wireless industry is Universal Mobile Telecommunications System (UMTS), which is a so-called 3G broadband, packet-based transmission of text, digitized voice, video, and multimedia with a potential of higher than 2 Mbps data rates. Once UMTS is fully implemented, computer and phone users can be constantly attached to the Internet as they travel and, with the roaming service, they have the same set of capabilities no matter where they travel. Users will have access through a combination of terrestrial wireless and satellite transmissions. The higher bandwidth of UMTS promises new services, such as video conferencing and Virtual Home Environment (VHE) in which a roaming user can have the same services to which the user is accustomed when at home or in the office, through a combination of transparent terrestrial and satellite connections.

The heart of the mobile telephone network is the mobile-services switching center (MSC), which has the capability of making tens of thousands of decisions each second, relying on powerful digital computers. The MSC's tasks are as follows:

- Acknowledging the paging of the user
- Assign the user a channel
- Broadcast user's dialed request
- Return the call
- Handle *handoff*
- Handle bookkeeping and account charging

The handoff process monitors the signal strength of both transmitter and receiver, allocating new channels as required. This is completely hidden to the user and presents technical challenges. It is the maturity of computer and software technologies that have made the rapid growth of mobile communications possible.

The personal communications services (PCS), is another addition to the mobile telephone services, which provides users with a single telephone number no matter where they go. PCS uses both connection-oriented and connectionless (Chapter 3) full digital technologies similar to digital cellular telephone systems, but with much smaller cell size ($<$100 m), to allow more users in a small area. Broadband PCS utilizes channels in the 1.8–2.3 GHz frequencies, while narrowband PCS utilizes channels in the 800–940 MHz frequencies. The smaller cell size and higher frequency result in very low transmitting power ($\sim$0.25 W), tiny transmitting antenna, and very lightweight telephone device. The narrowband PCSs are mainly used for high capacity and acknowledgment paging, while broadband PCS services for two-way messaging and digital voice message delivery. To make way for PCS, as many as 30,000 incumbent microwave systems must be relocated to a new spectrum. When fully deployed, PCS networks will use more than 100,000 cells across the United States.

The principal problem with mobile communication is the rapid variation in signal strength that occurs as the communicating parties move. This variation is due to fading, the varying interference of scattered radiation. Increasing the transmitter power is not a solution to fading in mobile communications, where transmitter power must be limited to prevent interference with neighboring cells. Generally, the superior communication quality provided by wired telephone services is not attained.

At extremely high frequencies, the effects of obstructions are significant because of the efficiency-lowering combination of high frequency and short wavelength. Factors such as street orientation (waveguide effect), tunnels, and foliage (attenuation) may significantly influence mobile communications, sometime with a variation as high as 18–30 dB, depending on street direction, tunnel length, and season. Therefore, installation of a mobile telephone system requires a large initial effort to determine the propagation behavior in the area covered by the network. Propagation planning incorporating a mixture of computer simulation and observation is necessary.

Low Earth Orbit (LEO) satellite systems may play a critical role in future wireless communications. The LEO satellites ensure strong signals and communication quality by allowing the beams projected onto the ground to be more tightly focused than is possible with geosynchronous communication satellites. Echos are minimized owing to the low earth orbit, and the receiving antenna of a LEO system can be small enough to be carried on a handheld subscriber unit. The services are available in remote places, not connected by regular cellular phones, such as the Antarctic and Arctic areas. The dual-mode handset units are often able to switch from conventional cellular telephony to satellite telephony when outside the reach of the terrestrial networks.

Several LEO-based systems, notably GlobalStar and Iridium, have faced unimpressive results due to the bulky size of the handset and high cost compared to the now dominating and popular cellular phone systems.

Finally, in the wireless communications system category, there are cordless phones for home and office use. A cordless phone consists of two parts—a base station that connects to the regular telephone network, and a handset that can be used around the house without a cord while the call is ongoing. The older cordless phone system, called cordless telephone first generation (CT1) is an analog system. The newer cordless telephone second generation (CT2) is digital. Based on the same architecture with which the handset communicates with the base station in the vicinity through the air interface, the base station acts as an interface to a public switched telephone network (PSTN) to set up a connection through the network to the destination address.

An experimental design for the CT2 system is to add a database to the system to track the location of all the handsets. This database maintains the identification number of the nearest base station of the handset. Every time the handset moves to another base station area, the database is updated. When there is an incoming call from the normal subscriber destined for the CT2 handset, the database will notify the PSTN of the location of the handset and the call will be routed to the appropriate base station. The base station will signal the right handset to ring, allowing better mobility than was possible with the old generation cordless phones (Figure 2.32).

2.7.2 Wireless Data Services

Demand for wireless data services is growing along with the popular interest in accessing Internet and other data communications services. People want to access the Internet

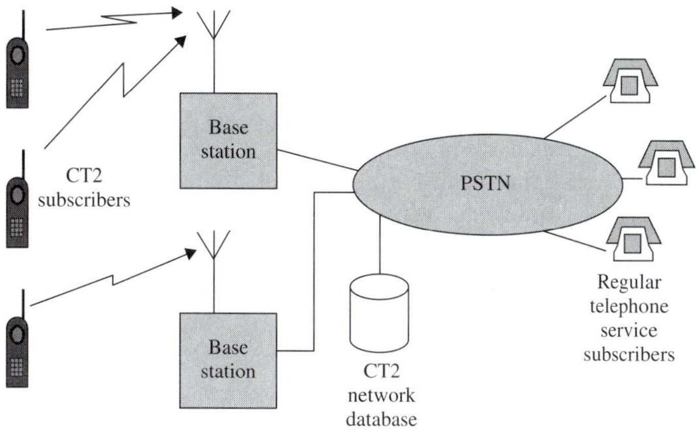

Figure 2.32 The cordless telephone second-generation (CT2) system enhances mobility by using a network database, which keeps track of subscribers' locations and is connected to a public switched telephone network.

from anywhere without being limited to a desktop and a phone jack. Packet radio relies on radio frequencies to transmit packetized data over private networks that in turn are connected to the public data networks. The connectionless packet radio system uses channels between 800 and 900 MHz and supports transparent subscriber roaming and call handoff throughout a national coverage area. Subscribers access the network with a laptop linked to a wireless radio modem. The biggest problem associated with private packet radio is low speed, with an actual throughput ranging from 2 to 8 bps that may show up as sluggish response times.

Microcellular spread-spectrum technology is another wireless data transmission that is gaining in popularity on university and corporate campuses, where Internet access is considered to be essential. Spread-spectrum providers install small radios on top of poles or buildings and transmit files by means of low-amplitude signals. Subscribers pay a low monthly fee and tap into the network from a laptop or personal digital assistant (PDA) equipped with a radio modem. The spread-spectrum system scatters data packets across randomly selected channels in the unlicensed 902–928 MHz band and operates at speeds 10–45 Kbps, which are higher than those provided by other wireless data services. Given spread spectrum's speed, downloads and searches can be done with little delay.

In 1992 a standard for wireless data communications was established. Cellular digital packet data (CDPD) is a connectionless digital data service that transmits packets over the idle capacity of an analog cellular network. It divides data files and transactions into small packets and sends them between voice transmissions on existing cellular channels. Since it is an overlay service, CDPD is less expensive to implement than a newer technology. All the cellular operators have to do is deploy new hardware and software throughout their networks and at base stations; it is not necessary to build an entirely new network. Users need only a cellular modem or cellular phone with modem capabilities. Any application that uses a modem's standard AT (advanced technology) command set should be able to support CDPD. Unlike other wireless technologies that use radio frequencies or satellite communications, CDPD leverages today's cellular infrastructure, including the current cellular network equipment, allowing customers to use existing cellular systems to send and receive data. CDPD also interfaces with landline data networks.

CDPD is based on TCP/IP (Chapter 5), which simplifies communications with landline networks that already run on this protocol. Instead of dialing in to the cellular system, CDPD subscribers use an IP address to access the network.

CDPD has several advantages over other wireless data services:

- CDPD operates at a theoretical speed 19.2 Kbps and an effective speed 10–12 Kbps. This is significantly higher than CDPD's major competitor, private packet radio, which typically runs at 2.4–4.8 Kbps effective speed.
- There is virtually no need to set up a call because CDPD is a packet-switched service. Switched cellular connections, by comparison, take about 30 seconds to set up.
- CDPD response times for database are fast. Queries are typically less than 5 seconds.
- CDPD is more secure. Because data are transmitted in packets over different cellular channels, electronic eavesdropping is very difficult. CDPD also offers built-in security features including authentication and data encryption.

CDPD is a good fit for any application involving short, bursty messages, such as point-of-sale credit card verification, vehicle dispatch, package tracking, telemetry, and brief e-mail exchanges. CDPD is less expensive for short, bursty messages, while circuit-switched cellular is more economical for lengthy files. CDPD service providers bill by the kilobyte rather than by connect time. So CDPD is more economical for the transmission of files less than 2–5 Kbyte long. Anything larger should probably be sent via switched cellular.

Despite delays in getting CDPD to market, interest has been growing. People who are already familiar with cellular technology would view CDPD as an upgrade to existing cellular networks and a reliable alternative with a viable future compared with other wireless network alternatives.

A group of six companies in the United States has come up with a specification that essentially creates a gateway between CDPD and circuit-switched cellular. This scheme allows customers to access the CDPD network from anywhere in the country, either through the cellular network or via a hardwired connection. Customers who want to double up their wireless data service must have IP addresses and subscriber devices with both circuit-switched cellular and CDPD capabilities. The cellular network will recognize when subscribers move out of the CDPD coverage area and transfer messages to them via circuit-switched cellular. Linking the two technologies could effectively give cellular service providers a way to deliver nationwide CDPD.

CDPD is still in its infancy and still underutilized. When more customers come online, response times may suffer. Nevertheless, cellular providers have very high hopes for CDPD because of its advantages. It is possible that CDPD will eventually become the primary technology used by mobile workers to transmit and receive data.

Figure 2.33 compares various mobile communications services in terms of mobility and cost. Many developing countries, lacking an existing wired communications infra-

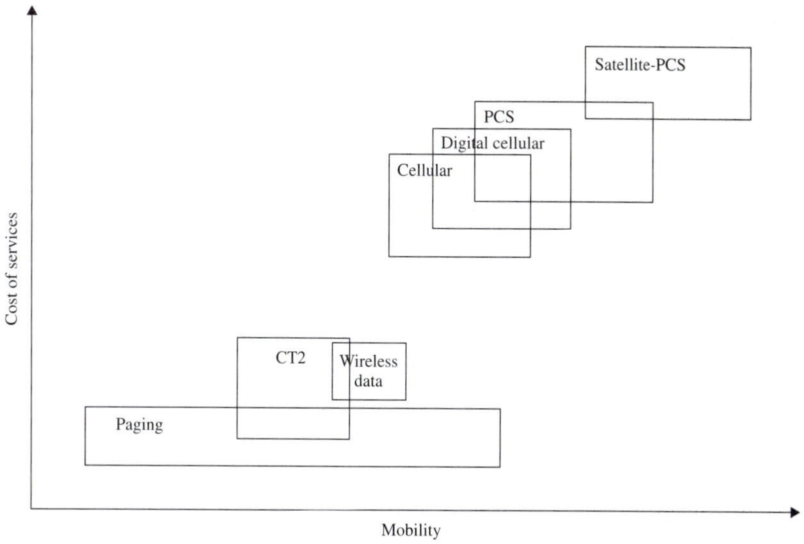

Figure 2.33 The mobility and cost of various wireless telephone services.

structure, bypass wired systems and start to deploy the digital cellular telephone and PCS wireless systems directly, to speed up the construction of a national communications system.

GSM is a standard that is more popular in Europe and many Asian countries. It competes with advanced CDMA technology and attempts to ensure interoperability without stiff competition, which works to the benefit of public in terms of both cost and service quality. GSM is a digital network that provides ubiquitous and robust wireless data connectivity available with transmission speeds of 9.6 kbps and up. General Packet Radio Service (GPRS) has been introduced as an intermediate step, sometimes also called as 2.5G, to lessen the impact of the delay in implementing 3G wireless systems. Using this standard, high-speed data over the current GSM and TDMA-based wireless network infrastructures may be transported efficiently. The features of GPRS include:

- GPRS packetizes the user data and transports it over Public Land Mobile Networks (PLMN) using an IP backbone and interfaces to other Public Data Networks (PDNs), including the Internet. Packet-switching (Chapter 3), which complements rather than replaces GSM circuit-switching data services, provides a seamless and immediate connection from a mobile station to the Internet or corporate Intranet.
- Multiple users can share radio channels because of the packet-switching with connection speeds of 14,400 bps (using 1 radio channel) to 115,000 bps (combining all 8 radio channels together), allowing short "bursty" traffic, such as web browsing, and e-mail, as well as large volumes of data. Line charges typically base on the actual amount of data transferred.

- GPRS uses GSM network for table look up in the Location Register database to obtain GPRS user profile data while GPRS signaling and data traffic do not travel through the GSM network.
- Perception of "always on" for continuous operation because of its fast connection setup, typically 0.5–1 second, and simultaneous voice and data communications.
- GPRS supports several Quality of Service (QoS) profiles so service providers can offer selective services to users.

Digital mobile phones are ubiquitous and wireless Internet/Intranet access is entering the adoption stage while broadband and video-over-wireless are just on the horizon. The other trends and development in wireless data communications include:

• Fixed wireless, also known as MMDS (Multipoint Multichannel Distribution Services), which makes use of the wireless spectrum for data transmission by shooting signals between carrier-owned base stations and transceivers affixed to customer sites within a 35-mile radius. Fixed wireless is as much as 30 times faster than dial-up modems with speeds from 384 kbps to 1 Mbps for download and 384 kbps to 512 kbps for upload. The fixed wireless service could be installed within five days and the fixed broadband wireless technology may soon be in a position to catch up with its Internet-access rivals, DSL and high-speed cable. To put fixed wireless access alongside its wireline connectivity, a high degree of reliability is needed. There are attempts to tackle MMDS' line-of-sight restrictions caused by buildings or terrain located between towers and receivers. This technology is promising, however, the customer premise equipment costs must come down in order to ultimately spur fixed wireless.

• Wireless LAN (WLAN) based on the 802.11b standard is another powerful wireless technology that is gaining popularity. WLANs can achieve wireless performance and throughput comparable to wired Ethernet at the speed of 11 Mbps and are on the verge of becoming a mainstream connectivity solution. A mobile user plugs a WLAN card into his/her notebook and desktop computer that communicates wirelessly to one of many access points that in turn connects to an ISP through high-speed connection. To date, wireless LANs have been primarily implemented in university campuses and some vertical applications such as manufacturing facilities, warehouses, and retail stores. The majority of future wireless LAN growth is expected in healthcare facilities, educational institutions, and corporate enterprise office spaces. WLAN is proliferating and finding its way into hotels, airports, restaurants, convention centers, etc.

• Bluetooth is a short-range wireless networking standard that allows all kinds of devices to communicate and transfer information. For example, with a Bluetooth-enabled laptop and PDA, scheduling and contact information can be synchronized on the fly without having to connect cables or align an infrared port. A Bluetooth-enabled headset can connect wirelessly with a cell phone or cordless phone because Bluetooth facilitates voice communication as well as data transmission; this eases the user's movement eliminating the need for wires to connect the devices. Bluetooth operates in the 2.4 GHz range, similar to many late-model cordless phones. The normal range for Bluetooth devices is 10 meters, but it can be extended as far as 100 meters. With a raw data rate of 1 Mbps, the standard is not designed to replace high-bandwidth LANs, although Bluetooth-enabled wireless LAN products are available.

The network formed by connecting Bluetooth devices is called a piconet. Several of these piconets can be linked, increasing bandwidth and allowing devices to communicate. Security concerns were paramount in the design of the Bluetooth standard. Built-in encryption and authentication between devices ensure that others cannot eavesdrop on the Bluetooth network. Bluetooth devices alter their transmission power in accordance with the distance between them, significantly reducing the range of the radio transmission when possible, thus reducing power consumption and security risk.

One of the primary benefits of Bluetooth is that manufacturers can incorporate it into many devices without having to design a special chipset. The Bluetooth chipset includes all of the components needed for Bluetooth connectivity, making an easy integration for device vendors.

Bluetooth may eventually change the way we use our cell phones, PDAs, laptops, and other digital devices. With a solid and secure basic technology, and support for a broad range of devices, the Bluetooth wireless standard promises to proliferate connections between devices and to render cable hassles a thing of the past.

The number of people with access to wireless Internet may exceed the number of fixed connections. Mobile advertising presents extensive opportunities to reach users anywhere and everywhere. But basic information by way of wireless Internet is no longer enough to satisfy customer demand. Subscribers want to have information customized to their specific needs. Numerous portals are striving to meet that demand with personalized and location based information. An example of daily advertising could be discounts to a movie, discounted airfares, lunchtime discounts at a restaurant, tickets to special events, etc. However, the amount and type of advertising needs to be controlled so that the user does not dismiss all advertising as spam.

On a Web-enabled cell phone, the user can query for directions without typing in street names because the precise location is automatically calculated at the time call is placed. Various ways of pinpointing a person's location are being tried, from Global Positioning Satellite systems built into handsets to sophisticated triangulation by the base stations that send and receive cell phone signals. The government wants to make sure that people who dial 911 on mobile phones can get emergency help. The FCC has made requirement for all carriers to have 911 location-tracking systems. As of this writing, wireless carriers are in the process of integrating that technology into their networks so every call can be tagged with an exact location to comply with government regulations.

2.7.3 Geographic Position Systems and Their Applications

The Global Positioning System (GPS) is a worldwide radio-navigation system that includes a constellation of 24 satellites at an altitude of 17,600 km (11,000 miles) and their ground stations. The satellites were originally launched by the U.S. government for military purposes. An advanced GPS device uses these satellites as reference points to calculate positions as accurate as to a few meters anywhere on the earth. With an advanced form of GPS, the differential GPS technique, one can make measurements on the earth's surface to an accuracy of one centimeter.

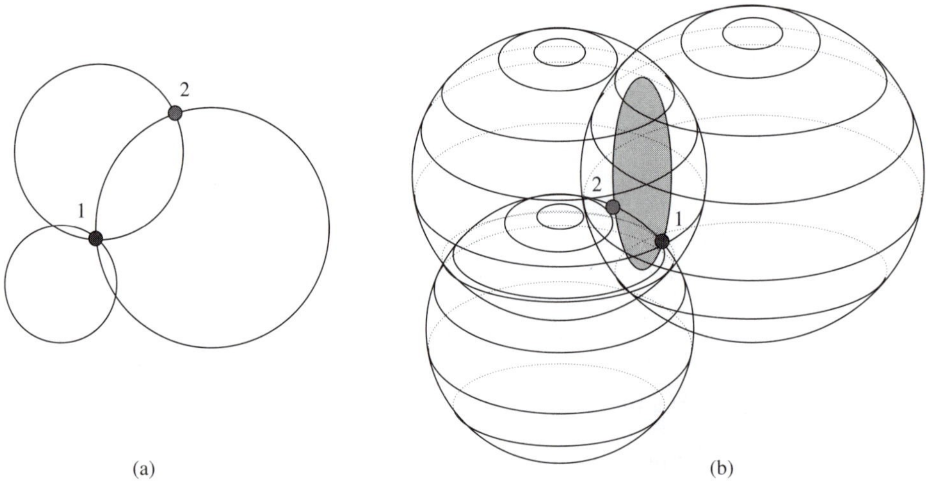

Figure 2.34 (a) Two circles in a plane have two intersecting points, 1 and 2. Point 2 might be easily removed by known conditions or by a third circle. (b) The intersection of two spheres forms a circle while three sphere have two intersection points, 1 and 2. Again, point 2 might easily be excluded by introducing a fourth satellite or by an educated guess. The center of the sphere is where the satellite is located.

GPS uses the principles of spherical trigonometry to determine a GPS receiver's location based on the distances from at least four GPS satellites. The idea can be explained as follows. In a two-dimensional plane, if you know the distances to two given points with known positions in the plane, you can draw two circles from these two points, using the distance to the known point (the satellite) as the radius. One of the two intersections of the two circles might be obviously ridiculous, and the remaining one would be your position in the plane. Nevertheless, a third known point (another satellite) will certainly give you a definite choice—point 1 (Figure 2.34a).

In three-dimensional space, however, spherical trigonometry must be applied. The intersection of two spheres, each of which has a GPS satellite as the center, forms a circle having as its radius the distance to the satellite. A third sphere centered on another satellite will intersect with the intersecting circle at two points, one of which is your position. With an educated guess, or by using a fourth satellite, your location can be determined (Figure 2.34b). Here the key is to determine the locations of the satellites and the distances to them, given the following conditions:

- Since the orbit for every GPS satellite is very precise, all the GPS receivers on the ground can tell where each satellite is in the sky, moment by moment, based on an almanac programmed into their computers.
- Each satellite is equipped with an extremely precise and expensive atomic clock.
- Each satellite constantly broadcasts a unique and very complex sequence of pseudorandom code (PRC) to identify itself and to serve as the synchronous basis for measuring time as well.

- A GPS ground device is equipped with a multiple-channel receiver that can receive signals from at least four satellites simultaneously. Each GPS receiver is also equipped with a less precise inexpensive clock and a pseudorandom code generator that is able to duplicate the same code received from any GPS satellite.

A GPS receiver receives signals from the closest four or more satellites and finds the time needed for the signal to travel from each satellite to the receiver. It then multiples the time by the speed of the light (the speed of signal travel) to get the distance to the satellite. Based on the positions of the three or four satellites and the distance to each of them, the location of the GPS receiver can be determined.

The GPS satellites are under constant monitoring by the U.S. Department of Defense. A very precise radar system is used to check each satellite's exact altitude, position, and speed, and to remove errors due to various factors such as gravitational pull from the moon and the sun, and even the pressure of solar radiation on the satellites. Accurate information about the satellites' positions is relayed back up to the satellites and is included in the timing signals they broadcast. All that remains unexplained is how to determine the time precisely. This is where the pseudo random code and the fourth satellite come in.

Since each satellite has its own unique PRC, the GPS receiver can easily identify the particular satellite and generate an identical PRC sequence synchronously based on its own clock. When the two PRC sequences are compared, the time difference is the time needed to travel from the satellite to the receiver (Figure 2.35).

Since light travels at the speed of 300,000 km (187,500 miles) a second, a time difference of one microsecond means a 30-meter distance. If every GPS receiver had an atomic clock, timing accuracy would not be a problem. However, a handheld GPS receiver selling for $200 obviously cannot include a $50,000–$100,000 atomic clock.

Since a timing error by the inexpensive receiver clock applies to all the measurements equally, the error can be corrected by introducing an extra satellite. We use two-dimensional figures to simplify the problem. Suppose that in a two-dimensional plane we find satellite A to be 65 ms (0.065 second) "away" while satellite B is 75 ms "away" according to the inaccurate receiver clock. Then two circles intersect at X, which may be or may be not the accurate position of the receiver (Figure 2.36a). Now if we find that the third satellite C (in the plane, though) is 70 ms "away," we can check to see whether

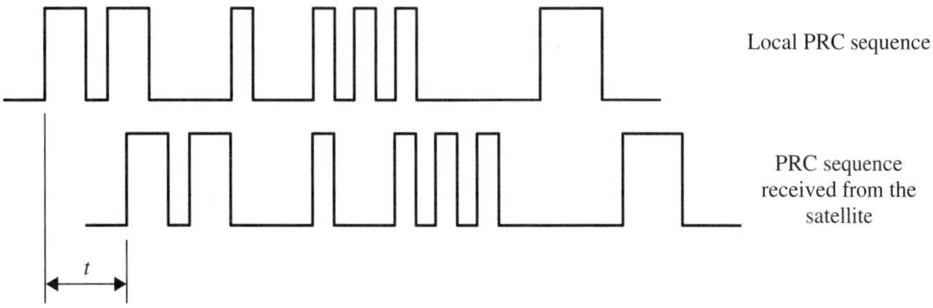

Figure 2.35 The time difference between the two pseudo-random code sequences is the time needed for the signal to travel from the satellite to the receiver.

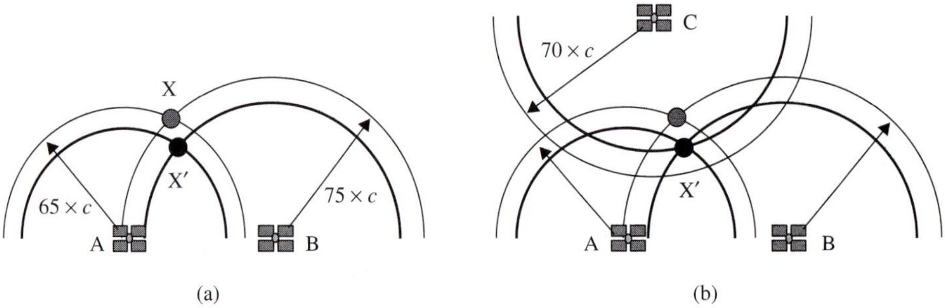

Figure 2.36 Using an extra satellite to remove the timing error and find the actual position.

the third circle crosses point X. If it does, there is no timing error; X is the accurate position of the receiver. Otherwise, by increasing or decreasing the radii of the three circles equally at the same time, a cross point X′ can be found at which all three circles meet. That X is the true position of the receiver, and the timing error, of course, also can be found (Figure 2.36b). In three-dimensional space, the *fourth* satellite, rather than the *third* for the two-dimensional illustration, is always introduced to correct the timing error.

With this clever approach and the extra satellite signals, an inexpensive GPS receiver not only removes the timing error and provides accurate position information, it also essentially turns itself to an atomic clock, which finds many uses of its own.

There are usually more satellites available than a receiver needs to determine its position. In practice, when two satellites are too close together, only one is used to minimize errors. Still, this is not the end of the story. Pursuant to the selective availability policy—an attempt by the U.S. government to prevent hostile powers from using GPS to aim weapons accurately—noise and inaccuracy are introduced into the satellite's timing and orbital data. Military GPS receivers can use a special decryption key to eliminate these barriers to accurate reception. Although imposing no major harm on most domestic applications, such noise and erroneousness constitute the largest single source of inaccuracy in the GPS system. The government claims that the selective availability policy has been phased out. In any event, an advanced GPS technique called differential GPS can effectively minimize intentional accuracies of the magnitude introduced in accordance with the defunct policy.

Differential GPS uses stationary receivers as additional references just for error information. Because these stationary receivers' positions are fixed and precisely surveyed, the reverse procedure is employed by a stationary receiver to determine the timing error based on its position. That is, the stationary receiver compares the ideal travel time and the actual travel time to find the error correction data. A stationary receiver must calculate the data quickly for every satellite it can receive, then encode the information into a standard format and transmit the data, one by one, to each satellite. When the distance between a stationary station and the regular GPS receivers is small—a few hundred kilometers, say—in comparison to the satellite altitude, the moving receiver can use the error correction data accordingly to correct is measurement to a particular satellite.

Differential GPS stationary stations were initially established by domestic organizations, such as surveying or oil drilling companies, which require highly precise data. Later, the U.S. Coast Guard and other national agencies started setting up GPS stations all over the world. The error correction signals are transmitted by the GPS stationary stations on the radio beacons in the 300 kHz range and can be received at no charge by anyone needing them.

GPS may have important impacts on everyday life, saving time, cutting costs, and giving mobility the average citizen never had before. It is possible some day that a a blind person's watch will tell its wearer, "You're standing at 500 Market Street, in San Francisco."

GPS allows ships and boats to navigate their way into harbors accurately. GPS can also detect ships in trouble in a storm, allowing rescue crews to avert potential disasters. Trucking companies are using GPS to monitor their fleets and decrease costs. There are 60,000 buses in this country and most of them will soon be equipped with GPS receivers, permitting a control center to know the bus locations, and monitor their movements closely and adjust schedules accordingly. The Federal Aviation Administration (FAA) is developing a GPS aviation system, which will guide airplanes in taking off, flying, and landing—replacing the ground-based navigation system that has been around since the 1940s.

Emergency services will improve their response times because GPS can provide a precise route to a patient's residence. Police dispatchers are using GPS to identify police cars closest to an incident. GPS will be used to automatically dispatch emergency medical helicopters to an accident. This ability is particularly critical when victims are unconscious and obviously unable to make a call for help. But the combination of a device attached to a car's air bag and the driver's cellular phone will enable GPS to instantly transmit the vehicle's location to the local 911 facility, which will locate the accident instantly.

GPS is also penetrating many new applications fields. In agricultural management, farmers have started doing "precision agriculture" by using GPS to micromanage fields right down to the thistle patch. GPS allows farmers to precisely monitor field variables by sampling soil, gathering crop yield data, and assessing pest infestation. In the geophysics area, the GPS satellite network has enabled ground-based observers to measure the earth's slow deformation more precisely than ever before.

Just as electricity, the automobile, and the computer forever changed the way people lived, so may GPS, along with other innovative network technologies, bring about major modifications in lifestyle. In a few years, people may wonder how we ever lived without GPS.

2.8 VIDEOCONFERENCING

Videoconferencing allows groups of people in different places to hold interactive meetings at which they can see and hear one another. This technology is probably the most exciting thing in data communications and even daily life. It is the most powerful tool for business, government, education, entertainment, and many other applications. An interactive multimedia videoconference system can provide a degree of interaction that offers the next best thing to being there.

Early videoconference systems utilized large pieces of expensive equipment to provide room-based videoconference service. Participants at each site gather in a specially equipped conference room around a table and watch monitors displaying similar rooms at the other sites. The static nature of this equipment, combined with its cost, makes it a poor vehicle for casual or person-to-person conferencing.

Desktop videoconferencing is a new kind of system that allows participants to sit at their own desks in their own offices and use their personal computers to call up other participants much as in standard telephony.

Based on the underlying communications channels, a videoconference system may be based on a LAN, a WAN, or an analog telephone line.

2.8.1 Desktop Videoconference Systems

Bandwidth is always the limiting factor associated with communication. Sending video through a communications channel requires a lot of bandwidth. The picture shown on a desktop computer monitor is made up of very small dots called pixels. A typical 15 in. color monitor can display an array of 1024 × 768 pixels, which can be comfortably watched by a single person. The number of pixels displayed, expressed as the number of rows by number of columns, is called resolution. In contrast to computer monitors, a TV set usually has a resolution about 300 × 400. For a digital line drawing figure, light and dark can be represented by one bit of data for 0 or 1. To display color or gray-scale images calls for about 256 different colors or gray-scale, consuming 8 bits (1 byte) of data for each single pixel. Generally speaking, larger numbers of available colors are compensated, to some extent, by the inadequate resolution. Since each pixel in a frame for a still picture corresponds to a byte of information that describes the pixel's color, a frame of 300 × 200 pixels has 60 Kbytes (480 Kbits) of video information. To play video games, many pictures, or frames, are needed.

Since the human eye is relatively slow with "persistence of vision," it perceives 20 or more sequential but slightly different still pictures as continuous movement, provided these pictures are displayed in a short period of time (one second). The frame rate is the number of still images displayed every second. A TV set displays 25 frames per second (fps) in most of Europe and China, and 30 fps in North America and Japan. Experience shows that a user's perception of the quality of video is strongly linked to the frame rate.

Suppose we want to send video that consists of a series of 300 × 200 pixel frames. This means that to send 30 pictures through the communications channel each second, we need a data rate of 1.8 Mbps, which presents a bottleneck, especially in many WAN environments.

A standard ISDN connection (Chapter 6), which is affordable for the desktop, can provide a data rate of only 128 Kbps. Ethernet, which can provide a raw speed of 10 Mbps, is the popular choice of communications channel for desktop videoconferencing. Because of packet assembling/disassembling, error checking, acknowledgment, and other overhead, however, the actual data rate is about 1.5 Mbps. Furthermore, Ethernet is a bandwidth-sharing network protocol. Like packet-switched circuits, Ethernet is at a disadvantage because data packets may not arrive in a timely manner. Typically there are other users who may consume about 70% of the bandwidth, leaving around 450 Kbps for a videoconference application, or less than 1 fps of video, which seldom provides good quality videoconferencing.

In such an environment, only data compression (Chapter 3) can make videoconferencing possible. The compression techniques used in most PC applications are called lossless, which means that if the original data are compressed, transmit/received, and then decompressed, the user will end up with exactly the same data. Lossless compression may get a video image down to half its original size—a 2:1 compression ratio—which still is not good enough.

Lossless compression is essential for the transmission and storage of software, documents, and other critical data, but the compression ratio can barely surpass 5:1. In lossy compression, a part (hopefully not critical) of the original data is changed or lost during compression and decompression; reasonably acceptable images can be obtained with compression ratios of 5:1 to 10:1 and even higher. Another form of compression is interframe compression, which sends a full frame then sends only information about how that frame has changed. Because many elements of picture, such as a wall in the background, do not change much from frame to frame, adding in interframe compression may give compression ratios of 25:1 or more in typical situations, while maintaining a pretty good image. Of course, the more motion in a picture, the harder it is to do interframe compression. The audio signal can comfortably be compressed by a factor of 5:1 to 10:1.

The most critical issues regarding compression are its interoperability with other videoconferencing facilities and its efficiency of operation on a heavily loaded network.

Compressed data bear no resemblance to the original form. Both conferencing parties must support the same compression algorithm for interoperability. ITU/CCITT H.261 is a well-defined and widely recognized video compression logarithm. The device that performs **co**mpression/**dec**ompression is called a codec. The codec, which is an integral part of any videoconference system, also acts as an interface device between the videoconference device and the network.

Compression/decompression is a very computation-intensive operation. The better the desired compression ratio, the longer the calculation takes. When the compression becomes too complicated, codec often replaces the communications channels as the bottleneck in videoconference systems. Use of an expensive piece of hardware is the ideal way of achieving required frame rate. Otherwise, the computer must be very powerful, to be able to run the software compression/decompress algorithm. The H.261 standard uses a compression scheme that first sends very coarse image information across the channel, then successively refines the picture as long as there is enough bandwidth left to perform this function.

2.8.2 Document Conference Systems

Early digital videoconference systems offered a white board, which is essentially a shared version of a simple "paint" program. Document conference systems provide true document sharing: that is, the accessing and sharing of information residing in a computer in a real-time group setting, from remote locations, even over simple telephone lines. Images, documents, design drawings, spreadsheets, and regular word processing files can be shared and exchanged to allow participants to work concurrently and interactively on participating computers.

Because a document conference system does not require line video, implementation is relatively simple and inexpensive. Many PC users already have the computer hardware and network connections needed to participate in document conferences:

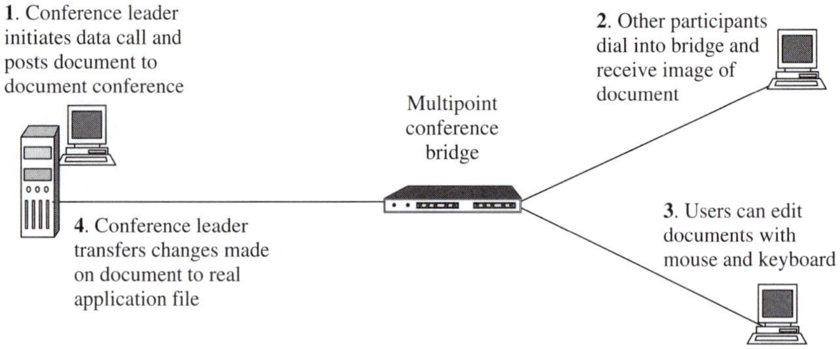

Figure 2.37 Document conference software provides true document sharing. Conference participates can mark up, edit, print, and save the documents they are collaborating on.

- Personal computers with fast modems
- Regular phones or speakerphones
- Two phone lines per PC—one for the phone and one for the modem
- Document conference software

A DSVD (digital simultaneous voice and data) modem or ISDN line–interface device may save a regular telephone line.

In a document conference, the group leader brings up an application with a document, and other participants can then change that document in real time even if they do not have the relevant application on their computers (Figure 2.37). The standard for interoperability of whiteboarding and document conferencing is T.120.

With an advanced version of a videoconference system, a dual-monitor setup can typically display on each monitor a combination of live video from the far end, a captured still image, and/or a computer application, while a single-monitor setup uses the picture-in-picture mode to show more than one type of display. While a still-image signal is being sent to the remote location, the motion video will momentarily freeze until the transmission is complete.

2.8.3 Videoconference Standards

Videoconferencing comprises the synchronous deployment of voice, data, and video across a network, which requires substantial collaboration of standardization efforts. Figure 2.38 shows the major ITU conferencing standards, and the list that follows describes a few of them.

- T.120 is a series of standards for common collaborative tools that regulate the interoperation of whiteboarding and document-sharing applications. T.120 addresses a far-ranging set of user needs with a comprehensive set of standards for applications involving "collaborative computing" or "data conferencing." These standards are independent of hardware, network, operating system, and application details.

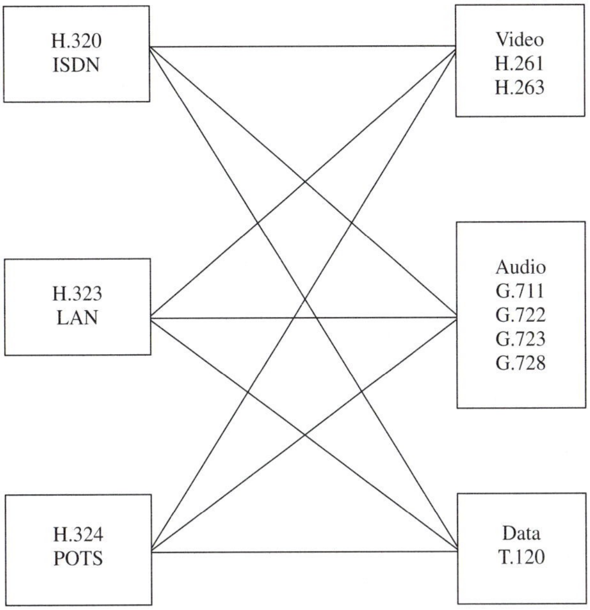

Figure 2.38 ITU conferencing standards for video, audio, and data in WAN, LAN, and public telephone network systems.

• H.320, also called P×64 (meaning multiples of 64 Kbps), is the established standard for videoconferencing over circuit-switched channels. It provides a standard for the coding, transmission, and decoding of audio and video over digital channels, typically over ISDN for videoconference room setups. H.320 is a family of standards that includes the H.261 video compression/decompression standard. Built into H.320 is the ability to translate between the NTSC video format used in the United States and Japan and the PAL video format used in most countries in Europe and in China.

• H.323 is the protocol suite that governs moving video and audio channels over a LAN connecting to the larger WAN and Internet environment. It is specifically designed to allow H.323-compatible products to communicate with H.320-compatible products through a gateway unit—a hardware connection point.

• H.324 describes a device entitled "multimedia terminal for low-bit-rate visual telephone services over the GSTN," where GSTN stands for general switched telephone network, or public switched telephone network (PSTN), also called the plain old telephone service (POTS). H.324 is an ITU-recommended standard family approved in 1996 for modems designed to handle point-to-point multimedia over analog telephone lines. H.324-compliant products serve in a range of formats, including stand-alone videophones and TV-based videophones, where the TV is used as a display device. Since H.324 is an umbrella standard, it establishes standards for audio, video, and data communications, as well as procedures for call setup and control by using other ITU recommendations. For a desktop PC, the H.324 specification requires at least a 28,800 bps v.34 modem that also meets the v.80 (video-ready) standard. Audio quality should meet telephone standards. H.324 has had an enormous impact on the teleconference market.

H.320 is the ITU-T standard for videoconferencing aimed at defining the minimum requirements all videoconferencing systems must support in a WAN environment. This minimum requirement ensures that all H.320-compliant systems will be able to communicate with each other. The system will differ in the optional requirements that can be implemented to improve the quality of the audio and video. Which optional requirements are implemented, and how well this is done, will be left to the system designers and manufacturers. Many videoconference operations involve calls with more than two parties. H.324 supports multipoint calls by having H.324 terminals connect through multipoint control units (MCUs). The H.324 recommendations specify methods for multipoint rate matching and synchronization.

The performance of a videoconferencing product is judged by the quality, speed, and continuity of the video and audio it presents. The following factors affect system quality.

- **Picture resolution**, which is critical to the quality of the picture, comes in two types: common intermediate format (CIF), with resolution of 352 × 288, and quarter-common intermediate format (QCIF), with 25% of the resolution of CIF (176 × 144). CIF is also occasionally referred to as FCIF, or full CIF.

- **Frame rate**, also referred to as the number of frames per second (fps), is the number of times the picture image is refreshed. H.320 systems can support frame rates of 7.5, 10, 15, or 30 fps. When the frame rate is low, the motion appears to be flickering. The higher the frame rate, the smoother the motion.

- **Preprocessing** is a complex process that reduces the amount of background recoding. If preprocessing is not used, the video encoder may spend much time encoding noise due to poor lighting, which will fool the system into acting as if there is motion in the background when in fact there is none. Preprocessing ensures that only real motion is encoded.

- **Postprocessing** can compensate for picture degradation due to fast motion. It may reduce the blocking and noisy effects caused by H.320 video codecs. Postprocessing can also be used to enhance the frame rate and reduce the flickering motion effect due to low frame rates.

- **Motion compensation** is performed at both the encoder and the decoder. Essentially, motion compensation encodes the moving block only rather than the entire video area for every frame. This refinement is especially important at lower bit rates. Motion estimation is performed at the encoder to determine what the motion vector should be. All H.320 systems must have the ability to decode a motion compensation signal. Video quality improvements are made possible by the encode ability, which is optional.

- **Audio quality** under H.320 is specified as follows:
  1. 48–64 Kbps narrowband
  2. 48–64 Kbps wideband
  3. 16 Kbps narrowband

  G.711 is telephone-quality audio with a 3 kHz bandwidth. G.722 produces stereo-quality audio at 7 kHz. At higher data rates, typically 256 Kbps and above, G.722 offers the user the best audio quality available. G.728 (16 Kbps) is narrowband audio, which may leave more bandwidth for video at low bit rate calls.

TABLE 2.3
The H.320 Audio Classes

	Class 1	Class 2	Class 3
Picture resolution	QCIF (176 × 144)	QCIF or CIF	CIF (352 × 288)
Frame rate, fps	7.5	10 or 15	30
Audio	G.711 narrowband 3 kHz	G.722 wideband 7 kHz	G.722/G.728 (Narrowband–low data rate) selectable
Pre- or Postprocessing	No	Optional	Fully supported
Motion compensation	No	Optional	Fully supported

Simply put, class 1 systems are required to meet only the minimum level of support; class 2 systems need to support part of the optional features in addition to the requirement to a class 1 system. Class 3 systems are required to support all the required functions plus all optional features. Table 2.3 summarizes the three classes of systems and shows the functions and quality provided by each. Unfortunately, even manufacturers who deliver class 3 systems are affected by manufacturers of products of inferior quality. For example, if a class 1 system connects to a class 3, the class 3 system will have to downgrade its capabilities to communicate with the class 1 system. Thus a high-quality system is forced to degrade its resolution. The class 3 system would be forced to display a QCIF resolution and to use 7.5 fps, as well as the lower quality G.711 audio for the call to work. A less capable system will always force the higher quality system to downgrade its performance for the sake of holding the conference. The best solution for a user is to require a flexible system with upgradable software, which provides investment protection by allowing the user to take advantage of new features without having to buy new hardware.

Data rates have a very big impact on perceived quality. At T1 speeds (1.544 Mbps), the video quality will be optimal. Realistically, for a H.320 system, 768 Kbps is a tradeoff between high-quality video and cost. Probably the most common data rate in use today is 384 Kbps, while 128 Kbps is becoming more popular with the availability of ISDN.

Since many H.320-compliant products offer a document-sharing application, the T.120 standard is often used to guarantee for interoperation in this area. With H.323 and H.320 in place, theoretically, a videoconference can be originated and terminated at any location equipped with hardware observing the ITU protocol sets.

Two non-ITU-approved "standards"—Intel's ProShare videoconference product family and the CU-SeeMe videoconference system of Cornell University—are worth mentioning. The CU-SeeMe is a network/Internet desktop videoconference package widely distributed as freeware over the Internet. In terms of "seats," it is probably the most popular videoconferencing solution in the world.

In general, if a videoconference product is not compliant with H.320, H.323, or H.324, it will not interoperate with products made by other vendors. Although this may not be a barrier to productive use of videoconferencing within a company, it is certainly inconvenient when one would like to communicate with the outside world.

Internet/intranet systems are inexpensive and widely available for videoconferencing. But the Internet does not guarantee good quality for audio and video because so many users are sharing the communications channels. The Resource Reservation Protocol

(RSVP) and Real-Time Transport Protocol (RTP) standards for guaranteed real-time Internet bandwidth might change the situation. However, most Internet/intranet videoconference systems expect to be given an IP address, thus, if the Internet/intranet connection is behind a firewall, there may be difficulties.

More and more organizations are enhancing their networks with ATM and fiber-based technologies, providing plenty of capacity for medium-quality videoconferencing. Although most videoconference systems use terrestrial lines, systems based on satellite transmissions are also available, particularly for distance-learning and business applications. Satellites offer the unique ability to reach multiple locations cost-effectively, and they are sometimes the only alternative in remote areas not served by terrestrial lines.

2.9 CHAPTER SUMMARY

This chapter covers

- Data and digital communications components
- Basic theories, concepts, and techniques of digital communications (digitization, modulations, etc.)
- Digital communications applications and the problems and challenges they present

We began by analyzing analog and digital signals using an intuitive approach, with reference to mathematical explanations where necessary. The Fourier series approach was shown to be utilized for the digital systems. Next, we discussed the current modem technology, including modulation and keying techniques.

Wireless communication system, which gaining popularity owing to its ease of use and better availability, was the subject of one section of the chapter. In addition to cellular technology, other technologies such as PCS and satellite systems were discussed in detail.

Among other applications of data communications, it is expected that videoconference systems will gain in popularity as higher bandwidths become available. The remainder of the chapter discussed the standardization and technologies involved in this enterprise.

2.10 PROBLEMS

2.1. What type(s) of wireless communication system require a direct line of sight between transmitters and receivers?

2.2. Why is it generally true that a digital signal requires a higher bandwidth than an analog signal?

2.3. Name the three components that completely describe an analog signal.

2.4. What are the three modes of communication in fiber optic communication?

2.5. Consider a periodic signal that completes one cycle in 0.01 second. What is the frequency of the signal?

2.6. What is the bandwidth of a signal that ranges from 10 kHz to 10 MHz?

2.7. Explain what the time period of a signal is. If there are two sine waves with frequencies of 20 and 40 kHz, what are their time periods? In general, if the frequency of signal Y is double the frequency of signal X, how are the time periods of X and Y related arithmetically?

2.8. Consider the signal $x(t) = 10 + \cos(20t) + \cos(400t) + \sin(5000t)$.
(a) What is the minimum (lowest) frequency of the signal $x(t)$ in hertz?
(b) What is the maximum (highest) frequency of the signal $x(t)$?
(c) What is the bandwidth of the signal $x(t)$?
(d) What is the bandwidth of the signal $y(t) = 100 \sin(10{,}000t) + x(t)$?

2.9. Let the Fourier coefficients of a Fourier series of a digital signal be given as follows:

$$a_n = \frac{1}{\pi n\,[\cos(\pi n/4) + \cos(\pi n) - \cos(3\,\pi n/2) + \cos(2\,\pi n)]}$$

$$b_n \frac{1}{\pi n\,[\sin(\pi n) + \sin(\pi n/4) - \sin(2\,\pi n) + \sin(3\,\pi n/2)]}$$

(a) Evaluate the tenth coefficients of the Fourier series (i.e., a_{10} and b_{10}).
(b) Find the rms (root-mean-square) amplitude of the tenth harmonic sqrt($a_{10}^2 + b_{10}^2$).
(c) Find the phase Φ of the tenth harmonic.
(d) For a bit rate of 9600 bps, and assuming 8 data bits per character, what would be the first three harmonic frequencies?
(e) How many harmonics of the signal can be transmitted on a line with cutoff frequency (or BW) of 4500 Hz, assuming that the bit transmission rate is kept at 9600 bps with 8 data bits per character?

2.10. Assuming the data used in Problem 2.9, plot the amplitudes of first 100 harmonics (versus time). Use a data bit rate of 1000 bps. (You must write a program to generate the data.)

2.11. Using the 8-bit ASCII character c (01100011) as the binary digital signal to be transmitted, determine the following:
(a) Fourier coefficients of the resulting Fourier series.
(b) The rms Fourier amplitudes of the first three harmonics of the binary signal.
(c) The third harmonic frequency of the binary signal if the bit rate is 9600 bps.

2.12. A periodic band-limited signal has only three frequency components: dc, 100 Hz, and 200 Hz. In the sine–cosine form, the signal is

$$x(t) = 12 + 15\cos(200\,\pi t) + 20\sin(200\,\pi t) - 5\cos(400\,\pi t) - 12\sin(400\,\pi t)$$

Express the signal in amplitude/phase form, that is,

$$x(t) = \frac{b_0}{2} + \sum_{n=1}^{\infty} d_n \cos(2\,\pi nft + \Phi_n)$$

where

$$\Phi_n = -\tan^{-1}\left(\frac{a_n}{b_n}\right) \qquad \text{and} \qquad d_n = \text{sqrt}(a_n^2 + b_n^2)$$

2.13. Use ASK, FSK, and PSK to modulate 1 0 1 1 0 1 1 1 0.

2.14. Calculate the baud rate and type the encoding for the following bit rates:
(a) 36,000 bps, 4 QAM
(b) 8000 bps, 32 QAM
(c) 3000 bps, 8 PSK

2.15. Calculate the bit rate and type of encoding for the following baud rates:
(a) 1000 baud, 32 QAM
(b) 2000 baud, ASK
(c) 5000 baud, FSK

2.16. Consider two amplitude, four-phase, 8-QAM encoding with 3-bit representation of signal amplitudes for the specifications shown in Figure P2.16.

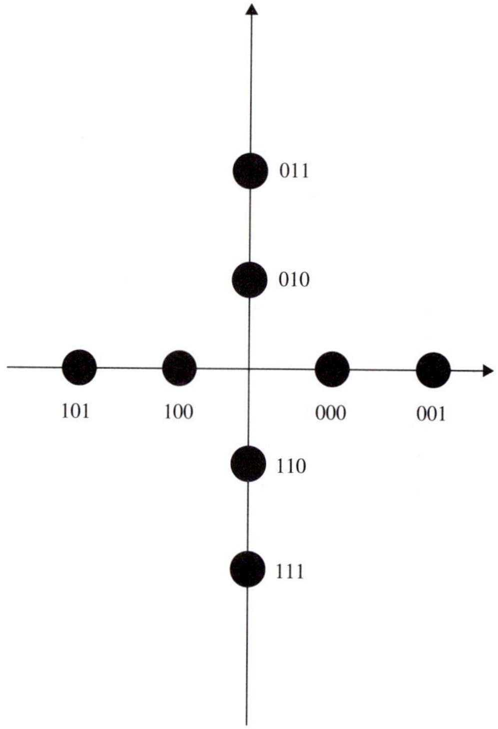

Figure P2.16

2.17. Show the time domain for the following bit pattern to be transmitted in one second. Also, indicate the baud rate.

0 0 0 0 0 1 0 1 0 0 1 1 1 0 0 1 1 1 1 1 0 1 0 1

2.18. Is it true that baud rate is the reciprocal of the shortest signaling element? Explain.

2.19. What is the theoretical maximum data rate for a communication channel having a 100 kHz bandwidth and operating at an *S/N* rate of 20 dB?

2.11 REFERENCES

BOOKS

Agrawal, G. P., *Fiber-Optic Communication Systems*, 2nd ed. New York: Wiley, 1997.

Bates, B., and D. W. Gregory, *Voice and Data Communications Handbook*, 2nd ed. New York: McGraw-Hill, 1998.

Duran, J., and C. Sauer, *Mainstream Videoconferencing: A Developer's Guide to Distance Multimedia*. Reading, MA: Addison-Wesley, 1997.

Faruque, S., *Cellular Mobile Systems Engineering*. Norwood, MA: Artech House, 1996.

Halsall, F., *Data Communications Networks & OSI*, 4th ed. Reading, MA: Addison-Wesley, 1996.

Hecht, J., *Understanding Fiber Optics*, 3rd ed. Englewood Cliffs, NJ: Prentice Hall, 1999.

Held, G., *Understanding Data Communications*, 6th ed. Indianapolis, IN: New Riders, 1999.

Hughes, L., *Introduction to Data Communications: A Practical Approach*. Sudsbury, MA: Jones and Bartlett, 1997.

Jones, C., *Yachtsman's GPS Handbook: A Guide to the Global Positioning System of Satellite Navigation*. Stillwater, MN: Voyageur Press, 1996.

Kaminow, I. P., and T. L. Koch, eds., *Optical Fiber Telecommunications IIIB*. San Diego: Academic Press, 1997.

Lee, C. Y., *Mobile Cellular Telecommunications: Analog and Digital Systems*, 2nd ed. New York: McGraw-Hill, 1995.

Lu, C., *The Race for Bandwidth—Understanding Data Transmission*. Redmond, WA: Microsoft Press, 1998.

Prasad, R., *Universal Wireless Personal Communications*. Norwood, MA: Artech House, 1998.

Ramaswami, R., and K. N. Sivarajan, *Optical Networks: A Practical Perspective*. San Diego: Academic Press, 1998.

Rappaport, T. S., *Wireless Communications: Principles and Practice*. Piscataway, NJ: IEEE Press, 1996.

Rosen, E., *Personal Videoconferencing*. Englewood Cliffs, NJ: Prentice Hall, 1996.

Sampei, S., *Applications of Digital Wireless Technologies to Global Wireless Communications*. Englewood Cliffs, NJ: Prentice Hall, 1997.

Schaphorst, R., *Videoconferencing and Videotelephony: Technology and Standards*. Norwood, MA: Artech House, 1997.

Shay, W. A., *Understanding Data Communications and Neetworks*. Boston, MA: PWS Publishing, 1995.

Stallings, W., *Data and Computer Communications*, 5th ed. Englewood Cliffs, NJ: Prentice Hall, 1996.

Stallings, W., *Business Data Communications*, 3rd ed. Englewood Cliffs, NJ: Prentice Hall, 1998.

Sveum, M. E., *Data Communications: An Overview*. Englewood Cliffs, NJ: Prentice Hall, 1999.

Walrand, J., *Communication Networks: A First Course*, 2nd ed. New York: WCB McGraw-Hill, 1998.

ARTICLES

Baqai, S., M. Woo, and A. Ghafoor, "Network Resource Management for Enterprise Wide Multimedia Services," *IEEE Communications Magazine* 34 (January 1996), 78–83.

Banerjee, S., D. Tipper, M. Weiss, and A. Khalil, "Traffic Experiments on the vBNS Wide Area ATM Network," *IEEE Communications Magazine* 35 (August 1997), 126–133.

Chen, J. C., K. M. Sivalingam, and R. Acharya, "Comparative Analysis of Wireless ATM Channel Access Protocols Supporting Multimedia Traffic," *Journal of Mobile Networks and Applications (MONET)* 3, no. 3 (1998).

Eleftheriadis, A., and T. Puri, "MPEG-4: An Object-Based Multimedia Coding Standard Supporting Mobile Applications," *Journal of Mobile Networks and Applications (MONET)* 3, no. 1 (1998).

Pahlavan, K., T. H. Probert, and M. E. Chase, "Trends in Local Wireless Networks," *IEEE Communications Magazine*, 33, no. 3 (1995), 88–95.

WORLD WIDE WEB SITES

Ohio State University cabling FAQs
http://www.cis.ohio-state.edu/hypertext/faq/usenet/LANs/cabling-faq/faq.html

Informative web site about data communications concepts
http://www.fit.qut.edu.au/selwyn/DavidChen/

Many data communications tutorials from TechWeb
http://www.data.com/Tutorials/

A data communications tutorial
http://www.sangoma.com/tutorial.htm

Another useful data communications site
http://www.cit.ac.nz/smac/dc100www/default.htm

One of the good videoconference resources on the World Wide Web is the Videoconference Resource Center, which provides a detailed roundup of products, services, and technical information
http://www.videoconference.com

3

THE OSI SEVEN-LAYER NETWORK MODEL

A communications system consists of many components and involves more than one communication party. A computer network comprises two or more nodes, typically computers, connected by underlying communication channels. A computer network is the next logical step beyond stand-alone computer systems, and a network can be constructed from a nesting of networks. This is called internetworking.

Networks are usually complex. Following the divide-and-conquer principle, a modern network is best designed, constructed, and described in a layered structure, or by using networking standards to connect layered protocols. Fundamentally, we use layered protocols to do the following:

- Reduce the complexity of network design and implementation.
- Provide for peer-to-peer layer interaction across the network.
- Allow modifications to be made in one layer without affecting others.

In a layered structure, a complex network system can be partitioned into many modules that belong to different layers. Changes in design and construction that happen in one layer have no significant impact on the functioning of other layers. For example, a change in the error checking algorithm in a lower layer will not affect the routing algorithm in a higher layer. In the meantime, an encryption algorithm in a presentation layer certainly relies on cooperation from the decryption function in the corresponding layer at the other end of the network—a typical peer-to-peer layer interaction, where a virtual connection is presented to all the network layers. In this sense, a network node is like a high-rise building. The components in each layer communicate with those in the corresponding layer of another node in the network via a virtual link. However, the data actually go all the way from the upper layers to the lowest layer of one system, traveling through the physical links, reaching the bottom layer of another system, and then reversing the process to climb all the way up to the components in the peer layer of the first node (Figure 3.1).

Protocols in a layered set work together to specify complete network functions. Each intermediate layer based on the protocols uses the layer below it to provide a service to the layer above. ISO's seven-layer open systems interconnect (OSI) model is an attempt to provide a international standard framework that describes the complete network protocol stack in heterogeneous computer network environments.

The OSI network architecture is partitioned into seven layers, from a physical layer at the bottom to the applications layer at the top, as follows:

Physical layer: Medium and signal format of raw bit transmission.

Data link layer: Access to and control of transmission medium.

Network layer: Format of individual data packets.

Transport layer: Delivery of sequences of packets.

Session layer: Management of connections between programs.

Presentation layer: Conversions for representing data.

Application layer: Detailed and application specific information about data being exchanged.

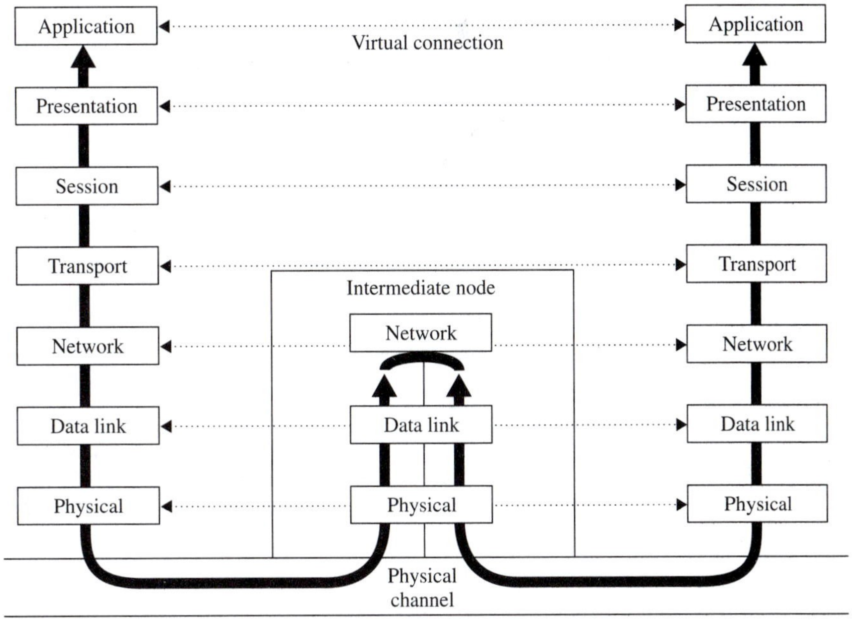

Figure 3.1 In the OSI seven-layer model, the peer-to-peer connections between two systems are virtual, while the actual data go from layer to layer within the system most of the time, bottom–up or top–down. The intermediate node may be a router, bridge, switch, or comparable unit.

In a networked host computer, the highest layers—the application layer and presentation layer—are usually part of the application programs. The middle layers are often part of the system software (operating system, device drivers, etc.), and the lowest layers are usually network-dedicated with data link and physical layer built into the network hardware, comprising interface cards and the like. Although the OSI model is not fully implemented in actual network systems, and complete compliance does not exist, the layered concept stands well as a reference and guide in network design.

3.1 PHYSICAL LAYER

The physical layer handles the transmission of raw bits over a communications channel. The protocols for the physical layer specify the medium used for the transmission (e.g., electronic vs optical or wireless), offer choices for the signal formats (e.g., serial vs parallel, synchronous vs asynchronous), and convert abstract raw bit streams into common codes understandable by all the connected parties. Chapter 2 discussed communications media and the most important theories about digital transmission. This chapter gives a more in-depth discussion of related topics from the perspective of a layered structure.

Physical level standards have been widely used for years in point-to-point and wide-area network applications. CCITT/ITU has established X.21–X.24 to specify the functions at the physical lever for leased circuits. Numerous other standards, such as EIA-232 and v.21–v.34, are widely used for various purposes at the physical layers. For LANs and newer high-speed network technologies, many new protocols have been added to the existing protocols, and these are discussed later.

3.1.1 Data Encoding

Digital data are rarely transmitted in their original format. Encoding is the procedure that converts the original signals or raw data into a different form for different reasons (better utilization of existing facilities, increasing efficiency, reducing error rate, synchronization, security, etc.). There are many encoding methods for data transmission purposes. Modulation, discussed in Chapter 2 is actually a form of encoding that is applied to the transmission of digital data over an analog communications line. In addition to data transmission, encoding techniques are widely used in error detection and correction, data compression, and data encryption, introduced in later chapters.

In the simplest and plain format, a bit stream of binary data may be transmitted by a series of lower volts for a 0 bit and a higher volts for a 1 bit; if all the signal elements have the same polarity, the signal is *unipolar*. Otherwise, a negative signal can mean 1 and a positive signal for 0, or vice versa. This encoding scheme is called polar. For historical reasons, 1 and 0 are sometimes called a mark and a space respectively. When binary data 1s and 0s are simply represented by two different voltages, the encoding is called non-return-to-zero level (NRZ-L) (Figure 3.2a). In a simple format such as NRZ-L, one bit of data (a 1 or a 0) is transmitted by a signal element. But multiple bits can also be transmitted by a single signal element, as discussed in connection with modulation tech-

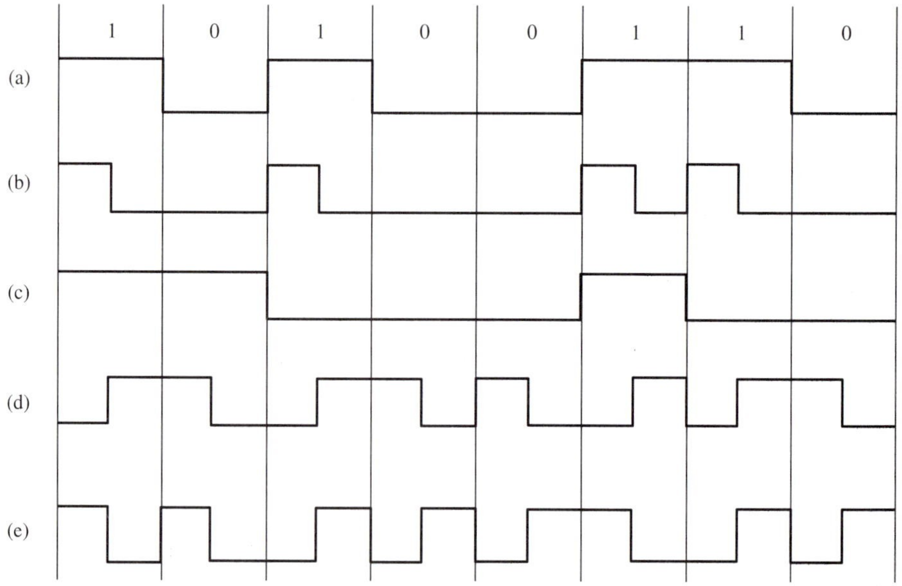

Figure 3.2 Various digital encoding schemes: (a) NRZ-L coding, (b) RZ coding, (c) NRZ-I coding, (d) Manchester coding, and (e) differential Manchester coding.

Hohulski

niques. (Section 2.5.4) ISDN (integrated services digital network) and HDSL (high-bit-rate digital subscriber line) use so-called 2B1Q (2 binary, 1 quaternary) coding to transmit 2 bits at a time, using different voltage levels to represent information (Chapter 6).

In NRZ-L code, the voltage level is constant during bit intervals. Because of its simplicity, NRZ-L coding is used universally for internal data transfers in a computer system.

Return-to-zero (RZ) coding makes a low-to-high voltage change (transition) at the beginning of a logic 1 followed by a high-to-low transition in the middle of the bit. A 0 bit has no signal transitions at the beginning (Figure 3.2b).

NRZ-L coding may cause problems in a communications system in which the two wire heads of a twisted-pair line are attached to a device. In the not uncommon event of accidental inversion of the wire heads, all 1s and 0s will be inverted, too. A variation of NRZ-L uses a signal transition at the beginning of a signal element to represent a 1 and no transition for 0. In other words, it compares the voltage (or polarity) of two adjacent signal elements, rather than measuring the absolute value of the signal element, to determine whether a 1 or a 0 being transmitted. If the two voltages are different, a 1 is transmitted, otherwise, a 0. This encoding scheme is called non-return-to-zero, invert-on-ones (NRZ-1) (Figure 3.2c). NRZ-I has no wire polarity problem and improves error rate under poor communication conditions. NRZ-1 is an example of differential encoding. Differential encoding has the drawback that if one single signal transition is missed, all the following 1s and 0s are inverted.

Both NRZ-L and NRZ-I maintain a constant voltage for the duration of a bit time. They are easy to implement. Because on average, a 1 and a 0 have the same chance to appear in the bit stream, and two adjacent bits have 50:50 chance of remaining unchanged, the voltage changes less frequently than the number of bits transmitted. Thus NRZ encoding makes efficient use of bandwidth. Neither NRZ-L nor NRZ-I provides synchronization capability. Hence the stream 000010000 may wrongly be interpreted as 10000 or 00001, depending on when the count is started.

Manchester coding and differential Manchester coding are introduced to add synchronization capability. In the Manchester coding scheme, a transition at the middle of each bit period is used for data as well as for clocking. A low-to-high transition represents a 1, and a high-to-low transition represents a 0 (Figure 3.2d). In differential Manchester coding, a 0 is indicated by the presence of a transition at the start of the interval, and a 1 is represented by the absence of a transition at the start of the interval. There is always a transition in the middle of the interval for clocking and synchronization purposes (Figure 3.2e). Differential Manchester coding has the edge over other differential encoding schemes—it is more resistant to noise.

Manchester and differential Manchester codes are biphase codes, which are widely used in the network industry. Ethernet and CSMA/CD (the IEEE 802.3 LAN standard physical layer) use Manchester coding with coaxial cable and twisted-pair wire media. Differential Manchester coding is used in IEEE 802.5 token ring LANs with shielded twisted-pair wiring (Chapter 4).

Biphase codes are self-clocking codes and easy to implement. They have no dc components and are more error resistant than NRZ codes. However, because they require at least one transition per bit interval and may have, half of the time on average, two transitions, the signal pulses are half the original width; hence they consume twice as much bandwidth as the straight NRZ encoding. This restricts the applications of a biphase coding scheme from higher speed networks, such as the 100 Mbps FDDI protocol (Chapter 7).

3.1.2 Multiplexing Schemes

A network may consist of millions of computers, and it is virtually impossible to completely connect them. However, complete connectivity may still be provided via sharing the physical links. Multiplexing (MUX) techniques, which allow more than one communications channel to be carried over a single physical connection, are just like the multitasking and time-sharing concepts used in computer science to share CPU resources. There are many elegant methods for multiplexing. In traditional telephone systems that use analog transmission, multiplexing is implemented by means of the frequency division multiplexing (FDM) method. For data communications and networks, time division multiplexing (TDM) allows time-sharing. For digital wireless telephone system, TDM is a maturing technology and code division multiplexing (CDM) is gaining popularity.

3.1.2.1 Frequency Division Multiplexing (FDM)

The idea for the FDM technique originated from radio broadcasting. The air is treated like a huge communications pipe. Different radio wave signals can be transmitted through the air without interfering one another because they use carrier waves with different frequencies. The difference of carrier frequencies is what separates one channel from another at the receiver end. FDM analog signals are transmitted over the physical link at different frequencies in much the same way as the signals for different television stations are transmitted at different frequencies through the air or via a television cable.

For the traditional analog telephone system, twelve 4000 Hz voice channels are grouped together and modulated and mixed (multiplexed) into the 60–108 kHz carrier band. In each 4000 Hz channel, only 3000 Hz (from 300 to 3300 Hz) is used for real signal, while two sidebands of 500 Hz each guard against possible interference between adjacent channels. A low- and a high-frequency filter at the receiver side separate the useful signal from the sidebands and carrier wave. In addition, the filter separates (demultiplexes) the signal from one channel from other channel signals (Figure 3.3).

The next FDM level for the telephone system is the 60-channel supergroup comprising five groups of signals ranging from 312 to 552 kHz. Each group is modulated as a single signal with a 48 kHz bandwidth in a supergroup. Then 5 supergroups (CCITT standard) or 10 supergroups (Bell System standard) are merged into a master group that accommodate 300 or 600 voice channels, respectively.

3.1.2.2 Time Division Multiplexing (TDM)

Digitized signals, such as voice, images, and data, are often transmitted through a time division multiplexing (TDM) system, in which a large data rate connection is divided into many small time slots. TDM is implemented by interleaving portions of each channel in time slots. In TDM, many communications channels share a single connection with a large data rate (sometimes inaccurately called bandwidth), which dedicates an individual band-

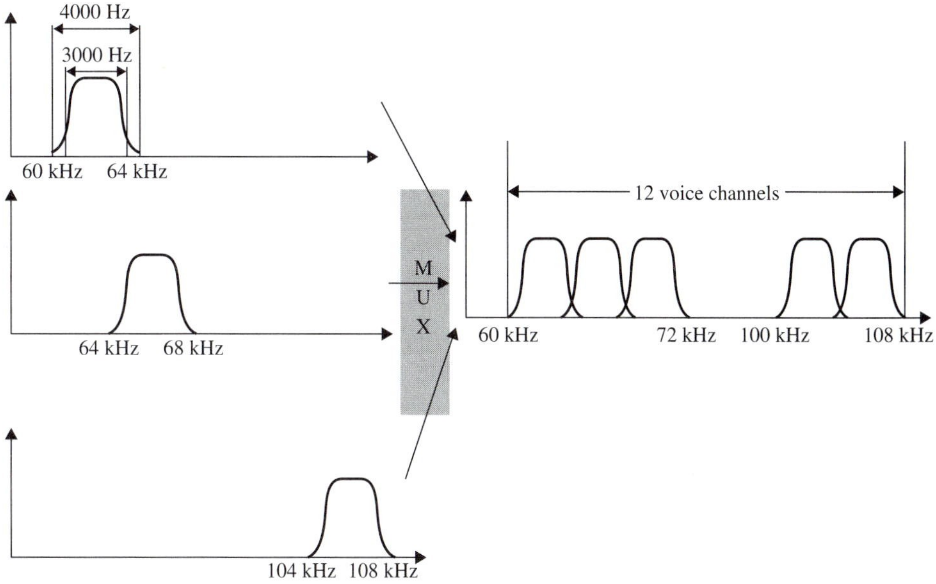

Figure 3.3 FDM divides the transmission bandwidth into smaller bands (subchannels).

width channel to each application or user. The time division multiplexer (MUX) cyclically scans the incoming data from the multiple input ports. Data are "picked" in bits, bytes, or block units and interleaved into frames on a single high-speed communications line. Thus, TDM systems take digital data as input. Consequently a channel bank that performs sampling quantization and encoding, using PCM (pulse coding modulation) to convert the analog signals into digital, is required to interface with the common carrier analog telephone network (Figure 3.4).

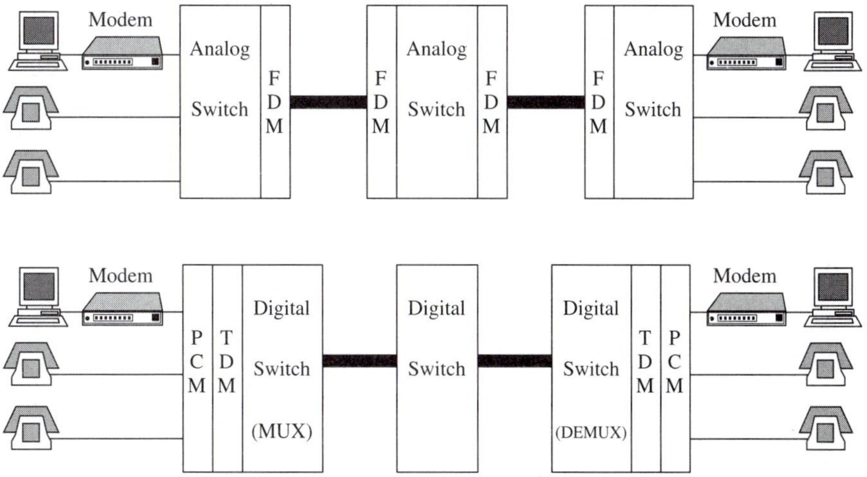

Figure 3.4 Comparison between FDM and TDM.

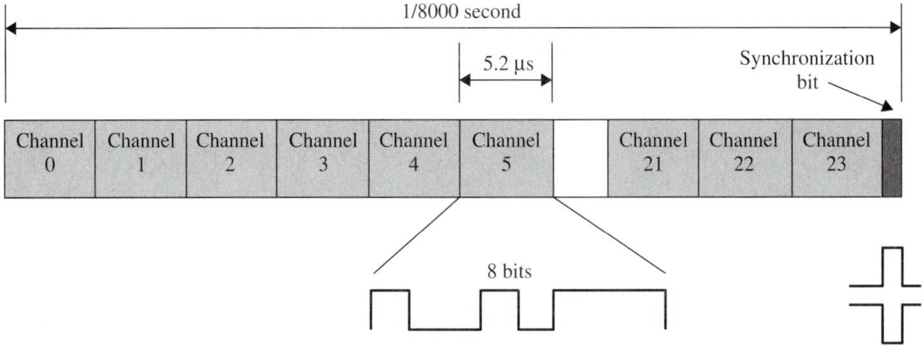

Figure 3.5 A T1 frame interleaves 24 channels of 8 bits each, plus one synchronization bit. The total data rate of the T1 is $8000(8 \times 24 + 1) = 1{,}544{,}000$ bps (1.544 Mbps).

The most widely used TDM system is the North American (AT&T) digital TDM hierarchy. The T1 system, first offered as tariff services in 1983, is at the bottom of the hierarchy. Common carriers transmit 24 voice channels together with a T1 digital channel. Because a voice channel uses a 4000 Hz bandwidth (including the two 500 Hz sidebands), a sampling rate of 8000/s is sufficient to transmit all the information in the original analog signal according to the Nyquist theorem. Each sampling takes 8 bits (256 levels), and 24 channels plus one synchronization bit are combined to form a 193-bit frame. A rate of 8000 fps, therefore, requires a T1 data rate of 1,544,000 bps (1.544 Mbps). In summary, a T1 frame has a duration of 1/8000 second, which contains 24 samplings plus one synchronization bit. Each sampling has a signal width of 5.2 μs and contains 8 bits of data for a single voice channel. Each voice channel has a data rate of 8000×8 (bits) = 64 Kbps (Figure 3.5).

The synchronization bit establishes and maintains synchronization between the sending and receiving devices and performs a function similar to the start bit in an asynchronous communications system and a synchronous byte in a synchronous transmission system. The bit alternates as 1 and 0 in each succeeding frame and forms the pattern 10101010 . . . , which can be easily scanned by the receiver to resynchronize when necessary. When 7-bit data format is used, the eighth bit is used for signaling and control, resulting in a 56 Kbps data rate. The formal name of T1 is DS-1, while a digitized voice channel (64 Kbps) is called DS-0. Two DS-1 lines can be combined to form a DS-1C (T1C) channel (3.152 Mbps), and four DS-1 lines are combined to form a DS-2 (T2) channel (6.312 Mbps). After that, 28 DS-1 lines or 7 DS-2 lines form the DS-3 (T3) channel (44.736 Mbps), and 6 DS-3 channels are multiplexed into a DS-4 (T4M) trunk (274.176 Mbps) to provide 4284 voice channels of 64 Kbps each (Figure 3.6).

When fewer than 24 voice channels are needed, fractional T1 service is provided. In Europe and China, the basic carrier data rate corresponding to T1 is E1, with a 2048 Kbps data rate. Furthermore, frame length, placement of the synchronization bit, and synchronization bit functions are also different. These differences highlighted the serious need for integrating international digital networks and resulted in the ISDN (integrated services digital network) standard (Chapter 6).

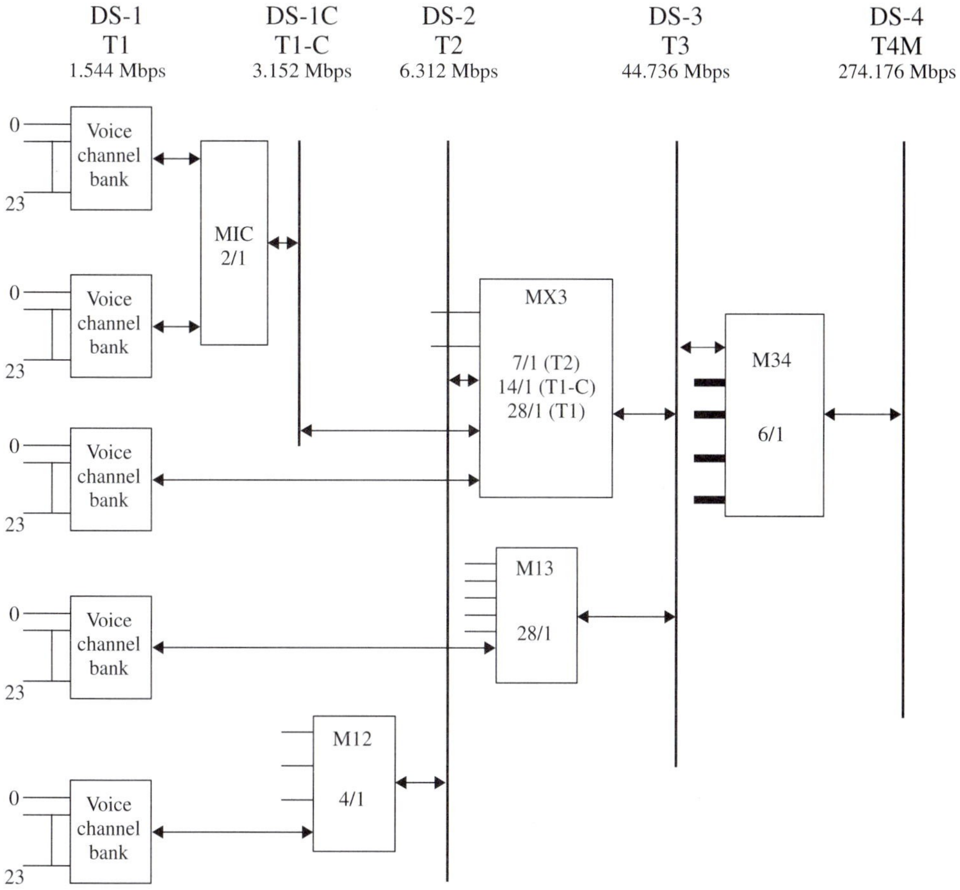

Figure 3.6 The North American digital hierarchy: DS-01, DS-1C, DS-2, DS-3, and DS-4M.

A multiplexer sends out control signals to input devices to keep the data streams synchronized. The control signals are generally issued from a master clock in the MUX. Because different communication devices operate with different physical transmission times and delays, the MUX may provide internal buffers for the time-independent data streams to allow the data to arrive at random intervals, to be multiplexed later into the high-speed link. This technique, called isochronous TDM, provides elastic buffers that can expand or contract to accommodate dynamic traffic flows.

TDM may be used together with FDM: that is, FDM may be used to multiplex multiple channels, which are then merged into a larger TDM channel. Like FDM systems, the conventional TDM wastes the bandwidth of the communication channel because time slots in the frames are often unused. For example, vacant slots occur with idle terminals and conversation gaps. Various efforts have been made, and some variations of TDM systems, such as the *statistical TDM*, have been developed to improve channel utilization. However, because of the bursty nature of communications, much expensive bandwidth was wasted carrying idle patterns in case of data and silence in case of voice. Furthermore, these channels represented strict boundaries that limited the performance of such high-bandwidth traffic as image transfer, LAN, and system network architecture (SNA) (Section 3.3) users. As communications became even more bursty, widespread, and mission critical, it became clear that the original TDM architecture was not quite appropriate for the new networking requirements.

3.1.2.3 Code Division Multiple Access (CDMA)

Thanks to the availability of very low cost, high-density digital integrated circuits, which reduce the size, weight, and cost of communication devices, an "old' multiplexing technology—code division multiplexing access (CDMA), proposed 40 years ago—is finding new life in cellular telephone communications.

CDMA has been used by the military in antijamming devices and to ensure secure communications because the spread-spectrum signal used by CDMA is difficult to detect or jam, and resists interference. An important issue for any cellular telephone system is its multiple-*access* capacity, meaning support for multiple, simultaneous users. That is, in cellular communications a large number of users must be able to share a common pool of radio channels, and any user can gain access to any channel, although a user is not always assigned to the same channel. A channel can be thought of as merely a portion of the limited radio resource, which is allocated, temporarily, for a specific phone call. CDMA uses an innovative multiple-access method to define how the radio spectrum is divided into channels and how channels are allocated to each user of the communication system.

The CDMA approach uses coding theory and complicated algorithms to spread information about multiple channels over the entire frequency spectrum. A CDMA cellular phone call starts with a standard rate of 9600 bps. It then spreads to a transmitted rate of about 1.23 Mbps. "Spreading" means that digital codes are applied to the data bits associated with users in a cell. These data bits are transmitted along with the signals of all the other users in that cell. When the signal is received, the codes are removed from the desired signal, separating the users and returning the call to a rate of 9600 bps. These codes, which are shared by both the mobile station—the cellular phone—and the base station, are called pseudo-random code sequences. All users share the same range of the radio spectrum.

Now we explain CDMA with the help of an example. Assume that a bit of 1 or 0 is represented by a vector (code) c_i or $\bar{c}_i$ for channel i to be multiplexed using CDMA. All the channels to be transmitted are *orthogonal* with respect to one another. This can be mathematically expressed as follows:

$$c_i \cdot c_j = \frac{1}{n} \sum_{k=1}^{n} c_{ik} c_{jk} = \begin{cases} 1 \text{ if } i = j \\ 0, \text{ otherwise} \end{cases} \tag{3.1}$$

The operator "·" is called the inner product. One way to implement the orthogonal relation is to use bipolar coding, that is, to use a special sequence of -1 and $+1$—vector or code—to represent a single bit of signal for each channel. For code $\bar{c}_i$, we simply replace every -1 with $+1$ and replace every $+1$ with -1 in the sequence for c_i. Apparently

$$c_i \cdot \bar{c}_j = \begin{cases} -1 \text{ if } i = j \\ 0, \text{ otherwise} \end{cases} \tag{3.2}$$

To preserve orthogonality for any pair of vectors from different channels, the number of pairs of identical elements must equal the number of pairs of different elements. Adding all the code sequences arithmetically will simultaneously transmit multiple channels. For channel i to transmit a 1, the code sequence for c_i is simply transmitted, and to transmit a 0 the code sequence for $\overline{c}_i$ is transmitted. To retrieve signals for channel i from signal S, where

$$S = C_1 + C_2 + \cdots + C_i + \cdots$$

and $C = c_i$ or $\overline{c}_i$, we find the inner product of c_i and S. Based on Equations (3.1) and (3.2),

$$c_i \cdot C_j = 0, \text{ when } i \neq j$$

we have

$$C_i \cdot S = c_i \cdot C_i = \begin{cases} +1, & \text{if } C_i = c_i \\ -1, & \text{if } C_i = \overline{c}_i \\ 0, \text{ if channel } i \text{ is silent} \end{cases}$$

In the following example, we use the vectors (+1, +1, +1, +1), (−1, −1, +1, +1), (−1, +1, −1, +1), and (−1, +1, +1, −1) to represent 1 for channels 1, 2, 3, and 4, respectively. Thus 0s are expressed as (−1, −1, −1, −1), (+1, +1, −1, −1), (+1, −1, +1, −1), and (+1, −1, −1, +1), respectively.

	Bit 1	**Bit 0**
Channel 1	(+1, +1, +1, +1)	(−1, −1, −1, −1)
Channel 2	(−1, −1, +1, +1)	(+1, +1, −1, −1)
Channel 3	(−1, +1, −1, +1)	(+1, −1, +1, −1)
Channel 4	(−1, +1, +1, −1)	(+1, −1, −1, +1)

Note that all the channels are orthogonal from one another, for example,

$$c_1 \cdot c_3 = \frac{-1 + 1 - 1 + 1}{4} = 0$$

$$c_2 \cdot c_4 = \frac{-1 + 1 - 1 + 1}{4} = 0$$

while

$$c_1 \cdot c_1 = \frac{+1 + 1 + 1 + 1}{4} = 1$$

$$c_4 \cdot c_4 = \frac{-1 - 1 - 1 - 1}{4} = -1$$

Suppose that at a given moment, channel 1 is transmitting 0, channel 2 is transmitting 1, channel 3 is not transmitting (silent), and channel 4 is transmitting 0. (Note that transmitting a signal 0 is different from not transmitting at all.) The resulting signal S would be

$$S = (-1, -1, -1, -1) + (-1, -1, +1, +1) + (0, 0, 0, 0) \\ + (+1, -1, -1, +1) = (-1, -3, -1, +1)$$

and we have

$$c_1 \cdot S = \frac{(+1, +1, +1, +1) \cdot (-1, -3, -1, +1)}{4} = -1$$

$$c_2 \cdot S = \frac{(-1, -1, +1, +1) \cdot (-1, -3, -1, +1)}{4} = +1$$

$$c_3 \cdot S = \frac{(-1, +1, -1, +1) \cdot (-1, -3, -1, +1)}{4} = 0$$

$$c_4 \cdot S = \frac{(-1, +1, +1, -1) \cdot (-1, -3, -1, +1)}{4} = -1$$

Increased privacy is inherent in CDMA technology. CDMA phone calls will be secure from the casual eavesdropper since, unlike an analog conversation, a simple radio receiver will not be able to pick individual digital conversations out of the overall RF radiation in a frequency band.

In the final stages of the encoding of the radio link from the base station to the mobile unit, CDMA adds to the signal a special "pseudo-random code" that repeats itself at finite intervals. Base stations in the system distinguish themselves from each other by transmitting different portions of the code at certain times. In other words, the base stations transmit time-offset versions of the same pseudo-random code. To assure that each time offset used remains unique, CDMA stations must remain synchronized to a common time reference.

This precise common time reference is provided by the Global Positioning System, the satellite-based radio navigation system discussed in Chapter 2. GPS is capable of providing a practical and affordable means of determining continuous position, velocity, and time to an unlimited number of users. For station identification mechanisms to work reliably and without ambiguity, base stations must be synchronized within a few microseconds. Any convenient mechanism can be used for this purpose, but the system was designed under the assumption that the GPS itself would be used. GPS is a system made of a family of low-earth-orbit satellites that broadcast a spread-spectrum signal and ephemeris information from which a sophisticated Kalman filter algorithm in a receiver can derive both a very accurate position and a very accurate time.

CDMA (and spread spectrum in general) was always dismissed as unworkable in the mobile radio environment because of what was called the "near–far problem." It was always assumed that all the stations transmitted constant power. In the mobile radio environment, some users may be located near the base station while others are far away. The propagation path loss difference between the extreme users can be many tens of decibels.

If there is, say, a 30 dB difference between the largest and smallest path losses, then there is a 60 dB difference between the signal-to-noise ratio (SNR) of the closest user and the farthest user, because these are the received powers. To accommodate the farthest users, the spreading bandwidth would have to be perhaps 40 dB, or 10,000 times the data rate. If the data rate were 10,000 bps, then $W = 100$ MHz. The spectral efficiency is abysmal, far worse than even the most inefficient FDMA or TDMA system. Conversely, if a more reasonable bandwidth is chosen, remote users will receive no service.

This observation was, for years, the rationale for not even attempting any sort of spread spectrum in any but geosynchronous satellite environments, where the path loss spread was relatively small.

The key to the high capacity of commercial CDMA is extremely simple. If, rather than using constant power, the transmitters can be controlled in such a way that the received powers from all users are roughly equal, then the benefits of spreading are realized. If the received power is controlled, the subscribers can occupy the same spectrum, and the hoped-for benefits of interference averaging accrue.

Maximum capacity is achieved if we adjust the power control so that SNR is exactly what it needs to be for an acceptable error rate. [If we set the SNR in Equation (2.8) to the target SNR, we can find the basic capacity equation for CDMA.]

3.1.2.4 Wave Division Multiplexing (WDM)

Dense WDM switches give users full system flexibility and use of conventional fiber for very high-speed transmission. Each optical channel is capable of carrying information over a single optical fiber at a rate of up to 10 Gbps. For a 128-channel dense WDM switch at full capacity, each fiber pair can deliver 1.28 terabits per second (tbps) bandwidth, the rough equivalent of 16 million simultaneous telephone calls.

Wave division multiplexing is similar in principle to frequency division multiplexing. With WDM, discrete colors of light become high-bandwidth channels for carrying data over a strand of fiber. WDM allows multiple channels to share the same fiber connection and boosts the speed and capacity of fiber optic links without installing more fibers to provide high-speed interconnection among widely scattered sites. WDM is also important as a means of setting up multiple data centers, and for purposes of disaster recovery and CPU redundancy.

The primary value of WDM lies in expanding the capacity of existing fiber networks without laying more fiber cables. All the user has to do is install wave division multiplexers at both ends of existing fiber spans. This is a step up from older fiber optic systems that require a fiber pair for each channel. WDM is particularly popular among carriers that otherwise would have to install thousands of miles of new fiber to boost capacity. In many situations, running new fiber is practically impossible because of rights of way.

WDM requires no repeaters over a 6-mile span. WDM switches are protocol independent. You can send ATM through them, FDDI, or any other type of frame. The multiplexers do not open up the packets or cells passing through them. They just send units down the line at up to 200 Mbps per channel. WDM might be useful for enterprise applications in which high bandwidth is essential, such as linking dual corporate data centers, multilocation manufacturing businesses, hospitals, and financial institutions. Some enterprise users consider WDM even if they do not have an existing fiber network.

3.2 DATA LINK LAYER

3.2.1 Asynchronous and Synchronous Communications

In a computer system including the I/O subsystems, most data transfers occur in parallel. For instance, the transfers from random access memory (RAM) to CPU, and from disk to memory, happen in parallel. Parallel transmission is fast because it allows 8, 16, 32, and 64 bits of data to be transferred at once. However, parallel transmission needs multiple circuits for the interface of transmitter and receiver. In addition, the expense of running the multiple-conductor cables over long distances is prohibitive. Therefore, parallel transmission is appropriate for short distance.

In contrast to computer systems, in digital communications and in network environments data are transmitted almost exclusively serially (i.e., over links in a single communications channel instead of a set of parallel lines). Many more devices may be connected to a serial link than to a parallel bus. Also, a serial connection may continue across multiple media (e.g., copper to fiber optic, then to copper), which is virtually impossible with a parallel bus. In a serial communications system, signals are sent through the communications line one at a time. When nonbinary logic is used, more than one bit of data may be transmitted at once by a serial communications system like the quadrature amplitude modulation (QAM) modem, described in Chapter 2.

Asynchronous serial transmission, the simplest communications format, is found primarily in point-to-point communications systems. A character or a byte of data is the unit of transmission between two devices. An asynchronous transmission system copes with timing and synchronization problems by splitting a character into a fixed number of bits and sending one character, or byte of data, at a time. To keep the communications synchronized, extra bits—a start bit and one or two stop bits—are added to the original data, creating a small frame. A single optional parity bit maybe added to the data, inserted between the data bits and the stop bit, for error checking. This parity bit may be

- Even parity: making the number of 1s in the 8 bits always an even number.
- Odd parity: making the number of 1s in the 8 bits always an odd number.
- Mark: always 1.
- Space: always 0.
- No parity: this bit is part of the data.

When the parity bit is used, at most 7 bits of data can be sent at a time. Parity check is performed on the data bits only, not the start and stop bits. Because 2 or 3 bits may be required to form the frame for each character (7 or 8 bits), a synchronous transmission is not very efficient. However, asynchronous communication requires little hardware support and has the advantage of simplicity. This mode is still often used in software for communications of microcomputers to host computers (e.g., Kermit). The most often used format is *N-8-1-1*, meaning no parity, 8 data bits, one start bit, and one stop bit. Figure 3.7 shows how the character G is transmitted in an asynchronous communications system. Note that in such a system, the least significant bit is transmitted first.

Synchronous transmission is more efficient and is appropriate for the fast serial transmission available in most high-speed digital networks. In a synchronous system, a stream

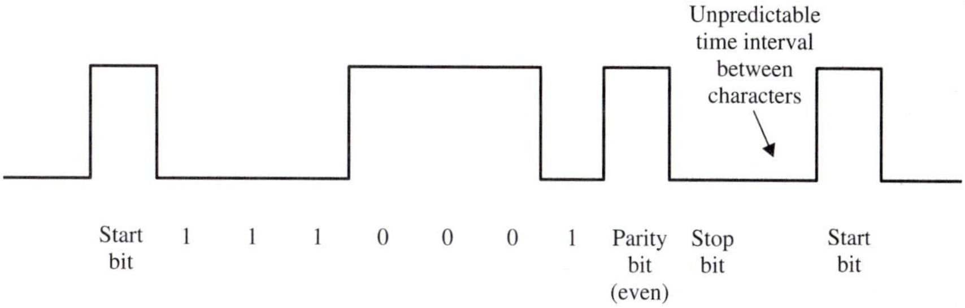

Figure 3.7 The ASCII character G (47_{16}) is transmitted in E-7-1-1 asynchronous transmission format with one start bit, one even parity bit, and seven data bits. The least significant bit is transmitted first. The zeros represent high voltage.

of bits or blocks of characters are transmitted without a start and stop bit for every character. Exact timing and synchronization are archived in three ways at different levels:

- A synchronization signal and a clock may be embedded inside the data signal and encoding mechanism as discussed before. Biphase encoding, such as Manchester and differential Manchester coding, is a good example of such synchronization schemes.

- Separate clock signal lines may be included in the interface between the DTE and DCE devices. Recall the EIA standard 232-D for serial line communications (Chapter 2). Pin 15 (or 24), transmit clock (TC) and pin 17 receive clock (RC) may be used along with pin 2, transmit data (TD) and pin 3, receive data (RD) respectively, to synchronize the transmission and reception of the data frame. Note that the synchronization is not between the transmitting modem and the receiving modem, but between the transmitting computer and its modem, and between the receiving computer and its modem. These clock signal lines are not used in asynchronous communications.

At the higher level, for a block of data (bits or characters), a preamble and postamble bit patterns or flags can be added in the front and end of the data block to form a frame. These flags contain control/status information and the information about the enclosed data pertaining to the sender and receiver. The data block may be bit-oriented or character-oriented, as we discuss in detail shortly. The exact format of the frame and the preamble/postamble depends on the network environment and communications protocols used. Usually, these starting and ending flags are multiple of 8 bits. For example, an Ethernet frame has a preamble 8 bytes long.

For large blocks of data, synchronous transmission is far more efficient than asynchronous. For example, for an Ethernet frame having a maximum length of 1500 bytes, the overhead for synchronization is only 8/1500.

3.2.2 Error Detection and Correction

Electromagnetic waves traveling over a transmission medium may encounter noise. A very common example is a lightning surge, which may cause corruption and loss of data. We see the effect of noise due to lightning, operation of common household appliances, and firing spark plugs from a passing vehicle as streaks and blank screens on our television screens. Both analog and digital signals are subject to noise that may result in changes to data and/or data loss. In computer communications, changing a bit may have severe effects on data. For example, the 7-bit binary ASCII code of character a is 1100001. If the code is changed to 1100011 during transmission, the data string no longer represents an a, but now represents a c. The presence of such errors may have adverse end effects on the data. A file containing the financial records of a company may have figures such as $1.# million dollars instead of the expected numeric value $1.3. Notice that this happened because of only a single incorrect bit (0100011 for # instead of 0110011 for 3).

Fortunately, single-bit errors are the most common type in data communication designs featuring simple error detection and correction techniques. However, multiple-bit errors or burst errors are possible too. A multiple-bit error would involve changing two or more bits during a transmission. So, for example, a 1100001 may change to 1000000. Burst errors are caused when a noise interferes with the transmission for a longer period of time, possibly causing a change in several consecutive bits. For example, if the sending bit pattern is 1000 0111 1001 1100, a burst error involving bits 3–9 (bit 0 being the rightmost bit) would result in the following received string: 1000 0100 0110 0100.

To avoid passing the effects of bit errors to higher OSI layers, the data link layer must detect any errors in a received message. Or, preferably, the bit error is fixed subsequent to the detection, thus correcting the error. One simple mechanism of error detection would be to transmit every bit twice. As soon as the receiver finds a pair of bits to be unequal, it can flag an error to higher layers. However, this redundancy technique effectively reduces the transmission efficiency by half, since there is one overhead bit for every bit transmitted.

A more practical method would be one for which overhead is less. Mainly, there are three redundancy-based techniques discussed in this section. The parity-based techniques append a single bit to a block of data to indicate the total number of bits; the parity bit can be even (for even parity) or odd (for odd parity). The second technique is based on the calculation of arithmetic checksums, using the binary ones-complement arithmetic. The calculated checksum is appended to the end of the message, and the sum of the data and the checksum is obtained upon receipt of the message. If the resulting sum is all 1s, then the message is considered to be error free; otherwise, there is an error in the message. The third redundancy method is based on using binary division of polynomials in modulo 2 to calculate the cyclic redundancy checksum (CRC).

Error detection at the data link layer is usually sufficient, and the frames in error may be retransmitted by using a protocol between the sender and receiver. However, some codes are designed to correct errors at the receiver end without need for retransmission. In one such code, called Hamming code, the even parity of specific bits in the data is included in such a way that in case of a single bit error, the checksum bits obtained at the receiver point to the erroneous bit.

3.2.2.1 Parity

Parity-based redundancy systems use either odd or even parity. The parity bit is obtained based on the count of 1s in the data block. The number of 1s, including the parity bit in the encoded data, must be even for even parity. For example, even parity of the ASCII character c is 1, since ASCII a is 1100001. Single-bit parity for a data block may not de-

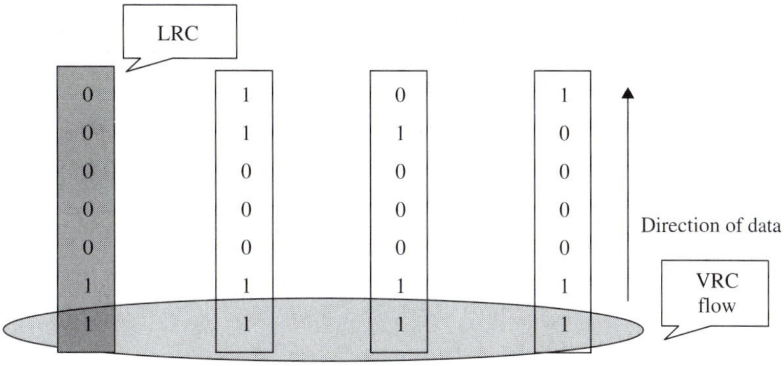

Figure 3.8 Parity using LRC and VRC.

tect an even number of errors (i.e., errors in even numbers of bits may go undetected). If our example of ASCII `a`, the number of errors had been odd (say the received bit pattern, with the leftmost bit as even parity, had been 11011001), the error would have been detected because of the odd number of 1s. To solve the problem of undetected errors when the number of errors is even, a two-dimensional parity mechanism may be used. Here the even parity for each vertical block of data is called the vertical redundancy check (VRC). Parity in the second dimension, called the longitudinal redundancy check (LRC), provides a double check. For transmitted data consisting of the ASCII character set `abc`, the corresponding check is shown in Figure 3.8.

The VRC/LRC technique shows how even parity in two dimensions may be used to accomplish single-bit error detection. Another redundancy technique that is based on even parity and allows for error correction is called Hamming code. In this code the number of redundancy bits required to correct any single error in m data bits is termed p. The resulting code is $m + p$ bits long. This code may be applied to data blocks of any length. However, p increases rapidly with the number of data bits. In general, $2^p \geq m + p + 1$. So for an ASCII character of seven bits ($m = 7$), p must be at least 4 bits, resulting in a total of 11 bits. The check bits (or redundancy bits) are positioned at 1, 2, 4, 8, . . . , 2^n, where n is an integer. The position number starts at 1 and continues until all the positions have been filled with either data bits or parity bits. Figure 3.9 shows the positions of bits in a 7-bit ASCII character.

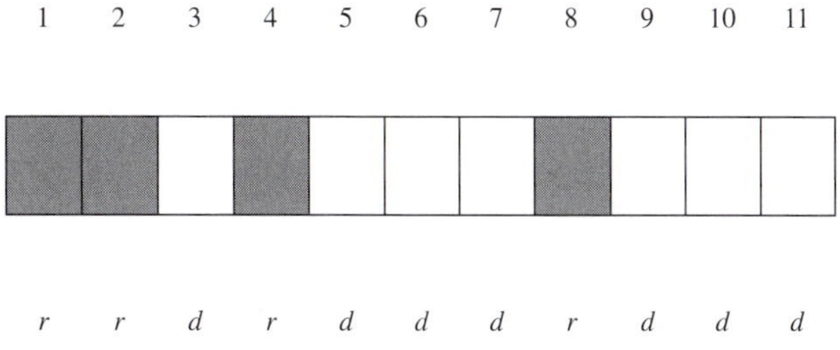

Figure 3.9 Positions of parity and data bits in Hamming code.

The r bits are obtained as even parity of d bits at specific positions. Actually the position numbers to be included in determining the parity is obtained by breaking down the d-bit positions as powers of 2:

$$3 = 1 + 2$$
$$5 = 1 + 4$$
$$6 = 2 + 4$$
$$7 = 1 + 2 + 4$$
$$9 = 1 + 8$$
$$10 = 2 + 8$$
$$11 = 1 + 2 + 8$$

and so on.

The r bit at position 1 is obtained by locating 1s in the foregoing breakdown. Wherever a 1 appears at the right-hand side, the corresponding data bits are used in determining the r bit. We see that r at position 1 is obtained as even parity of data bits at positions 3, 5, 7, 9, and 11. Similarly for r at position 2, we locate 2s at the right-hand side. The data bits used in this case will be at positions 3, 6, 7, 10, and 11. The process is repeated for the r bits at positions 4, 8, and so on.

Now let us apply the technique to get the r bits for the 7-bit data string 1 1 0 0 0 0 1 with respective positions at 3, 5, 6, 7, 9, 10, and 11. Put together, it will be 1(3) 1(5) 0(6) 0(7) 0(9) 0(10) 1(11), where the numbers in parentheses are the bit positions. The r bit at position 1 will be an even parity of 1, 1, 0, 0, and 1, giving a 1. Similarly, the even parity at position 2 will be an even parity of 1, 0, 0, 0, and 1, giving a 0. The r bits at positions 4 and 8 will both be 1s. The resulting bit sequence will be (1) (0) 1 (1) 1 0 0 (1) 0 0 1, in which the r bits are in parentheses.

Error correction at the receiving end is done by calculating c bits, which are r bits as obtained just shown, but this time the bit at position r is also included in the calculation. Let us say that the received bit pattern is 1 0 0 1 1 0 0 1 0 0 1. Notice that there is an error at position 3. The c bit at position 1 will be an even parity of bits at positions 1, 3, 5, 7, 9, and 11 (i.e., even parity of 1, 0, 1, 0, 0, 1 equal to 1). Similarly the c bit at position 2 will be a parity of 0, 0, 0, 0, 0, 1 equal to 1, and the c bit at position 4 will be a parity of 1, 1, 0, 0 equal to 0. The c bit at position 8 will be a parity of 1, 0, 0, 1 equal to 0. Now arranging the c bits as positions 8, 4, 2, and 1, we get 0 0 1 1, which is the binary combination of 3 pointing to the erroneous bit in the received message.

Notice that the Hamming code technique as presented here works well as long as there is only one bit in error. The method must be modified to correct multiple errors in the data, resulting in higher number of r bits. The main idea is to produce enough overlap in the parity bit calculations to allow error correction to take place at the receiving end. Since, however, the number of r bits increases dramatically to allow multiple error correction, this method is seldom practical for multiple error correction.

3.2.2.2 Arithmetic Checksum

In the arithmetic checksum method, the sender divides the sending data unit into equal segments of n bits. Then ones-complement arithmetic is used to add the segments together to get the result in n-bit form. This sum is complemented and appended to the data as the checksum field. The two fields, being complements of each other, must give all 1s when added together.

The receiver keeps receiving the n-bit data units and adds them to the checksum received at the end. If there is no error, the result will be all 1s. In the event of an error, the data unit is rejected. As an example, consider four 4-bit data units as 1000 (unit 1), 1101 (unit 2), 0101 (unit 3), and 1110 (unit 4). The checksum for these would be obtained by subsequent one complement additions, in which the final carry is added to the resulting binary sum: $1000 + 1101 = 0110$, followed by $0110 + 0101 = 1011$, followed by $1011 + 1110 = 1010$. For an error-free transmission, the resulting arithmetic checksum should be 0101, which gives all 1s when added to the last sum, 1010. Notice that if there are errors in any of the received data units or the checksum, the sum results will not give all 1s.

3.2.2.3 Cyclic Redundancy Checksum (CRC)

Like other data link layer error detection methods, the CRC of data is generated at the transmitter end by means of a hardware mechanism that involves sequential circuits using shift registers and flip-flops. A modulo-2 division is implemented by means of a generator polynomial $G(x)$ of degree n over the message polynomial $M(x)$ representing the message. The n-bit frame check sequence or the checksum is the remainder modulo-2 division of $M(x)$ by $G(x)$. However, n bits of 0 are appended at the end of $M(x)$.

As an example, consider $M(x) = X^7 + X^6 + X^5 + X^2 + X$, where $M(x)$ represents the message bit sequence 1 1 1 0 0 1 1 0. For $G(x) = X^4 + X^3 + 1$, the degree is $n = 4$, resulting in the division of 1 1 1 0 0 1 1 0 0 0 0 0 by 1 1 0 0 1. Figure 3.10 illustrates the division. The 4-bit remainder is obtained as 0 1 1 0, which when appended to the message gives us 1 1 1 0 0 1 1 0 0 1 1 0.

The CRC hardware implementation using $G(x) = X^4 + X^3 + 1$ consists of two exclusive-or gates and a couple of shift registers, as illustrated in Figure 3.11. Initially, the content of the shift registers is 0. When the data stream to be transmitted is coming sequentially, the results of the exclusive-or gates are pushed back into the shift registers one bit at a time. For each bit transmitted, the shift registers shift left one bit. At the end of the operation, the remainder is kept in the shift registers, which is attached to the end of the data (payload) and transmitted without change.

At the receiver side, the same device is used. The shift registers are set to zero initially. The incoming data is fed into the exclusive-or gate and pushed back into the shift registers through the exclusive-or gates. The difference is that after the incoming data including the attached CRC remainder is received, the shift registers contain all 0s. Otherwise, transmitting errors occurred.

3.2.3 Framing and Flow Control

Asynchronous transmission consists of the movement of short sequences of fixed-length data. To correctly recover the data, synchronization is needed between the sender and the receiver, and since the data length is fixed, clocks of reasonable tolerance can be used to achieve it. This simple synchronization mechanism does not work well for long data se-

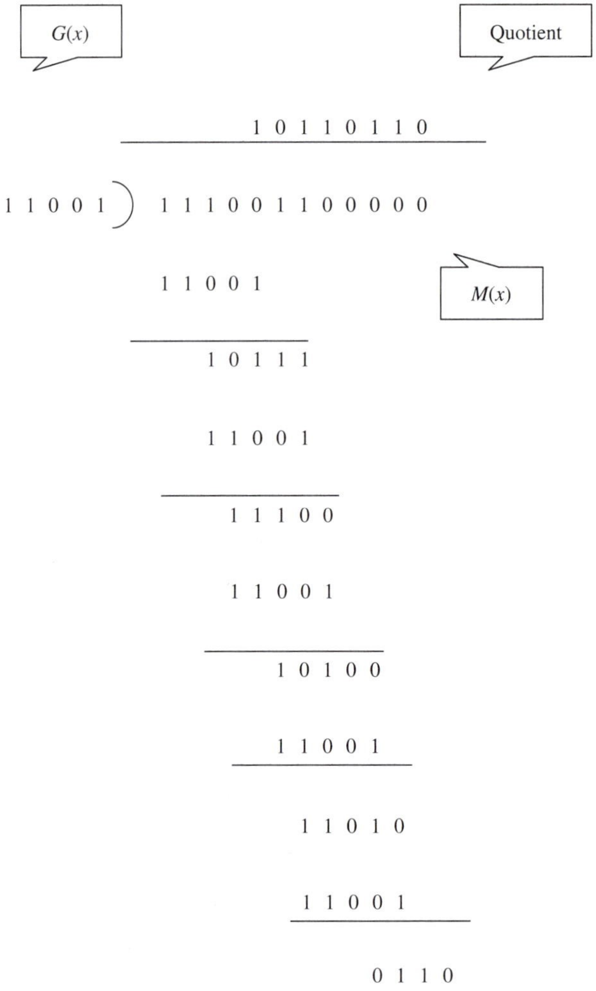

Figure 3.10 Evaluation of CRC using modulo-2 division.

quences, however. Thus in practice, sequences of bits called packets are used to increase transmission efficiency. The transmission of long packets requires improved synchronization of the receiver, which in turn requires special framing structures. The frames help indicate the start and end of packets for the receiver, allowing an arbitrarily long sequence of data. Two types of transmission may be done with a built-in synchronization and flow control mechanism: bit-oriented and character-oriented transmission. The two methods differ in the way they frame the bits. Bit-oriented transmission uses bit sequences to represent characters, and character-oriented transmission uses characters.

The bit-oriented transmission mechanism is illustrated in Figure 3.12. A special bit pattern 0 1 1 1 1 1 1 0 called a flag is used to indicate the frame's start and end. The frame contains the address, control, CRC, and data, in addition to the start and end flags. The length of each field depends on the protocol used. High-level data link control (HDLC), a popular bit-oriented protocol, is presented later in Section 3.2.4.

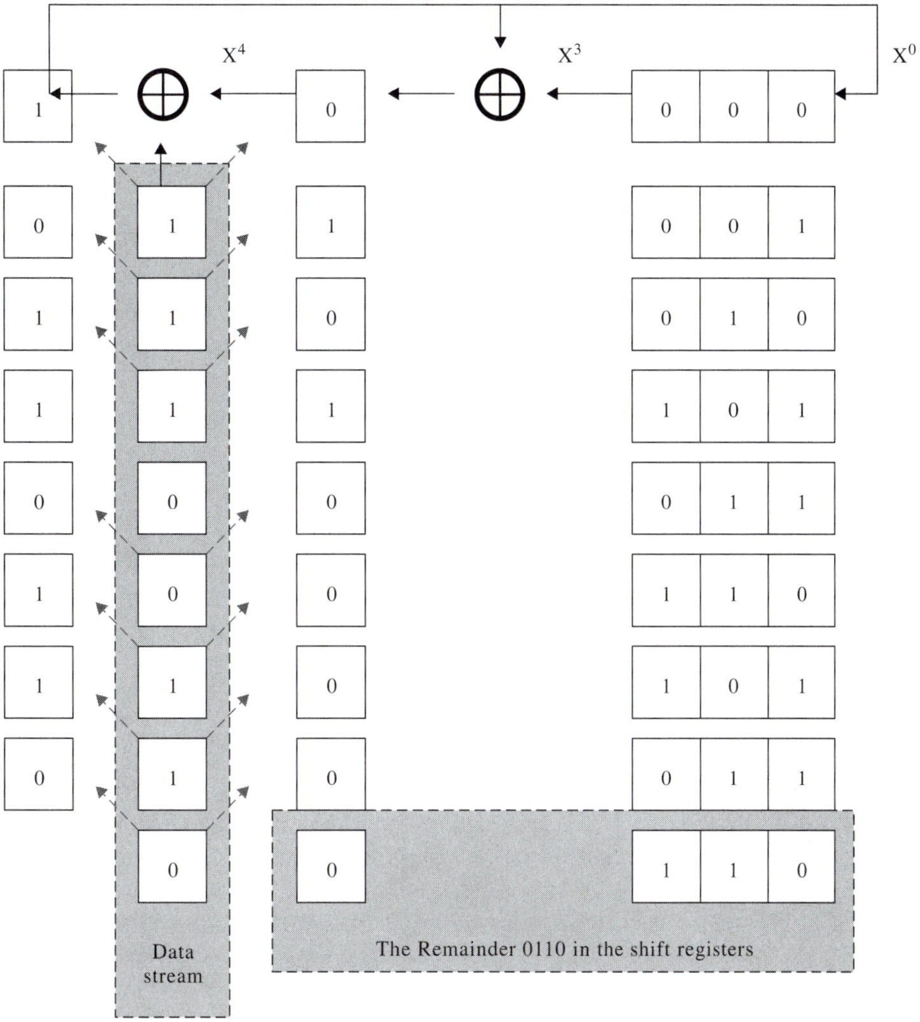

Figure 3.11 The CRC hardware implementation using $G(x) = X^4 + X^3 + 1$.

In character-oriented transmission, the packets are assumed to consist of an integral number of (8-bit) bytes. A typical frame structure is shown in Figure 3.13. The frame start is indicated by a special synchronization character SYN followed by a DLE (data link escape) character and an STX (start-of-text) character. The end of the frame is indicated by DLE, followed by an ETX (end-of-text) character. Two bytes of CRC and a byte of SYN are appended at the end to complete the frame.

Flag	Address	Control	Data (0 or more bytes)	CRC	Flag

Figure 3.12 Use of a bit-oriented protocol to construct a frame.

SYN	DLE	STX	Header	Data (0 or more bytes)	DLE	ETX	CRC	SYN

Figure 3.13 Use of a character-oriented protocol to construct a frame.

3.2.3.1 Bit Stuffing and Character Stuffing

A flag in the bit-oriented protocol determines the start and end of a frame. However, the data and other fields may also have the same sequence. Unless special measures are taken at the transmitter end, the receiver may take the wrong sequence as the end-of-frame flag. To avoid such an error, the sender end must be designed to ensure that no bit sequence identical to the flag will occur anywhere else in the data. If any such sequence is noticed, it may be modified by the process called bit stuffing of 0 bits. For example, if any sequence of five is found at the sender side after the start flag, a 0 is inserted, as shown in Figure 3.14. The reverse of stuffing, called de-stuffing, is done at the receiver end if a 0 is found in the received data after five consecutive 1s to recover the original data.

Just as in bit stuffing in bit-oriented protocols, it is necessary to perform character stuffing at the sender to avoid confusion at the receiver end. Character stuffing consists of replacing every occurrence of the pattern DLE in the frame between the two legitimate DLEs by DLE DLE. This prevents the pattern DLE ETX from appearing anywhere in the frame except at the end. A de-stuffing mechanism at the receiver simply removes a DLE from the frame if two consecutive DLEs are received.

3.2.3.2 Flow Control

Flow control defines both the way in which many frames are sent and tracked and how the stations do error control. Error control defines both how a station checks frames for

Flag		Flag
01111110	0 1 1 1 1 0 1 1 0 1 1 1 1 1 1 0 1 1 1 1 1 0 1 1 1 1 0	01111110

(a)

Stuffed bits

01111110	0 1 1 1 1 0 1 1 0 1 1 1 1 1 *0* 1 0 1 1 1 1 1 *0* 0 1 1 1 1 0	01111110

(b)

01111110	0 1 1 1 1 0 1 1 0 1 1 1 1 1 1 0 1 1 1 1 1 0 1 1 1 1 0	01111110

(c)

Figure 3.14 (a) Original frame at the sender, with flags. (b) Stuffed frame. (c) De-stuffed frame at the receiver.

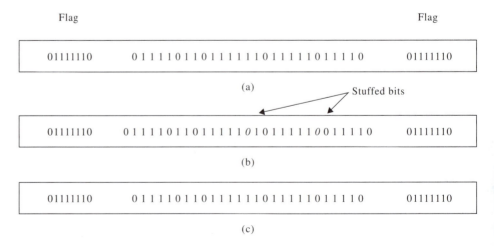

errors and what it does if it finds them. A common error control approach is to have the receiver send a message to the transmitter indicating that an error has occurred in the preceding transmission. The message is effectively a request to resend the erroneous frame. Thus this type of error control is often called automatic repeat request (ARQ).

In general, the flow control protocols ensure that all the related frames arrive at their destination accurately and in order. Many physical limitations of the stations must be taken into consideration to ensure a proper flow control. Buffer space, processing capability, and transmission line errors are some limitations that must be considered in obtaining proper flow control. For example, the receiver's buffer capacity is usually limited, and if a sender keeps transmitting the frames, the buffers at the receiver may in time fill up, resulting in buffer overrun, which can loosen and damage the frames. One method to stop the sender from overrunning the receiver's buffer is to insert special ASCII XON/XOFF characters in the data. Thus, a receiver may send an XOFF message to the sender along with its own data to stop the flow of incoming data. When the receiver is ready again, it can send an XON message to resume the transmission. It may be noticed that this method is more applicable to character-oriented asynchronous transmission.

Bit-oriented or frame-oriented transmission requires better organization than the character-oriented kind, since the information is sent and received in larger pieces rather than bytes or characters. Two well-known protocols that provide the flow control in synchronous transmission are the stop-and-wait and sliding window protocols.

3.2.3.3 Stop-and-Wait Protocol

In the stop-and-wait protocol, the receiver buffer space is assumed to be limited, such that the sender may not continue to transmit until an okay signal (an acknowledgment) is received. The acknowledgment is a frame sent by the receiver to the sender to indicate a successful transmission. The sender sends one frame and then waits for an acknowledgment from the receiver before resuming transmission, hence the name stop-and-wait protocol. The sender does not proceed with the next frame until a positive acknowledgment, indicating an error-free transmission, has been obtained from the receiver. If the acknowledgment message indicates an error in the transmission just completed, the sender transmits the same frame again instead of getting a new one from its buffer.

The receiver checks each received frame for errors (using CRC, checksum, etc.), and if no error is found, the recovered data are passed to the upper layer and a positive acknowledgment is sent to the sender. In the case of an error, the receiver sets the error field to 1 in the acknowledgment message, and the frame is retransmitted.

While this protocol seems to work, it has some shortcomings. For instance, it does not allow the sender to perceive transmission errors that have resulted in frame loss. What happens if the frame sent by the sender is lost on the line? Obviously if the receiver never got it, no acknowledgment for that frame will be generated, and sender has no way of knowing of the problem. Similarly, if an acknowledgment is lost on the line or damaged, the error field will be changed to indicate (incorrectly) an unsuccessful transmission. Somehow both sender and receiver must learn about this system break down.

Fortunately, the solution to these problems is a timer mechanism at the sender end and a capability for the receiver to determine the presence of duplicate frames in the case of an acknowledgment loss. The timer mechanism allows the sender to time-out after a certain time and retransmit the frame. If the cause of time-out was frame loss, then the receiver gets the frame and normal operation resumes. However, if the cause was acknowledgment loss or damage, the receiver will end up with the same frame twice, resulting in duplicate frames. Since there can be only one pending frame at one time, a duplicate can be detected easily by checking the sequence number bit that alternates between 0 and 1. (The protocol is also called the alternate bit protocol for this reason.) If two consecutive frames at the receiver have the same sequence, acknowledgment loss or damage must have occurred, and the receiver simply discards one frame.

The mechanism just described seems to take care of the inherent problems in the stop-and-wait protocol. We see that the protocol works properly under different error conditions. Its efficiency may be very low, however, since there is only one frame pending at a time. For example, on a 10 Mbps line with frame and acknowledgment lengths of 1000 bits, if the signal propagation delay from sender to receiver is 100 μs (approximately 20 km), the efficiency of the stop-and-wait protocol will be

$$\frac{1000/10^7}{(1000 + 1000) / 10^7 + 2 \times 100 \times 10^{-6}} = 0.25$$

Thus the line running at 10 Mbps can be used effectively up to 2.5 Mbps only, which is quite low. For higher distances, this efficiency is even smaller. Also, note that the effect of processing delays at both ends is neglected in this evaluation.

3.2.3.4 Sliding Window Protocol

In the sliding window protocol, an extension of the stop-and-wait protocol, a sequence of packets may be sent and received simultaneously. The idea is to "keep the pipe full in both directions." However, there is a limiting factor. The number of pending frames between transmitter and receiver may not exceed the limit allowed by the sequence number. That is, if the sequence number consists of three bits, then the frames 0–7 may be generated by one end. The same sequence number must be used for the next group of frames. To ensure that the frames belonging to the first group can be distinguished from frames belonging to the second, the sender cannot send more than 8 frames at a time.

To keep track of sent and received frames, windows are implemented that open and close as frames are being sent or received. In the sender window shown in Figure 3.15,

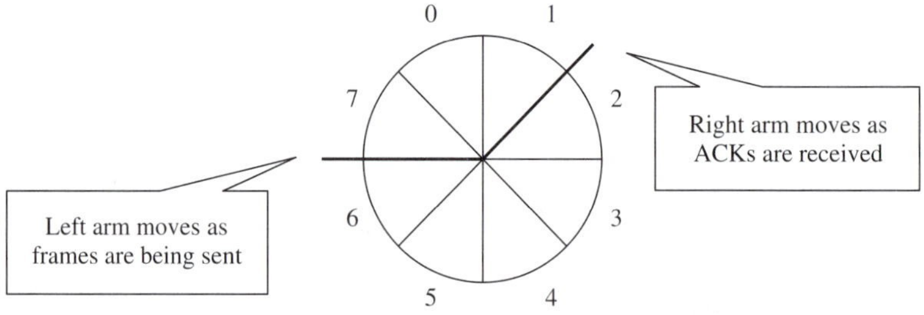

Figure 3.15 Sender's sliding window with two moving arms.

the left arm moves clockwise when a frame is being sent. The right arm moves in the same direction to indicate that an acknowledged frame is being received.

The receiver's window is designed in a similar fashion: the left arm moves when a frame is being received and the right arm moves when an acknowledgment is being sent.

It may be noticed that when this sliding window mechanism is used, the frames in transition are not always frame numbers 0–7. The sequence number is simply a continuous increment in modulo 7. Also, to allow error-free operation of the protocol, the maximum number of pending frames may not exceed the half of window size (e.g., 4 for a window size of 8). To prove this, let us consider a transmission scenario allowing 5 frame transmissions.

Assuming the start sequence number from 0, the transmitter may send frames 0–4 and may record these in the sending window. The receiver also receives the frames and adjusts the window accordingly to expect frames 5, 6, 7, 0, and 1 (i.e., 5 more new frames). Now suppose that the acknowledgment for frames 0–4 is lost. The sender will send these frames again. The receiver, expecting frames 5 onward to 0, will discard frames 1, 2, 3, and 4 as frames lost earlier and will accept frame 0 as one of the new set of expected frames. Since frame 0 was from the old set of frames, the protocol has failed.

The ARQ mechanisms used in the sliding window protocol are go-back-*n* and selective repeat. When a packet is lost during transmission, the go-back-*n* mechanism asks the sender to retransmit all the frames in the current window. In the selective-repeat ARQ, the receiver asks for selective retransmissions of damaged frames. At first, it appears that the selective-repeat ARQ is better than the go-back-*n* ARQ. However, the efficiency of these methods largely depends on the line error rate and number of retransmissions required.

3.2.4 High-Level Data Link Control (HDLC)

Many bit-oriented protocols were developed over the years, to secure the inherent advantages of this approach. The most notable are LAP series by ITU and HDLC by ISO. However, these protocols are mostly dependent on the synchronous data link control (SDLC) developed at IBM in 1975. Upon the recommendation of IBM, the ISO adopted this protocol as the standard in the form of HDLC. There are three types of station defined in HDLC: primary, secondary, and combined.

A primary station acts like a line controller and may poll the secondary stations for data transmission. A control message is sent online, and a secondary station may respond if it has any data to transmit. An unbalanced mode of transmission may occur if more than one secondary station responds, an event that may be followed by another round of polling. A combined station may act like both primary and secondary stations. Notice that the intent to transmit by secondary stations should be resolved in a multipoint (or multidrop) configuration, in which stations are connected on the same line. However, in a point-to-point configuration the intention may not exist, since there will be only one primary and one secondary station.

The modes of transmission in HDLC are normal response mode (NRM), asynchronous response mode (ARM), and asynchronous balanced mode (ABM). The NRM is a normal primary secondary operation, in which the secondary station must have permis-

Flag (1)	Address (≥ 1)	Control (1)	Information (0 or more)	FCS (2 or 4)	Flag (1)

(a)

Flag	Address	Control (0, NS, P/F, NR)	Information	FCS	Flag

(b)

Flag	Address	Control (10, S, P/F, NR)	FCS	Flag

(c)

Flag	Address	Control (11, M, P/F, M)	Management	FCS	Flag

(d)

Figure 3.16 HDLC frame formats: (a) standard, (b) I frame, (c) S frame, and (d) U frame.

sion from primary before transmission can occur, as explained earlier. In ARM, a secondary station may initiate a communication by sending control packets to the primary. Unlike the primary station, however, the secondary may not control the line and issue commands. In ABM, only combined stations are connected. Therefore, any station may initiate transmission and may change its role as primary and secondary station.

Three types of frame are defined for HDLC: information frame (I frame), supervisory frame (S frame), and unnumbered frame (U frame) as shown in Figure 3.16. The number of bytes in each field is given in parentheses in Figure 3.16a. We note that the main difference between the I frame and the S frame is that the information (or data) part is not included in the S frame. Therefore an S frame acts like a control frame for exchange of control information between the stations. A U frame is a management frame that is used like an S frame to exchange control information.

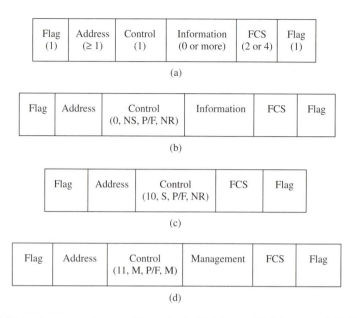

In an HDLC frame, the flag is set to 01111110. Bit stuffing is used to avoid confusion with the end-of-frame flag. The FCS is either a 2- or a 4-byte CRC. The control field identifies the three frame types, starting with 0 for an I frame, 10 for an S frame, and 11 for a U frame, respectively. The 3-bit NS and NR fields in the control field indicate the sent frame sequence number and the expected frame sequence number, respectively. Hence, NR is used in a way to acknowledge the frames up to frame number (NR − 1). Notice that NR exists in both I and S frames. The inclusion of an acknowledgment on an information frame is called piggybacking; that is, the acknowledgment is not a separate frame by itself but gets a free ride by being part of an I frame. The 3-bit S control in the S frame is used for supervisory purposes. For instance, it may indicate the type of error control as go-back-*n* ARQ or selective-repeat ARQ, as explained earlier. Similarly the five *M* bits (2,3 combination) allow 32 types of management information to be exchanged in the U frame. More types of management information can be expressed by using the information field. Some examples of management information include set normal response mode (SNRM), request disconnect (RD), and reset (RST).

The stations use the P/F bit in the frame for polling and responses. When the primary station sets the bit to 1, it means that a secondary station with address NR has been polled. In response, the secondary station may transmit frames with the P/F bit set to 0. When the primary station transmits its last frame, it sets the P/F bit to 1 to indicate the final frame.

3.3 NETWORK LAYER

It is important that messages be delivered from source to destination in minimal time. This goal can be achieved relatively simply for a point-to-point network, since the shared link directly connects the two stations. However, the task is more complex for a multipoint network, in which there may be one or more active stations between the two ends. In addition, there may be more than one path connecting the two ends, and forwarding and routing strategies must be devised for each case. The network layer of OSI architecture deals with the connection of two ends via a switching mechanism to allow the use of network links in a predetermined manner. The two services used are called connection-oriented services (CONS) and connectionless network service (CLNS). In connection-oriented service there are three main phases of communication. In the first phase a connection is established between the sender and the receiver, followed by the second phase consisting of data transfer. The connection may be terminated by either side in the third phase when the data transfer is complete or for some other reason.

There are no connection establishment and termination phases in a connectionless service. Rather, the stations transfer the data directly. The packets forming the data may take different routes to reach the destination, resulting in different packet delays and unordered arrivals at the destination. In this case the higher layers must perform error checking to ensure an error-free and reliable end-to-end service. It may be noted that the CONS provides better network service than CLNS because in connection-oriented service a path is established between the two ends and the packets must follow the same path, minimizing the chances of loss and preventing any unordered delivery of packets. However, the delay will depend on the line conditions. Once the connection has been established in CONS, the network is responsible for delivering the data packets in order, without causing any extra delays. However, the connection establishment and termination phases cause the extra delay.

3.3.1 Subnet Concept

An internetwork consists of many smaller networks capable of performing switching, routing, and forwarding on their own. These smaller constituent networks of an internetwork are called subnetworks or subnets. Each subnet is capable of operating on its own. In addition, interconnection of subnets is possible by means of special switching devices called intermediate systems (ISs) by ISO. Two types of IS are routers connecting the subnets at the network layer and switches (sometimes referred to as bridges) connecting the subnets at the second layer. A router may allow forwarding and routing of packets between different subnets operating under different protocols. However, a bridge operates between the two subnets operating under the same protocol. A bridge does not change or modify the contents of a received packet, it simply acts as an address filter that picks up a packet on one subnet and forwards it to another where the destination can be found.

The routers and switches may be configured to support either the CONS or CLNS. There are protocols implemented for these devices to support the internetwork traffic between subnets, as we shall see later.

3.3.2 Overview of Switching Techniques

A dedicated data circuit may operate at speeds of several million bits per second and even at gigabit rates. But such an expensive connection must be justified by a continuous need to transmit data between computers. A network may connect hundreds, of computers, or even millions, as in the case of the Internet. It is just impossible to connect all computers directly to one another. The interconnecting links must be shared between many computers. That is where multiplexing comes into play. For a multiplexed network, there are two ways of switching between communications channels: circuit switching and packet switching. Circuit switching is similar to traditional telephony in that a dedicated connection is provided between the two parties just as a dedicated telephone connection is necessary for the duration of a conversation.

Figure 3.17 illustrates a variety of switching mechanisms for transporting data across networks. At the left is the circuit switching technique, which provides transmission with the strongest connectivity—a constant traffic, but minimal processing demands on the network access devices. On the opposite side of the spectrum is the datagram packet switching technique, which demands more technical complexity and processing efforts but provides more flexible, dynamic, robust, and statistically sharable transmission with variable bit rates. Located between the two extremes are virtual circuit packet switching, frame re-

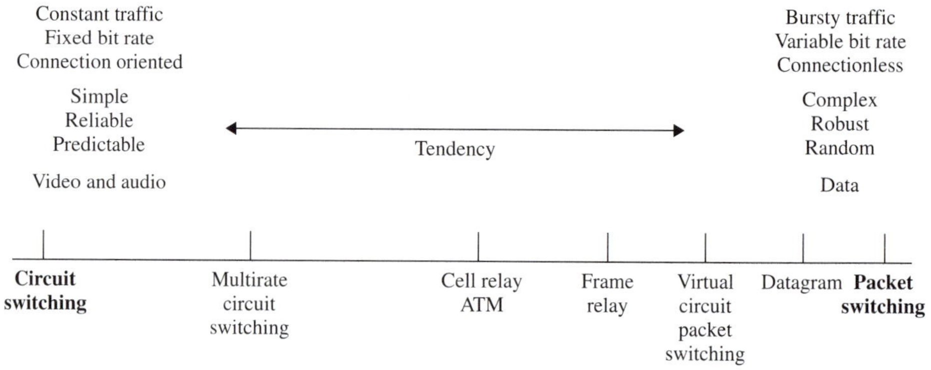

Figure 3.17 The major switching techniques, arrayed on a spectrum indicating their appropriate service, their properties, and applicators.

lay, cell relay (ATM), and multirate circuit switching. These relatively new telecommunication products are discussed in later chapters.

Data communications systems typically have bursty traffic; that is, the channel is sometimes quiet but very busy when, say, a large file is to be transferred in a relatively short period of time. When heavy traffic is anticipated, it is desirable for many communication channels to be bundled together, or multiplexed, to increase channel utilization; thus quick reconfiguration of the switching system from one channel to another must be possible. Circuit switching systems prove to be inefficient when it comes to quick reconfiguration due to bursty traffic because setting up and tearing down circuits takes hundreds of milliseconds rather than a few microseconds.

Packet switching is specially designed to accommodate the bursty multiprocess communication commonly found in computer networks and distributed computing environments. Frames and packets—chunks of data—originated from one computer are prefixed with headers containing address and routing information that identifies the source and destination computers. Then the packets are sent to the physical channel. Each switch, or router, examines the header of a packet, then decides where to forward it or, if it is within the same physical network that includes the switch or router, delivers to the final destination. Whereas for obvious reasons two networks connected by a circuit switch must operate at the same speed, packet switching can connect networks operating at different speeds. However, because of the store-and-forward nature of packet switching, packet switching often causes variation in delay.

Packet switching systems can recover from failure in less time and with less effort than are required in circuit switching. Also, packets may take different paths when a route becomes unavailable or too crowded. This makes packet switching more robust than circuit switching; however, packets may not arrive in the order originally sent. Disordering of packets may occur when a packet is retransmitted because of a detected error or when the packets traverse different paths (with different delays) in the network.

Since packet switching systems may consist of different networks operating at different speeds, buffers are introduced for flow control. When the buffers become full in an abnormal situation, congestion occurs and switches may be forced to discard packets that cannot be stored or delivered. When this happens, chances are that congestion will become worse. Detection, prevention, and avoidance of congestion are important topics in packet switching technology. Virtual circuit switching is a service designed to avoid disorderly packets, thus reducing the chances of congestion. In a virtual circuit switching system, the network appears to deliver information from one node to another via a dedicated circuit. In reality, no real circuit exists. The data are still delivered in the form of packets in the right order but with variable delay, thus the name virtual circuit. Usually the traffic paths are fixed in a virtual circuit system, but when the traffic is too bursty, routing selection still presents a serious challenge.

The first packet switching system, Arpanet, was developed by the U.S. Defense Advanced Research Projects Agency (DARPA) in 1969. The system used PDP-8 minicomputers made by Digital Equipment Corporation as packet switches, which were connected by dedicated 50 Kbps telephone lines. Similar projects were started in other countries at the same time. Since then, a great many private and public packet switching networks, notably the X.25 system with speeds varying from 56/64 Kbps to 1.5/2 Mbps (T1/E1), have been deployed.

Packet-switched communication channels share their total bandwidth among all users who want to access it. The communications channel is packet-switched because each time a user sends a packet of data, the packet goes into the shared channel with everyone else's data. The channel is designed to "switch" packets instead of circuits. When the channel is being used simultaneously by a large number of users (i.e., when it is "heavily loaded"), sending the data will take more time than is needed when the channel is being used by only a few users ("lightly loaded"). This is the feature that distinguishes circuit-switched from packet-switched communications channels. In the circuit-switched case the full bandwidth is always available to the user. In the packet-switched case a user must share the bandwidth with others.

3.3.2.1 Circuit Switching

Just like telephone systems, circuit switching requires a dedicated transmission path between source and destination. Since the line is dedicated for the user, there is continuous transmission of data. It is the users' perogative to pass the traffic in any manner they want. They can send bursty, stream, or interactive traffic because usually there is enough speed to support these varying types of traffic. If the network is not capable of handling fast traffic, the stations will know about it during connection establishment phase. Once the line has been established, that path will remain in effect for the entire conversation, and the network is not responsible for accommodating changes in demand by the user. However, if the network is experiencing heavy delays or if the destination station is busy, the path connection may be refused by means of a busy signal sent to the source station during the call setup phase.

A circuit switching system stores no data at the intermediate nodes. So, the only overhead expected is the call setup time. After the call setup, the message is transmitted in its entirety with no delays at the nodes. However, the user is responsible for message errors and loss of sequence in the packets.

3.3.2.2 Virtual Circuit Packet Switching

Packet switching was introduced for business communications some 20 years ago, in the form of IBM's systems network architecture (SNA), using X.25 as defined by the International Standards Organization (ISO), the International Telecommunications Union (ITU), and also Arpanet, the progenitor of the Internet. With the widespread deployment of packet switching devices around the world, the practice of breaking streams of data into discrete blocks (called frames or packets) and statistically multiplexing them over a transmission path is well proven and is considered to be a mainstream technology.

With traditional time division multiplexing and circuit switching, a dedicated bandwidth is allocated to each path (circuit) through the network on a static basis, for the duration of the call. For example, in a traditional voice call, just as much bandwidth is used to transmit the silence (when listening and pausing between words and sentences) as the sounds when parties are actually speaking. In virtually all data there is similar "silence" between transmissions. In fact, data transmissions are typically much more "bursty" than voice conversations. Thus, with dedicated bandwidth, the transmission facilities are essentially unused a large percentage of the time.

By contrast, statistical multiplexing means that paths (virtual circuits) are defined through the network. However, no bandwidth is allocated to the paths until actual data (real information) are ready for transmission. Then, the bandwidth within the network is dynamically allocated on a packet-by-packet basis. If, for a short period of time, more data need to be transmitted than the transmission facilities can accommodate, the switches within the network buffers (store) the data for later transmission. If the oversubscription persists, congestion control mechanisms must be invoked, as discussed later.

Packet switching was developed during the early stage of digital transmission, and error rates were substantial. Every effort was made to catch errors at the lower layers to reduce the effort needed to recover from errors. As a result, a significant amount of overhead in the form of redundancy and additional processing capacity was built into packet switching facilities. With today's high-speed communications systems, this overhead becomes unnecessary, and the redundancy and complexity become counterproductive. With a much cleaner communications channel, insignificant errors can be easily handled by the higher layers above packet switching in the end communications devices without the involvement of intermediate network nodes. Such improved communication conditions have led to newer and simpler packet switching technologies, such as frame relay (Chapter 6).

3.3.2.3 The X.25 Protocol

The X.25 standard for packet switching is a lower three-layer equivalent of the OSI model. This protocol, based on a physical layer, a link layer, and packet layer, is standardized by the ITU-T and is defined as an interface between data terminal equipment (DTE) and data circuit-terminating equipment (DCE). Widely used in public networks that are part of wide-area networks, X.25 makes use of the physical layer standard X.21, but in many cases uses other standards such as EIA 232. The link layer protocol, called LAP-B (Link Access Protocol—balanced), provides for reliable transfer of data across the physical link.

With X.25, packet layer data are transmitted as packets over virtual circuits, which may be permanent or dynamically established. A permanent virtual circuit is a fixed virtual circuit assigned by the network with no call setup overhead. One of the important aspects of X.25 is its service. The DTE is the connecting device that allows up to 4095 simultaneous virtual circuits with other DTEs over a single physical link. Flow and error controls in X.25 are identical to the ones used in HDLC.

3.3.3 Routing Strategies

The cost of moving data between stations on a network depends on several factors such as delay, the number of intermediate nodes, and the distance between them. When the outgoing link costs are known, a router can figure out the best path for a packet. Several algorithms exist for performing these calculations. Two well-known strategies are distance vector routing and link state routing. In distance vector routing the routers exchange cost information about neighbors with one another. They also share the complete routing table, and the inputs received by other routers are used to update the current tables. The link cost is considered to be one. Thus, the cost of sending data from one router to another in five hops would be 5.

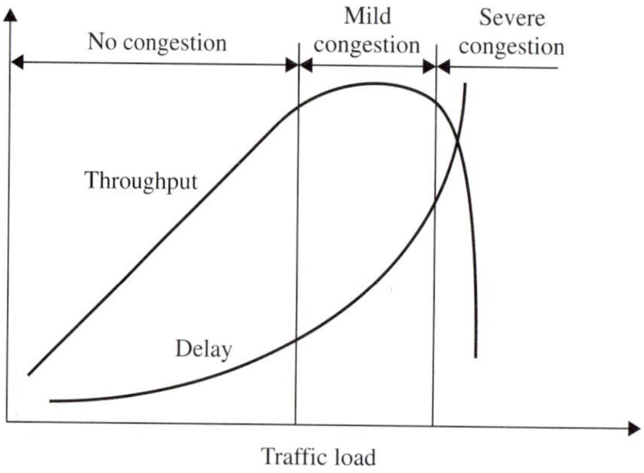

Figure 3.18 The relation between throughput, delay, and congestion. When congestion becomes severe, both throughput and delay deteriorate.

A practical example of this type of routing may be found on the Internet, which is an interconnection of subnets connected via routers and gateways. A router at start-up creates a routing table for the subnets it is connected to and then sends this information to other neighboring routers. Similarly, the other routers send their information to the first router and, after an initial setup phase, the updated network tables are distributed to every router.

Link state routing involves sharing the routing information with every other router in the network. Instead of sending the entire routing table, only the information about neighbors is sent. The routers send periodic updates to each neighboring router, which in turn sends the information to each of its neighbors, and so on. This process of routing is also called flooding. It may be noticed that this type of flooding may cause unnecessary duplicates. However, the routers are guaranteed to have most up-to-date original information from the originating router.

Comparing the two methods, we notice that the cost is expressed in terms of hop count in distance vector routing, whereas in link state routing it is expressed in terms of the weighted value based on traffic, link state, and security levels. Also, information exchange in distance vector routing (say 30 seconds) is more frequent than in state vector routing (say 30 minutes).

The distance vectors consist of hop distances and router identifications for each destination network.

Shortest-path algorithms have been a favorite topic of study for network designers and engineers for long time. In evaluating the shortest paths, however, most routers use one of two famous algorithms: Dijkstra's algorithm and the Bellman–Ford algorithm. Both these algorithms use graphs made up of nodes and arcs to calculate the shortest path between two nodes. A shortest-path tree from each router may be developed and sent to possible destination routers. Finally, the routers use the shortest-path tree to construct their routing tables.

3.3.4 Congestion Control

We have discussed the various routing and multiplexing schemes. The main challenge in implementing these strategies is achieving the maximum throughput with controlled delay. However, throughput and delay are directly related in a network. When the throughput is increased, the delay is likely to increase too, since more data are going across the network. This relation is shown in Figure 3.18. It may be noticed that when the network is lightly loaded there is no congestion in the network, since delay is limited. As the load on the network is increased, delay increases, and a region of mild congestion is reached. As the offered load is increased further, a period of severe congestion is reached, whereupon the network throughput actually drops instead of increasing. The throughput increases dramatically in this region, and the network becomes unstable. The task of congestion control mechanisms is to avoid this region.

Congestion avoidance and recovery mechanisms may be used to control the network to prevent complete collapse. Congestion avoidance methods are used when there is no congestion in the network. If the load increases, measures are taken in the network to keep operations in the same region. Congestion recovery mechanisms are used to recover from severe congestion. These procedures are typically used when the network starts to drop packets as a result of congestion.

3.4 TRANSPORT LAYER AND SESSION LAYER

The transport layer of OSI is responsible for providing reliable, cost-effective data transport from source computer to destination computer. The transport should be independent of the physical network currently in use. The goal of this layer is to provide service to users, which are normally processes in the application layer. There are two types of transport service, connection oriented and connectionless. In both cases, connections have the same three phases: connection establishment, data transfer, and connection termination. Although the data link and network layers also do these tasks, the transport layer provides reliable end-to-end control on the data. This is the last error check on data before delivery to the application layer.

To facilitate the task of the application layer, the transport layer also provides services to enhance the functioning of the client–server architecture. The transport layer primitives allow two or more applications to make connections to one another. Because of the importance of network software design and reliable-end-to-end data transport, we discuss the details of this layer in Chapter 5.

The session layer of OSI is very small in practical networks today. Most of the session layer tasks are usually built into applications. However, this layer is responsible for session management tasks such as checking for user logon to a remote time-sharing system. Another function of the session layer is to manage dialogue control, for example, to allow stations to share a channel.

3.5 PRESENTATION LAYER AND APPLICATION LAYER

The presentation layer of OSI, being closer to user applications, is concerned with the syntax and semantics of the transmitted information rather than with the reliable transmission of data from one point to another. Data encoding, compression, and security are some of the issues handled at this layer. For instance, different computers may have different codes for representing characters (ASCII, EBCDIC, or Unicode), integers (signed or unsigned), floating-point values, and other data structures. To ensure a smooth exchange of data between computers, the presentation layer is responsible for managing the abstract data structures and converting them from one form to another.

The application layer is responsible for providing the user interface for the network. After all, the network is of no benefit to the end user without user-friendly interfaces and software for data communication. A number of network applications are available that provide user-friendly network interfaces. Among them the most popular is the World Wide Web browser, found on almost all computing platforms, which allows hypertext file transfers via a markup language. We discuss some applications later in this section.

3.5.1 Data Compression

The compression and coding of data is important because it saves storage space and transmission time. Although the original data are already coded in digital form for computer processing, often more efficient coding (i.e., using fewer bits) can be obtained by using compression utilities. For example, run-length encoding replaces strings of repeated characters (or other units of data) with a single character and a count. However, the compressed data must be decompressed before being used again.

There are many compression algorithms and utilities. The standard Unix compression utility is called "compress," though GNU's "gzip" is better. Other compression utilities include "pack," "zip," and "pkzip."

The performance of a compression scheme is largely characterized by its compression ratio, the simple fraction that compares the size of data after and before being compressed. So, if a certain compression algorithm takes a 1000-byte chunk of data and returns compressed data of 500 bytes, it has a compression ratio of 1:2. Most compression schemes perform differently for data of different kinds. However, in general, running a compression algorithm on already compressed data will not reduce the size of the data. Thus compression ratios refer to an average compression ratio that is typically attributed to an algorithm.

There are two types of data compression. Compression schemes like those used in zip files and GIF files are lossless compression schemes. That means when the algorithm is used to compress data that are later to be uncompressed, the exact original data will be recovered. This is an extremely important characteristic for application programs. Typical lossless compression schemes are gzip and pkzip. However, having lossless compression of images is not very important. The visual effect of a picture may be of slightly lower quality, but the overall impact usually is not noticeably diminished.

When several similar files are to be compressed, it is usually better to join them together into an archive of some kind (using "tar," for example) and then compress them, rather than joining individually compressed files. This is because common compression algorithms tend to build tables based on the data from their current input, developed from the file the respective algorithm has compressed. These data do not need to be reproduced to compress subsequent files in the same archive, and thus the algorithm can perform a better compression ratio.

3.5.1.1 Huffman Coding and JPEG

It is the JPEG (Joint Photographic Experts Group) standard, developed jointly by ITU, ISO, and other standards organizations, that makes image compression possible. This standard uses Huffman coding for data compression, in addition to another form of arithmetic coding. Huffman coding exploits the property that not all symbols in a transmitted frame occur with the same frequency. In this method, fewer bits are used to encode the most common characters occurring in a data transmission than are used for less frequent characters. The result is a variable number of bits per character.

A bit-oriented transmission mechanism is used. First, the character stream is analyzed to determine the character types and their relative frequencies. A Huffman code tree is obtained that is essentially a binary tree with branches assigned the value 0 (for left branch) or 1 (for right branch). The code binary word used for each character, shown in the leaf nodes, is determined by tracing the path, proceeding from the root node to each leaf and forming a string of bits. An example code construction for the character string `AAABBC` is shown in Figure 3.19. Notice that the number of bits reduces to 10 as opposed to 48 bits required for ASCII representation. However, for the receiver to learn that a reduced character set has been used, extra transmission may be necessary. Obviously, the advantage of Huffman coding is more evident when the frequency distribution of characters being transmitted is wide and there are long characters strings with repeated letters. Code performance is determined by the average number of bits per symbol. Using the example of Figure 3.19, if the probability of occurrence of letters `A`, `B`, `C`, and `D` are 0.5, 0.3, 0.1, and 0.1 respectively, the average number of encoded bits per symbol would be 1.7 bits: $1(0.5) + 2(0.3) + 3(0.1) + 3(0.1)$.

JPEG is a lossy mechanism that encodes images. The mechanism is sometimes called a lossy sequential mode. The major steps in this mode are block preparation, application of a discrete cosine transform (DCT) to each block, quantization, and encodings. In block preparation, the image is transformed into pixel representation that may or may not cause

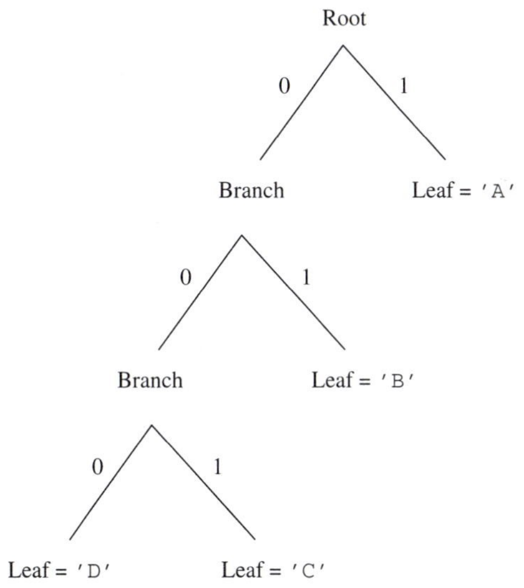

Figure 3.19 Traversing a Huffman tree from the root gives the binary string 111 0101 001 for the character string `AAABBC`.

a reduction in resolution. The pixels are considered to be grouped as blocks, and a DCT is obtained for each block in form of DCT coefficients. The coefficients indicate the level of spectral power present at each spatial frequency.

In the quantization phase, which is a lossy phase, a quantization table is used to transform DCT coefficients. The table allows the high-frequency coefficients to become zeros, which are eventually wiped out during encoding. In the last phase, a run-length encoding of each block is done by means of a zigzag pattern on the blocks. However, the run-length encoding is applied only on zeros that are grouped together as a result of quantization. This process actually compresses the data, as discussed earlier. The individual coefficient elements are encoded by means of the Huffman code. Finally a list of numbers that represents an image is obtained.

3.5.1.2 Run-Length Compression

One technique that uses a simple logic to shorten the expression of consecutively occurring data characters is called run-length compression. For example, `3A2B2C3B` will replace the character string `AAABBCCBBB`. For binary values the runs could be even longer, resulting in a better compression ratio. For example, an image may consist of runs having 25 zeros followed by a one or 55 zeros followed by a one. Binary codes may be given for variable length runs. For instance, a run of 15 zeros may be encoded by 1111, thus resulting in equivalent codes of 1111 1010 and 1111 1111 1111 1010, respectively.

The run-length compression method may be used to compress digital images by comparing pixel values that are adjacent and coding only the change in values. This technique is very effective for images with homogeneous regions, and may achieve compression ratios of 8:1 or higher. For images with even small variations, however, compression may actually increase the image byte size needed to represent the variations.

3.5.1.3 LZW Compression

Lempel and Ziv (LZ) designed a compression method using encoding segments. The segments of the original text are stored in a dictionary that is built during the compression process. When a segment of the dictionary is encountered later during scanning of the text, it is replaced by its index from the dictionary. The dictionary is the central point of the algorithm. It has the property of being prefix-closed (every prefix of a word of the dictionary is in the dictionary). Furthermore, a hashing technique makes its implementation efficient. The modified version of the LZ method is called the Lempel–Ziv–Welsh method in acknowledgment of several improvements introduced by Welsh. The algorithm is implemented by the "compress" command under the Unix operating system.

In LZW compression, the dictionary is initialized with all strings of length 1, the characters of the alphabet. When we have just read a segment *W* of the text and "a" is the next symbol (just following the given occurrence *W*), we proceed as follows.

If *W*a is not in the dictionary, we write the index of *W* in the output file, and add *W*a to the dictionary. We then reset *W* to "a" and process the next symbol (following "a"). If *W*a is in the dictionary, we process the next symbol, with segment *W*a instead of *W*. Initially, the segment *W* is set to the first symbol of the source text.

The decoding method is symmetric to the coding algorithm. The dictionary is recovered while the decompression process runs. It is basically done in this way. Read a code c in the compressed file, write into the output file the segment *W* having index c in the dictionary, and add the word *W*a to the dictionary, where "a" is the first letter of the next segment.

Two common commands, "gunzip" and "gzip" use variations of Lempel–Ziv compression. The first takes a list of files on its command line and replaces each file whose name ends with .gz, .z, .Z, -gz, -z, or _z and begins with the correct magic number with an uncompressed file without the original extension. This command also recognizes the special extensions .tgz and .taz as shorthands for .tar.gz and .tar.Z, respectively. When compressing, "gzip" uses the .tgz extension if necessary instead of truncating a file with a .tar extension.

The command "gzip" uses Lempel–Ziv coding to reduce the size of the named files. Each file is replaced by one with the extension .gz, whenever possible, while keeping the same ownership modes and access and modification times. (The default extension is z for MSDOS). The amount of compression obtained depends on the size of the input and the distribution of common substrings. Typically, text such as source code or English is reduced by 60–70%. Compression is generally much better than that achieved by LZW (as used in "compress"), Huffman coding (as used in "pack"), or adaptive Huffman coding ("compact"). Compression is always performed, even if the compressed file is slightly larger than the original. The worst-case expansion is a few bytes for the gzip file header, plus 5 bytes every 32K block, or an expansion ratio of 0.015% for large files. Note that the actual number of disk blocks used almost never increases.

3.5.1.4 Fractal Compression

Fractal compression is a promising patented technology for image compression that is claimed to be superior to JPEG. It is a form of vector quantization and a lossy compression with slow compression and fast decompression. The technique is usually traced to IBM mathematician Benoit B. Mandelbrot and the 1977 publication of his seminal book *The Fractal Geometry of Nature*. The book put forth a powerful thesis: "Traditional geometry with its straight lines and smooth surfaces does not resemble the geometry of trees and clouds and mountains. Fractal geometry, with its convoluted coastlines and detail ad infinitum, does." The promise of using fractals for image encoding rests on two suppositions:

1. Many natural scenes (e.g., clouds), possess details within a detailed structure.
2. An iterative function system (IFS) can be found that generates a close approximation of a scene.

However, fractal compression in its original form was not very convincing. The algorithm failed to incorporate the diversity of real-life pictures. To capture the diversity of real images, partitioned IFSs (PIFS) are employed. In a PIFS, the transformations do not map from the whole image to the parts, but from larger parts to smaller parts. An image may vary qualitatively from one area to the next (e.g., clouds, then sky, then clouds again). A

PIFS relates the areas of the original image that are similar in appearance. Using Jacquin's notation, the big areas are called domain blocks and the small areas are called range blocks. Every pixel of the original image must belong to (at least) one range block. The pattern of range blocks is called the partitioning of an image. A fractal-compressed image is an encoding that describes the following:

1. The grid partitioning (the range blocks)
2. The affine transforms (one per range block)

The decompression process begins with a flat gray background. Then the set of transformations is repeatedly applied. After about four iterations, the attractor stabilizes. The result is not (usually) an exact replica of the original, but reasonably close.

3.5.1.5 Facsimile Compression

Compression ratios of 1:2 may be obtained on text files by using the original Huffman coding. A better compression may be obtained for scanned images through facsimile (fax) compression, which is a modified Huffman coding. The method uses two tables for encoding called a termination codes table and a makeup codes table. The code words in these two tables provide bit combinations for white and black runs in the scanned image. Each scanned line consisting of black and white picture elements is encoded to incorporate runs, thus reducing the size of the scanned page. For example, an uncompressed scanned page produces about 2 million bits, requiring about 6 minutes of transmission at 4800 bps. With a compression ratio of just 1:2, the transmission time is reduced to half.

3.5.1.6 MPEG and MPEG-2

The standards of the Motion Picture Expert Group, MPEG (or MPEG-1) and MPEG-2, are used to compress the video and audio components of movies. However, MPEG is most effective on video, which takes more bandwidth and also contains more redundancy than audio. The objective of MPEG-1 is to produce video recorder quality output using a bit rate of about 1.2 Mbps. The compression is challenging, since with NTSC the uncompressed video may require as much as 472 Mbps. The next standard in MPEG family, MPEG-2, was originally designed for compressing broadcast quality video to 4–6 Mbps to make it fit in the PAL or NTSC broadcast channels and HDTV.

MPEG audio compression is done by using a fast Fourier transformation (FFT) to sample waveforms. FFT changes the signal from the time domain to the frequency domain. Spatial and temporal redundancies are dealt with in video compression. Spatial redundancy is handled by using JPEG to code each frame separately. Temporal (frame-to-frame) redundancy is tricky because it may require keeping track of changes in the image. This is accomplished by pegging block-by-block differences to the last frame (the predictive or P frame) and block-by-block differences to the last and the next frames (the bidirectional or B frames). However, in MPEG-1, D frames are also utilized for block averages used for fast forward. In addition, intracoded frames (I frames) are inserted for self-contained JPEG still pictures.

3.5.2 Encryption and Decryption

Network security was not accorded much importance in the initial phases of development. Now, as millions of people are using networks for banking, shopping, and other financial transactions, network security is a major problem. Most security failures are intentionally caused by network hackers rather than resulting from system malfunction.

The solution to security problems lies in an encryption and decryption system. Encryption is a process by which the plaintext at the sender's end is transformed to a ciphertext by means of an encryption key. The ciphertext is then transmitted. A decryption process (inverse of encryption) is used at the receiver's end to recover the plaintext. It is expected that through this process, the message will be secured against intruders, who can hear and copy the ciphertext but cannot decrypt it without the decryption key. It is therefore assumed that only the sender and the receiver know the decryption key.

Traditionally, network security has been considered to be part of the OSI presentation layer. However, many people place it with the application layer also. In our opinion, both approaches are valid. In TCP/IP suite of applications, where the architecture hierarchy is not as layered like the OSI model, it may make more sense to make security part of the application layer. However, in a strict seven-layer OSI model, the presentation layer handles the task of network security. We discuss this subject in detail later (Chapter 10), introducing the common encryption algorithms and showing how they can be used to provide network security.

3.5.3 Network Applications

The TCP/IP application layer is considered to be equivalent to the combined session, presentation, and application layers of the OSI model. Many applications based on TCP/IP have been developed over the years. Some of the popular ones are telnet, FTP (File Transfer Protocol), SMTP (Simple Mail Transfer Protocol), SNMP (Simple Network Management Protocol), and HTTP (Hypertext Transfer Protocol). These applications allow the underlying networks to be very transparent to the user, and the user may perform many useful tasks on the network, such as remote login, file transfer/retrieval via the World Wide Web, and sending and reading e-mail.

The basic idea in network applications is to have a client–server setup, in which an application program (client) issues requests for services to another application program (server), which provides the requested service. The client and service programs are typically located on two different computers connected via a network.

A graphical user interface (GUI) allows interaction between the client and server programs via graphical icons available to the user as part of the client program. The GUI facilitates the use of network applications, and no knowledge of network operation is necessary for a person to use the network. A typical example is the famous WWW application that may be used through a browser that provides a GUI interface to the user. A user may be using several ftp-like commands in the form of service requests/responses without realizing it. The available hypertext links make it possible to transfer files, images, and programs.

In the Unix environment, X-Windows provides the user with a graphical interface that is capable of running several network applications simultaneously. In addition, the protocol allows the applications to be run over the network; that is, a user may start an X-Windows application requiring specialized graphics at the remote end. This feature of the protocol facilitates running software packages, with built-in GUI to execute the commands, over a network. For instance, a remote user may run the "ghostview" program through X-Windows to display a postscript file.

3.6 NETWORK PERFORMANCE

Today's client–server applications can be very chatty and bursty. For example, in applications based on a standard query language (SQL), client and server communicate by exchanging SQL calls over the network as if it were the system bus of a stand-alone machine. The simplest query could generate hundreds of small network packets. Even if individual network components improve in reliability, and the bandwidth is increased, the network may not function reliably. It is important to understand how the overall network performance is affected. What parameters affect performance, and how can performance be improved?

A network connection may be viewed as a pipe, as shown in Figure 3.20. As the bandwidth is increased, the pipe width increases, while the length increases with increases in delay. The amount of data present on the line may be found as the product of bandwidth times delay. For example, if a line bandwidth is 10 Mbps and the delay from source to destination is 10 ms, the amount of data found on line will be 100 kilobits.

If the bandwidth is increased 10 times (or the delay is increased 10 times), the amount of data will increase 10-fold. The more data on the network, the higher the required sequence number at the data link layer to keep the pipe full. For instance, with a packet size of 1000 bits transmitted along a path with delay of 10 ms, 100 packets will fit on the cable in one direction if the transmission speed is 10 Mbps. With these numbers, it would be impossible to keep the pipe full with 3-bit sequence numbers of frames at the data link layer. However, if the delay is reduced to 1 ms and the transmission speed is reduced to 2 Mbps, the number of frames will be 2 in one direction and 4 for both directions. Now, if a sliding window protocol is used, the 3-bit sequence number should keep the pipe full in both directions. The second scenario gives a better channel utilization that approaches almost 100% under error-free conditions.

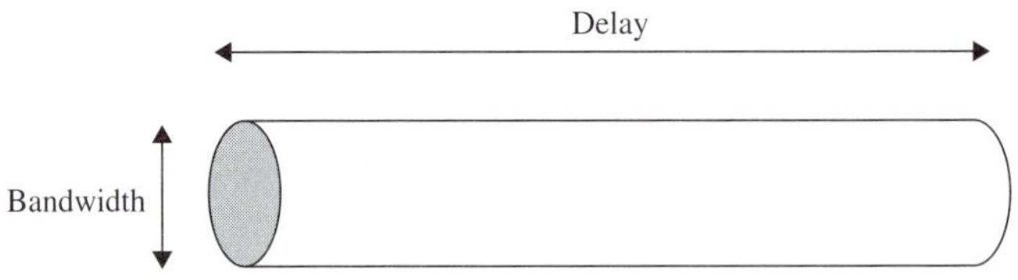

Figure 3.20 Speed and delay for a network connection can be viewed in terms of a pipe.

3.6.1 Delay

Delay refers to the amount of time needed for an offered frame to be delivered across the network to a remote user device. Since a network consists of complex interconnections and devices, the end-to-end delay may be very difficult to calculate in the closed form. However, approximate analytical and simulation models may be obtained that closely replicate the actual situation.

The two scenarios presented in our earlier example are cases of performance study that show the effects of propagation delay and the sliding window protocol on channel utilization. Other types of delay affect network performance as well, however. These delays are processing delay, caused by node processing, transmission delay, caused by limited media bandwidth, and the queuing delay, caused by the buffering of packets at the stations (Chapter 8). In addition, any platform (cell or switch) that does not have firewalling between users is vulnerable; a single overactive photovoltaic cell could cause delay for all users passing through the congested switch. This delay could build, causing protocol time-outs and retransmissions, further adding to the congestion. Note that trying to prevent congestion by simply discarding traffic early and often lowers reliability; thus, the end result is still lower performance.

3.6.2 Throughput

The throughput of a network is defined as the amount of data that can be transferred per unit time. Notice that this rate is not necessarily equal to the offered media capacity in bits per second. In fact, throughput and delay are often at odds with each other. As the throughput is increased owing to higher offered load, the delay increases too. Over time, however, increasing the offered load causes an uncontrolled excessive delay, resulting in network congestion, whereupon throughput falls instead of increasing. Careful studies show that to keep the network stable, bandwidth utilization should be below 50%. However, this figure largely depends on the type of network being used, as we shall see in later chapters.

3.6.3 Bandwidth Utilization

"Utilization" of a resource is defined as the percentage of time it is busy. In case of computer networks, we like to see resources being used most of the time. However, as we have discussed, higher throughput is associated with higher delay, and increasing the offered load beyond a certain point may in fact reduce throughout. Thus bandwidth utilization should not serve as an independent metric. Higher utilization does not necessarily mean a better network design. On the other hand, it may sometimes reflect a need to increase the network bandwidth.

3.6.4 Error Rate, Congestion, and Network Reliability

High application performance requires both reliability and low delay. Many switches simply admit all traffic into the network, without regard to instantaneously available bandwidth in the network. If congestion occurs, the server module buffers will fill, and the delay will increase. Increased delay lowers throughput. Once the delay experienced by users has reached a critical level, the server module starts setting forward- and backward explicit congestion notification bits (FECN/BECN) on all frames to notify end devices. Unfortunately, the reality is that most end devices ignore this notification. If the congestion continues to grow, and the buffer is about to overflow, the DE bit is used to decide which frames are to be discarded first. When frames are discarded, throughput is lowered even more, as protocol windows shrink in size and frames are retransmitted.

A much better solution is to prevent congestion whenever possible. If there is no spare capacity in the network, it does not make sense to blindly send in information at the access rate, only to have it add to the congestion and cause additional discards. The optimal technique would be to lower the rate of transmission for the virtual circuit, down to the committed information rate if necessary, and buffer the excess burst at the edge of the network.

Some network managers attempt to overcome discards by overprovisioning, so that the trunk bandwidth is much greater than the expected load. Aside from being expensive, this tactic creates other problems. First of all, subscribers will quickly learn that it is better to pay for an artificially low Committed Information Rate (CIR), even when traffic patterns dictate a higher CIR. This results in poorer performance for all. Second, as more users and traffic are added to the network, this artificial reserve will diminish and users will become upset that they no longer get the level of performance they have grown accustomed to. In addition, overprovisioning bandwidth still does not rectify the problems due to high delay.

Many frame relay architectures not only lack features aimed at congestion avoidance but also react to it in a way that unfairly penalizes some users. Most architectures simply allow all PVCs to transmit at their access rates, marking frames in excess of the CIR as discard eligible. In this architecture, virtual circuits compete for bandwidth, and those that generate the most frames the quickest get the majority of the bandwidth. While this arrangement is easy for a vendor to implement, it creates potentially serious problems for the user. This is especially true for the platforms that use shared trunk buffers without effective congestion avoidance. As these buffers fill, all users experience delay. When one of these many server modules becomes overloaded, it sets the FECN and BECN warning bits for all users, regardless of which circuit is actually causing the congestion by exceeding its CIR. The result is that devices capable of responding will throttle their traffic, even if the problem did not originate with them.

As congestion continues to build, frames marked DE eventually will be discarded. In addition to having serious consequences for the throughput of the higher level protocol, this method of discarding frames is unfair. Many devices, such as most routers, never mark traffic as DE. Some other end devices mark all noncritical frames as DE, in the hope that if a network does perform discarding, the more critical frames (those not marked DE) will be delivered. Even so, some frames in some devices that were transmitting within their CIRs will be discarded improperly.

At a given delay, throughput varies according to the reliability of the network. Lack of network reliability, in the form of lost frames, has a serious impact on all applications and protocols. Applications that individually acknowledge each frame or frame burst

(Novell Netware, SNA, TCP/IP, etc.) will wait for a time-out and then retransmit, thus reducing throughput and response time. Furthermore, when a sliding acknowledgment window is used, such as with TCP/IP and the IBM implementation, the discarding of a single frame will also cause retransmission of the entire window and the reduction of the window size down to one. Fortunately, the use of relatively error-free digital lines means that network-induced bit errors will have virtually no impact on throughput. However, if congestion is allowed to form, then the resultant discards will dramatically affect performance.

Both low delay and low error rate are needed to achieve maximum throughput. Low delay is important in all protocols, especially Novell Netware, SNA, and DECnet. Frame discards are very disruptive on all protocols. By having both low delay and high reliability, one minimizes the retransmission of redundant frames, waiting for acknowledgments or protocol time-outs, and the reduction of window sizes. If congestion continues to grow to the critical level, the network will then start to discard based on DE. However, the server module will not limit the discarded frames to those of the offending user. Any frame marked DE, whether by the user or by the network, is subject to discard. As it turns out, in an effort to prevent session loss, the IBM front-end processors mark all user data frames as DE and never mark control frames that way. Thus, the SNA traffic will also be subjected to discards, even if it was within its CIR. Once again, users suffer because of the actions of others.

3.7 CHAPTER SUMMARY

This chapter, dealing with the concepts of computer networks and data communications, discussed the seven-layer OSI model in detail. This chapter is a foundation for later chapters because references are often made to the OSI model in connection with different topics. Starting at the physical layer, different encoding methods were introduced, including the popular Manchester and differential Manchester methods. Since, however, Chapter 2 also deals with this subject, these approaches were not emphasized here. In addition to encoding methods, we discuss multiplexing and switching techniques.

This chapter performs the important task of linking the physical layer to the next layer, the data link layer. We showed how physical transmission (at the bit level) can be carried out in the presence of errors and noise. Error detection and correction methods were introduced with examples. Automatic repeat request (ARQ) techniques, the heart of the data link layer, for error recovery was discussed.

Another important concept covered in this chapter is framing. It was shown how frames are obtained by using either character-oriented (binary synchronous) protocols (or bit-oriented (HDLC) protocols. The example of X.25 served as an illustration of a bit-oriented protocol. Higher OSI layers (network to application) were mentioned for the sake of completeness. Later chapters provide a much more detailed coverage of these layers. However, this chapter treated some important topics related to higher layers, such as network routing/congestion control, compression schemes, network security, performance, and applications.

3.8 PROBLEMS

3.1 What technique is used to support more than one simultaneous speech conversation? Explain.

3.2 What are handshake signals, for what type of communication between a sender and a receiver are they used? How is the procedure carried out?

3.3 Explain the difference(s) between the serial and parallel transmission.

3.4 Categorize the following types of networks as LAN, MAN, WAN, or interconnection of WAN (Internet).

Network description	Network type
A company with its headquarters computer in London and branch office computers throughout Asia, Europe, and North America	
A club consisting of two workstations and one printer	
A city traffic control system managed by a network of 700 computers	
A network used by a commercial service providers such as America Online	
A campus network consisting of about 500 computers	

3.5 Is television an example of half-duplex operation?

3.6 Determine the total number of links needed for an N node connected as (a) mesh topology, (b) star topology, and (c) ring topology.

3.7 Categorize the following as simplex, half-duplex or full duplex modes of transmission.

Transmission description	Transmission mode
A heated argument between two drivers involved in an accident	
Transmission from radio at AM 1240	
Channel 13 television news at 11 P.M.	
A reversible vehicular traffic lane, like the commuter lane of Chicago's Kennedy Expressway, which runs west to east in morning and east to west in the evening	
A computer-to-monitor connection	

3.8 Match the following services to one or more of the seven OSI layers.

Service	OSI layer name
Reliable end-to-end transmission of message	
Breaking a transmitted bit stream into frames	
Determining which route through a subnet to choose	
Frame error and recovery	

Character conversion from EBCDIC to ASCII
Electrical and mechanical interface details
Segmentation and reassembly of messages
Encryption of data for security purposes
E-mail delivery system
Division of database update task through different dialogue units
File management and transfer

3.9 XYZ Inc. and ABC Inc. are involved in making network routers, bridges, and switches. After competing in the business for 10 years, the firms started to work toward some type of partnership. First, the companies' technicians met and figured out how to standardize their products. The recommendations were taken back to the first-level and second-level managers of the respective companies, and the technicians met again to discuss the details. Finally, the presidents of the companies met and worked out the business details. Discuss the similarity between this arrangements and the OSI seven-layer architecture. Does it fit the description of the seven-layer model? Give reasons to support your answer.

3.10 You need to come up with the design of a network architecture similar to the OSI model, but you can have only four layers in the design instead of seven. Explain each layer of your design and assign an appropriate name to it.

3.11 Suppose three terminals are connected to a statistical time division multiplexer and that each produces output as shown here (0 indicates no output). Construct the frame sent.

Terminal 1: A 0 0 B C

Terminal 2: 0 0 A C D

Terminal 3: C B C 0 D

3.12 A cable TV system has 100 commercial channels, all of them alternate programs with advertising. Is this more like TDM or like FDM? Explain.

3.13 Is it correct to say that in phase modulation each bit or group of bits is associated with a phase change of the carrier signal? Explain. Does this modulation technique provide the fastest transmission speed (bits per second)?

3.14 Consider Figure P3.14, a graph of voltage levels observed over time on a cable being used to transmit a binary digital signal.

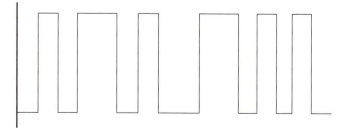

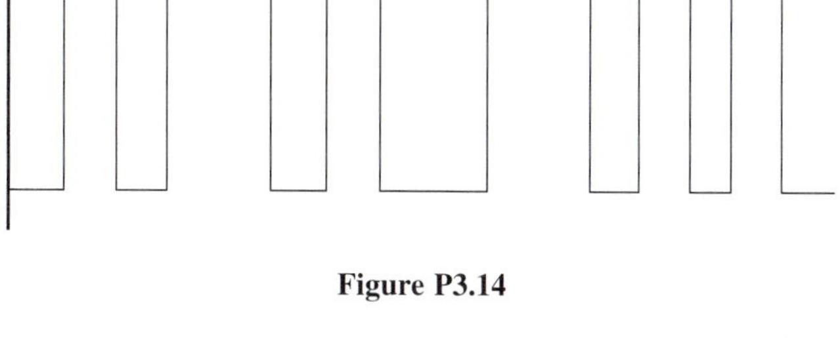

Figure P3.14

(a) Assuming that Manchester encoding is being used, what is the sequence of 0s and 1s represented by the signal on the cable?
(b) What is the bit sequence for the case of differential Manchester encoding?

3.15 What is the encoding method used in LANs set up according to IEEE Standard 802.3? What is the baud rate?

3.16 Why should the window size be less than half the maximum sequence number in a sliding window protocol? For example, W must be less than 4 for a 3-bit sequence number. Explain, and illustrate showing the sliding windows of transmitter and receiver.

3.17 Is a data link layer protocol necessary even if there were no bit errors or losses at the physical layer? Why or why not?

3.18 An analog signal is plotted for one second as shown in Figure P3.18. Assuming a PCM sampling rate of 16 samples/s and eight levels of encoding (0, . . . , 7), answer parts a–c.
(a) The resulting bit stream.
(b) The bit rate in bits per second.
(c) The baud rate.

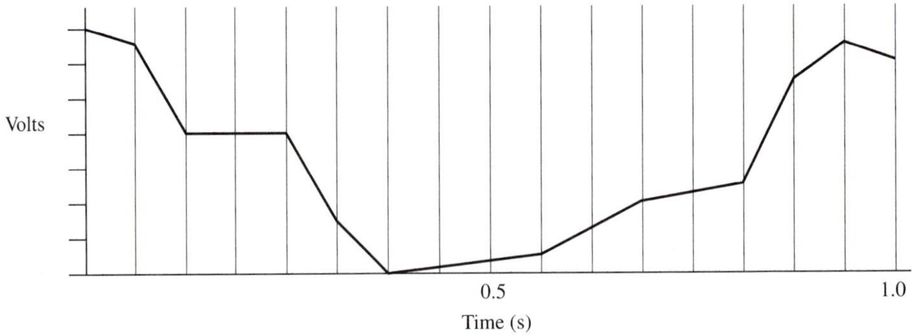

Figure P3.18

3.19 Assuming that each is correctly implemented, should the Selective Repeat Protocol (SRP) perform much better than go-back-n on a fiber optic link between two nodes about a half-mile apart? Explain.

3.20 Hardware or software may be used to perform error detection. If this function is implemented in hardware, does it then become part of the physical layer? Explain.

3.21 Consider the go-back-n algorithm with a window size of 7. Draw the sender and receiver windows and also describe the actions of both sending and receiving protocols, specifying the buffer contents in the following cases. What is the current state of each protocol after responding to the events specified?
(a) Station A sends frames 0–6. Station B receives them in order, but frame 4 was damaged.

(b) Station A sends frames 0–6 and station B receives them in order. Station B sends one data frame to A (which A receives correctly) after receiving frame 4 but before receiving frame 5.
(c) Same as in part b, but the data frame sent to A is damaged.

3.22 In stop-and-wait ARQ, what happens if a Negative Acknowledgment (NAK) is lost in transit? Why is there no need for such acknowledgments to be numbered?

3.23 Following is a data stream encoded using the Hamming code. Find all the parity bits and determine whether an error is present. If there is an error, locate the bit position in stream and rewrite the correct bit stream.

1 0 1 1 0 1 1 1 0 1 0 1

3.24 A generator function for CRC is given as $x^5 + x^2 + 1$.
(a) What is the generator function in binary form?
(b) What is the checksum (CRC) for the following message in binary and in polynomial forms?

1 1 0 1 0 1 1 0 1 1

3.25 If an ASCII character H is sent and the character I is received, what type of error has occurred (single-bit, multiple-bit, burst, etc.)? Explain.

3.26 If the data unit is 1 1 1 1 1 1 and the divisor is 1 0 1 0, what dividend (not remainder or quotient!) would be found at the transmitter from modulo-2 division in a CRC evaluation?

3.27 Assuming odd parity, what is the checksum bit for 1 0 0 1 0 1 1?

3.28 A router using HDLC has the following data block (without the flags) ready to be sent to the receiving router.

0100111110111100111111010101111111100100000010111110100

Carefully show how this sending router processes this data block to keep it distinct from adjacent data blocks. Assume the block has the data link layer header and trailer (except flags) attached already and show what needs to be done to complete the frame.

3.29 How would a receiving router using HDLC operate on the initial portion of the following received bit stream (including flags) before trying to read the frame header or check for errors?

0111111001101111111001111110001000000111110111110011111100111110110 . . .

3.30 In HDLC, what does each of the following received bit patterns represent?
(a) 0 1 1 1 1 1 1 0 1 0 1 0 1 0 1 0 . . .
(b) 0 1 1 1 1 1 1 1 0 0 1 1 0 1 0 1 1 . . .
(c) 0 1 1 1 1 1 1 0 1 1 1 1 0 1 0 1 0 1 1 1 0 1 1 1 . . .

3.31 Consider a 100 Kbps link that is 200 km in length. Each packet is 1000 bits long. The header and trailer together are 40 bits long. Assume piggybacked ACKs, a propagation speed of 200 m/μs and that a packet in either direction is lost, damaged, or stolen with probability 0.1.

(a) What is the throughput (bps) of this channel using stop-and-wait? Assume that the time-out value is set to exactly when the piggybacked ACK should return.

(b) What is the throughput (bps) of this channel using go-back-*n* with a window size of $W = 15$.

3.32 Is bisync (binary synchronous) an example of a bit-oriented protocol? Explain.

3.33 In the binary synchronous protocol, when a control code needs to be sent as part of a data block, what character code is used as a suffix.

3.34 Is X.25 an example of a bit-oriented protocol? Explain.

3.35 A colony is set up on moon. The 10 Mbps link from the earth to the lunar colony measures about 242,000 miles. Assume that the signal propagation speed is 186,000 miles per second.

(a) Calculate the minimum round-trip time (RTT) for the link.

(b) Using the RTT as the delay, calculate the delay × bandwidth product for the link.

(c) If a camera on the lunar base sends 25 Mbyte image file to the earth as a sequence of 1 Kbyte packets, how many bits are needed for the sequence number if we assume the use of a sliding window protocol?

(d) Assume that the data packets are 2500 bits long and acknowledgments are piggybacked (with negligible header). The error probability is 7%. What is the maximum achievable efficiency in the following cases?

(i) Stop-and-wait protocol (ABP) with processing delay of 0.1 ms at both sender and receiver ends.

(ii) SRP with window size 8 and negligible processing delay.

(iii) Go-back-*n* with window size 64 and negligible processing delay.

3.36 Assuming a call setup time of 0.5 second for a circuit switching scheme, what are the total data transfer times when 4064 bits of data must be transferred from station A to station B with (a) circuit switching and (b) datagram packet switching? The stations are separated by 5 hops (1 hop = 10 km), the signal propagation speed is 1000 km/s and the line capacity for each hop is 9600 bps. Assume that each packet is 1024 bits long with only 8 bits of overhead.

3.9 REFERENCES

BOOKS

Bertsekas, D., and R. Gallager, *Data Networks*, 2nd ed. Englewood Cliffs, NJ: Prentice Hall, 1992.

Black, U. D., *Data Communications and Distributed Networks*, 3rd ed. Englewood Cliffs, NJ: Prentice Hall, 1993.

Comer, D. E., and R. E. Droms, *Computer Networks and Internets.* Englewood Cliffs, NJ: Prentice Hall, 1999.

Forouzan, B., *Introduction to Data Communications and Networking*. New York: WCB McGraw-Hill, 1998.

Gibbs, M., and T. Brown *Absolute Beginner's Guide to Networking*, 2nd ed. Indianapolis: SAMS Publishing, 1994.

Keshav, S., *An Engineering Approach to Computer Networking*. Reading, MA: Addison-Wesley, 1997.

Nance, B., *Introduction to Networking*, 4th ed. Indianapolis, IN: QueCorporation, 1997.

Peterson, L. L., and B. S. Davie. *Computer Networks: A System Approach*. San Francisco: Morgan Kaufmann, 1996.

Shay, W. A., *Understanding Data Communications and Networks*. Boston, MA: PWS Publishing, 1995.

Spragins, J. D., J. L. Hammond, and K. Pawlikoski, *Telecommunications Protocols and Design*. Reading, MA: Addison-Wesley, 1991.

Tanenbaum, A. S., *Computer Networks*, 3rd ed, Englewood Cliffs, NJ: Prentice Hall, 1996.

Walrand, J., *Communication Networks: A First Course*, 2nd ed. New York: WCB McGraw-Hill, 1998.

WORLD WIDE WEB SITES

The Computer Network news, rumors, reviews, downloads, CNET Radio, and free members-only services
http://www.cnet.com/

Yahoo's listing on networking
http://dir.yahoo.com/Business_and_Economy/Companies/Computers/Communications_and_Networking/

Another networking listing from Yahoo
http://dir.yahoo.com/Computers_and_Internet/Communications_and_Networking/

Professor Raj Jain's web site (Ohio State University) with information on recent topics in networking
http://www.cis.ohio-state.edu/~jain/

The online network book by Professor Y. Yemini (Columbia University)
http://www.cs.columbia.edu/netbook/

4

LAN TECHNOLOGIES

Not long ago LANs (local-area networks) were a luxury, and many small organizations and companies were reluctant to adopt this technology. Now, it has become almost a necessity for small offices, and experts predict that within a few years, LAN setups will find their way into homes. The situation is similar to PC developments since the early 1980s. With the cheaper, faster, and more efficient machines developed over the last decade, most homes have at least one PC now. Because of the growing importance of LAN technologies, we discuss several aspects of LAN technologies in detail. Following an overview, we present the advantages of having a LAN. Then we move on to a discussion of LAN protocols, architectures, services, and operating systems.

4.1 LAN OVERVIEW

There are many changes going on in the field of computing. Network computing, making use of PCs, workstations, LANs, and WANs, is challenging traditional centralized data processing architectures. A network is a system that interconnects computers and devices, relying various media for the efficient use of shared resources and distributed information. A LAN, on the other hand, is a data communications network that is geographically limited, typically to a 1 km radius, allowing easy interconnections of terminals, computers, and peripheral devices, such as daily-use printers, fax machines, and copy machines. The large number of existing microcomputers close together geographically has been one of the major factors in the development of LANs, which focus on the technology of connectivity.

There are various ways to organize a network. Two popular and widely supported ones for LAN are the bus and token ring topologies, introduced in Chapter 1. In a bus network, devices are hooked up to a shared bus. In a hub network, the devices are connected to a central hub, which shares the same bandwidth between the devices. Both networks can become inefficient when two or more machines transmit at the same time, resulting in a collision. When a collision occurs on the bus, all machines on the network have to wait and retransmit. To solve the problem, bus and hub networks can be divided into segments: if two machines on a segment communicate with each other, their traffic stays inside the segment. If a collision occurs on a segment, only the machines on that segment need to retransmit. Thus, if network users can be divided into natural workgroups, segmentation can keep each workgroup's traffic from jamming the network.

A hub network can also be switched, rather than shared or segmented, to let each connected device get its own segment, consisting of a clear, collision-free channel. For example, a 12-port, shared Ethernet hub provides a total bandwidth of 10 Mbps for everyone to share, while a 12-port, switched Ethernet hub may provide 10 Mbps per port, for a total 120 Mbps. With the development of intelligent switching technology for hub networks, intelligent hubs can now monitor network traffic and provide a platform for automatic troubleshooting. By isolating the port, the manager can allow other devices to continue using the network while the problem is being fixed. Most intelligent hubs can do the same thing automatically.

The evolution of LAN technologies includes the following well-known phases.

Ethernet: A coaxial cable local-area network first described by Metcalfe and Boggs of Xerox PARC in 1976; specified by DEC, Intel and Xerox, also referred to as DIX (using the initials of these three companies), and IEEE 802.3. Now it is recognized as the industry standard. This 10-Mbps tried-and-true standby is often bundled with computers and offers inexpensive, highly compatible expansion cards.

Fast Ethernet: A newer 100 Mbps technology, fast Ethernet is an extension of the standard 10 Mbps Ethernet. With its high bandwidth and low cost, it offers the lowest cost–bandwidth ratio among LANs on the market.

Fiber distributed data interface (FDDI): This 100 Mbps, ANSI standard for LAN is defined as X3T9.5. The underlying medium is fiber optics, and the topology is a dual-attached, counterrotating token ring. FDDI rings are normally constructed in the form of a "dual ring of trees." A small number of devices, typically infrastructure devices such as routers and concentrators rather than host computers, are connected to both rings—these are said to be "dual-attached." Host computers are then connected as single-attached devices to the routers or concentrators. The dual ring in its most degenerate form is simply collapsed into a single device. In any case, the whole dual ring is typically contained in a computer room. This network topology is required because the dual ring passes through each connected device and requires each such device to remain continuously operational (the standard allows for optical bypasses but these are considered to be unreliable and error prone). Devices such as workstations and minicomputers that may not be under the control of the network managers are not suitable for connection to the dual ring. As an alternative to a dual-attached connection, the same degree of resilience is available to a workstation through a dual-homed connection, which is made simultaneously to two separate devices in the same FDDI ring. One of the connections becomes active while the other one is automatically blocked. If the first connection fails, the backup link takes over with no perceptible delay. This fault-tolerant technology can maintain connectivity even if a hub port or cable fails.

Copper distributed data interface (CDDI): CDDI is FDDI running over conventional copper cables instead of fiber optic cables. All FDDI connections, single-attached or dual-attached, can be either fiber or copper.

Asynchronous transfer mode (ATM): ATM, or "fast packet," is a method for the dynamic allocation of bandwidth by means of a fixed-size packet, called a cell. This emerging 155 Mbps technology is designed to carry multimedia traffic such as data, voice, and video.

There are some other LAN technologies as well, with rates in the kilo- and megabyte ranges. LocalTalk, a 235 Kbps local-area network, uses Apple Computer's own technology across an Ethernet network, and EtherTalk, a 10 Mbps local-area network is used to extend AppleTalk's networking capability across an Ethernet network. Both are based on the proprietary LAN protocol AppleTalk, developed by the Cupertino firm for communication between Apple products (e.g., Macintosh) and other computers.

Each LAN technology is typically associated with a particular type of cabling. Fast Ethernet is designed to run on category 5 twisted pair (100Base T fast Ethernet), ATM can run either on fiber optic or category 5 unshielded twisted pair. Ethernet usually runs over category 5 twisted pair (10Base T Ethernet).

4.2 PROTOCOLS AND STANDARDS

Protocols are designed to manage the flow of data on communication channels. Most of the protocols designed at the data link layer are used in LANs. For example, sliding window protocols for flow control, error detection/correction mechanisms, polling/selection, contention, and time slot mechanisms are used in LANs. We discuss here some of the industry standards that are in common service. We also illustrate how these industry standards employ the protocols to implement the data link layer tasks.

4.2.1 IEEE Standards

The widely accepted standards of the Institute of Electrical and Electronics Engineers are the most popular of all. The standards group of IEEE, called the 802 committee, is divided into subcommittees, each of which addresses a specific LAN architecture, as follows.

802.1: The high-level interface standard addresses matters related to network architecture, management, and interconnection. In addition, it deals with issues related to the higher OSI layers (above the data link layer).

802.2: Logical link control (LLC) and media access control (MAC) are two sublayers within 802.2 that are equivalent of the OSI data link layer.

802.3: Carrier sense multiple access with collision detection (CSMA/CD) standards cover a variety of architectures that are generally based on the Ethernet as originally proposed by Metcalfe and Boggs.

802.4: The token bus network standard describes how the token bus network operates.

802.5: The token ring network standard describes how the token ring network operates.

802.6: The metropolitan-area network (MAN) standard describes the operation of networks covering bigger distances (about 200 miles) than conventional LANs. Another MAN standard, the distributed queue dual bus (DQDB) by the American National Standards Institute (ANSI), is an adaptation of the IEEE MAN standard.

802.7: The broadband Technical Advisory Group provides guidance to other groups that are involved in establishing broadband LAN standards.

802.8: The Fiber Optic Technical Advisory Group provides guidance to other groups that are involved in establishing LAN standards using fiber optic cable.

802.9: Integrated data and voice networks (Iso-Ethernet) standards cover the architecture for networks that carry both voice and data, such as integrated services digital networks (ISDNs).

802.10: A LAN security standard addresses the implementation of security capabilities such as encryption/decryption network management and data transfer.

802.11: The wireless LAN standard covers multiple transmission methods for wireless transmissions that include broadcast frequencies such as spread-spectrum radio waves and microwaves.

802.12: The demand priority access method (VG-AnyLAN) is one of the newer groups that is involved in developing specifications for 100 Mbps speed over twisted-pair wires.

802.14: Cable-TV Based Broadband Communication Network Working Group

802.15: Wireless Personal Area Network (WPAN) Working Group

802.16: Broadband Wireless Access (BBWA) Working Group

802.17: Resilient Packet Ring Working Group (RPRWG) for use in Local, Metropolitan and Wide Area Networks for transfer of data packets at rates scalable to many Gbps.

We explain next some of the LAN standards that are widely used in the industry.

4.2.2 CSMA/CD, Ethernet, and IEEE 802.3

Ethernet is considered to be the most user-friendly, inexpensive type of LAN technology available today. The vast majority of computer vendors equip their products with 10 Mbps Ethernet attachments, making it possible to link other computers with an Ethernet LAN. As the 100 Mbps standard becomes more widely adopted, computers are going to be equipped with an Ethernet interface that operates at both 10 and 100 Mbps. The ability to link a wide range of computers is an essential feature of LANs. Most LANs must support a wide variety of computers purchased from different vendors, a mandate calling for the high degree of network interoperability provided by Ethernet. The first Ethernet system was developed in 1976 (Figure 4.1). Following that, IEEE Standard 802.3 was developed, to provide an open set of specifications for building Ethernet systems. The technology was made easily available to anyone. This openness, combined with ease of use, inexpensive implementation, and robustness, resulted in a large Ethernet market and is one of the main reasons for Ethernet's very wide implementation in the computer industry.

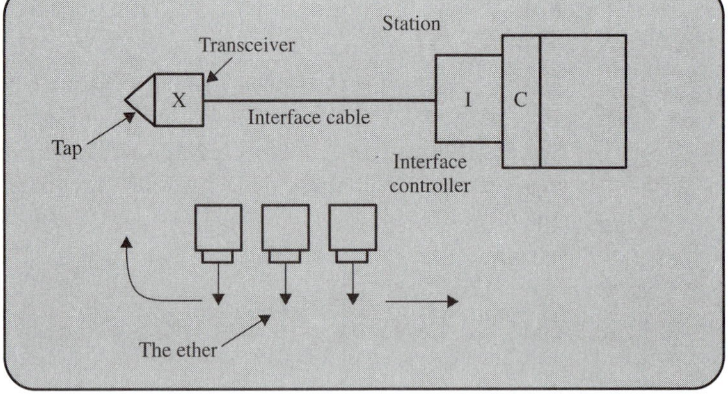

Figure 4.1 A 1976 drawing of the first Ethernet system by its coinventor, Dr. Robert M. Metcalfe.

The Ethernet system consists of three basic elements: the physical media (the ether) used to carry Ethernet signals between computers; a set of rules, embedded in each Ethernet interface, that allow multiple computers to fairly arbitrate access to the media; and an Ethernet frame that consists of bits used to carry data, control, and address information.

The station consists of an Ethernet controller, usually called a network interface card (NIC), that is connected to the Ethernet cable via a transceiver cable. The transceiver, which is also called a medium attachment unit (MAU), performs the CSMA/CD functions of carrier checking and detecting collisions on the line. It also acts as a connector that attaches a station to the Ethernet cable via a tap.

Since there is no central controller, each station on an Ethernet LAN operates independently of all other stations. All stations are attached to a shared broadcast medium called a "bus" (for channel). Ethernet signals are transmitted serially, one bit at a time, over the bus to every attached station. Before sending the data, a station listens to the channel, and when the channel is idle the station transmits its data in the form of an Ethernet frame, or packet. After each frame transmission, all stations on the network must contend equally for the next frame transmission opportunity. Access to the shared channel is determined by the medium access control (MAC) mechanism embedded in the Ethernet interface located in each station. The medium access control mechanism is based on carrier sense multiple access with collision detection (CSMA/CD).

The CSMA/CD protocol calls for each connected station to perform "carrier sensing"—that is, to wait until there is no signal carrying data on the channel. However, there may be a carrier (or ether) on the channel indicating an idle channel. Since other station interfaces on the network go through the same process, this method is called "carrier sense, multiple access." If, while one station interface has just started the data transmission using the carrier, another station has detected an idle channel and starts transmission, the frames may collide, invalidating both transmitted frames. Since signals take a finite time to travel on the channel, a collision may occur because it is possible that two stations will simultaneously detect an idle channel. When a collision occurs, the stations involved must sense it, stop the transmission, and resend the frames. This is called collision detection.

Figure 4.2 shows an Ethernet frame as specified by IEEE 802.3. The frame consists of several fields including address fields, a variable size data field that carries from 46 to 1500 bytes of data, and an error-checking field that checks the integrity of the bits in the frame to make sure that the frame has arrived intact. The preamble and SFD (start frame delimiter) synchronize the receiver by using a 64-bit sequence of alternating 1s and 0s, ending with 11 (i.e. 101010 . . . 11). The "length" field indicates frame length. However, the original Ethernet frame as specified by DEC, Intel, and Xerox (also called the DIX frame) has a type field instead of the length field that specifies the protocol type above the MAC sublayer. Since the data part of the 802.3 frame consists of the upper layer data provided from the logical link control (LLC—802.2 layer), there is no need to provide

Preamble	SFD	Destination address	Source address	Length	Data	32-bit CRC

Figure 4.2 IEEE 802.3 frame.

the upper layer protocol type. The data part must contain at least 48 bytes to make the collision detection process work. Note that if the actual data from higher layer is too small, extra padding bits may be needed. The maximum data size is 1500 bytes.

The two address fields in the frame carry 48-bit addresses, called the destination and source addresses. IEEE controls the assignment of these addresses by administering a portion of the address field. This 48-bit address is also known as the physical address, hardware address, or MAC address. The unique 48-bit address is preassigned to each Ethernet interface upon manufacture. As each Ethernet frame is sent onto the channel, all Ethernet interfaces look at the first 48-bit destination address field. Each interface compares the destination address of the frame with its own address. The Ethernet interface with the same address as the destination address in the frame reads the entire frame and delivers it to the higher layer software running on that computer. All other network interfaces stop reading the frame when they discover that the destination address does not match their own address. Computers attached to an Ethernet can send application data to one another using high-level protocol software, such as the TCP/IP protocol suite.

Ethernet was designed to be easily expandable to meet the growing networking needs of a site. To allow extension of Ethernet systems, vendors sell devices that provide multiple Ethernet ports. These devices are known as hubs, since they provide the central portion, or hub, of a media system. There are two major kinds of hub: repeater hubs and switching hubs. Each port of a repeater hub links individual Ethernet media segments together to create a larger network that operates as a single Ethernet LAN. The total set of segments and repeaters in the Ethernet LAN must meet the particular Ethernet specifications. The second kind of hub provides packet switching, typically based on bridging ports. We discuss the details of LAN switching architectures in Chapter 7. The high-speed Ethernet networks, such as 100 Mbps Ethernet, gigabit Ethernet, Iso-Ethernet (IEEE 802.9), and 100 VG-AnyLAN (IEEE 802.12) are discussed in Chapter 6. The details of cabling and interface are provided later in this chapter. Here we outline the major 10 Mbps Ethernet configurations used in the industry.

Thick coaxial ethernet, type 10Base5: Uses 10 Mbps Ethernet media with baseband signaling and maximum segment lengths of 500 m. A maximum of five segments can be connected with a total of 2.5 km and not more than four repeaters in a path. The standard asks for a minimum intertap distance of 2.5 m, allowing a maximum of 1000 stations in all.

Thin coaxial ethernet, type 10Base2: Uses 10 Mbps Ethernet media with maximum segment lengths of 185 m. The connectors and cables in this case are NICs, thin coaxial cable, and a T-shaped device called a BNC-T connector. The transceiver circuitry is moved to the NIC in this case.

Twisted-pair Ethernet, type 10Base-T: Uses 10 Mbps Ethernet media over twisted-pair cable instead of coaxial cable. A maximum segment length of 100 m is supported. The segments are connected to a hub to form a star topology. The intelligent hub implements all the networking operations, replacing the function of all transceivers.

Fiber optic Ethernet, type 10Base-F: Uses 10 Mbps Ethernet media over fiber optic cable.

4.2.3 Token Bus and IEEE 802.4

Collisions are considered to be normal in Ethernet operation. The system may not be suitable for a high-traffic situation if excessive numbers of collisions are anticipated, however, because the higher number of retransmissions may drastically reduce the throughput causing further increases in the delay. The token bus standard (IEEE 802.4) combines features of Ethernet and token ring to provide a deterministic (or predictable) delay under heavy loads without causing collisions.

The token bus system operates on a bus topology and is considered to be suitable for industry applications in which factory automation and process control are desired. A station with the token has right to transmit on the bus, while others may listen to the transmission to determine whether it is meant for them. The physical bus is connected as a logical ring; that is, stations are connected as a logical ring in that the token is passed among stations. If a station wants to send data, it must wait for the token. To implement the protocol, each station must complete the following steps.

- Wait for the token.
- When the token has been received and there are data to be transmitted, then
  (a) Transmit the data.
  (b) Wait for an acknowledgment.
  (c) Pass the token to the next node.

Obviously, if a station does not have any data to transmit, the token is not used but is passed to the next station. The token is passed from one station to another based on station addresses. For example, it can be passed in ascending or descending MAC address order. Figure 4.3 shows token passing on a bus in descending order. Notice that it is not necessary for all the stations on the bus to be active at all times.

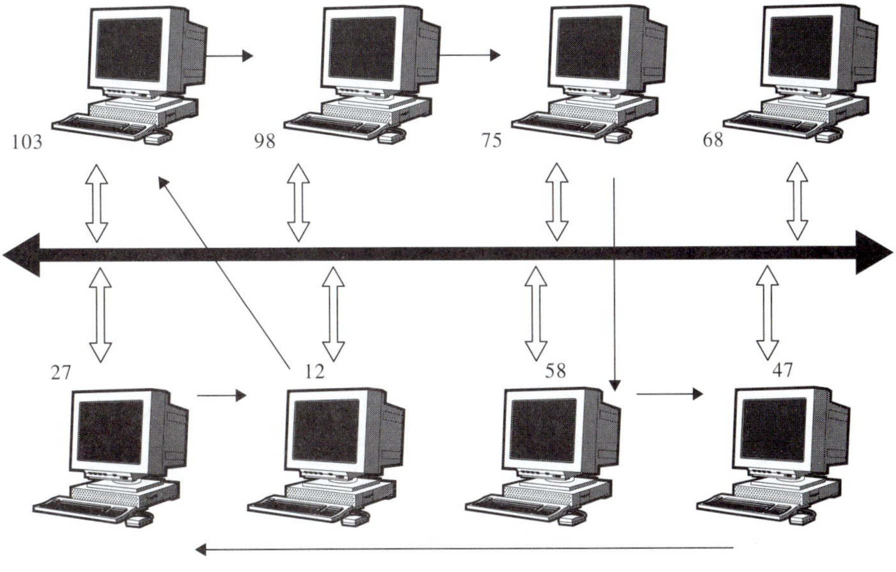

Figure 4.3 Logical ring formed on token bus with station addresses in descending order.

The token bus protocol allows for new stations to be added to the logical ring via periodic "solicit-successor" messages issued by the stations on the logical ring. This message contains the sending station's address and the address of that station's current successor node. Stations on the bus inspect the two addresses, and a station that has an address that falls between these two addresses will respond to the message. For example, assume the station with address 75 in Figure 4.3 broadcasts a "solicit-successor" message with two address fields, 75 and 58. Now, if a station with address 68 is interested in transmitting, it will respond to the message. However, if there are two or more stations responding to the message, the contention must be resolved by following an orderly process.

The protocol specifications include scenarios of failure and error recovery. For instance when an active station in the logical ring fails or shuts down, the sending station must recognize the situation by sending a "who is next" message. The successor of the failed station recognizes the address of its predecessor in the "who is next" message and responds to it.

In addition, a rotation timer is used to control the timing of token passing. When a sending station does not receive the token back in a prescribed interval, the sending station transmits the token again. If there is no response the second time, recovery measures, as described earlier, are taken.

4.2.4 Token Ring and IEEE 802.5

The stations in a token ring network take turns in participating in data transmission, very much like the token bus system, except that the transmission medium is a ring instead of a bus. The ring supports data rates of up to 16 Mbps and consists of a series of 150-Ω twisted-pair sections connected to one another via a hub called a concentrator or MAU (multistation access unit). The NIC on each station connects to a MAU that can support up to eight stations. A ring is formed by connecting the output port of one station to the input port of the next station in the ring.

Token passing is done from one NIC to another in sequence until a station with data to send is encountered. If the token is free, the station captures it and then sends a data frame. The data frame rotates in the ring, and each station examines the destination address field in the frame to determine whether a match exists. The intended recipient recognizes its address and copies the message while allowing it to proceed in the ring until it reaches the station that sent it.

The sender, after recognizing the source address field and checking to confirm its receipt, accepts the frame and discards it. Figure 4.4 shows the token ring frame format. The different fields are start delimiter (SD) or flag, access control (AC) for priority, frame control (FC) for frame type, destination address (DA), source address (SA), data, CRC, end delimiter (ED), and frame status (FS). The access control field of the frame consists of a token bit that is set to indicate that the frame is a data frame. The data field contains

SD (1)	AC (1)	FC (1)	Destination address (6)	Source address (6)	Data (≤4500 bytes)	CRC (4)	ED (1)	FS (1)

Figure 4.4 Token ring frame with nine fields; field lengths in bytes in parentheses.

the data, which may be up to 4500 bytes supplied as an IEEE 802.2 (LLC) protocol data unit (PDUs).

The IEEE 802.5 committee has come up with a new token ring architecture using the impending switched token ring standard called 802.5r, dedicated token ring (DTR). The DTR standard defines requirements for new end stations and concentrators, and it specifies a protocol for full-duplex operation. This new architecture coexists with current token ring equipment and the token passing access protocol. As a result of this new standard from IEEE 802.5, DTR-enabled adapters and concentrators are available from virtually all the major token ring vendors. These products have the benefits of early interoperability testing and of token ring's outstanding reliability and network management.

The DTR concentrator consists of C-ports and a data transfer unit (DTU). The C-ports provide basic connectivity from the device to the token ring stations, classic concentrators, or other DTR concentrators. The DTU is the switching fabric that connects the C-ports within a DTR concentrator. In addition, DTR concentrators can be linked to each other over a LAN or WAN via data transfer services such as asynchronous transfer mode (ATM).

A LAN segment on a classic concentrator can be linked to the rest of the network by attaching a DTR C-port (operating in station-emulation mode) to any available lobe on the device. This small segment is then given the entire 16 Mbps token ring bandwidth. DTR-enabled end stations operate identically to token ring stations when attached to a classic concentrator. Using the new access protocol, called transmit immediate (TXI), they also can operate in full-duplex mode when attached to a DTR concentrator.

4.2.5 Commercial LAN Systems

AT&T's Information System Network (ISN) employs three networks inside a central controller, each operating at 8.64 Mbps. The contention bus handles network access. The other two networks handle the transmit and receive operations, while providing high-speed interface to devices. Using a central controller with four remote concentrators, ISN has the capability to support up to 1680 end-user devices. It also provides support into T1 digital trunks, Ethernet, and other digital services.

Among other popular commercial LAN systems, IBM's token ring is worth mentioning. This ring allows several smaller rings to be attached to a backbone ring attached through the bridges. In addition to providing the connectivity, the bridges may provide speed translations between rings operating at speeds that may vary between 4 and 16 Mbps. The physical ring provides for IEEE 802.5 unidirectional point-to-point transmission of signals to and from up to 250 stations attached to one ring.

4.2.6 Wireless LAN and IEEE 802.11

Wireless LAN systems eliminate the need for wiring between the stations. The networks are called cellular packet networks supported by a standard. Two of the emerging standards are Hiperlan and 802.11. These standards call for the use of wireless transmitters equipped in the wireless NICs at the stations that communicate with one another via a hub. The transmitter implements the physical layer by converting the bits into radio waves operating at frequencies regulated, in the United States, by the Federal Communications

Commission (FCC). The radio waves are usually either TDMA or CDMA, depending on the standard used.

The MAC of Hiperlan is called nonpreemptive multiple access (NPMA) because it in fact provides nonpreemptive access to high-priority traffic and fair access to traffic of the same priority. The cycles in NPMA have three phases: priority resolution, contention resolution, and transmission. The priority resolution phase guarantees that only the highest priority stations will transmit during that cycle. The stations with same priorities contend during the contention phase by transmitting bursts of random lengths. A station survives if it senses an idle channel right after the end of a burst.

The IEEE 802.11 protocol offers two services, namely, contention free and contention. The contention-free service is designed to set up bounded delay transfers. Time is divided into frames consisting of contention-free and contention phases. During the contention-free phase, a control station polls the other stations, which then transmit their packets one at a time. During the contention period, the stations transmit by means of a carrier sense multiple access with collision avoidance (CSMA/CA) protocol, which is similar to a CSMA/CD protocol but favors a station that waits longer.

4.3 LAN HARDWARE

A LAN is made up of hardware, software, and transmission media. However, in building a LAN, a wide variety of options are available. One need to choose from media of at least three types, as well as a variety of network operating systems, network methodologies, computer platforms, and client operating systems. In this section we talk about the different kinds of hardware that are used to design a LAN.

4.3.1 Connecting Components: Cabling, Connectors, Transceivers, Repeaters, and Network Interface Cards

The lowest level of the seven-layer network structure is a hardware base, which defines the physical link between computers, peripheral devices, and any other resource in a network. In addition to computers that run everyday applications, networking software that makes the whole network functional, and network interface cards (NICs), there are some communication devices and technologies needed. Among them are cabling, connectors, transceivers, and repeaters, the basic LAN components for data transmission.

4.3.1.1 LAN Transmission Media

A variety of transmission media are used to provide LAN connection, as discussed in Chapter 2. Two types of physical transmission medium are used in today's networks: copper and fiber. The most common copper transmission media are twisted-pair and coaxial cables. The transmission media of fiber are fiber optic cables. Figure 4.5 shows some cable types.

Twisted-pair copper cables (10BaseT) are the earliest ones to be used, and the longest. They are relatively inexpensive and easy to connect. The two most common types of

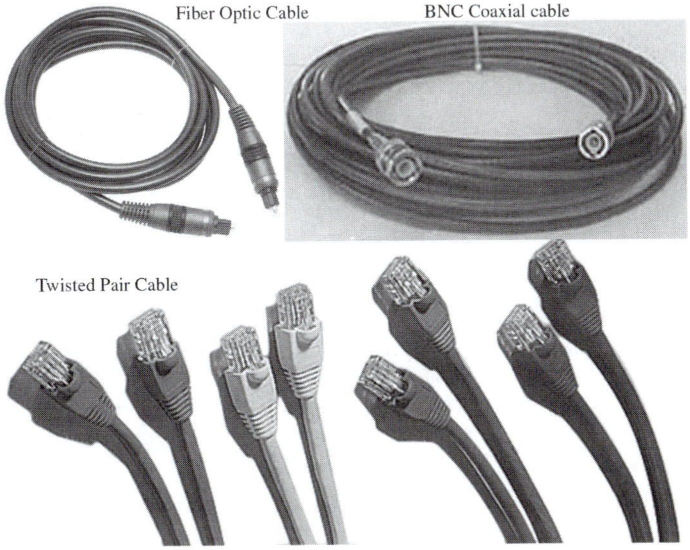

Figure 4.5 Network media types.

twisted-pair copper cable used in networking today are unshielded twisted pair (UTP) (Figure 4.6) and shielded twisted pair (STP). The difference between UTP and STD is that unshielded copper cable is subject to natural electrical interference, which can cause network errors; shielded cable costs more, but is more stable. Twisted-pair cables are usually wired in a star pattern with a repeater hub in the center of the star and workstations at the end of the tentacles.

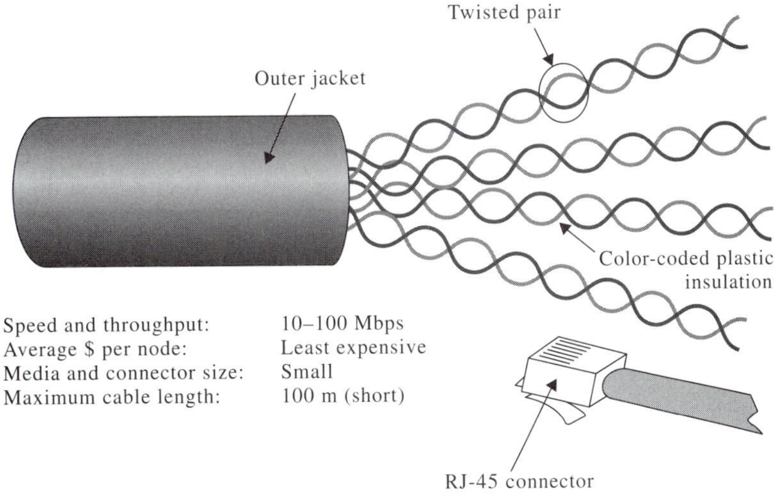

Figure 4.6 Unshielded twisted-pair (UTP) cable.

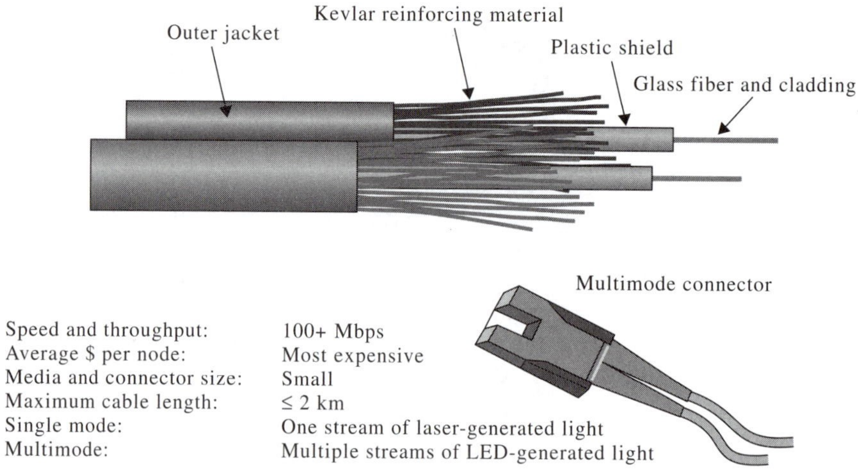

Figure 4.7 Components of fiber optic cable.

Coaxial cables are mostly used in bus topology networks. They come in two types: thin coaxial cable (10Base2) and thick coaxial cable (10Base5). Coaxial cables are the earliest high-performance cables. Thick coaxial cables are very stable and robust, while thin coaxial cables are cheaper and easier to install. A computer or other network device can be attached to the bus cable through a T-connector at the back of its NIC. One terminator is attached to each end of the bus. This configuration may become irritating because when problems occur, they are very difficult to isolate. The usual reasons for adopting copper cables are their low cost and ease of installation and maintenance.

Fiber optic is a new connecting technology that is more expensive but free of electrical interference and able to cover longer connection distances than any other type of cable. Fiber optic cable has been usually limited to high-speed and high-security installations because both materials and installation are expensive. As the engineering problems involved in using fiber optics have been resolved, this technology has gained in popularity.

Based on these three major types of transmission medium (twisted-pair copper cables, coaxial cables, and fiber optic cables), corresponding connecting or cabling technologies are developed.

The main factor limiting the use of fiber optic cable is not the cost of the cable itself but the cost of installation. The labor and expertise required to install and terminate fiber optic cable account for most of the expense. Some elements of a fiber optic cable installation are shown in Figure 4.7. The fiber optic connector is so important that its performance affects network performance directly. Because of the cost of termination, fiber optics is often limited to use as a network backbone. But no other network transport medium can match the bandwidth, scalability, or physical transmission capabilities of fiber optics.

4.3.1.2 Ethernet Devices

Ethernet uses a transmitter–receiver (usually called a transceiver: Figure 4.8), a physical device that connects a host interface (e.g., an Ethernet controller) to a local-area network.

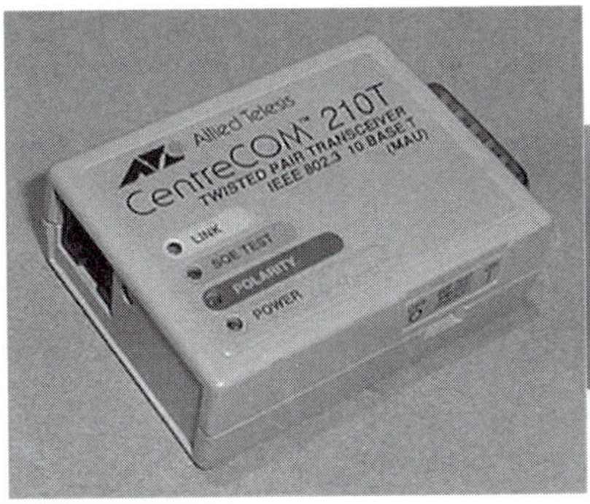

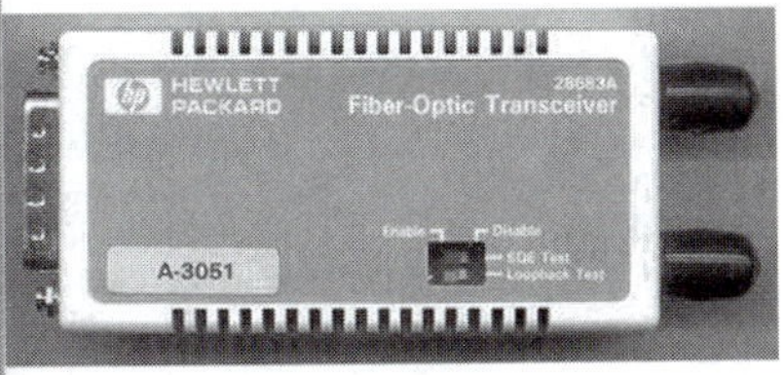

Figure 4.8 Transceiver for thick Ethernet cable (10Base5).

Ethernet transceivers contain electronics that apply signals to the cable and sense another host's signals and collisions. A tap located on top of the transceiver provides connection to the Ethernet cable. The station is connected to the transceiver by a 15-wire cable also called an attachment unit interface (AUI), which performs the physical layer interface functions between the station and the transceiver. Each end of an AUI terminates in a DB-15 (15 pin) connector, as shown in Figure 4.9.

The thin Ethernet implementation is also called 10Base2, as defined by the IEEE 802 committee. The same implementation is sometimes also referred to as Thin-Net, Cheap-net, and thin-wire Ethernet). A BNC-T connector, as shown in Figure 4.10, provides the connection to the cable. The main advantage of thin Ethernet is its reduced cost due to lighter weight cable. Also, installation in office environments is easier because the wire is more flexible than the thick Ethernet kind. Because of its shorter range (185 vs 500 m for thick Ethernet) however, thin-wire Ethernet is used mainly in small offices.

An alternate implementation that is more common is inclusion of the transceiver in the NIC, which provides a direct connection to a BNC-T connector as shown in Figure

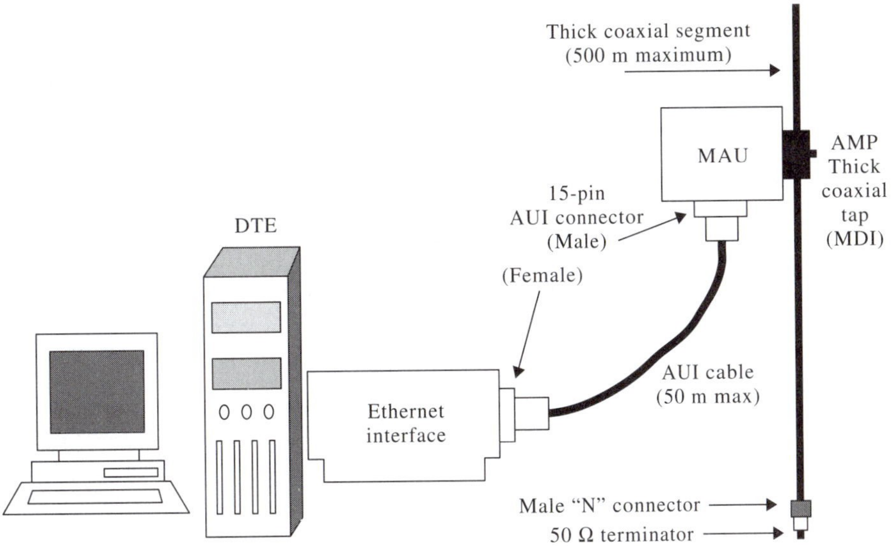

Figure 4.9 Thick Ethernet cable connection.

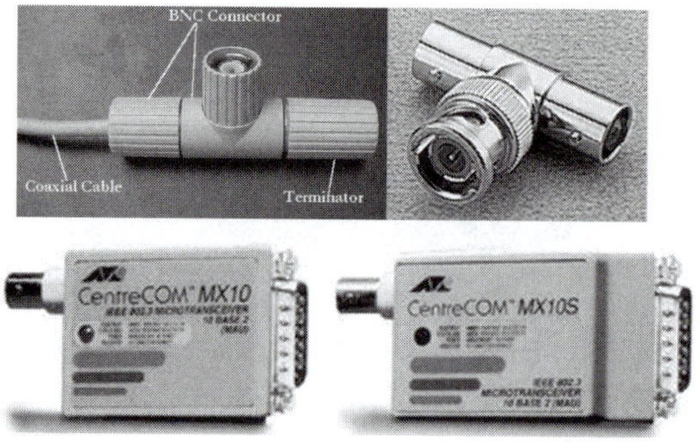

Figure 4.10 A thin Ethernet transceiver with a BNC-T connector.

4.11. This implementation eliminates the need for an AUI cable to connect the transceiver and the NIC.

Finally, the 10BaseT is a star LAN topology utilizing the Ethernet standards. A UTP is used instead of coaxial cable. The data rate of 10 Mbps and a maximum distance of 100 m (from star hub to station) are supported by the standard. Figure 4.12 shows the connection between the station and the hub using UTP cable and RJ-45 connectors (Figure 4.6) that look very much like telephone connectors. The twisted-pair network hub takes care of all networking operations, with a port provided for each station.

4.3.2 Workstations and Network Servers

Besides the LAN transmission components already discussed, the other two most important and end-user-related LAN components are the workstations on which everyday applications and network software are run, and the network server—that provides services to clients. A workstation is a general-purpose computer designed to be used by one person at a time and offering higher performance than normally is found in a personal computer, especially with respect to graphics, processing power, and multitasking ability.

A network server is a computer that provides services for other computers connected to it via a network. The most common example is a file server that has a local disk and services requests from remote clients to read and write files on that disk, often using Sun's Network Filing System (NFS) protocol or Novell NetWare on IBM PCs (Figure 4.13).

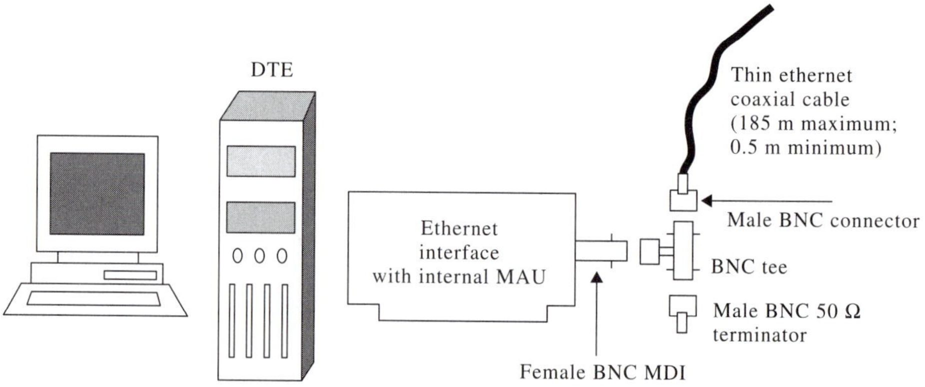

Figure 4.11 Thin Ethernet cable connection.

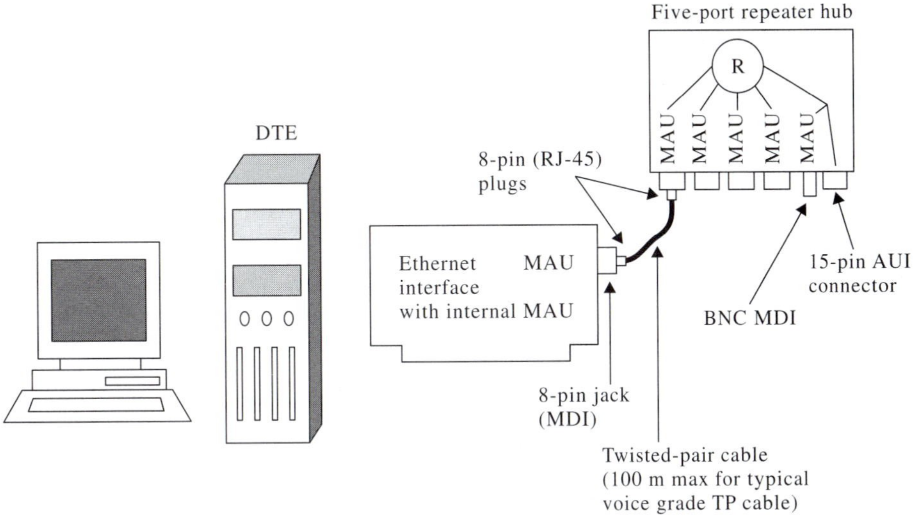

Figure 4.12 Twisted-pair connection.

With the development of technologies and servers, software that formerly could be run only on mainframes may now run on network servers. A typical network server should have at least the following features: systems management software, automatic server recovery, remote maintenance, and predictive diagnostics.

The two types of server, super and midrange servers, support massively parallel processing and generally are limited to uniquely designed processors, specific operating systems and services software, and PC-based servers, which can run general-purpose operating systems and the corresponding services and software.

The super and midrange servers have high performance, large memory storage capacities, tremendous expandability and manageability, support for symmetric multiprocessing, bus architectures designed to maximize concurrency, sophisticated cache designs,

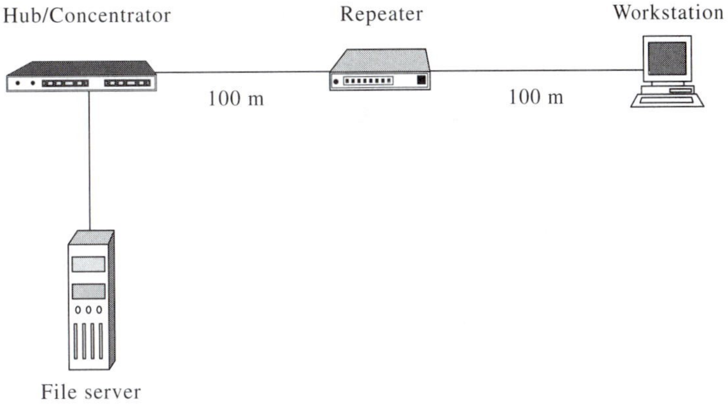

Figure 4.13 File server/workstation configuration in a LAN setup.

and numerous fault-tolerant features designed to provide maximum server availability. However, many of these features can also be achieved at least partially on PC-based servers at less cost.

A network server must provide for error checking and correcting in the memory to maintain the integrity of the data in random access memory and on disk. In addition, the server must utilize components, such as intelligent drive controllers and redundant subsystems, for swapping the data without bringing down the server. We talk about the network storage systems in more detail in the next section.

4.3.3 Network Storage Systems and Hierarchical Storage Management (HSM)

The dramatic price decrease for hard disk storage, and the unprecedented growth of networks and new applications such as multimedia, caused network storage systems to become very popular. The proliferation of data storage options poses challenging problems for network and system administrators. Initially, system administrators believed in supplying more hardware to satisfy user requirements for more disk space. However, recently the trend has been to explore and deploy hierarchical storage management (HSM) solutions for client/server distributed computing environments. A typical implementation from ANL is shown in Figure 4.14.

The HSM system consists of three components.

- The front end, which implements the file system and exports it as an NFS file system to the HSM server
- The ADSM (ADSTAR distributed storage manager) server, which keeps the database of all files and controls the tape system
- The tape system, with tape drives and typically around a terabyte (1 TB) of tape

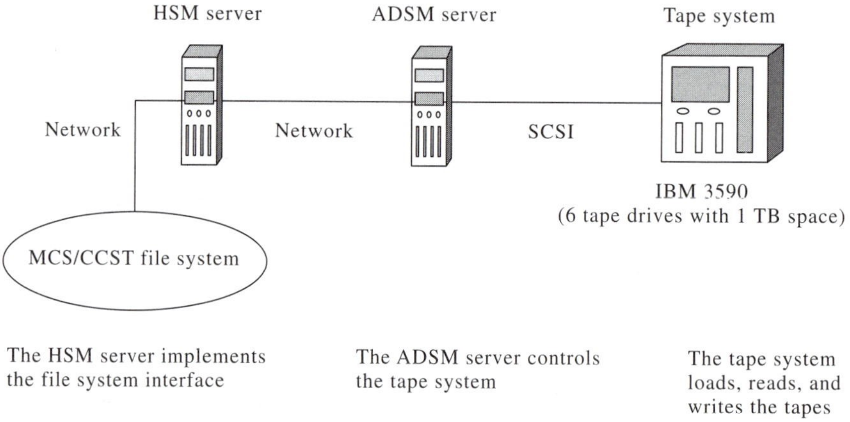

Figure 4.14 Atypical HSM implementation used at Argonne National Laboratories utilizing a tape system and an ADSM server.

Generally speaking, HSM is a storage management strategy whereby data are automatically "migrated" from one storage medium to another based on a set of policies. Migrated files appear to be "local" to the system and, when accessed, they are automatically recalled from the migration store. The system works as follows.

1. The user puts a file into HSM space, which causes it to be stored on the HSM disk.
2. When the file is stored on the HSM disk, it also is copied onto HSM tape.
3. At some point, the HSM disk will hit a "high water mark," meaning that the disk is almost full. The HSM system then removes files from the disk to free up space.
4. The files that were removed can still be referenced via a standard file system.
5. If the user references a file that is on HSM tape but not on the HSM disk, the file is copied onto the HSM disk.
6. Whether a file is on disk or on tape is completely transparent. However, there is a long delay while a file is copied in from tape.

The ADSM server is like an average Unix file system, but files that have not been used recently are moved from the active disk into a tape system. When those files are referenced again, they are copied from tape back on to disk. Effectively, this lets the user archive tape files that would be useful occasionally for reference and gives a place to store very large data sets that would not fit into available space on a home file system.

HSM was originally thought of as a panacea to all storage management problems, and vendors promised dramatic reductions in total cost of ownership. With experience, IS personnel found that they traded in one set of problems for another, and mission-critical data were not being protected or managed effectively. It was also noticed that the dynamics of the mainframe–data center environment are completely different from the distributed client–server environment. Issues such as heterogeneous platforms and operating systems, network bandwidth, and varying tape speeds were not considered before deployment. As a result, users found that the HSM solution touted as a "corporate standard" in reality managed storage on only one of the dozen or so different platforms that typically exist in a distributed environment.

In HSM, data staging is defined as the process whereby the migrated data is moved from one level of storage hierarchy to the next within the migrated store. Staging can be one or many levels. The operation of staging is transparent to the user and is governed by staging policies. Staging policies dictate when the data will be staged from one storage medium to the next one in the hierarchy.

HSM complements backup storage but does not replace it. By strategically deploying HSM, organizations can not only protect the network against accidental data loss but also manage network storage resources more effectively.

One key advantage of deploying the solution from one vendor is satisfying the requirement for "peaceful" coexistence between backup/archiving and HSM. If the solution is not integrated, users might find that backing up a file causes recall or de-migration, or that the migrated store cannot be backed up using the backup product. Conversely, a valid candidate for migration might be ignored if accessed by the backup software.

Next we discuss about some typical network storage systems used in the industry. Network storage is storage for large distributed computing environments. The following properties can assure reliable network operation.

- A network must be highly accessible (i.e., able to survive even major component failures).
- It must support geographically separated clients of diverse architectures.
- It should be deployable and manageable as a single domain, independently from computing resources.
- It should scale over a wide range of capacity, performance, and connectivity.
- It should be secure against intrusion.

4.3.3.1 Magnetic Disk

A magnetic disk is a flat rotating disk covered on one or both sides with magnetizable material. Data are stored in concentric rings called "tracks" and read and written by a disk drive. The two main types are the hard disk and the floppy disk.

4.3.3.2 Optical Disks: Magneto-optical Disks, CD, CD-ROM, WORM

A magneto-optical disk is a plastic or glass disk coated with a compound (often TbFeCo) with special properties. Reading is done by bouncing a low-intensity laser off the disk. Originally the laser was infrared, but frequencies up to blue may be possible; the shorter the wavelength, the higher the possible density. The polarization of the reflected light depends on the polarity of the stored magnetic field. To write, a higher intensity laser is used to heat the coating material to its Curie point, allowing its magnetization to be altered and "frozen" as it cools.

There are several reasons for using CDs with read-only memory: the dropping price of CD-ROM drives and CD-ROM media, the large storage capability, the duration and high quality, and the compatibility of CD-ROM drives and CD-ROM media.

WORM (write once, read many) is the acronym that describes any type of storage medium to which data can be written to only a single time, but can be read from any number of times. Typically this is an optical disk, and information is permanently recorded on its surface by laser etching. WORM media have a significantly longer shelf life than magnetic media and thus are used when data must be preserved for a long time.

4.3.3.3 Tapes

Tapes are slow storage devices with least expense in terms of dollars per megabytes. A newer type of tape device includes the Sony Corporation's Advanced Intelligent Tape (AIT). This technology features a 3.5 in., 8 mm drive that teams with Sony's new metal evaporative media. The first-generation AIT drive features a native capacity of 25 GB with a native transfer rate of 3 MB per second (MBps). Media load time is less than 7 seconds, and the average file access time has been clocked at less then 27 seconds. The only element the drive lacks is compatibility with other similar devices.

Being incompatible with existing 8 mm formats, AIT technology departs from the conventional wisdom in the area of future growth of the technology. Conforming to the 8 mm technology, proponents suggest, would lower capacity and performance, limit recording density increases, and surrender the advantages of the new head and media technologies. As the design currently stands, AIT drives will accept only the special cassettes and class. The abilities to achieve maximum storage efficiency, performance, and reliability are the paramount criteria for tape systems, along with a timely upward migration capability.

The AIT architecture represents a direct challenge to current midrange to high-end tape subsystems. Chief among these is digital linear tape (DLT), the longitudinal recording technology developed by DEC and now offered by Quantum. Quantum recently introduced its newest generation, the DLT 7000, which has shipped evaluation units thus far. The DLT 7000 holds 35 GB per cartridge in native mode and offers a 5 MBps transfer rate with a 200,000-hour mean time between failures. The other key competitor to the AIT architecture is the long-expected Mammoth 8 mm drive from Exabyte. Mammoth provides a 20 GB capacity with a 3 MBps transfer rate.

With the capacity of some databases expected to increase to 800 petabytes (1 PB = 100 quadrillion bytes) in the next few years, storage area networks (SANs) are emerging as another way for organizations to give users fast access to data without taxing the LAN. Uncoupling storage systems from host server bus connections and consolidating storage in high-speed networks at the server "back end" are the key elements of SANs (also known as system area networks), according to supporters of the concept.

Many system analysts believe that separating and consolidating network data storage and their concomitant functions onto a SAN not only eases storage management, but also enhances the performance of mission-critical LAN traffic. Given that the concept is evolving, users of full-blown SANs are still uncommon. But the architecture does have its pioneers.

4.3.3.4 RAID Technology

Redundant arrays of inexpensive (independent) disks (RAID) was originally a project at the computer science department of the University of California at Berkeley, under the direction of Professor Katz, in conjunction with Professors John Ousterhout and David Patterson. It featured a prototype disk array file server having a capacity of 40 GB and a sustained bandwidth of 80 MBps. The server is being interfaced to a 1 Gbps local-area network. A new initiative in the project was to construct a geographically distributed storage system spanning disk arrays and automated libraries of optical disks and tapes.

RAID technology has been developed to address three areas of disk storage:

1. Large capacities
2. Increased input/output performance
3. Reliability through redundancy

RAID uses multiple smaller disks that function as one large drive and provide for data recovery if a single drive fails in most cases. The task of maintaining large amounts of high-availability storage is attained through different approaches, as explained next.

A RAID device can be configured to act like a large, single logical drive. This is usually done with the aid of specialized RAID software or hardware, such as a RAID controller. The RAID mechanism acts as an intermediary between the multiple disk drives and the operating system. Because the RAID mechanism allows simultaneous read/writes to all the drives in the array and sometimes uses memory buffering of I/O requests, an overall increase in the I/O performance for read/write operations is associated with RAID technology. The RAID mechanism can be configured to provide data redundancy through mirroring, which is storing two copies of the original data, or through a scheme that uses "parity drives." Parity striping is a form of partitioning that allows storage of multiple copies of data across multiple drives, thus preventing the failure of one drive from causing data loss.

RAID technology is available in both hardware and software implementations. The reason for deciding on one approach over the other generally boils down to cost vs performance. Hardware RAID can offer special-purpose processors, dedicated circuits for parity generation, and large dedicated caches for high performance. Hardware RAID can be fully redundant, can offer battery backup for RAM caches, and can contain the necessary software packages within the storage subsystems controller firmware. These features, however, can make hardware implementations more expensive than software RAID.

Software-driven RAID systems divide data for striping, generate parity information, and monitor components by running a program on a host processor. This approach is less expensive to implement than hardware but is slower, offers fewer options for redundancy, and can affect the performance of other applications on the system. Some RAID vendors sell both hardware and software products.

The RAID solutions protect against a disk drive failure through a variety of RAID levels that implement parity recovery mechanisms. Spare disk drives are often supported to implement automatic rebuilding of failed drives. Removable disk canisters are typically utilized to allow easy replacement of a failed disk drive. However, other points of failure must also be discussed when one is considering the reliability of a system. For example, redundant load-sharing power supplies, redundant fans, redundant controllers, and dual-host attachments also should be included in the array design.

RAID employs the technique of striping, which involves partitioning each drive's storage space into units ranging from a sector (512 bytes) up to several megabytes. The stripes of all the disks are interleaved and addressed in order. In a single-user system where large records, such as medical or other scientific images, are stored, the stripes are typically set up to be small (about 512 bytes) so that a single record spans all disks and can be accessed quickly by reading all disks at the same time. In a multiuser system, better performance requires establishing a stripe wide enough to hold the typical or maximum size record. This allows overlapped disk I/O across drives.

The most popular RAID levels and uses are as follows.

RAID 0: Data striping with no data protection. This technique has striping but no redundancy of data. It offers the best performance but no fault tolerance.

RAID 1: Data mirroring with totally redundant data. Two or more copies of data are written to two or more different disks at the same time. Data may be read from either disk, based on device availability. Although reliability is high, so is cost, since twice the amount of disk storage must be purchased.

RAID 2: Data striping plus an error correction scheme written to additional drives. Data are striped at the bit level; multiple error-correcting disks provide redundancy. This RAID level is not commercially implemented.

RAID 3: Bit- or byte-level data striping with data parity information stored on a dedicated drive. Data are striped at the byte level, and one drive is set aside for parity information. This approach works well for large files where large blocks size and sequential I/O are used. Since only one drive is used for parity, the cost is less than that of a RAID 1 implementation.

RAID 4: Block- or record-level data striping with data parity information stored on a dedicated drive. Data are striped in blocks, and one drive is set aside for parity information.

RAID 5: Block- or record-level data striping with data parity information distributed on all drives. Data are striped in blocks, and parity information is rotated among all drives in the array.

RAID levels 1, 3, and 5 are the most commonly used. RAID 1 is used most often for environments with high random I/O rates and high levels of write activity. Since RAID 1 is fully redundant and, therefore, expensive, it is usually limited to small-capacity RAID systems or segments of an entire system.

RAID 3 is generally used for environments with large sequential transfers, It is less expensive than RAID 1 because a single parity drive can support several data drives. RAID 3 is frequently used in applications such as video, audio, and document imaging.

RAID 5 is generally used for environments with high random I/O rates and frequent read activity. A RAID 5 system can provide good sequential transfer by buffering reads and writes in cache. Write caches can be used to minimize the performance effects of the multiple drive accesses required for RAID 5 writes.

Effective data management can go a long way toward improving RAID performance where large file transfers are common. By understanding the size of the storage stripe written and the percentage of the time writes will occur on a stripe, users can adjust RAID levels for best results. In the future, RAID subsystems will be able to use a GUI interface to talk to the host directly via a network card. This will allow instant feedback about types of I/O transfer occurring, and each subsystem can adjust accordingly.

With the evolution of RAID, a new technology called adaptive RAID is available. Adaptive RAID combines significant firmware and hardware advances to make the RAID technology faster, easier to use, and network extensible. There are several products on the market. One of them allows the user to define sophisticated operations on preconfigured logical drives. Rather than simply shifting from an optimal to a degraded level, it is possible to convert from a degraded level to a lower optimal level on the fly, thereby improving the performance of a degraded system. Another feature allows the administrator to transparently remove or insert a physical drive from the system at any time. Data are redistributed to the logical drives accordingly. With a firmware-driven system, this feature may be used regardless of operating systems. Another product allows for hot spares, dynamic data migration, and intelligent subsystem monitoring. This new technology also allows automatic conversion between RAID levels and automatic restriping of logical drives without stopping system usage.

4.4 LAN SERVICES AND LAN OPERATING SYSTEMS

4.4.1 Network File System

Network File System (NFS) is a software that mounts remote file systems across homogeneous and heterogeneous operating systems. For example, NFS offers a good solution for a heterogeneous system using two or more of the following: Windows 95/98/NT, Unix, IBM mainframe or midrange computers, or any other different operating system platform.

NFS allows administrators to build network systems in which all the file systems on the network appear to the user as a single local file system. NFS is implemented by means of the Remote Procedure Call (RPC) protocol. All NFS operations are implemented as RPC procedures. RPC is a method of interprocess communication over a network. An RPC operates much like a traditional local procedure call. The fundamental difference is that the calling procedure executes in one machine, and the called procedure executes in another.

One of the reasons for the widespread implementation and commercial success of NFS is that it is an open system standard. This means that the protocol is specified in publicly available documents in sufficient detail to permit implementation from the official description.

NFS consists of a client–server system. An NFS server can export local directories for remote NFS clients to use. NFS redirects all system calls related to file operations to the appropriate system. The client is the redirector that channels the file operation requests to the host system, while the host or server administers the access rights to different files and directories. NFS runs over IP, using UDP (commonly). There are NFS implementations that works using TCP as the network transport service.

The Network File System was developed and implemented by Sun Microsystems. Since its initial release in 1984, Sun has introduced NFS version 2 in 1985, NFS version 3 in 1993, and NFS version 4 in 2000. Although NFS was first implemented under the Unix operating system and many Unix concepts were integrated into the protocol, it remains platform independent. Currently, NFS exists on all major computer and operating system platforms, including VMS, DOS, Windows, Unix, and MVS. NFS version 2 is defined in Internet RFC standard 1094, for file services on TCP/IP networks on the Internet. It is also a component of the X/Open Computing Application Environment (CAE). NFS version 3 and 4 are documented in RFC 1813 and 3010 respectively.

Some of the most important NFS design principles include the following.

Network protocol independence: NFS is built on the top of a network-independent Remote Procedure Call (RPC), which gives NFS flexibility to run on multiple transport protocols.

Portability: Because NFS is machine and operating system independent, it is easily ported to multiple operating system and hardware platforms.

Fast recovery from failure: NFS is designed to recover quickly from system failures and network problems, so that users have minimal disruption of service when failures occur.

Security: The NFS architecture enables the utilization of multiple security solutions, so that appropriate solutions can be chosen for each distributed file-sharing environment.

Performance: NFS is designed for high performance so that users can access remote files as quickly as they can access local files.

The NFS protocol is defined as a set of RFCs, their arguments, results, and effects. The NFS procedures enable clients and servers to communicate with each other. File and directory operations such as read, write, rename, remove, mkdir, and rmdir are part of the list of procedures that allow NFS to manipulate files and directories.

The RPC mechanism enabled the design of NFS as an independent operating system. Every NFS client wishing to connect to the server uses the RPC request that contains arguments intended for that procedure call. The request is encoded into external data representation (XDR) and sent to the transport layer. The TCP/IP or (more likely) User Datagram Protocol/Internet Protocol (UDP/IP) packets actually communicate between client and server. The TCP protocol is better for congested wide-area networks because of its superior retransmit features and congestion control. The UDP protocol is designed for exchanging smaller packets quickly and is best in networks that seldom experience congestion and lost messages. The client selects the transport protocol in the mount command.

There are many advantages of NFS. One major advantage is that files on the remote hosts can be accessed in exactly the same way as local files. Also the users need not know where any file is physically located. Another advantage is the saving of disk space, because data that consume large amounts of space can now be stored into a single server. Unfortunately, NFS has some drawbacks as well. For large networks NFS access to the files can be slow because the system is dependent on TCP/IP or UDP/IP. Another major concern about NFS for a long time has been security. However, special authentication services have been developed that offer improvements in this area.

4.4.2 Network Directory Services

A directory service translates unique network names into unique addresses. It is a naming scheme that applies to all network items such as users, printers and file servers, and its information is distributed throughout the network. To both the end user and the network administrator, a directory allows the network to appear as a single unit instead of as a collection of servers. For the user, logon occurs only once to access the data, applications, and services to which the user is entitled. For the administrator, there is now one central location to manage.

As an example, consider adding a new server for all 3000 users of your WAN. Without a directory service, you would have to add all 3000 user IDs to the new file server.

With directory services, you just hook the server to the WAN and announce its presence. Users log on to a location in the directory tree rather than to a specific file server.

The directory not only simplifies network management but also makes distributed computing possible. As the defining element of distributed computing, directory services are usually the point at which security, applications, and the other network services converge to create a seamlessly distributed computing environment.

Directory services, which have received more attention as organizations have been obliged to deal with the problems caused by building large, complicated, and heterogeneous networks, are becoming a necessary component of LANs for tracking users and resources in a distributed environment. Administration is easier with a directory because it is not necessary to define external aliases for giving users access to multiple servers and applications. Defining interdependent aliases can be a time-consuming task that is confusing to manage. Enterprise computing is not a question of how well applications integrate with the server, but of how well the network is integrated. A set of software services aims at hiding differences between multivendor products in a distributed, client–server environment.

The directory service's function is easing administration for large, multiserver internetworks. One of the most fundamental, yet least understood, components of today's directory services are specialized databases designed to store information about network resources. Ultimately, their purpose is to provide users with transparent access to all network resources and users. Although vendor implementations vary, the X.500 is the only standard for directory services to date.

Novell implements X.500 based on a directory tree. Microsoft's implementation is known as a domain; its basic unit of administration is a flat-file (as opposed to hierarchical) database that generally reorients a single workgroup. Each workgroup, however, is administered independently.

4.4.3 Network Printing and Fax Services

A network printer has different requirements from a stand-alone printer. An ideal modern network printer usually has the following characteristics.

1. A powerful RISC-equipped controller that boosts productivity, provides fast first-page-to-print time, and high throughput for complex documents.
2. Compatibility with a broad line of network operating systems, whether LAN-based or host-based.
3. A secure mailbox option that ensures privacy in a shared environment. Support of an industrial network management standard such as open SNMP for efficient print management is also important.

For many reasons, applications developers often fail to plan and build centralized control and management of the output medium, from printing and faxing to e-mail, paging, voice mail, and WWW servers. A solution would allow any user to output pages to any device on a wide-area network, including managing the process, outputting pages to multiple devices, and even getting notification, feedback, and full-job tracking and control.

Software may provide centralized management of and unified access to output processes and destinations and also may automate the transformation of documents into a variety of formats. It may allow users to send documents to multiple printers, fax machines, and e-mail accounts simply by dragging and dropping icons on the desktop. Users may pause, delete, and resume their print jobs and move them from one queue to another.

Printing on many networks is a confusing mix of operating system print services, third-party utilities, and long print queues. The most common solution for network printing is the external adapter, known as a print server. Print servers hook one or more parallel or serial printers to a network and accept oncoming jobs from a variety of protocols and from a number of users' operating systems.

Actually, printer servers function like scaled-down routers. Incoming print job packets are passed off to appropriate protocol stacks, where they are stripped of their headers and routed to the correct printer, at which point nearly all print servers offer about the same performance. Different vendors' models tend to have unique advantages and disadvantages. Some printer servers have a bidirectional interface to allow properly equipped printers to return status messages to users. Nearly all external print servers include a basic SNMP management program to verify that various printers are running.

4.4.4 Backup

Now that client–server LANs are handling mission-critical applications, businesses realize that having the network down for even a few days can translate into losses. Putting together a disaster recovery strategy for a company network is not for just mainframe systems any more. Backup and recovery mechanisms are finding their way in today's LANs.

One of the key features of enterprise backup products is their ability to set up data protection and tape rotation schemes, and to run backup and restore operations unattended. Most of the products provide an option for customizable rotation schemes as well as predefined grandfather–father–son rotation schemes, in which full backups are performed weekly and monthly, while incremental or differential backups are performed daily.

Many tape backup products have only two modes of writing to tape—append or overwrite. However, there are variations of this method. For example, a product tracks files over a configurable period of time and writes those that do not change to tape in an "archive set." Then it appends incremental backups after the archive set. Archive sets are never overwritten. At each tape set rotation, the incremental data sets are overwritten at the end of the tape with new archive sets, and then writing continues with additional incremental backups to the end of the tape. When a file has been saved in an archive set for a configurable number of times, say three, it is considered to be "safe" and is not backed up again unless it changes.

There are some important issues to consider in network backup and storage management systems. We present here a summary of these issues.

4.4.4.1 Ensuring Data Availability

In a perfect world, data would be available to whoever needs it, 100% of the time. The goal, therefore, of any storage management system, is total and constant data availability. Ensuring data availability begins with identifying the critical data across the network and establishing and enforcing consistent policies and procedures to reliably protect and manage all data across the enterprise.

While this was relatively easy to do in the traditional data center environment, distributed heterogeneous networks have created enormous challenges, forcing IT managers to make difficult decisions and trade-offs between reliability, performance, and cost. The one thing that is clear is that virtually all systems, from corporate desktops to network servers, contain valuable data that would be costly (sometimes impossible) to replace. In nearly all but the best managed environments, distributed corporate data are at risk and need to be subject to the same level of protection and management found in the traditional data center environment, including not only routine backup, but off-site protection, automated archiving, and hierarchical storage management.

4.4.4.2 Cost Containment

In assuming responsibility for enterprise storage management, the first challenge stems from the origins of today's enterprise networks, most of which were established outside IT control and consequently lack standards for hardware, software, or configuration. Because of this, extensive knowledge and expertise is required locally to manage each network, resulting in skyrocketing costs as the networks expand. To achieve economies of scale in management costs, many organizations want to build a standard storage management infrastructure from which they can provide storage management as a service to internal customers. Establishing storage management standards reduces costs associated with operator training, support, and maintenance. In designing the storage management infrastructure, it is important to consider the overall performance requirements for high-volume backup and, most importantly, time-critical disaster recovery. Management of storage growth, and capacity planning through archiving and HSM technologies, must also be considered. The system needs to scale evenly with additional storage capacity and additional sites.

4.4.4.3 Scalable Performance

There are two fundamental storage management models: centralized, where distributed data are pulled to a single location across the network; and distributed, where storage devices and processing are placed as close to the data as possible, often on the same machine, to improve performance and eliminate or reduce network traffic and bandwidth requirements. In a hybrid or multitier architectures, which combine both centralized and distributed models, multiple servers at a site or campus are managed by a small number of medium-scale "storage servers," which in turn can be managed remotely as a group.

A distributed architecture typically provides the best performance, since the data movement can be restricted to high-speed paths such as SCSI, often within a single machine, rather than "pulled" or "pushed" across slower LAN or WAN connections. This is especially critical in disaster recovery scenarios that must come into play when critical operations have been brought to a halt. It is not uncommon for local backup and recovery performance available from a centralized model to be 10 times greater than remote performance. Sometimes, depending much more on the storage hardware in use and network bandwidth availability, a distributed architecture also offers enhanced scalability. Thus, since additional components can be easily added to the system without negatively impacting the workload of existing components, constant performance is ensured.

4.4.4.4 Continuous Reliability

The best design and highest performance matter little if the system fails to meet its objectives when needed most urgently. In examining system reliability, simply look for the weakest link. High-availability computing technologies such as RAID, SFT-III, and server clusters attempt to identify "weak links" and provide redundancy and fault tolerance to eliminate critical failure points.

An enterprise storage management system should be designed in the same manner, to ensure that the failure of any component will not prevent the entire system from operating. This is one of the major drawbacks of a highly centralized architecture and a major strength of a properly designed distributed system. If processing, storage, and management are centralized, physical disasters and network hardware or software failures can impact data availability for the entire enterprise. Aside from distributing processing responsibility, the distributed components should themselves provide fault-tolerant characteristics such as redirection between storage devices, automatic media management, and alert notification.

And so we have distributed, autonomous systems with no easy way to manage them. Doesn't this leave us right where we started? Fortunately, in just the past several years, systems and network management technologies in general have finally started to address the needs of managing today's distributed network environment. These systems have evolved from basic SNMP-type "polling" and queries to threshold-based intelligent agents and probes providing greater scalability and reducing network traffic. In addition, sophisticated event correlation technologies have become available to relate information from a variety of distributed components and elicit the appropriate type of action without operator intervention. This technology is being moved from the central management console out to the devices being managed and the applications themselves. The result is a new class of truly network-aware storage management products that can be easily and efficiently managed from a central location anywhere on the network.

4.4.5 LAN Operating Systems

There is no doubt that connecting computers and other devices together makes the computing environment more powerful, efficient, and cost-effective. At the same time, it is also true that a cluster of networked computers and devices is more complicated than an individual computer. There are many more facilities to be managed in a network than in a stand-alone personal computer. It is more challenging for a system administrator to master the software and data processing in a network than in a centralized computer system. In a network environment, the operating system does more than the basic low-level jobs—typically, scheduling tasks, allocating storage, handling the interface to peripheral hardware, and presenting a default interface to the user.

In general, a network operating system (NOS) is an operating system that includes software to communicate with other computers via a network. There are two major LAN NOS design paradigms: peer-to-peer design and client–server design. In peer-to-peer network architecture, any computer may make a request of any other computer. There is no dedicated file server, nor is any machine dedicated to only one task. Peer-to-peer is usually considered to be suitable for wide-area networks only. The largest network in the world, the Internet, works with the peer-to-peer method for internetworking. In the client–server network architecture, client machines make requests of another machine called the server.

LAN operating systems from different vendors have their own architectures associated with various services and applications. No LAN operating system can be the best fit for all users. For instance, NetWare is given very high marks as a file and print server, while NT is ranked higher as an applications server. The product choice depends on the type of user requirements. We now present a brief overview and comparison of several popular LAN operating systems.

4.4.5.1 NetWare

NetWare is Novell's proprietary networking operating system for the PC machines. NetWare uses the IPX/SPX, NetBEUI, or TCP/IP network protocols. It supports MS-DOS, Microsoft Windows, OS/2, Macintosh, and Unix clients. NetWare for Unix lets users access Unix hosts. NetWare 2.2 was a 16-bit operating system, later versions are 32-bit operating systems. Netware 6 (3Q,2001) uses latest version of Novell's file system called Novell Storage Services (NSS3.0). It uses a 64 bit interface to support large files and volumes (up to 8 TB each).

IPX/SPX was developed by Novell and is the native protocol for NetWare in LANs. While TCP/IP is adopted and supported as a LAN protocol by most other companies and software, the necessity for Novell to support IPX/SPX is important.

Novell provides four ways to bring NetWare into the TCP/IP world:

1. A NetWare file server acts like a Unix server. Running NetWare's Network File System (NFS) lets Unix workstations hook up to a NetWare server and access the files on a server's hard drive.

2. A NetWare file server works as a router for any IP network. A NetWare server will behave like an IP router if configured by users. The server can use two network interface cards to route between multiple IP networks or between a NetWare network using the IPX protocol and a Unix network running TCP/IP.

3. NetWare LAN WorkPlace and LAN WorkGroup allow a DOS workstation to talk to a Unix server. LAN WorkPlace allows users to send data across an IP network connection to a distant NetWare server.

4. NetWare can pass NetWare packets across an IP network with the IP tunnel feature. Using the IP tunnel feature, a NetWare server can communicate with a second NetWare server across an IP backbone. The tunnel feature sends NetWare IPX packets across the IP backbone embedded in IP packets.

Novell has also introduced NetWare/IP to integrate the TCP/IP. NetWare/IP can serve as a gateway between NetWare file servers without using the basic IPX protocol, as before. Unlike IP tunneling, the traditional NetWare method of routing IPX packets over TCP/IP, NetWare/IP in its native form runs over TCP/IP between servers and clients.

Besides running NetWare over TCP/IP, NetWare/IP can join two or more IP and IPX servers. NetWare/IP also includes many of the TCP/IP transport features found in Novell's LAN WorkPlace product, such as WinSock support, duplicate IP-address media (Ethernet, token ring, FDDI, and ARCnet) via Open Data Link Interface drivers.

Originally designed for file and print services, NetWare included such features as high performance, easy administration, inexpensive and efficient hardware, and interoperability. For achieving these goals, NetWare has minimal software interaction between applications and hardware. It uses the unprotected ring 0 portion of memory for maximum performance and relies on well-behaved NLMs (NetWare Loadable Modules) to prevent system crashes.

One feature that makes NetWare outstanding is its network directory services (NDS). In addition to providing file and print services to large numbers of users, NDS, which is based on a small, fast kernel, guarantees speed. Novell has added another design goal to its new version, namely, easier administration. NetWare provides true robust and scalable enterprise directory services for enhanced administration capabilities. The network structure is based on a set of X.500 conventions. But NDS still lacks the capability to change the directory structure once it has been defined. The new version of NetWare also provides several administration tools for configuring and monitoring the network. Also, it continues to support RAID levels 0 and 1 in both mirrored and duplex configurations.

NDS allows for a gradual transition to a tree, allowing IS to make a big server change without disturbing clients. IntranetWare has a greatly improved installation program, which is capable of detecting installed hardware. Although it is not as friendly as some GUI-powered programs, it is just as easy to install as Windows NT.

NDS is a true enterprise directory service, with a replicated, distributed, hierarchical architecture independent of any single server—and, thus, with no single point of failure.

With the current development, many application software programs run in networking environment. For example, most complex database applications work across networks, running with special versions of the database engine on various platforms. Because of the tight integration that database vendors have achieved with it, NetWare can provide the database engines with performance that most other NOS cannot provide.

NetWare products come with both a DOS-based and a Windows-based administrative program to manage the NDS database. With these programs, administrators can easily add, edit, and delete NDS objects from the directory tree. The NDS database can be split into replicas and placed on different servers. Any section of the database can be replicated to provide layers of redundancy. If a server that contains part of the NDS database goes down, a replica of that section on another server can take over until the first server comes back online. This partitioning can either be automatic or explicitly controlled.

Novell provides four types of replicated partition: master, read/write, read-only, and subordinate reference. There can be only one master partition replica for each partition in an organization; these are the main partitions and typically the first ones referenced. Read/write replicas are similar to masters, but there can be many of them for each partition. Read-only replicas are self-explanatory, and subordinate reference replicas are created either automatically by NetWare or explicitly by administrators for referencing replicas located on other servers. All these replicas are transparent to the users.

One of the limitations of NetWare is that it can run only on the Intel or Intel-compatible processors, while other network operating systems, like NT, can run on Intel, DEC Alpha, and MIPS PowerPC processors.

The whole process of setting up the NDS database is rather cumbersome in comparison to the experience with StreetTalk. Banyan's **V**irtual **N**etworking **S**oftware (VINES) is easier in that it has only three levels, but this also limits an administrator's ability to model the organization's structure. The added complexity of NDS is the price the user pays for more flexibility. NDS is extremely extensible. Everything under NDS—including users, organizational units, and organizational roles—is considered to be an object, and each object has various attributes.

The graphical installation utility in NetWare 5 has made the installation somewhat easier. To make installation even easier, the CD that holds the operating system is bootable and works with most CD-ROM drives. The entire installation is Java-based and requires only about 30 minutes to complete, compared with the hour or more needed to install earlier versions. The hard disk requirement is about 35 MB from about 15MB required for version 4.11.

4.4.5.2 Windows NT

Windows NT, or NT, is Microsoft's 32-bit multitasking operating system developed from what was intended to be OS/2 3.0 before Microsoft and IBM ceased joint development of OS/2. NT was designed for high-end workstations, servers, and corporate networks. Unlike Windows 3.1, which was a graphical environment that ran on top of MS-DOS, Windows NT is a complete operating system, no longer based on MS-DOS. It has true multithreading, built-in networking, security, and memory protection.

NT is based on a microkernel—that is, a modern object-oriented kernel with 32-bit addressing for up to 4 GB of RAM. Its virtualized hardware access fully protects applications. It has installable, high-performance, robust file systems (e.g., FAT, HPFS, NTFS) with built-in networking and multiprocessor support and full UNICODE support. NT is also designed to be hardware independent. Once the machine-specific part, the hardware abstraction layer (HAL), has been ported to a particular machine, the rest of the operating system should theoretically compile without alteration. Older versions of NT needed at least a 386 processor or equivalent, at least 12 MB of RAM (preferably 16 MB) and at least 75 MB of free disk space.

NT is a plug-and-play OS. The core OS components are the kernel, the user, and the graphics device interface (GDI). The 32-bit NT kernel provides many advanced OS services, including multiple threads of execution, thread scheduling and object contention synchronization, memory management, memory-mapped file I/O, and a multitude of high-performance protected-mode 32-bit preemptive I/O subsystems.

On top of the core 32-bit kernel, NT has several very high performance modular and installable file systems. To compensate for the resource limitations associated with the Windows environment, NT introduced a series of compromises intended to offer improvements. When a 32-bit application terminates for any reason, NT identifies it and immediately releases all system resources that had been created by that application.

NT Server also provides an automated IP address pooling service that automatically allocates IP addresses to clients without user intervention, regardless of the user's location on the network. NT Server's network administrative scheme relies on its domain directory services. To simplify the maintenance of separate user accounts on different servers, users can establish trust relationships between domain directories to allow them to have a single network-wide login.

But NT's domain service considered to be less powerful than Novell's Network Directory Services, discussed in Section 4.4.5.1. Windows NT uses ring 3 memory to load applications. A thread running in ring 3 can be terminated without bringing down the entire system, but performance might be affected.

Because NT supports extremely large numbers of applications through its OLE (object linking and embedding) protocol, as well as because of the ease of use of its familiar GUI and greatly improved performance, it is more attractive than other NOSes. With the built-in support of TCP/IP, a mature and most effectively implemented network protocol, and the built-in NetWare OLE called distributed component object module (DCOM), NT is becoming a prominent NOS for the Internet applications.

MS-NT server directory service is a domain-based naming service. The directory, which supports only the NT Server platform, organizes servers and users into domains for the purpose of management. It does not support enterprise-wide logical naming. Moreover, it relies on trust relationships between domains for interdomain operation, which increases the amount of administration required. The domain model, however, is advantageous in decentralized organizations, where individual business units want management autonomy yet enterprise-wide integration.

Since NT 3.1 was released in 1993, Microsoft has steadily improved file, print, and application services and has continually increased its installed base. The domain architecture that NT is based on makes the system difficult to administer over multiple sites. However, with Windows NT Server 4.0, Microsoft included a very impressive array of tools and applications to create a comprehensive Internet solution. By including Internet Information Serve (IIS), NT owners could serve up HTTP, FTP, and Gopher all from one easy-to-use package. Furthermore, NT came with FrontPage to create and manage Internet and intranet sites.

Microsoft has put a lot of effort into making NT an attractive choice for current NetWare users. However, migrating NetWare servers to NT remains frustrating. On one hand, Microsoft provides an array of tools to help the task. File and Print for NetWare (FPNW) provides support for existing NetWare clients. On the other hand, these tools do not always work right the first time.

Now we discuss some of the major disadvantages in NT 4.0. Primarily, NT's current domain-based security model is a nightmare. First, it is not hierarchical; that is, there is no easy way to search and grasp the relationships between domains. The second major problem is the lack of an object model. Users, printers, databases, and so on are just objects with different attributes. This fact precludes any kind of central management and makes extensibility very difficult.

The third problem with NT 4.0's directory services, interoperability, is not isolated to NT, but also appears in Novell Directory Services (NDS) and Solaris's NIS+. Because not one directory service captured the market during the past decade, no directory services are designed to work with the others.

Microsoft's Windows 2000, originally named NT 5.0 provides a stronger infrastructure for building distributed applications and solve some problems for current Microsoft customers. Microsoft, however, must still provide adequate functionality for customers to proceed with an upgrade or to migrate from other platforms, such as Novell's NetWare. Many tests show quite a bit of improvement. Following is the summary of some of the features of Windows 2000.

The server has been improved to provide support for ATM and the layer 2 Tunneling Protocol, and various quality-of-service specifications, such as the **Res**ource Reservation Protocol, or RSVP. Microsoft has also implemented its Routing and Remote Access Server (RRAS) add-on to the product.

One key add-on to the RRAS feature set is Network Address Translation (NAT), which eases the burden of writing distributed applications. However, core services such as the Active Directory may still need a few improvements.

Perhaps the most important feature of Windows 2000 is the Active Directory. However, Active Directory still relies heavily on the Windows NT domain architecture, which introduces some limitations. In particular, since each directory object still requires a globally unique name, Active Directory's implementation of hierarchical organizational units appears to lack hierarchy.

This "flat" directory implementation is also evident in the user list dialogues, which are used to perform Active Directory administration. With Active Directory there are now three different group types—Domain local, Global, and JScript (Microsoft's implementation of JavaScript) in user login scripts. Contrary to some reports, however, login scripts are not integrated into the Active Directory; rather, they are still implemented on the file systems and require additional replication management to perform adequately. It has been also reported that if two administrators update the same directory object on different domain controllers within the space of a synchronization interval, the second administrator's changes will be applied and the first administrator's changes will be overwritten. This is true even if both administrators change different attributes of the same directory object, such as a user or group object.

Microsoft has announced Windows XP as successor of Windows 2000 and the Windows 9X lines of operating systems. It has been claimed the most reliable NOS yet, having a restore back feature without losing the current work. It also has the capability to support remote technical assistance and an Internet connection firewall. To control the problems caused by non-standard drivers, Microsoft plans to provide signatures for third party approved drivers. In addition, Microsoft has also announced a suite of .net enterprise servers capable of building, deploying, and managing Web based solutions and services. The enterprise server consists of several servers such as application center, biz talk, commerce, exchange, mobile information and SQL etc.

4.4.5.3 IBM LAN Server and OS/2

OS/2, which was jointly developed by IBM and Microsoft, is a single-user multitasking operating system that works with up to 16 MB of memory and can run several existing MS-DOS applications one at a time. The older version of OS/2 triggers a special "virtual mode" allowing it to run multiple MS-DOS applications simultaneously. OS/2 is strong on connectivity and the provision of robust virtual machines. It can support Microsoft Windows programs in addition to its own native applications. It also supports the Presentation Manager graphical user interface. OS/2 supports hybrid multiprocessing (HMP), which provides some elements of symmetric multiprocessing (SMP), using add-on IBM software called MP/2.

The system takes advantage of the object-oriented capabilities of OS/2 and OS/2 WARP to provide a strong new set of graphical administration tools. For moderate file and print service loads, LAN Server has revved its performance considerably. Its support of the new symmetric multiprocessing version of OS/2 2.11 allows it to handle CPU-intensive applications better than ever, assuming the user has a multiprocessor system.

IBM has enhanced LAN Server's enterprise-wide networking capabilities by including an improved TCP/IP protocol stack (easily configurable as the default) and aliasing features that allow easier management of resources across workgroups and domains.

Despite OS/2 Warp's forward-looking object foundation, its current usability seems both backward and bothersome. With Windows 98/2000 as an alternative, few users put up with OS/2 Warp's massive learning curve long enough to learn to like it, much less love it. Consequently, it is not regarded as a serious competitor to the Windows and NetWare systems. However, OS/2 WARP still lives up to its reputation for multitasking rings around the competition.

4.4.5.4 Plan 9

Plan 9 is an operating system developed at Bell Labs by many researchers who had been intimately involved with Unix. Plan 9 is superficially Unix-like but features far finer control over the name space on a per-process basis and is inherently distributed and scalable.

Plan 9 is divided according to service functions: CPU servers concentrate computing power into large multiprocessors, file servers provide repositories for storage, and terminals give each user of the system a dedicated computer with bit map screen and mouse on which to run a window system. The sharing of computing and file storage services provides a sense of community for a group of programmers, amortizes costs, and centralizes and hence simplifies management and administration. The pieces communicate by a single protocol, built above a reliable data transport layer offered by an appropriate network, which defines each service as a rooted tree of files. Even for services not usually considered to be files, the unified design permits some simplification. Each process has a local file name space that contains attachments to all services the process is using and thereby to the files in those services. One of the most important jobs of a terminal is to support its user's customized view of the entire system as represented by the services visible in the name space.

4.4.5.5 Comparing Network Operating Systems

With its familiar Windows interface, NT is easier than NetWare to install, configure, and manage. Also, because it is nondedicated, you can use it to run applications, an excellent option for a workgroup server. NT's code, however, is not as well optimized and battle-tested as NetWare's. This performance hit is somewhat offset by NT's scalability. NT runs on a variety of processors and supports multiple processors out of the box.

NetWare provides the power necessary to run an organization's corporate services and applications and provide them to the entire Internet. Through NDS, NetWare offers a solid way to control these services within the enterprise. But NDS also prohibits smaller workgroups or departments from having their own servers. For raw throughput and a highly reliable file system, nothing beats NetWare. Management is handled from a client and includes a Windows utility for administering users.

On a medium-sized network, NT Directory Services can prove simpler to manage than Netware's NDS. But if the network has more than a handful of servers and is geographically dispersed, NDS is certainly the better choice. Also, NDS is more scalable than NT, letting users add as they grow.

When it comes to fault tolerance, the capability of a NOS to gracefully handle inevitable software and hardware errors can save data, time, and money. Redundant disks, adapters, and servers can provide the level of protection needed in the event of a system failure. Although NetWare and NT have many of the same fault tolerance features, those implemented in NT do not quite measure up to their NetWare counterparts. For the file system, NetWare offers protections against single-disk faults (hot fix and mirroring), single-disk channel faults (duplexing), and single-server faults (server mirroring). NT-based server redundancy solutions, however, provide adequate protection.

With Windows NT as a workgroup server, a department has more control over its environment, and NT requires less technical prowess to administer. Unlike NetWare, it is easy to bring up or down several NT servers without disrupting the rest of the network.

Another LAN operating system called LANtastic, by Artisoft, Inc., has been gaining popularity recently. The company claims that LANtastic is the best NOS available for businesses with a mix of PC operating systems. This operating system allows sharing of files, printers, CD-ROMs, and applications among the PCs that run any combination of Windows or DOS operating systems.

Finally, we turn our attention to Sun's Solaris NOS. The Solaris's advantages are its stability and scalability in terms of processors, users, and clients on a network. The very same copy of Solaris can run, and stay running, on a single-processor desktop machine and on a mainframe-bashing server equipped with 64 processors that handles a database measured in terabytes and controls huge network traffic. Its weakness is its difficulty of use (because of the somewhat difficult-to-use Unix operating system).

Irrespective of the choice of an NOS, the smooth transition from an older version of NOS to a modern NOS requires three things. First, user accounts must be transferred intact. Second, file system data, including permissions and file attributes, must be moved across. Finally, existing client software must be manipulated in such a way that back-end transition occurs during a weekend, without inconveniencing hundreds or thousands of clients.

4.5 CHAPTER SUMMARY

This chapter deals with LAN issues such as LAN protocols, access schemes, cabling, server issues, and operating systems. We have discussed and compared popular types of LAN including IEEE 802.3 Ethernet, token bus and token ring. Various protocol issues, including media access control, were discussed in connection with the IEEE logical link control (LLC) model. The LAN protocols discused use the OSI model to deal with the lower two layers.

Storage and server issues discussed in this chapter dealt particularly with RAID systems. It was shown how the RAID systems could allow a development of reliable LAN storage. Other systems discussed are NFS and NDS. In addition, active directory implementation aspects from the NetWare and Windows NT perspectives were discussed, and network operating systems were compared.

In comparison to Chapter 3, the emphasis here is on implementation practices of networking. This focus will also be obvious in Chapter 7, which deals with different switching mechanisms, but only after Chapters 5 and 6 have clarified some routing and WAN issues (network and transport layers of OSI).

4.6 PROBLEMS

4.1 The standard 10 Mbps Ethernet requires a minimum frame size of 64 bytes (46-byte payload) and a maximum frame size of 1518 bytes (1500-byte payload).

(a) Explain why we need to have both a minimum and a maximum size for Ethernet frames.

(b) Explain why we pick these two particular sizes. In your calculation, use the following parameters: each Ethernet cable has a maximum length of 500 m; there are at most four repeaters for each Ethernet; the signal travels at the speed of 200,000 km/s in the cable.

4.2 Using the delay–bandwidth product, find out how much data can be present on an Ethernet segment at one time. Note that the Ethernet segment is limited to 100 m at 10 Mbps. Assume a signal propagation speed of 200,000 km/s.

4.3 Identify each part (marked with "?") in the Ethernet setup of Figure P4.3.

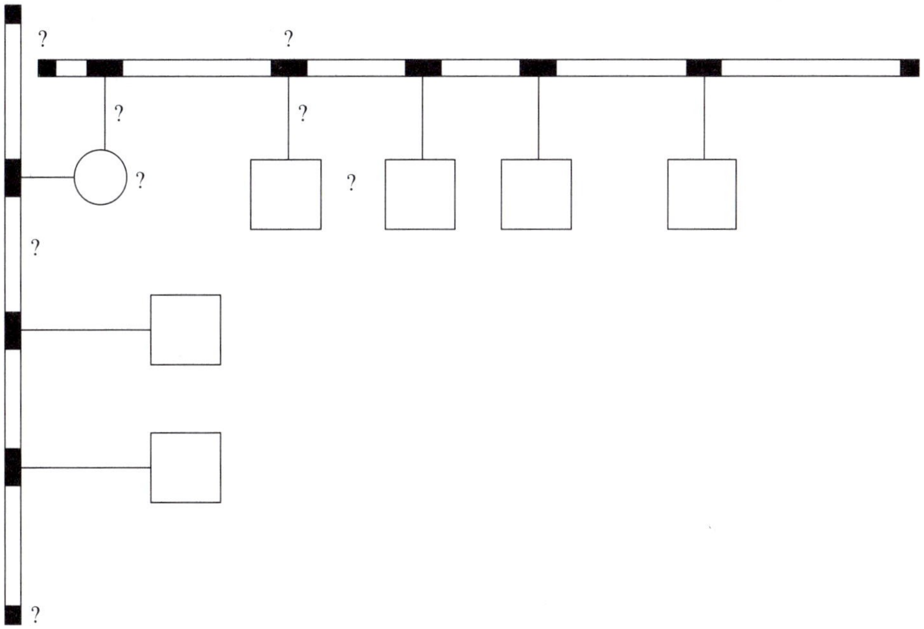

Figure P4.3

4.4 In an 802.3 frame structure using the 10 Mbps baseband standard, if the data portion is 1000 bytes, what would be the length of the pad portion? What percentage of the frame is used for data in this case? Why is the minimum frame size set to 64 bytes?

4.5 You are assigned to install network wire for a seven-story building. Each floor has the same floor plan, with 15 rooms. Each room contains a socket to connect the computers with a separation of 4 m, between the sockets, both horizontally and vertically, between the floors. How much cable would be needed for the following topologies, assuming that you can run cables between any pair of sockets, horizontally, vertically or diagonally?
(a) Star with a single router in the middle
(b) 802.3 bus
(c) Ring without a wire center

4.6 What happens in a token bus if a station accepts the token and then crashes immediately?

4.7 A 4 Mbps token ring has a token holding timer value of 10 ms. What is the longest frame that can be sent on this ring?

4.8 Consider building a CSMA/CD network running at 1 Gbps over a 1 km cable with no repeaters. What should be the minimum frame size?

4.9 Assume that a station in a token bus system transmits five priority 6 frames, three priority 4 frames, two priority 2 frames, and a single priority 0 frame in a round of token passing. If the bus is operating at 100 Mbps and the data transferred for each priority class is 1000 bits, what is the percentage of bandwidth used by each priority class?

4.10 At a transmission rate of 5 Mbps and a propagation speed of 200 m/μs, to how many meters of cable is the 1-bit delay in a token ring interface equivalent?

4.11 In a 802.5 token ring, the sender removes its frame from the ring after the transmission is complete. What modifications would be necessary in the standard to have the receiver remove the frame instead? What would the consequences of this change be?

4.12 Assuming that the order of token passing is 1-6-2-5-3-4-1-6-2 . . . in a six-station 10 Mbps token bus system with end-to-end propagation delay of 100 μs, what is the fraction of utilized bandwidth for one round of token passing if each station transmits 1000 bits of data with 100 bits of overhead? What percentage of bandwidth is wasted owing to the order of token passing? Predict the effect on utilized bandwidth of changing the token passing order to 1-2-3-4-5-6-1-2 . . . .

4.7 REFERENCES

BOOKS

Braginski, L., and M. Powell, *Running Microsoft Internet Information Server 4.0*. Redmond, WA: Microsoft Press, 1998.

Cady, D., *Administering NetWare 5*. New York: McGraw-Hill, 1999.

Carlo, J. T., M. Siegel, et al., *Understanding Token Ring Protocols and Standards*. Norwood, MA: Artech House, 1998.

Doering, D., and T. Simpton, *A Guide to Novell Netware 5: Network Administration*. Boston, MA: International Thomson Publishing, 1999.

Frisch, A., *Essential Windows NT System Administration*. Sebastapol, CA: O'Reilly & Associates, 1998.

Hancock, B., *Advanced Ethernet/802.3 Network Management and Performance*. 2nd ed. Digital Press, 1995.

Held, G., *Token-Ring Networks: Characteristics, Operation, Construction and Management*. New York: John Wiley, 1994.

Held, G., *Ethernet Networks: Design, Implementation, Operation, and Management*, 2nd ed. New York: Wiley, 1996.

Held, G., *Ethernet Networks*, 3rd ed. New York: Wiley, 1998.

Hughes, J. F., and B. W. Thomas, *NDS for NT: Installing, Configuring, and Designing Novell Directory Services*. Foster City, CA: IDG Books Worldwide, 1998.

IBM International Technical Support Organization (IBM Redbooks). *Inside OS/2 LAN Server 4.0*. Armonk, NY: IBM Corporation, 1995.

IBM International Technical Support Organization (IBM Redbooks), *Using ADSM Hierarchical Storage Management*. Armonk, NY: IBM Corporation, 1996.

IBM International Technical Support Organization (IBM Redbooks), *Configuring and Implementing the IBM Fibre Channel RAID Storage Server*. Armonk, NY: IBM Corporation, 1999.

Institute of Electrical and Electronics Engineers, *IEEE Information Processing Systems Local Area Networks*, Part Four: *Token Passing Bus Access Method and Physical Layer Specifications*. Piscataway, NJ: IEEE Press, 1990.

Kaplan, A., and M. S. Nielsen, *NT 5: The Next Revolution*. Scottsdale, AZ: Coriolis Group, 1998.

King, R. R., *Mastering Active Directory*. Alameda, CA: Sybex, 1999.

Kuo, P., and J. Henderson, *Novell's Guide to Troubleshooting NDS*. Foster City, CA: IDG Books Worldwide, 1999.

Nemzow, M., *Fast Ethernet Implementation and Migration Solutions*. New York: McGraw-Hill, 1997.

Northrup, A., *Introducing Microsoft Windows 2000 Server*. Redmond, WA: Microsoft Press, 1999.

Plas, J., *Deploying Windows NT 4 in the Enterprise*. Indianapolis: SAMS Publishing, 1997.

Quinn, L. B., and R. G. Russell, *Fast Ethernet*. New York: Wiley, 1997.

Quinn-Andry, T., and K. Haller, *Designing Campus Networks*. New York: Macmillan, 1998.

Raid Advisory Board, *The Raidbook*, 7th ed. Scottsdale, AZ: Coriolis Group, 1999.

Ramos, E., A. Schroeder, and A. Beheler, *Data Communications and Networking Fundamentals Using Novell NetWare (3.12)*. Englewood Cliffs, NJ: Prentice Hall, 1995.

Reed, A., *Implementing Directory Services: Microsoft Active Directory, Novell NDS, and Cisco/Microsoft Directory-Enabled Networks*. New York: McGraw-Hill, 1999.

Riley, S., and R. A. Breyer, *PC Week: Switched and Fast Ethernet*," 2nd ed. Emeryville, CA: Ziff Davis Press, 1996.

Siever, E., *Linux in a Nutshell*, 2nd ed. Sebastapol, CA: O'Reilly & Associates, 1999.

Spinney, B. L., *Ethernet Tips and Techniques*. 3rd ed. Englewood Cliffs, NJ: Prentice Hall, 1997.

Spurgeon, C., *Ethernet Configuration Guidelines*. Menlo Park, CA: Peer-to-Peer Communications, 1996.

Spurgeon, C., *Practical Networking with Ethernet*. Scottsdale, AZ: Coriolis Group, 1997.

Stern, H., and M. Loukides, eds., *Managing NFS and NIS*, Sebastapol, CA: O'Reilly & Associates, 1991.

Stokes, N., A. Buecker, J. Friedrichs, and V. Moroian, *Getting to Know OS/2 Warp 4*. Englewood Cliffs, NJ: Prentice Hall, 1996.

Trulove, J., *LAN Wiring: An Illustrated Guide to Network Cabling*. New York: McGraw-Hill, 1997.

Welsh, M., and L. Kaufman, *Running Linux*. Sebastapol, CA: O'Reilly & Associates, 1995.

ARTICLES

Gray, J. N., et al., "Parity Striping of Disk Arrays: Low Cost Reliable Storage With Acceptable Throughput," 16th International Conference on VLDB, Australia, August 1990.

Katz, R. H., D. A. Patterson, and G. A. Gibson, "Disk System Architectures for High Performance Computing" *IEEE Proceedings* 78, no. 2 (February 1990).

Ng, S. W., "Improving Disk Performance via Latency Reduction," *IEEE Transactions on Computers* 40, no. 1 (January 1991), 22–30.

WORLD WIDE WEB SITES

Charles Spurgeon's Ethernet Web Site provides extensive information about Ethernet (IEEE 802.3) local area network (LAN) technology. Including the original 10 megabit per second (Mbps) system, the 100 Mbps Fast Ethernet system (802.3u), and the gigabit Ethernet system (802.3z)
http://www.ots.utexas.edu/ethernet/

Raid Technology information from Indiana University
http://www.uwsg.indiana.edu/usail/peripherals/disks/raid/index.html

Illustrated overview of RAID by Advanced Computer & Network Corporation
http://www.acnc.com/raid.html

Intelligent storage solutions by Storage Computer offers a trademarked RAID 7 technology that it says offers complete fault tolerance at the same performance of single-disk storage
http://www.storage.com/

Storage area network solutions by Clarion
http://www.clariion.com/

Operating systems links of Yahoo
http://dir.yahoo.com/Computers_and_Internet/Software/Operating_Systems/

Novell's NetWare web site
http://www.novell.com/netware5/

Microsoft's NT-Server web site
http://www.microsoft.com/ntserver/

5

TCP/IP AND THE INTERNET

As we move from LANs to WANs, many new issues arise. A wide-area-network consists of computers that are connected like a network graph with many alternate paths between source and destination. The access methodology is no longer limited to the physical and data link layers. The network for a computer may be represented by a cloud. Which path should the traffic take through such a shifting of network environment? Path determination occurs at the OSI network layer with the help of routers. The path determination function enables a router to evaluate the available paths to a destination and to establish the preferred handling of a packet.

Routing services use network topology information when evaluating network paths. This information can be configured by the network administrator or collected through dynamic processes running in the network. The network layer interfaces to networks must provide best-effort end-to-end packet delivery services to its user, the transport layer. The network layer sends packets from the source network to the destination network based on routing tables.

After the router has determined which path to use, it can proceed with switching the packet: taking the packet it accepted on one interface and forwarding it to another interface or port that reflects the best path to the packet's destination.

Internetworking functions of the network layer include network addressing and best path selection for traffic. Network addressing uses one part to identify the path used by the router and one part for ports or devices on the net. The user traffic is forwarded between the nodes by means of routing tables that are obtained and updated by the routing protocol programs running at the routers. Two methods of routing involve distance vector and link state evaluations. Network discovery for distance vector involves the exchange of routing tables between the neighboring nodes. For link state, routers calculate the shortest paths to other routers with a broadcast of routing information to allow the updates of other routing tables.

An internet (with lowercase i) is a network of heterogeneous networks (network clouds) connected via internetworking devices such as routers and gateways. The Internet (with uppercase I) is the name of the worldwide network that connects millions of computers as small subnetworks. Every subsidiary net observes a minimal set of protocols to allow data to travel smoothly among computers attached to the Internet even though the heterogeneous networks may have different architectures, run different operating systems, use different data formats, and run with different speeds. These conditions call for a careful design that will allow connection of computers at the OSI application layer. However, the important tasks of data routing and providing reliable and error-free end-to-end connectivity are addressed at the network and transport layers, the heart of internetworking technology.

In this chapter, we start with a general overview of Internet architecture and discuss the Internet addressing and routing protocols. Next, we discuss two well-known Internet transmission protocols, the User Datagram Protocol and the Transmission Control Protocol. We also discuss the routing issues as related to internetworking and the Internet. We discuss the standard Internet services such as ftp and smtp, in addition to popular applications such as gopher, wais, and www. We also discuss the domain name system and directory services that form the foundation of the Internet.

5.1 INTERNET ARCHITECTURE

5.1.1 Internet Addresses

There are millions of computers connected to the Internet. To allow each host to communicate with others connected to the network, an identifier is necessary. This identifier is also referred to as the Internet address of the host. The most common Internet protocol, called TCP/IP, assigns to each host a 32-bit integer address called the Internet address or IP address, which is different from a host's physical address. The Internet address is chosen carefully to make routing efficient. An IP address encodes the identification of the network to which a host attaches as well as the identification of a unique host on that network. The bits of IP addresses for all hosts on a given network share a common prefix.

Conceptually, each address is a pair (netid and hostid), where *netid* identifies a network, and *hostid* identifies a host on that network. There are three types of Internet address, class A, class B, and class C, as shown in Figure 5.1. Class D addresses are used for multicasting and class E is reserved for future use.

Class A addresses, which are used for only large networks ($2^7 = 128$ networks), have about 2^{24} hosts per network: 7 bits for netid and 24 bits for hostid. Class B addresses, which are used for intermediate-sized networks having between 2^8 (256) and 2^{16} hosts, allocate 14 bits to the netid and 16 bits to the hostid. Class C addresses have only 2^8 hosts, allocating 21 bits for netid and only 8 bits for hostid. Class D addresses have 28 bits for multicasting.

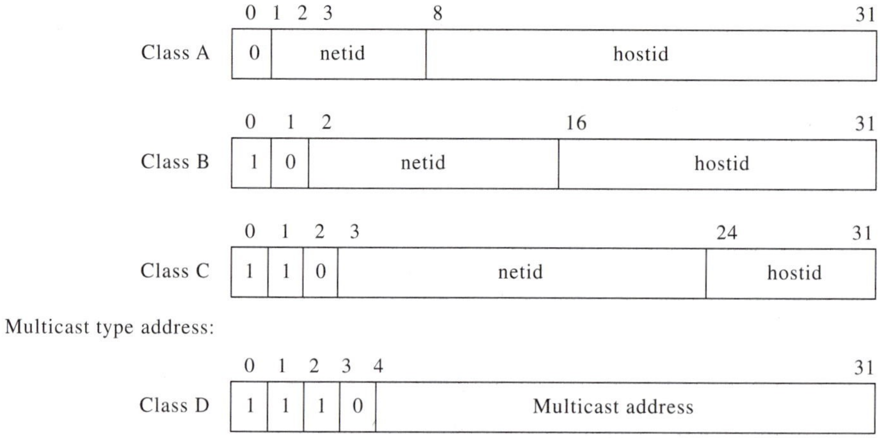

Figure 5.1 Internet address classes. Class E with the left-most 4 bits as 1 and the fifth bit as 0, is reserved for future use.

5.1.2 Gateway Addressing

The computers that are connected internally to a localized network as well as to an intermediate computer to pass the data to other networks are called Internet gateways or Internet routers. A gateway has two physical interfaces, and an IP address is required for each physical interface. Thus, we notice that an IP address specifies a connection to a network rather than to an individual machine. A machine that has *n* connecting networks will have *n* IP addresses.

5.1.3 Network and Broadcast Addressing

IP addresses with hostids consisting of all 1s are reserved for broadcasting. A broadcast address, or directed broadcast address, refers to all the hosts on the network, since knowledge of network address is required. During start-up, a machine does not know its IP address. Thus until it learns its IP address, it uses the local network broadcast address, or limited broadcast address, which is a 32-bit string of 1s.

Internet addresses can refer to networks as well as to hosts. An IP address with hostid 0 refers to "this" host, and netid 0 means "this" network. Using netid 0 is important when a host wants to communicate over a network but does not yet know the network IP address. The host uses network id 0 temporarily, and other hosts on the network interpret the address as meaning "this" network.

5.1.4 Dotted Decimal Notation

According to convention, 32-bit IP addresses are written as four decimal integers, one for each byte, separated by decimal points. For example,

10001101 11010001 10000011 00001010

is written as

141.209.131.10

5.1.5 Loopback Addressing

The class A network address 127.0.0.0 is reserved for loopback and is designed for testing and interprocess communication on the local machine. If a program uses the loopback address to send data, the protocol software in the computer returns the data without sending any traffic across the network.

A packet sent to the address 127 should never appear on any network. A host or gateway never propagates routing or reachability information for network number 127; it is not a real network address.

5.1.6 Weaknesses in Internet Addressing

Since the addresses refer to connections, not to the host, if a host moves from one network to another, its IP address must change. When any class C network grows to more than 255 hosts, it must have its address changed to a class B address. Changing network addresses can be incredibly time-consuming, and the process is difficult to debug, as well.

Because routing uses the network portion of the IP address, the path taken by packets traveling to a host with multiple IP addresses depends on the address used.

5.1.7 Mapping of Physical and IP Addresses

In a TCP/IP network, each machine is assigned an IP address and a physical address. The goal of the Address Resolution Protocol (ARP) is to provide low-level software that hides physical addresses and allows higher level programs to work with Internet addresses only. Finally, the communication must be carried out by a physical network using whatever physical address scheme the hardware supplies.

For example, when machine A wants to communicate with machine B, machine A has the IP address of B. Now ARP at A will map the IP address of B to physical address of B.

5.1.7.1 ARP Scheme

When host A wants to resolve an IP address, say Ib, it broadcasts a special packet that asks the host with IP address Ib to respond with its physical address, say Pb. All hosts, including B, receive the request, but only B will recognize that the IP address in the broadcast message is the same as its own IP address. B then sends a reply that contains its physical address, Pb. When A sends a broadcast message, it contains physical address of A. Thus B will receive Pa and can directly reply to A. When A receives the reply, it uses the physical address Pb to send the Internet packet directly to B.

5.1.7.2 ARP Cache

According to the ARP scheme just explained, whenever a host needs the physical address of another machine, it broadcasts a request. But broadcasting is too expensive because it requires every machine on the network to process the broadcast packet. To reduce communication costs, hosts that use ARP maintain a cache of recently acquired IP-to-physical address bindings to avoid having to use ARP repeatedly. Whenever a host receives an ARP reply, it saves the machine's IP address and corresponding hardware address in its cache for future. When transmitting a packet to any other host, it looks into the cache to check the physical address of the destination host. If the physical address is found, there is no need to broadcast the request.

In the same way, when a machine requests any other machine's physical address, the request itself contains the physical address of the requesting machine, which all other hosts in the network may store in their caches.

When a new machine appears on the net, it broadcasts its own IP–physical address binding so that all other machines can store the bindings in their caches for future use and thus avoid running ARP.

5.1.7.3 ARP Implementation

ARP is divided into two parts:

- A part that determines physical addresses when sending packets
- A part that answers requests from other machines

Part 1. When a host needs to send a packet to a destination host, it looks in the ARP cache to check the binding between IP address and physical address. If the binding is available, the host extracts the physical address from the cache and uses it to send the data. Otherwise it simply broadcasts a request.

Broadcasting an ARP request to find an address mapping can be nontrivial, considering some possible obstacles.

- The target machine could be down or just too busy to accept the request. If so, the sender might not receive a reply, or the reply might be delayed.
- The request itself may fail and never reach the destination. In this case, the sender should retransmit.
- The host machine may be obliged to allow other application programs to proceed while address resolution is being done. If this happens, the sender must store the outgoing data.
- If applications are allowed to run, they may generate requests for the same IP address for which a physical address request has been sent. Therefore multiple requests should not be broadcasted.
- Suppose machine A has a binding for machine B in its cache and now machine B changes hardware and so also its address. The cache should be refreshed periodically to accommodate new changes in the network. Cache refresh is controlled by using a timer. The timer for an entry should be reset whenever an ARP broadcast arrives containing a binding.

Part 2. Whenever an ARP packet arrives from the network, the receiving host extracts the sender's IP address and physical address. It then looks into the local cache to determine whether a binding for the sender IP address exists. If there is such a binding, the host updates the cache entry. Then it processes the rest of the ARP packet.

If the incoming ARP packet is a request, the receiving machine must verify that it is the target of the request. If so, the ARP software forms a reply by supplying its physical

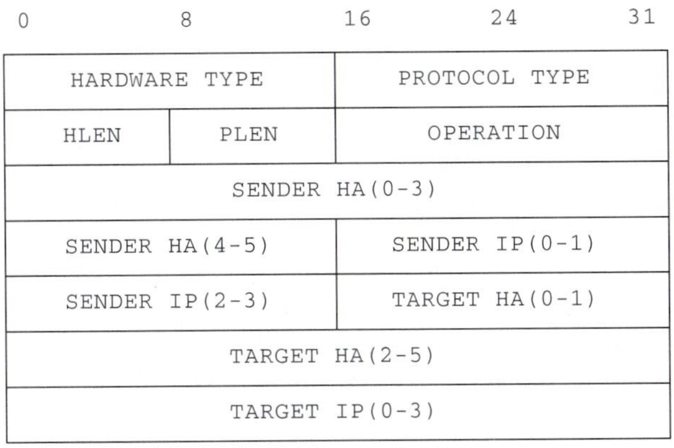

Figure 5.2 An ARP message used on Ethernet.

hardware address and sends the reply directly back to the requester. The receiver also adds the sender's address pair to its cache if the pair is not present. If the IP address in the ARP request does not match the IP address of the receiver, the request is ignored.

Another type of incoming ARP packet is the reply for a past request from the receiver. In dealing with the reply packet, first the cache is updated for the address binding. Then the receiver tries to match the reply with a previously issued request. Between the time the machine broadcasts its ARP request and receives the reply, application programs or higher level protocols may generate additional requests for the same address; part 1 of ARP implementation software takes care of not sending multiple requests for the same IP address. All requests for the same IP address are stored in a queue, and when a reply comes for that IP address, the ARP software removes items from the queue and supplies the address binding to each. If a machine does not issue a request for the IP address in reply, it is ignored.

5.1.7.4 ARP Protocol Format

ARP packets do not have a fixed format to allow arbitrary physical addresses and arbitrary protocol addresses. They are designed to be useful with a variety of network technologies.

For example, Figure 5.2 shows a 28-byte ARP message format used on Ethernet. This format shows the included Ethernet (physical or hardware) addresses of 48 bits or 6 bytes and the IP protocol addresses of 4 bytes. In the ARP message shown in Figure 5.2, `HARDWARE TYPE` specifies a hardware interface type for which the sender seeks an answer; it contains 1 for Ethernet. The `PROTOCOL TYPE` specifies the type of high-level protocol address the sender has supplied; which is 0×800 for IP protocol addresses. The `OPERATION` has value 1 for an ARP request, 2 for an ARP reply, 3 for a RARP (discussed in the next section) request, and 4 for a RARP response.

5.1.8 Reverse Address Resolution Protocol (RARP)

A diskless machine uses a TCP/IP protocol called RARP (Reverse Address Resolution Protocol) to obtain its IP address from a server. RARP is adapted from the ARP protocol.

A RARP message is sent from one machine to another encapsulated in the data portion of an Ethernet frame. An Ethernet frame carrying a RARP request has the usual preamble, Ethernet source and destination addresses, and packet type fields in the front of the frame. The frame type contains the value 0×8035 to identify the contents of the frame as a RARP message. The data portion of the frame contains the 28-byte RARP message, which has the same format as ARP message.

The sender broadcasts a RARP request that specifies itself as both sender and target machine, and supplies its physical network address in the target hardware address field. All machines on the network receive the request, but only those authorized to supply the RARP service process the request and send a reply; such machines are known as RARP servers. For RARP to get through, a network must contain at least one RARP server.

Servers reply by filling in the target protocol address field, changing the message type from request to reply, and sending the reply back directly to the machine making the request. The sender receives replies from all the RARP servers, even though only one is needed.

5.1.8.1 Timing RARP Transactions

Since RARP uses the physical network directly, it must handle the response and retransmit the request itself. In general, RARP is used in local-area networks, where probability of failure is low. If a network has only one RARP server and it cannot handle the load, packets may be dropped.

5.1.8.2 Primary and Backup RARP Servers

More than one server can be used in a network to make the system more reliable. If one server is down or heavily loaded, another can response to the requests. A disadvantage of using many servers is that when a request is broadcast, the network becomes overloaded as all the servers attempt to respond. On the Ethernet, many servers make the probability of collision high.

As a solution to the foregoing problem, each machine is assigned one primary RARP server. In normal conditions only the primary server responds to the requests. All nonprimary servers receive a given request but merely record its arrival time. If the primary server is unavailable, the original sender machine will time out and will retransmit the same request. When a nonprimary server receives a second copy of a RARP request soon after the first, it responds.

In another solution, each nonprimary server that receives a request computes a random delay and then sends a response.

5.2 INTERNET PROTOCOL (IP) AND DATAGRAMS

LANs have become established as the networks for local communication within buildings or single sites. But LANs are limited with respect to the distance they cover and the number of computers and other network devices they can support. Also, because data sharing

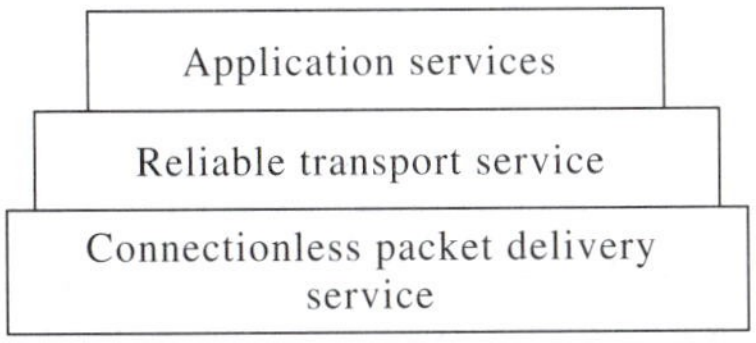

Figure 5.3 TCP/IP sets of services.

and broadcasting are prominent features of LAN technology, the amount of data a LAN can transmit and data security and management become important issues when a LAN attains large size. This means that to create a large network serving a whole enterprise or multiple sites, different LANs need to be interconnected. TCP/IP was developed not only to create LANs, but also for internetworking multiple LANs. Today, these protocols are the primary building blocks for the Internet.

TCP/IP provides three sets of services, as shown in Figure 5.3. The service types are connectionless, reliable, and application-oriented. Connectionless service is described as an unreliable, best-effort packet delivery system analogous to the service provided by network hardware that operates on a best-effort delivery paradigm. This service is called unreliable because delivery is not guaranteed. A packet may be lost, duplicated, delayed, or delivered out of order, but the service will not detect such conditions, hence cannot inform sender or receiver. The service is called connectionless because each packet is treated independently of all others. A sequence of packets sent from one machine to another may travel over different paths, and some may be lost while others are being delivered.

Connectionless service is said to use best-effort delivery because the Internet software makes an earnest attempt to deliver packets. The Internet does not discard packets normally; unreliability arises only when resources are exhausted or underlying networks fail.

Some features of TCP/IP are as follows.

- IP specifies the exact format of all data passing across the TCP/IP network. Thus, the protocol defines the basic unit of data transfer used throughout a TCP/IP internet.
- IP software performs routing, which means that it decides the path over which data will be sent.
- IP includes a set of rules that characterize

  How hosts and gateways should process packets

  How and when error messages should be generated

  Conditions under which packets can be discarded

The basic transfer unit on a net, called an Internet datagram or IP datagram, consists of a header and data.

Transmission of an IP datagram between two machines on a single physical network does not involve routers. The sender encapsulates the datagram into a physical frame, binds the destination IP address to a physical hardware address, and sends the resulting frame directly to the destination. Because the Internet addresses of all machines on a single network include a common network prefix, and because extracting that prefix can be done with a few machine instructions, testing whether a machine can be reached directly is extremely efficient.

Indirect delivery is more difficult than direct delivery because the sender must identify a router to which the datagram is to be sent. The router must then forward the datagram on toward the destination network. Routers in a TCP/IP internet form a cooperative, interconnected structure. Datagrams pass from router to router until they reach a router that can delivery the datagram directly.

The usual IP routing algorithm employs a routing table on each machine that stores information about possible destinations and how to reach them. Because both hosts and routers route datagrams, both have IP routing tables. Whenever the IP routing software in a host or a router needs to transmit a datagram, it consults the routing table to decide where to send the datagram. After taking into account everything about routing, we can set down the following IP routing algorithm:

```
RouteDatagram (Datagram, Routing Table)
Extract destination IP address, D, from the datagram and
compute the network prefix, N;
If N matches any directly connected network address
  deliver datagram to destination D over that network,
  (This involves resolving D to a physical address,
  encapsulating the datagram, and sending the frame.)
else if the table contains a host-specific route for D
  send datagram to next-hop specified in the table
else if the table contains a route for network N
  send datagram to next-hop specified in the table
else if the table contains a default route
  send datagram to the default router specified in the table
else declare a routing error;
```

5.2.1 IP Datagram Format and Types of Service

Figure 5.4 shows the IP datagram format. The contents of the bit sets can be outlined as follows.

Bits 0–3 (`VERS`): The first 4 bits contains the version number of the IP protocol that was used to create the datagram. This field is used to verify that the sender, receiver, and any gateways in between agree on the format of the datagram. All IP software checks the version field before processing a datagram. Machines reject the datagrams with protocol versions that differ from theirs.

Bits 4–7 (`HLEN`): These 4 bits contain the datagram header length, measured in 32-bit words. All fields except IP options and padding have fixed lengths. A common IP datagram header without options or padding fields has a typical header length of 5 words.

Bit 8–15 (`SERVICE TYPE`): This field is broken into five subfields, as shown in Figure 5.5. The values 0–7 indicate datagram precedence, allowing the sender to indicate the importance of each datagram. Bit sets D, T, and R specify the type of transport: D for low delay, T for high throughput, and R for high reliability. The DTR bits

0	4	8	16	19	24–31
VERS	HLEN	SERVICE TYPE	TOTAL LENGTH		
IDENTIFICATION			FLAG	FRAGMENT OFFSET	
TIME TO LIVE		PROTOCOL	HEADER CHECKSUM		
SOURCE IP ADDRESS					
DESTINATION IP ADDRESS					
IP OPTIONS (IF ANY)					PADDING
DATA					
...					

Figure 5.4 IP datagram format.

may be used as a hint to the routing algorithm. For example, if a gateway knows more than one possible route to a given destination, it can use the type of transport field having characteristics closest to those desired.

Bits 16–31 (TOTAL LENGTH): The length of the IP datagram is measured in bytes and includes both header and data. The data area size is total length minus header length times 4. Since the field is only 16 bits long, the maximum size of an IP datagram is 64KB.

5.2.2 Datagram Encapsulation and Fragmentation

The underlying physical network transports datagrams. Each datagram travels in a distinct physical frame, for efficient Internet transportation. The technique of carrying one datagram in one physical frame is called encapsulation. To the underlying network, a datagram is like any other message sent from one machine to another.

Ideally the entire IP datagram should fit into one physical frame. A field in the frame header identifies the data being carried. The Ethernet uses the type value 0×0800 to specify that the data area contains an encapsulated IP datagram. The Ethernet limits transfer to 1500 bytes of data frames, while some other networks may allow a different size data. For instance, proNET-10 allows 2044 bytes per frame, and 2044 bytes is called this network's maximum transfer unit (MTU). However, MTU sizes can be quite small ($\leq$128 bytes). Since a datagram may travel across many types of physical network as it moves across the Internet, IP must select a maximum datagram size to ensure that each datagram will always fit into one frame.

0–2	3	4	5	6–7
PRECEDENCE	D	T	R	UNUSED

Figure 5.5 The service type subfields.

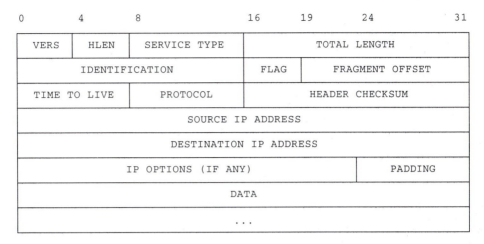

Limiting datagrams to fit the smallest possible MTU in the Internet makes transfers inefficient when those datagrams pass across a network that can carry larger size frames. By the same token, allowing a datagram to be larger than an internet's minimum network MTU means that some datagrams will not fit into a single network frame.

TCP/IP software chooses a convenient initial datagram size and arranges a way to divide large datagrams into smaller pieces when they need to travel over a network with smaller MTU. The smaller pieces are called fragments, and the process of dividing datagrams into smaller pieces is called fragmentation.

Fragmentation occurs at a gateway somewhere along the path between the datagram source and its ultimate destination. Fragment size is chosen so that each fragment can be shipped across the underlying network in a single frame. IP represents an offset of data as a multiple of 8 bytes. So fragment size is also chosen to be the multiple of 8 bytes nearest to the network MTU. The last piece of a fragmented datagram will be smaller than the other pieces. Fragments must be reassembled to produce a complete copy of the datagram before it can be processed at the destination.

Each fragment contains a datagram header that duplicates most of the original datagram header, except for a bit in the flag field, followed by as much data as can be carried in the fragment of a limited MTU.

5.2.3 Reassembly and Fragmentation Control

In a TCP/IP network, once a datagram has been fragmented, the fragments travel as separate datagrams all the way to the final destination, where they must be reassembled. Two disadvantages of this technique are as follows:

- Even if some of the physical networks encountered after the point of fragmentation have large MTU capability, only small fragments will traverse them.
- If any fragments are lost, the datagram cannot be reassembled. Thus, the probability of datagram loss increases when fragmentation occurs.

The Identification field in an IP frame contains a unique integer that identifies the datagram. When a datagram is fragmented, most of the fields are copied into the fragments. The identification field is one that must be copied so that the destination will know which arriving fragments belong to which datagram.

Machines sending IP datagrams must generate a unique value for the identification field for each unique datagram. Retransmissions of IP datagrams contain the same identification number.

The fragment offset field specifies the offset in the original datagram of the data being carried in the fragment, measured in units of 8 bytes, starting at offset zero. To reassemble a datagram, the destination must receive all the fragments, from offset 0 to highest offset. The destination must process with care because fragments do not arrive in order, and there is no direct communication with the gateway that did the fragmenting.

The low-order 2 bits of the 3-bit flags field control fragmentation. They are for the purpose of debugging and testing Internet software. The first bit specifies whether the datagram is to be fragmented. It is actually a "*do not fragment*" bit. Thus setting it to 1 means that the datagram will not be fragmented.

The second low-order bit specifies whether the fragment contains data from the middle of the original datagram. It is called the "*more fragments*" bit. The last fragment will have this bit turned off, while others will have it on.

5.2.4 Other Fields

5.2.4.1 Time-to-Live (TTL) Field

The TTL field specifies how long, in seconds, a datagram is allowed to remain on an Internet. Whenever a machine injects a datagram into an Internet, it sets a maximum time that the datagram should survive. Gateways and hosts that process datagrams must decrement the TTL field as time passes and remove from the network any datagrams whose time has expired.

Whenever a TTL field reaches zero, the gateway discards the datagram and sends an error message back to the source. Keeping TTL fields in datagrams guarantees that datagrams cannot travel on the Internet forever. Even if routing tables become corrupt and gateways route datagrams in a circle, the TTL guarantees their destruction when the limit is exceeded.

5.2.4.2 Protocol Field

The value in the protocol field specifies which high-level protocol was used to create the message being carried in the "data" area of a datagram. The value of "protocol" specifies the format of the data area. Values in this field corresponding to high-level protocols are standards that are controlled to guarantee agreement across the entire Internet.

5.2.4.3 Header Checksum

The checksum is used to ensure the integrity of the header values. It is formed by treating the header as a sequence of 16-bit integers (in network byte order), adding them together using one's-complement arithmetic, and then taking the one's complement of the result. When the header checksum is being computed, the header checksum field is assumed to contain zero initially. However, this applies only to the header values, not to the data.

5.2.4.4 Source and Destination IP Addresses

The IP frame contains 32-bit IP addresses of the datagram's sender and intended recipient. These specify the original source and the final destination of the datagram.

5.2.4.5 IP Options

"IP options" is not a required field in IP datagrams. It is included for network testing and debugging. The length of this field varies depending on which options are selected. When options are present in datagram, they appear contiguously, with no special separators between them. Each option consists of a byte of option code, a byte of length, and data bytes for that option (Figure 5.6).

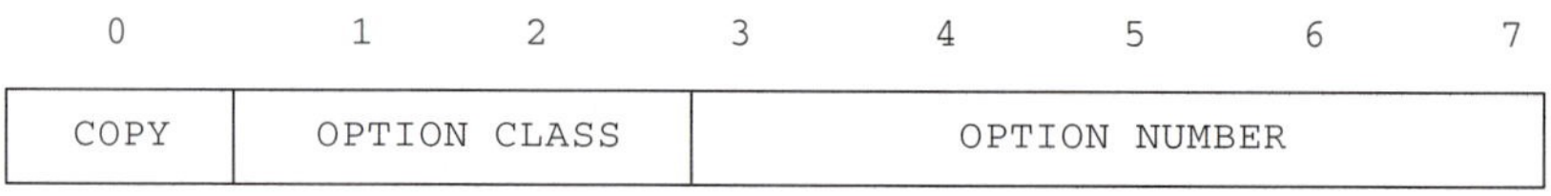

Figure 5.6 Option byte field in an IP datagram.

The copy bit in the option byte, when set to 1, specifies that the option should be copied into all fragments. When it is set to 0, the copy bit means that the option should be copied into the first fragment only, and not into all fragments. Table 5.1 gives the meanings of the other fields in the byte.

5.2.4.6 Record Route Option

The 3-byte option field allows the source to create an empty list of IP addresses and arrange for each gateway that handles the datagram to add its IP address to the list.

Whenever a machine handles a datagram that has the record route option (Figure 5.7) set, the machine adds its address to the record route list. To add itself to the list, a machine first compares the pointer and length fields. If the pointer is greater than the length, the list is full, so the machine forwards the datagram without inserting its entry. If the list is not full, the machine inserts its 4-byte IP address at the position specified by the pointer, and increments the pointer by 4. When the datagram is received, the recipient must extract and process the list of IP addresses.

5.2.4.7 Source Route Option

The source route option provides a way for the sender to dictate a path through the Internet. The option format is shown in Figure 5.8. IP supports both strict and loose forms of source routing.

TABLE 5.1
Option Class and Numbers in the OPTION Byte

Option Class	Option Number	Length	Description
0	0	—	End of option list. Used if options are not part of the header.
0	1	—	No operation.
0	2	11	Security and handling restrictions.
0	3	Variable	Loose source routing. Used to route a datagram along a specified path.
0	7	Variable	Record route. Used to trace a route.
0	8	4	Stream identifier. Used to carry a SATNET stream identifier.
0	9	Variable	Strict source routing. Used to route a datagram along a specified path.
2	4	Variable	Internet timestamp. Used to record timestamps along the route.

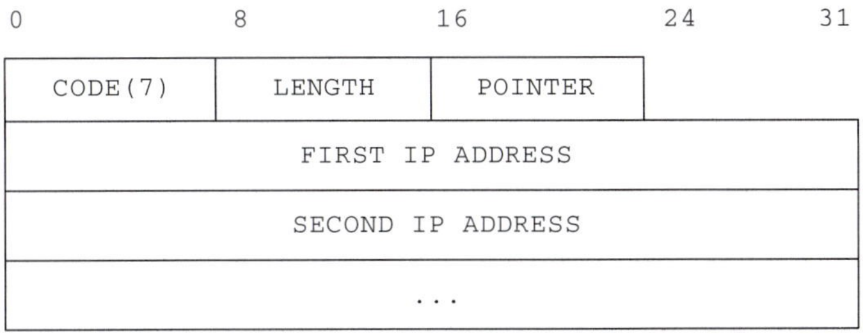

Figure 5.7 The record route option: the **code** field contains option number and option class, the **length** field specifies the total length of the option as it appears in the IP datagram, including first 3 bytes, and the **pointer** field specifies the offset within the option of the next available slot.

Strict source routing: The addresses specify the exact path the datagram must follow to reach its destination. The path between two successive addresses in the list must consist of a single physical network; an error results if a gateway cannot follow a strict source route.

Loose source routing: The datagram must follow the sequence of IP addresses, but allows multiple network hops between successive addresses on the list are allowed.

Both source route options require gateways along the way to overwrite items in the address list with their local network addresses. Thus, when the datagram arrives at its destination, it contains a list of all addresses visited, exactly like the list produced by the record route option.

Each gateway examines the pointer and length fields to see if the list has been exhausted, resulting in a pointer that is greater than the length. When this happens, the gateway routes the datagram to its destination as usual. If the list is not exhausted, the gateway follows the pointer, picks up the IP address, replaces it with the gateway's address (corresponding to the network over which it routes the datagram), uses the address obtained from the list to route the datagram.

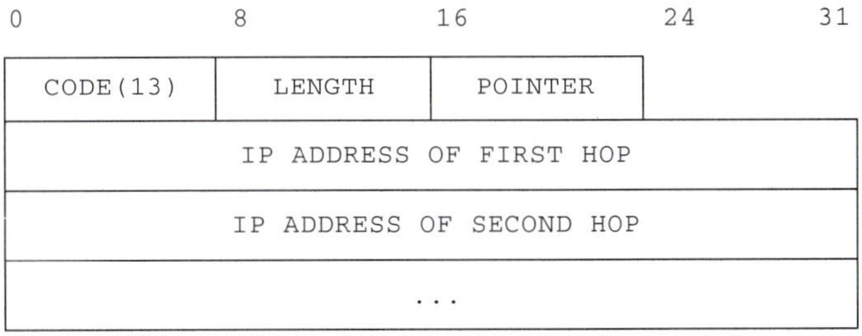

Figure 5.8 The source route option.

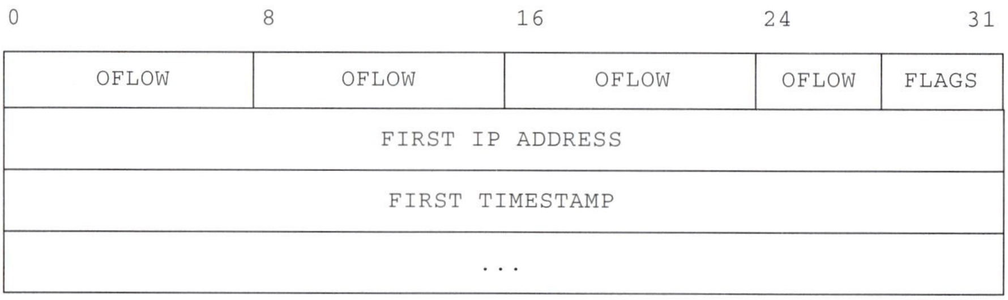

Figure 5.9 The timestamp option.

5.2.4.8 Timestamp Option

The timestamp option works the same as the record route option. It contains an initially empty list, and each gateway along the path from source to destination fills in one item in the list. Each entry in the list contains two 32-bit items, the IP address of the gateway that supplied the entry, and a 32-bit integer timestamp. The overflow field ("oflow") is the number of gateways that could not supply a timestamp because the option was too small. The possible flags values are shown in Table 5.2.

Timestamps give the time and date at which a gateway handles the datagram, expressed as milliseconds since midnight, Universal Time. Timestamps issued by independent computers are not always consistent even if represented in Universal Time; each machine reports time according to its local clock, and the clocks associated with the transmission of a given datagram differ.

5.2.5 Error and Internet Control Message Protocol (ICMP)

To allow gateways in the Internet to report errors or provide information about unexpected circumstances, ICMP protocols are added to the TCP/IP family. ICMP is a required part of IP and must be included in every IP implementation. ICMP messages travel across the network in the data portion of IP datagrams. ICMP software communicates with higher level protocols or application programs transmitting the control messages.

ICMP is not restricted to gateways; an arbitrary machine can send an ICMP message to any other machine. Thus a way exists to report errors to the original source. Although

TABLE 5.2
Flag Values in the Timestamp Option

Flag Value	Meaning
0	Record timestamps only; omit IP addresses.
1	Precede each timestamp by an IP address.
3	IP addresses are specified by sender; a gateway records a timestamp only if the next IP address in the list matches the gateway's IP address.

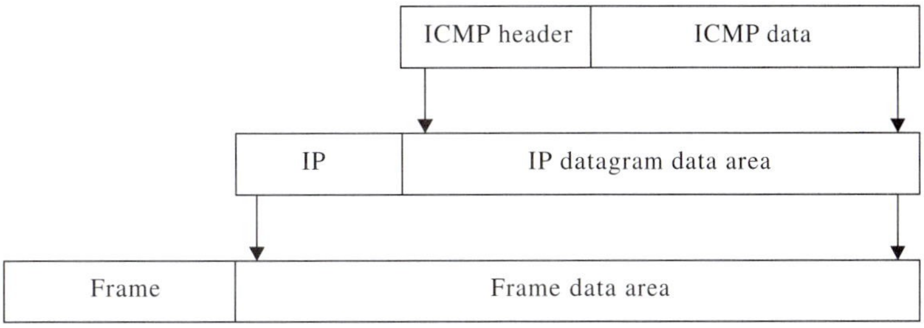

Figure 5.10 ICMP encapsulation.

the protocol specification outlines the intended use of ICMP and suggests possible actions to take in response to error reports, ICMP does not fully specify the action to be taken for possible error. The source must relate errors to individual application programs and take action to correct the problem.

Most errors stem from the original source, but some do not. There are reasons for restricting ICMP reporting to the original source. The datagram contains only fields that specify the original source and the ultimate destination; it does not contain a complete record of its trip through the Internet. Because gateways can establish and change their own routing tables, there is no global knowledge of routes. Thus, when a datagram reaches a given gateway, it is impossible to know the route by which it has arrived. If the gateway detects a problem, it cannot know the set of intermediate machines that processed the datagram, so it cannot inform them of the problem. The ICMP encapsulation and message format are shown in Figures 5.10 and 5.11, respectively. Table 5.3 gives the meanings of type fields in an ICMP frame.

5.2.5.1 Echo Request/Reply Message (Type 8/0)

Type 8/0 messages are used to determine whether a given network or machine is reachable (Figure 5.12). A host or gateway sends an ICMP echo request to a specified destination. Any machine that receives an echo request formulates an echo reply and returns it to the original sender. The request contains an optional data area; the reply contains a copy of the data sent in the request. Receipt of a reply successfully verifies following:

- IP software on the source machine is routing the datagram.
- Intermediate gateways between the source and destination are operating and are routing the datagram correctly.

0	8	15	31	
TYPE	CODE	CHECKSUM	OPTIONAL	IP HEADER AND FIRST 64 BITS OF DATAGRAM

Figure 5.11 ICMP message format.

TABLE 5.3
Meanings of the Type Field in an ICMP Message

Type Field	ICMP Message Type
0	Echo reply
3	Destination unreachable
4	Source quench
5	Redirect (change a route)
8	Echo request
11	Time exceeded for a datagram
12	Parameter problem on a datagram
13	Timestamp request
14	Timestamp reply
15	Information request (not used now)
16	Information reply (not used now)
17	Address mask request
18	Address mask reply

- The destination machine is running, and both ICMP and IP software packages are working.
- Routes in gateways along the return path are correct.

On many systems, the command users invoke to send ICMP echo requests is called ping. Ping sends a series of ICMP echo requests, captures responses, and provides statistics about datagram loss.

5.2.5.2 Destination Unreachable (Type 3)

Whenever an error prevents a gateway from routing or delivering a datagram, the gateway sends a "destination unreachable" message (Figure 5.13, Table 5.4) back to the source and then drops the datagram. "Network unreachable" implies routing failures; "host un-

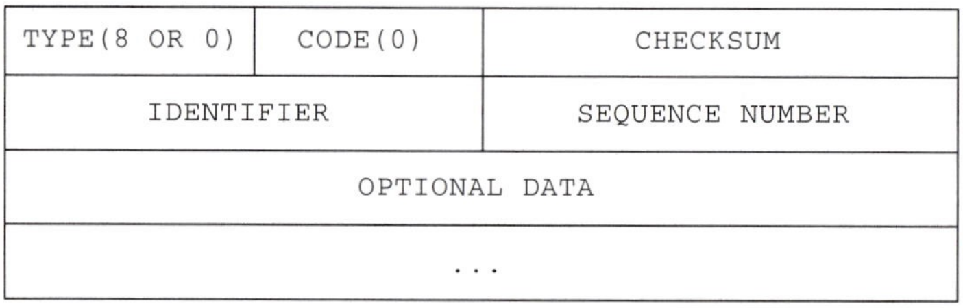

Figure 5.12 The type 8/0 ICMP message format: type field, 8 (request) or 0 (reply); identifier and sequence number, fields used by senders to match replies to requests; optional data, a variable length field that contains data to be returned to the sender, which are the same data received in the case of echo reply.

0	8	16 31
TYPE(3)	CODE(0-5)	CHECKSUM
UNUSED (MUST BE ZERO)		
INTERNET HEADER + FIRST 64 BITS OF		
...		

Figure 5.13 The type 3 ICMP message format.

reachable" implies delivery failures. Since the message contains a short prefix of the datagram that caused the problem, the source knows exactly which address is unreachable.

5.2.5.3 Source Quench (Type 4)

A datagram that arrives too quickly for a host or gateway to process is enqueued in the memory temporarily. If the datagrams are part of a small burst, such buffering solves the problem. If the traffic continues, the host or gateway eventually exhausts memory and must discard additional datagrams that arrive. Such a machine would issue the ICMP "source quench" message—a request for the source to reduce its current rate of datagram transmission (Figure 5.14). Usually, congested gateways send one source quench message for every datagram that they discard.

The host that receives source quench messages from a machine lowers the rate at which it sends datagrams until it stops receiving source quench messages; it then gradually increases the rate as long as no further source quench requests are received. The datagram prefix field contains a prefix of the datagram that has triggered the source quench request.

TABLE 5.4
Meaning of the Code Field in an ICMP Type 3 Message

Code Value	Meaning
0	Network unreachable
1	Host unreachable
2	Protocol unreachable
3	Port unreachable
4	Fragmentation needed and DF set
5	Source route failed
6	Destination network unknown
7	Destination host unknown
8	Source host isolated
9	Communication with destination network administratively prohibited
10	Communication with destination host administratively prohibited
11	Network unreachable for type of service
12	Host unreachable for type of service

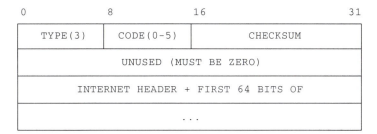

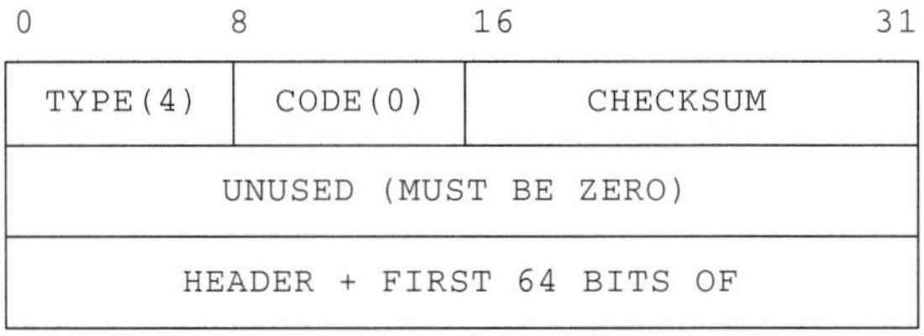

Figure 5.14 The type 4 ICMP message.

5.2.5.4 Route Change Request (Type 5)

Gateways exchange routing information periodically to accommodate network changes and keep their routes up to date (Figure 5.15, Table 5.5). Gateways are assumed to know correct routes; hosts begin with minimum routing information and rely on gateways to update their respective routing tables. When a gateway detects a host using a nonoptimal route, it sends the host an ICMP message called a *redirect*, requesting that the host change its routes. The gateway also forwards the original datagram on to its destination. Generally, gateways send ICMP redirect requests only to hosts, not to other gateways.

The ICMP redirect scheme allows a host to boot even if it knows the address of only one gateway on the local network. The initial gateway returns ICMP redirect messages whenever a host sends a datagram for which there is a better route. The host routing table remains small but still contains optimal routes for all destinations in use. The gateway internet address field in a type 5 message contains the address of a gateway that the host is to use to reach the destination mentioned in the datagram header.

5.2.5.5 Time Exceeded for a Datagram (Type 11)

Whenever a gateway discards a datagram because its hop count has reached zero or because a time-out occurred during the wait for fragments of a datagram, it sends an ICMP "time exceeded" message back to the datagram's source, using the format shown in Figure 5.16.

Code value 0 indicates that the TTL (time-to-live) count has been exceeded, and code value 1 indicates that fragment reassembly time has been exceeded. Fragment reassembly refers to the task of collecting all the fragments from a datagram. When the first fragment of a datagram arrives, the receiving host starts a timer and considers it an error if the timer expires before all the pieces of the datagram arrive. Code value 1 is used to report such errors to the sender; one message is sent for each type 11 error.

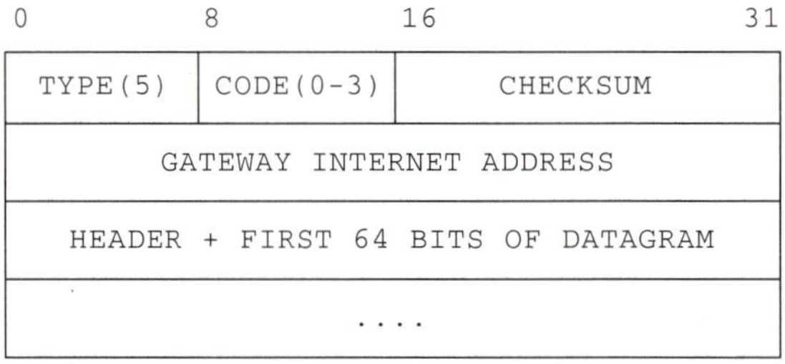

Figure 5.15 The type 5 ICMP message.

TABLE 5.5
Meaning of the Code Field in a Type 5 ICMP Message

Code Value	Meaning
0	Redirect datagrams for the net (now obsolete)
1	Redirect datagrams for the host
2	Redirect datagrams for the type of service and net
3	Redirect datagrams for the type of service and host

5.2.5.6 Parameter Problem (Type 12)

When a gateway or host finds problems with a datagram not covered by previous ICMP error messages (e.g., an incorrect datagram header), it sends a parameter problem message to the original source (Figure 5.17). Such errors may occur when arguments to an option are incorrect. The sender uses the POINTER field in the message header to identify the byte in the datagram that caused the problem. Code 1 is used to report that a required option is missing.

5.2.5.7 Timestamp Request/Reply (Type 13/14)

The timestamp messages are used for synchronization purposes to obtain the time from another machine. A requesting machine sends an ICMP timestamp request message (type 13: Figure 5.18) to another machine, asking that the second machine return its current value for the time of day. The receiving machine returns a timestamp reply (type 14: Figure 5.19) back to the machine that made the request.

The identifier and sequence number fields are used by the source to associate replies with requests. Remaining fields specify times, given in milliseconds since midnight, Universal time. The originate timestamp field is filled in by the original sender just before the packet is transmitted, receive timestamp is filled immediately upon receipt of a request, and transmit timestamp is filled immediately before the reply is transmitted.

5.2.5.8 Information Request/Reply (Type 15/16)

The information request/reply messages are obsolete and hence not used any more. They were used formerly to allow hosts to discover their Internet address at system start-up.

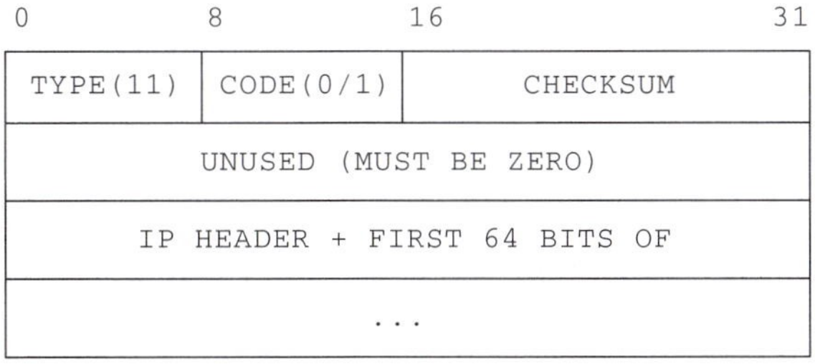

Figure 5.16 The type 11 ICMP message.

0	8	16 31
TYPE(12)	CODE(0/1)	CHECKSUM
POINTER	UNUSED (MUST BE ZERO)	
IP HEADER + FIRST 64 BITS OF		
...		

Figure 5.17 The type 12 ICMP message.

5.2.5.9 Address Mask Request/Reply (Type 17/18)

A machine may send an address mask request message to a gateway to learn about the subnet mask used for the local network. In reply it receives an address mask (Figure 5.20). The machine making the request can send the message directly if it knows the gateway's address. If it does not know the gateway's address, it will broadcast the message.

5.2.6 IPv6

We pointed out some addressing weaknesses of IP version 4 in the preceding section. The primary motivation behind the development of IPv6 is to support the continued growth of the Internet. Once the IP header had to be changed to improve the addressing, other weaknesses also became candidates for fixing. However, since IPv6 is still undergoing modification, we will talk only about the major features that are likely to stay.

The most important difference in IPv6 is the addressing scheme. Addresses of 16 bytes (128 bits, or 32 hex digits) provide for 2^{128} hosts on the Internet, an amount considered to be much more than we will ever need—approximately 7×10^{23} IP addresses per square meter on the entire earth, including water. The reason for this overkill is the fear that even if the address space is not utilized efficiently, further revisions of IP will be necessary.

The address format is shown with groups of 4 hex digits separated by colons. A valid IPv6 address can be in one of the following forms:

FDEC:0000:0000:1234:7654:FFEE:ABCD:1111

FDEC::1234:7654:FFEE:ABCD:1111

::192.41.20.41

0	8	16 31
TYPE(13/1)	CODE(0)	CHECKSUM
IDENTIFIER		SEQUENCE
ORIGINATE TIMESTAMP		

Figure 5.18 The type 13 ICMP message.

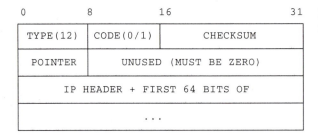

RECEIVE TIMESTAMP
TRANSMIT TIMESTAMP

Figure 5.19 The type 14 ICMP message.

The pair of colons in the address signifies the missing 0s in the address. A leading pair of colons followed by the decimal dot notation shows the conventional IPv4 addresses.

The IPv6 header consists of the following:

VERS: These 4 bits are for version number (6).

PRIORITY: These 4 bits control the quality of service by indicating non-time-sensitive and time-sensitive classes of data.

FLOW LABEL: A 3-byte label to provide quality-of-service control for a particular type of data.

PAYLOAD LENGTH: A 2-byte field to define up to 65,536 bytes of payload in the datagram.

NEXT HEADER: A 1-byte field that replaces the IP options and protocol fields of IPv4. If options are needed, they are included in one or more special headers following the IP header. The next header field includes the special header value for identification. If there are no special headers, this field simply acts as the header for higher level protocol. Fragmentation in IPv6 is handled also by using the optional headers. When fragmentation is necessary, the entire original datagram (including the original header) is divided into pieces and placed in the payload section of fragments. A new base header is used for each fragment, with an inserted fragment extension header.

HOP LIMIT: A 1-byte field that corresponds to the TTL field of IPv4.

SOURCE/DESTINATION ADDRESS: These two 16-byte addresses are the last two fields before the optional extension headers and data.

0	8	16	31
TYPE(17/18)	CODE(0)	CHECKSUM	
IDENTIFIER		SEQUENCE	
ADDRESS MASK			

Figure 5.20 The type 17/18 ICMP message.

5.3 ROUTING PROTOCOLS

Routing is the process by which datagram routing tables are built at nodes or routers. Forwarding consists of taking a packet, looking at its destination address, and then sending the packet to its destination in accordance with the table. It is important to realize the role routers play in forwarding and routing. Routers are capable of supporting multiple independent routing protocols and maintaining routing tables for several routed protocols concurrently. This capability allows a router to deliver packets from several routed protocols over the same data links. In addition, formation of the table is a result of routing protocol implementation at routers. We first explain the process of routing table formation using distance vector and link state routings. Later in this section we will discuss how the Internet routing protocols operate.

5.3.1 Routing Tables and Routing Mechanisms

A routing table has columns for the destination network (or node) with the corresponding costs and the next router address to reach a destination. Additional information may vary depending on the routing protocol being used. There are several algorithms for calculating the routes. However, the most popular are distance vector and link state routing. The distance vector routing approach determines the direction (vector) and distance to any link in the network. The link state (also called shortest-path-first) approach re-creates the exact topology of the entire network (or at least the partition in which the router is situated).

5.3.1.1 Distance Vector Routing

In the distance vector-based routing algorithm, also known as the Bellman–Ford algorithm, routers pass periodic copies of a routing table to other routers to indicate the changes in the topology. Each router receives a routing table from its direct neighbor. For example, in Figure 5.21, router B receives information from router A. Router B adds a distance vector number (such as a number of hops), increasing the distance vector, then passes the routing table to another neighbor, router C. The same step-by-step process occurs in all directions between direct-neighbor routers. In this way, the algorithm accumulates network distances and is able to maintain a database of network topology information. Distance vector algorithms do not allow a router to know the exact topology of the network.

To use distance vector routing, each router begins by identifying its own neighbors. In Figure 5.22, the port to each directly connected network has a distance of 0. As the distance vector network discovery process proceeds, routers discover the best path to destination networks based on information from each neighbor. For example, router A learns about other networks based on information it receives from router B. Each of these other network entries in the routing table has an accumulated distance vector to show how far away that network is in the given direction.

When the topology in a distance vector protocol network changes, routing table updates must occur. As with the network discovery process, topology change updates pro-

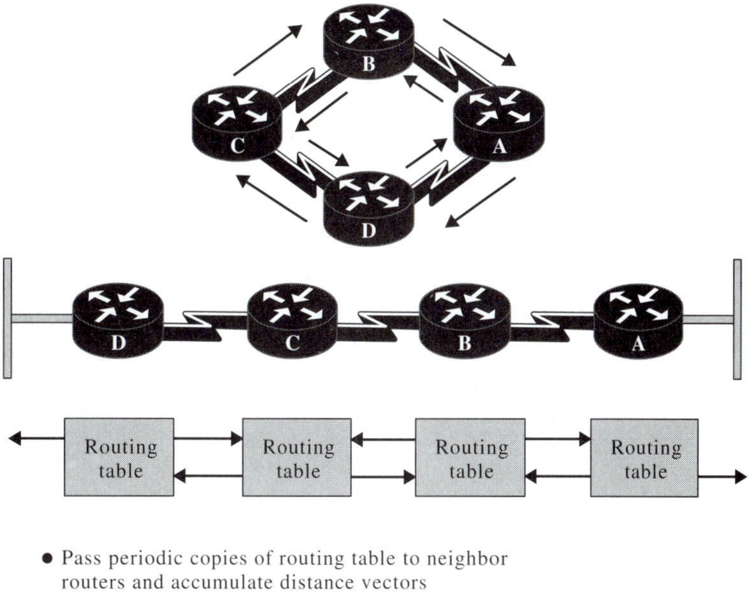

Figure 5.21 Distance vector routing.

ceed step by step from router to router. Distance vector algorithms call for each router to send its entire routing table to each of its adjacent neighbors. Distance vector routing tables include information about the total path cost (defined by its metric) and the logical address of the first router on the path to each network it knows about.

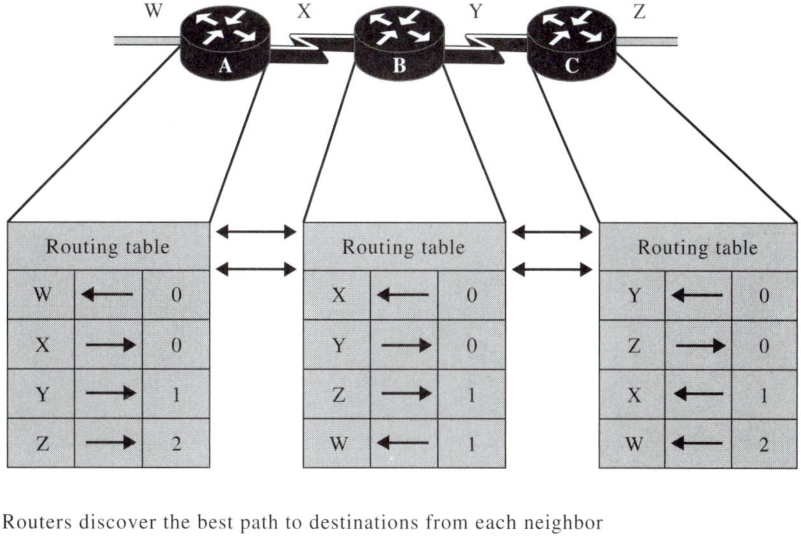

Figure 5.22 Distance vector routing operation.

5.3.1.2 Link-State Routing

The second basic algorithm used for routing, the link state algorithm, also known as shortest-path-first (SPF) algorithms, maintains a complex database of topology information. Whereas the distance vector algorithm has nonspecific information about distant networks and no knowledge of distant routers, a link state routing algorithm maintains full knowledge of distant routers and how they interconnect.

Link state routing uses link state advertisements (LSAs), a topological database, the SPF algorithm, the resulting SPF tree, and finally, a routing table of paths and ports to each network (Figure 5.23). Designers have implemented this link state concept in open-shortest-path-first (OSPF) routing, which we discuss later.

Link state algorithms rely on using the link state updates. Whenever link state topology changes, the routers that first become aware of the change send information to other routers or to a designated router that all other routers can use for updates. However, it is important that the routers agree on some common network state. To this end, each router does the following.

- Keeps track of its neighbors: the neighbor's name, whether the neighbor is up or down, and the cost of the link to the neighbor.
- Constructs an LSA packet that lists its neighbor router names and link costs. This includes new neighbors, changes in link costs, and links to neighbors that have gone down.
- Sends out this LSA packet so that all other routers receive it.

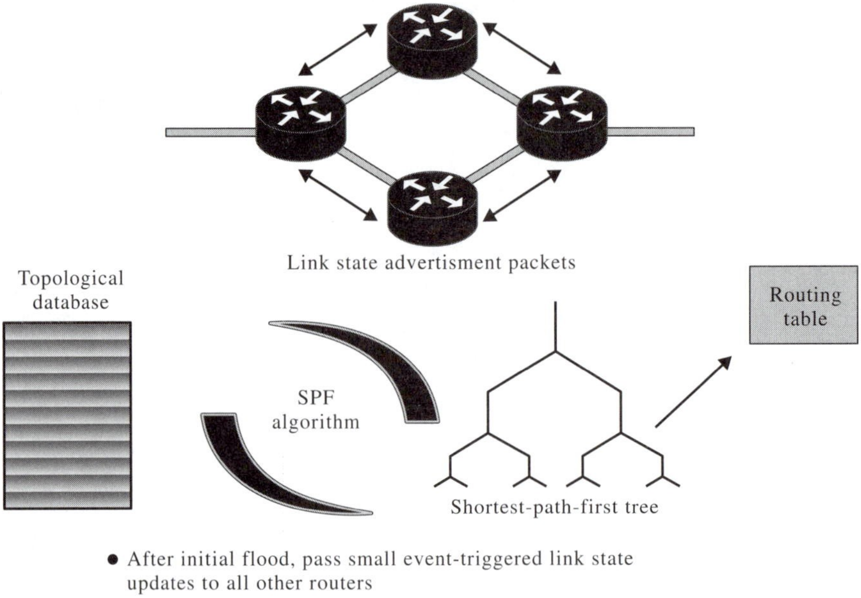

Figure 5.23 The link state routing concept.

- Records each LSA packet it receives in its database immediately, to ensure that it has the most recently generated LSA packet from every other router.
- Using accumulated LSA packet data to construct a complete map of the internetwork topology, proceeds from this common starting point to rerun the SPF algorithm and compute routes to every network destination.

Each time an LSA packet causes a change to the link state database, the link state algorithm recalculates the best paths and updates the routing table. Then every router takes the topology change into account as it determines the shortest paths to use for packet switching.

The routing table updates in the link state routing method (Figure 5.24) use the following process.

- Routers exchange LSAs with each other. Each router begins with directly connected networks for which it has direct information.
- Each router, in parallel with the others, constructs a topological database consisting of all the LSAs from the internetwork.
- The SPF algorithm computes network reachability, determining the shortest path first to each other network in the link state protocol internetwork. Each router constructs this logical topology of shortest paths as an SPF tree. With itself as root, this tree expresses paths from the router to all destinations.

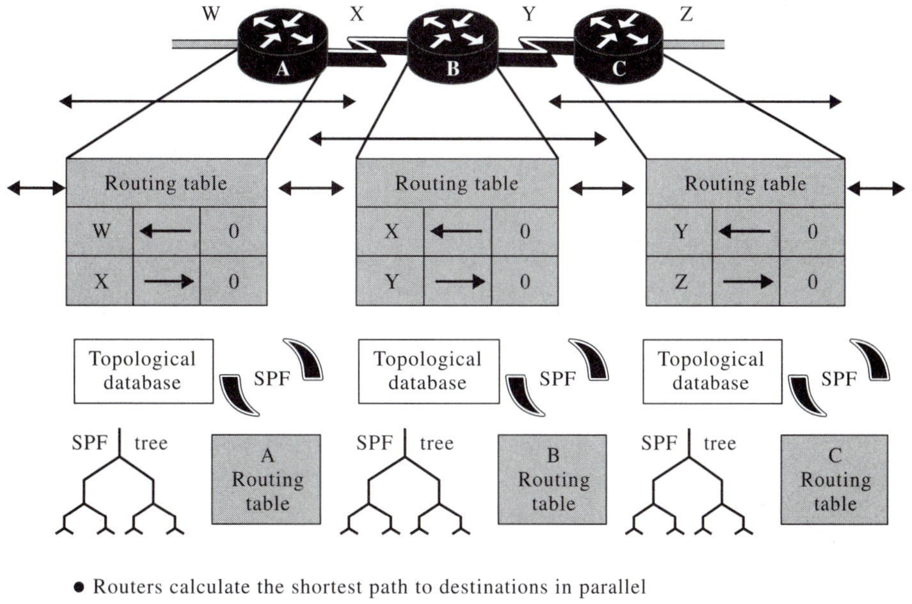

Figure 5.24 Routing table updates in link state routing.

- Each router lists its best paths and the ports to these destination networks in the routing table. It also maintains other databases of topology elements and status details.

5.3.1.3 LAN-to-LAN Routing

The network layer must interface with various lower layers because the data eventually must pass through them. Routers must be capable of seamlessly handling packets encapsulated into different lower level frames without changing the packets' layer 3 addressing.

Figure 5.25 shows an example of this with LAN-to-LAN routing. In this example, packet traffic from source host 4 on Ethernet network 1 needs a path to destination host 5 on network 2. The LAN hosts depend on the router and its consistent network addressing to find the best path. When the router checks its router table entries, it discovers that the best path to destination network 2 uses outgoing port To0, the interface to a token ring LAN.

Although the lower layer framing must change as the router switches packet traffic from the Ethernet on network 1 to the token ring on network 2, the layer 3 addressing for source and destination remains the same. The figure shows that despite the different lower layer encapsulations, the destination address remains Net 2, Host 5.

5.3.1.4 LAN-to-WAN Routing

In addition to LAN-to-LAN traffic, the network layer must interface with various lower layers for LAN-to-WAN traffic. As a network grows, the path taken by a packet might encounter several relay points and a variety of data link types beyond the LANs. For ex-

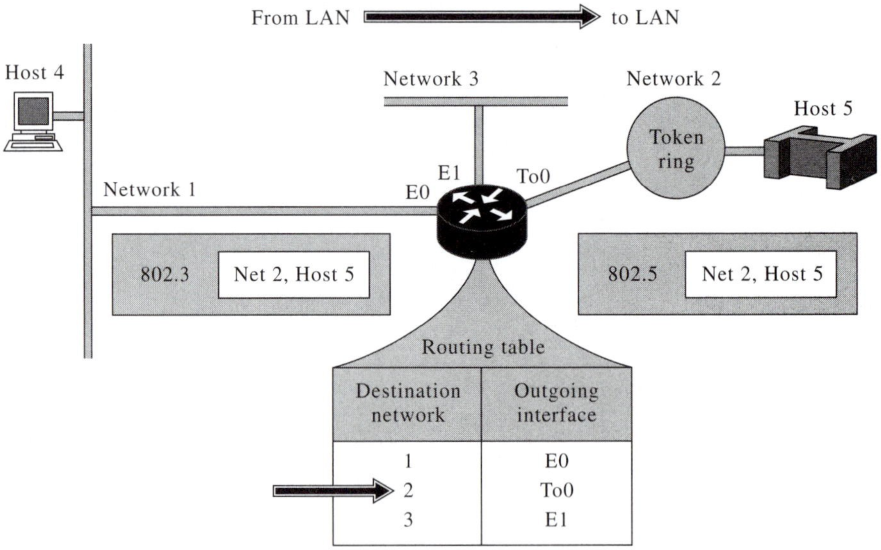

Figure 5.25 LAN-to-LAN routing.

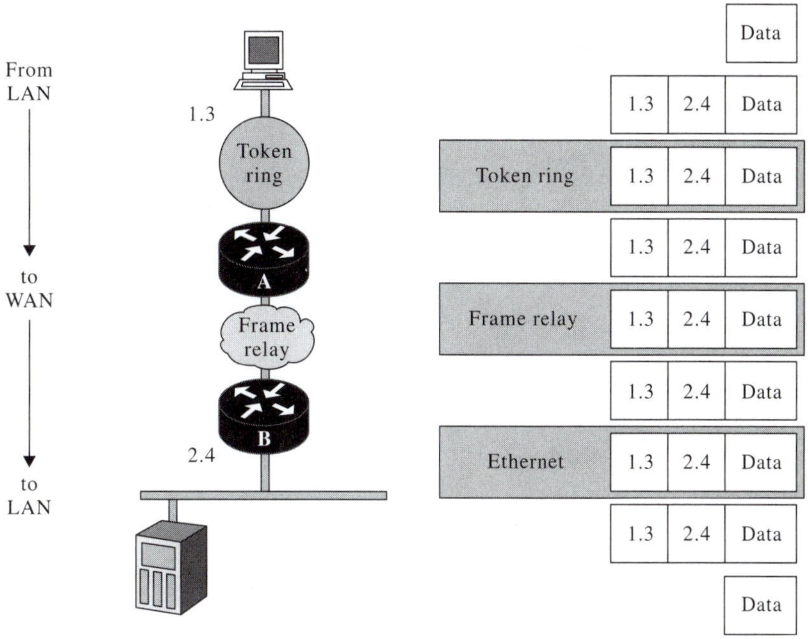

Figure 5.26 LAN-to-WAN routing.

ample, in Figure 5.26, a packet from the top workstation at address 1.3 must traverse three data links to reach the file server at address 2.4, at the bottom.

The complete process of sending the frames from network 1.3 to network 2.4 is the following:

- The workstation sends a packet to the file server by encapsulating the packet in a token ring frame addressed to router A.
- When router A receives the frame, it removes the packet from the token ring frame, encapsulates it in a frame relay frame, and forwards the frame to router B.
- Router B removes the packet from the frame relay frame and forwards the packet to the file server in a newly created Ethernet frame.
- When the file server at 2.4 receives the Ethernet frame, it extracts and passes the packet to the appropriate upper layer process.

The routers enable LAN-to-WAN packet flow by keeping the end-to-end source and destination addresses constant while encapsulating the packet at the port to a data link that is appropriate for the next hop along the path.

5.3.1.5 Routing Policies

There are two main policies used to form the routing tables, manual and automatic. In the manual policy, the system administrator sets a router's table at start-up. This is called static routing. Whenever the router receives a packet with a particular destination address, it looks at the table for the next hop. If the destination address is not found in the table, the router forwards the packet to a default router connected to it. The system administrator, while setting up the routing table, also makes the default router entry available. The default route is therefore manually defined by the system administrator as the path to take, when no route to the destination is known.

In the second policy, the routers accept information from other routers to periodically update their entries. This type of routing policy is called dynamic routing.

Static routes are explicitly configured and entered into the routing table. When the routers use a combination of static and dynamic policies, static routes take precedence. A static route allows manual configuration of the routing table. No dynamic changes to this table entry will occur as long as the path is active.

A static route in the network reflects some special knowledge of the networking situation known to the network administrator. Routing updates are not sent on a link if defined only by a static route, thereby conserving bandwidth.

5.3.2 Subnet Routing

As mentioned earlier, we can create networks of 2^{24}, 2^{16}, and 2^8 nodes with class A, class B, and class C networks, respectively. These are not small numbers for many situations. With any Internet class, subnetting is introduced to allocate a part of the host address space to network addresses and leave the remaining part to other physical networks. This adds flexibility and administrative benefits. Subnetting can get the most out of the limited 32-bit IP address space and reduce the size of the routing table in a large internetwork. Note that subnetting cannot apply to a single physical network.

If subnetworks are intended as part of the subnetting process, a network-wide *netmask* should be decided first. The netmask determines which bits in the IP address space represent the subnetwork addresses and which bits represent host addresses. The netmask also determines how many subnetworks will be created and how many nodes are included in each subnetwork. A netmask is specified in decimal dot format in the file `/etc/netmasks` (or `/etc/inet/netmasks`) in a Unix operating system, using local files for name service. In the netmasks file, a line containing both the network address to be partitioned and the intended netmask specifies the netmask for the particular class network. A 1 at a particular position in the netmask indicates that that particular bit in an IP address ought to be the network address. A 0 indicates that the bit belongs to the host address. The netmask can be applied to an IP address simply by using the bitwise logical AND operator. This operation separates the network and subnetwork address from the complete IP address.

For example, for a class B network with network address 131.31.0.0, the two left-most octets are assigned to the network address, and the right-most octets are assigned to the host number. Therefore, there may be as many as 65,534 computers connected in the single class B network. However, subnetting allows an entire class B network to be partitioned into 254 subnetworks with up to 254 host computers on each, simply by specifying a netmask 255.255.255.0. This netmask indicates that not only the first two octets, but also the third octet serve as the network address and only the fourth octet is for the host addresses. If the netmask 255.255.255.0 is applied to the IP address 131.31.126.3, the result is the IP address 131.31.126.0:

255.255.255.0 and 131.31.126.3 = 131.31.126.0

In binary form, the operation is as follows:

	11111111.11111111.11111111.00000000	Netmask
and	10000011.00011111.01111110.00000011	IP address
	10000011.00011111.01111110.00000000	Subnetwork address

Now the system will look for a network address of 131.31.126.0 instead of the network address 131.31.0.0. Since there are 254 values in the third octet of the IP address, excluding all 0s and all 1s that are reserved, 254 subnetworks can be created with 254 host computers in each, specified by the fourth octet of the IP address. To create only two subnetworks with 32,766 hosts in each, a subnet netmask of

255.255.128.0 or 11111111.11111111.10000000.00000000

should be used.

Usually, netmask bits should be contiguous, but alignment on byte boundaries is not necessary. At least two bits must be reserved for the host number because all 0s and all 1s are reserved for network and broadcasting.

As an example, consider a class A address 15.16.193.6 with a subnet mask of 255.255.248.0. The corresponding subnet address would be 15.16.192.0 and the broadcast address would be 15.16.199.255. Notice that the mask provides 21 consecutive 1s starting from the left (also written as 15.16.193.6/21). This mask allows for a total of 2048 (or remaining 11-bit combinations) to be used for subnet addresses. Overlapping the mask bits on IP address gives the corresponding subnet address. The broadcast address is obtained by using all 1s within the subnet address excluding the masked bits. So, the bit positions 0-10 would form the broadcast address as the leftmost 21 bits are masked resulting in 15.16.199.255 as the broadcast address. Another approach to understand the broadcast address would be to first figure out the first and last usable addresses in the subnet and then use the last address with all 1s as the broadcast address. In this example, the total number of usable addresses would be 2047 (8 times 256 less 1), with the first and last usable addresses as 15.16.192.1 and 15.16.199.254, respectively. Thus the broadcast address would be 15.16.199.255. Note that the next address 15.16.200.0 is not part of the current subnet since bit 12 is changed in this case, which is supposed to be masked.

Following are some more examples with addresses arranged in the format, IP Address/mask bits, subnet address, broadcast address, first usable address, last usable address

172.16.129.201/20, 172.16.129.192, 172.16.129.207, 172.16.129.193, 172.16.129.206

192.168.30.101/19, 192.168.30.96, 192.168.30.127, 192.168.30.97, 192.158.30.126

10.16.193.6/21, 10.16.192.0, 10.16.199.255, 10.16.192.1, 10.16.199.254

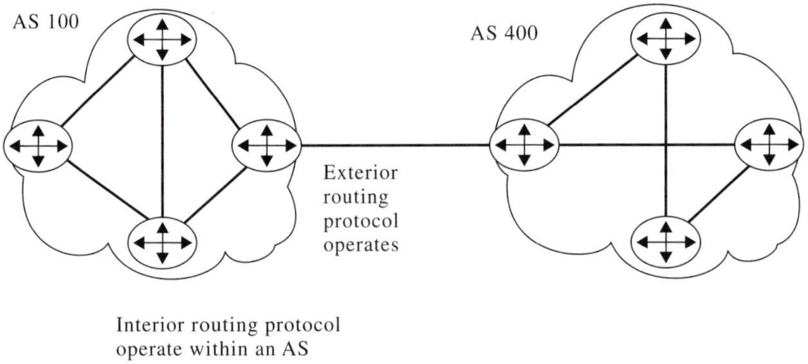

Figure 5.27 Interior and exterior routing protocols.

5.3.3 Internet Routing

The routing protocols used on the Internet fall into two categories: protocols within an autonomous system (AS) and protocols that operate between the autonomous systems (Figure 5.27). An autonomous system consists of routers that present a consistent view of routing to the external world. A network information center (NIC) assigns a unique autonomous system to enterprises. This autonomous system is a 16-bit number. A routing protocol such as Routing Information Protocol (RIP) or Cisco's Interior Gateway Routing Protocol (IGRP) requires that a specific and unique autonomous system number be assigned during configuration.

5.3.3.1 Routing Information Protocol (RIP)

RIP, a distance vector routing protocol, was originally specified in RFC 1058.

A router is initially configured to operate using RIP so that it can learn about other paths in the network as it talks to other routers. Initial configuration must also include information about the networks that are directly connected to the configured router. This allows updating of routing information from the directly connected networks. The router must refer to entries about networks or subnets that are directly connected. Each interface must be configured with an IP address and mask. The initial setup software learns about the IP address and mask information from configuration information provided as input. When the router is operational, the distance vectors are transmitted between the neighbors to exchange the routing information.

In RIP, hop count is used as the metric for path selection. The maximum allowable hop count is about 15, indicating that RIP may be limited to fairly small networks. Routing updates are broadcast every 30 seconds by default. A RIP frame consists of (net-address, distance) pairs. However, the net-address consists of 16-byte (net-family, net-address) pairs, out of which 14 bytes are used for the net-address. In the current RIP (version 1), only 4 bytes are used as the IP address. Version 2 would allow subnetting that may require the use of other bytes. Other fields in the frame are used for version number, and a 1-byte command is reserved for control purposes.

5.3.3.2 Open Shortest Path First (OSPF)

An OSPF works within an AS just like RIP. However, it is based on the link state routing that is alternately known as SPF, as mentioned earlier. OSPF adds many important features to the link state routing protocol. Among the most notable features are authentication of routing messages, scalability via addressing hierarchy, and load balancing. RFC 1583 contains a detailed description of OSPF link state concepts and operations.

In OSPF, the advertised costs from routers are not used in updating the routing tables unless the 8-byte authentication information included in the frame is as expected. This authentication may be a password that the involved routers had agreed upon. The use of this type of authentication can keep unintended updates out of the routing table. For instance, a misconfigured router may advertise an inaccurate cost of 0 to every router in the network. This would soon bring the network to a standstill as all routers updated their tables wrongly and sent their data to the misconfigured router. An authentication will prevent this kind of disaster.

OSPF allows a domain or an AS to be partitioned into smaller areas. With this partition, a router does not need to know how to reach every network in the AS. Only the information for a router in the nearby area is needed. The hierarchy of addressing allows a reduction in the amount of information that must be transmitted to and stored at the routers. OSPF also allows multiple routes to the same destination, thus distributing the load evenly to optimize performance.

The OSPF header consists of the source address (4 bytes), the area ID (4 bytes), checksum (2 bytes), authentication (8 bytes), and authentication type (1 byte). In addition, the OSPF version number (1 byte), OSPF type (1 byte), and message length (2 bytes) are included.

5.3.3.3 Border Gateway Protocol (BGP)

Exterior Gateway Protocol (EGP) is the older Internet protocol used on gateways that connect autonomous systems. A gateway is considered to be a router that can connect two or more autonomous systems. However, in general terms the gateway is actually a computer that may provide a connection between two computers at the application layer.

EGP assumed the Internet to have a treelike topology, which severely restricted its application. The Internet may have evolved from a single backbone, but the tree analogy no longer holds, and thus EGP was modified as BGP.

BGP assumes the Internet to have an arbitrary connection of autonomous systems that may be connected via multiple backbones representing different service providers. In BGP, a "speaker" that acts like a spokesperson represents each AS. In addition, there is a border gateway that provides access within an AS, which may or may not be the speaker.

The BGP speaker advertises access information for the networks within its AS and the other networks that can be reached through it, in case it is also a border gateway. Unlike the distance vector or link state protocol, BGP advertises the complete path as an enumerated list of autonomous systems to reach a particular network.

It may be noted that the complexity of the BGP is of the order of the number of border gateways in the network. The border gateways in turn may be used as a transit to access networks. For example, all the Internet traffic meant for networks in Michigan would have to pass through one of the border gateways located at Ann Arbor. Thus the problem of finding a path to a network will reduce to the problem of finding a path to the border gateway that will further forward the data to the correct network. Knowledge of this pattern helps in building scalable networks.

5.4 USER DATAGRAM PROTOCOL (UDP)

UDP, a TCP/IP protocol suite that provides connectionless datagram transmission, is a simple protocol that exchanges datagrams without acknowledgments or guaranteed delivery, requiring that error processing and retransmission be handled by other protocols. UDP is connectionless and is considered to be unreliable. Although UDP is responsible for transmitting messages, no software checking for segment delivery is provided at this layer; hence the description "unreliable." In this section we provide a brief overview of UDP. A complete description is given in RFC 768.

To distinguish between the many programs running on a single machine, UDP uses a port number corresponding to each program. Each UDP message, in addition to data information contains the port information for source and destination machine. Port numbers are used to keep track of different conversations crossing the network at the same time. Application software developers agree to use well-known port numbers to do specific tasks that are defined in RFC 1700 (Table 5.6). For example, any conversation bound for the FTP application uses the standard port number 21. Conversations that do not involve

TABLE 5.6
Some Well-Known Port Numbers in UDP

Decimal	Keyword	UNIX Command	Description
0	—	—	Reserved
7	ECHO	echo	Echo
9	DISCARD	discard	Discard
11	USERS	systat	Active Users
13	DAYTIME	daytime	Daytime
15	—	netstat	Who is up, or NETSTAT
17	QUOTE	qotd	Quote of the day
19	CHARGEN	chargen	Character generator
37	TIME	time	Time
42	NAMESERVER	name	Host name server
43	NICNAME	whois	Who is
53	DOMAIN	nameserver	Domain name server
67	BOOTPS	bootps	Bootstrap protocol server
68	BOOTPC	bootpc	Bootstrap protocol client
69	TFTP	tftp	Trivial file transfer protocol
111	SUNRPC	sunrpc	Sun Microsystems Remote Procedure Co
123	NTP	ntp	Network Time Protocol
161	—	snmp	SNMP net monitor
162	—	snmp-trap	SNMP traps
512	—	biff	Unix comsat
513	—	who	Unix who daemon
514	—	syslog	System log
525	—	timed	Time daemon

0	16 31
UDP SOURCE PORT	UDP DESTINATION
UDP MESSAGE LENGTH	UDP CHECKSUM
DATA	
...	

Figure 5.28 The UDP frame format.

an application with a well-known port number are assigned numbers randomly chosen from within a specific range. These port numbers are used as source and destination addresses in the TCP and UDP segments.

Some ports are reserved in both TCP and UDP, and applications might not be written to support them. Port numbers have the following assigned ranges:

- Numbers below 255 are for public applications.
- Numbers from 255 to 1023 are assigned to companies for salable applications.
- Numbers above 1023 are unregulated.

UDP uses IP as its underlying protocol. It does not use acknowledgment or ordering of messages. Hence it is called an unreliable connectionless delivery service. An application using UDP accepts full responsibility for handling the consequences of unreliability, including message loss, duplication, delay, out-of-order delivery, and loss of connectivity. Protocols that use UDP include Trivial File Transfer Protocol (TFTP), Simple Network Management Protocol (SNMP), Network File System (NFS), and the domain name system (DNS).

The UDP frame consists of 16-bit source and destination ports with 64K port numbers. The other fields include the checksum (16 bits) and the data length (16 bits), allowing a maximum of 64 KB of data (including the header). The minimum value of length field is 8 (i.e., the length of a UDP header in case of no data). The frame format is shown in Figure 5.28.

5.5 TRANSMISSION CONTROL PROTOCOL (TCP)

TCP is one of the most widely used full-duplex transport protocols on the Internet. It provides a reliable and in-order byte stream between two ends as it frees applications from having to worry about missing or reordered data. In addition, it also provides a flow control mechanism for the byte streams. The flow control mechanism allows the receiver to limit the number of bytes it accepts from the sender. It is an end-to-end control mechanism that is helpful in controlling the buffer overrun at the receiver. It is important to see how flow control differs from congestion control, in which the idea is to prevent the injection of too much data into the network, which would cause the network links and switches to be overloaded.

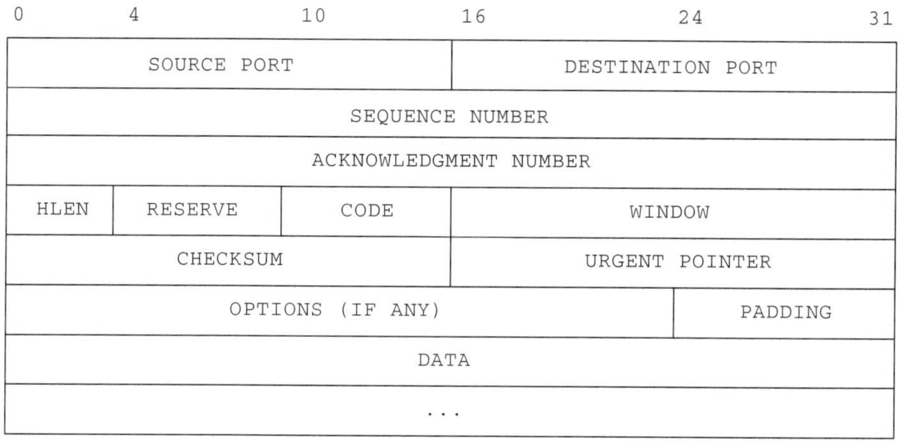

Figure 5.29 The TCP frame format.

TCP is a connection-oriented protocol meaning that the data transfer is preceded by a connection establishment phase, and a connection termination phase follows the data transfer.

5.5.1 TCP Headers and Services

TCP is a reliable, connection-oriented transmission protocol based on IP. It allows multiple application programs on one machine to communicate to one another through a demultiplexing operation. The demultiplexing operation makes it possible for two or more application programs running on same or different hosts to simultaneously carry out a data transfer. For this purpose, it uses port numbers just like UDP. The connection is considered to be an abstraction consisting of a virtual circuit between two applications running on usually two different machines. The host machine address and port number for each machine serve as the end points of such a virtual circuit.

The frame format of a TCP message is shown in Figure 5.29. The source port and destination port fields contain the TCP port numbers that identify the application programs at the two ends of the connection. The sequence number field identifies the position in the sender's byte stream of the data in the segment. The acknowledgment number field identifies the number of the byte the source expects to receive next. The field called HLEN contains an integer that specifies the length of the segment header measured in 32-bit multiples. It is needed because the options field varies in length, depending on which options have been included. Thus the size of the TCP header varies accordingly. The 6-bit reserved field is reserved for future use. The window field contains the buffer size that limits the data TCP software is willing to accept every time it sends a segment.

Some segments carry only an acknowledgment while some carry data. Other carry requests to establish or close a connection. TCP software uses the 6-bit code bits field to determine the purpose and contents of the segment. The 6 bits are interpreted as follows:

URG-bit (left-most): Urgent pointer field is valid.

ACK-bit: Acknowledgment field is valid.

PSH-bit: This segment requests a push.

RST-bit: Reset the connection.

SYN-bit: Synchronize sequence numbers.

FIN-bit: Sender has reached the end of its byte stream.

Like UDP, TCP combines static and dynamic port binding, using a set of well-known port assignments for commonly invoked programs (see Table 5.7).

TABLE 5.7
Some Well-Known Port Numbers in TCP

Decimal	Keyword	Unix Command	Description
0			Reserved
1	TCPMUX		TCP multiplexer
5	RJE		Remote job entry
7	ECHO	echo	Echo
9	DISCARD	discard	Discard
11	USERS	systat	Active users
13	DAYTIME	daytime	Daytime
15		netstat	Network status program
17	QUOTE	qotd	Quote of the day
19	CHARGEN	chargen	Character generator
20	FTP-DATA	ftp-data	File-Transfer-Protocol (data)
21	FTP	ftp	File Transfer Protocol
23	TELNET	telnet	Terminal connection
25	SMTP	smtp	Simple Mail Transfer Protocol
37	TIME	time	Time
42	NAMESERVER	name	Host name server
43	NICNAME	whois	Who is
53	DOMAIN	nameserver	Domain name server
77		rje	Any private RJE service
79	FINGER	finger	Finger
93	DCP		Device Control Protocol
95	SUPDUP	supdup	SUPDUP Protocol
101	HOSTNAME	hostnames	NIC host name server
102	ISO-TSAP	iso-tsap	ISO-TSAP
103	X400	x400	X.400 mail service
104	X400-SND	x400-snd	X.400 mail sending
111	SUNRPC	sunrpc	Sun Microsystems Remote Procedure Call
113	AUTH	auth	Authentication service
117	UUCP-PATH	uucp-path	UUCP path service
119	NNTP	nntp	Usenet News Transfer Protocol
129	PWDGEN		Password generator protocol
139	NETBIOS-SSN		NetBIOS session service
160–223	Reserved		

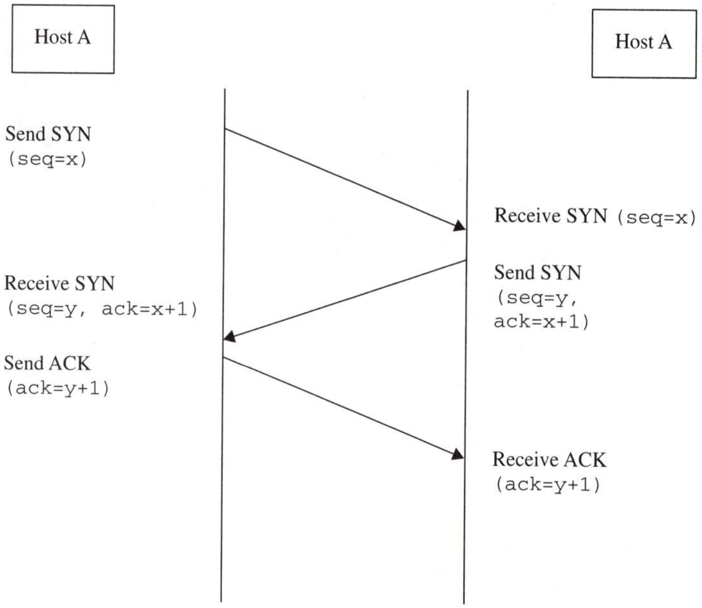

Figure 5.30 TCP three-way handshake and open connection.

5.5.2 Connection Establishment and Termination

Connection establishment in TCP is done via a three-way handshake between the sending and receiving hosts. Figure 5.30 shows this handshake operation. The sending host (host A) sends a SYN signal to the receiving host (host B) indicating that it is ready to transmit and would like to synchronize the byte sequence numbers. After host B has received the SYN signal from host A, host B sends a 2nd SYN signal back to host A with a sequence number, say `y`, and an ACK field equal to `x + 1`. The sequence `y` in the SYN signal indicates the starting sequence number of the backward traffic from host B to A. When host B receives this SYN signal from host B, it sends an ACK signal to B indicating its expectation of byte sequence `y + 1` onward.

Connection termination occurs in TCP when either party does not receive an IP datagram for twice the time a packet might live on the Internet (120 seconds). Sending the FIN signal terminates the connection. It is necessary to wait for twice the maximum time because the host terminating the connection must receive the ACK for its FIN signal. In case the ACK is lost, the retransmitted FIN signal will close the port connection. However, if the sending host does not wait for this period and another TCP connection using the same port number is established, a delayed FIN signal may erroneously terminate the connection.

5.5.3 Flow Control and Window Size Advertising

Flow control in TCP is provided by means of a window mechanism. Window size refers to the number of messages that can be transmitted while awaiting an acknowledgment. It determines how much data the receiving station can accept at one time. After a host has advertised the window size number of bytes, it must receive an acknowledgment before any more messages can be sent. So, with a window size of one, each segment must be acknowledged before another segment is transmitted. This results in inefficient use of bandwidth by the hosts. A larger window size allows more data to be transmitted pending acknowledgment.

TCP uses an acknowledgment mechanism with the ACK number referring to the octet expected next, calling on the sliding window protocol, which works very much like the one used at data link layer. The "sliding" part of "sliding window" refers to the fact that the window size is negotiated dynamically during the TCP session. A sliding window results in more efficient use of bandwidth by the hosts. TCP provides sequencing of segments with a forward reference acknowledgment. Each datagram is numbered before transmission. At the receiving station, TCP reassembles the segments into a complete message. If a sequence number is missing in the series, that segment is retransmitted. Segments that are not acknowledged within a given time period result in retransmission.

5.5.4 TCP Time-Out and Retransmission

TCP guarantees a reliable delivery of data even in the case of transmission losses and delays. This is possible because each segment of data is retransmitted if an ACK is not received within a certain period of time. The time-out period is set as a function of the round-trip time. However, owing to wide variation of the round-trip times (RTTs) between hosts on the Internet, choosing this period may call for some discretion. The simplest technique involves a recursive estimation of the RTT based on the current estimate and the sample RTT. Therefore

$$\text{estimated RTT} = a \times \text{estimated RTT} + b \times \text{sample RTT}$$

here $a + b = 1$, and a and b are mainly chosen to smooth the estimated RTT. Notice that if b is large, the changes in RTT are tracked. A large a is more stable but not quick enough to adapt to real changes in the RTT. When an estimate for the RTT is obtained, the time-out period is simply taken to be twice as long.

There are at least two algorithms that modify the foregoing simple algorithm for the computation of the RTT. In the first, called the Karn/Partridge algorithm, the sender does not obtain a sample RTT once a retransmission has occurred. This eliminates the possibility of an incorrect sample RTT if the retransmission and ACK signals cross over. The second algorithm, proposed by Jacobson and Karles, uses a weighted difference between the sample and estimated RTTs to obtain a new value for the estimated RTT. The change in time-out is then obtained from the estimated RTT and the deviation.

5.6 INTERNET STANDARD SERVICES

Much of the impetus for the development of TCP/IP is the need for internetworking services—the ability of end users to communicate through a local machine to a remote machine or to remote end users. The traditional TCP/IP services are supported by the appropriate protocols, such as the following.

- The File Transfer Protocol (FTP), which allows the transfer of files from one computer on the Internet to another computer on the Internet.
- The Network Terminal Protocol (telnet), which provides a means a user on the Internet to log onto any other computer on the network.
- The Simple Mail Transfer Protocol (SMTP), which allows users to send messages to one another on the Internet.
- In a TCP/IP Internet environment, IP routers form the active switches that managers need to examine and control. The prevalent network management protocols, SNMP and SNMPv2, are based on a management information base (MIB).

In this section we discuss about some of these standard services provided on the Internet. The MIB and SNMP services are discussed later in Chapter 9, Network Management.

5.6.1 File Transfer Protocol (FTP)

To use FTP, TCP port 21 is fixed as the command channel and TCP port 20 as the data channel. In Unix systems, the protocol consists of ftp and the server daemon process ftpd. FTP differs from other file transfer programs in many respects. The most prominent differences include the use of separate channels for control information and data and the fact that FTP data transfers do not run in the background (i.e., work without a spooler).

FTP differs from other client–server applications in that it establishes two virtual connections, each being bidirectional, between the hosts. One connection is used for data transfer, while the other is used for control information. The control connection is used for commands, replies, and process updates. While data transfer takes place across the data connection, the control connection issues reassurances to the user interface with respect to the beginning of file transfers, expected duration times, completion of data transfers, and successful data transfers.

5.6.2 Simple Mail Transfer Protocol (SMTP) and Examples

SMTP is a standard protocol for mail transfer in the TCP/IP protocol suite. SMTP focuses on the message-passing mechanism of the underlying mail delivery system across a link from one machine to another. TCP port 25 is defined for SMTP. Like FTP and telnet, SMTP is desirable for its simplicity. It incorporates many features of FTP.

In Unix systems, SMTP is implemented by the program sendmail. This program can communicate with mail services other than SMTP and, to some extent, also operates as a gateway between different mail systems. It is possible for sendmail to function not only as an SMTP server, but also as an SMTP client. However, users never use sendmail directly, but mostly use pine or mail, which controls and simplifies the processing of a message. Sendmail is activated only to forward the message. If forwarding is not immediately possible, the message is entered into an output queue. Regular attempts are then made to forward the message from the queue to the destination.

A configuration file "`sendmail.cf`" is used to control sendmail, which makes the program very adaptable. In addition to the definition of the local-mail-forwarding program and many other uses, this file contains commands for converting the address for the connected mail system. File aliases, which may be used to create distribution lists and for forwarding requests, are edited and then converted into an indexed database by means of the command new aliases. Alias names may also be used to send a message to a program, for example, to set up an automatic answering service. The prototype networks provide complete e-mail services.

However, sendmail does not specify the following:

How the mail system accepts mail from one machine to another

How the mail system accepts mail from a user

How the user interface presents the user with incoming mails

How mail is stored, or how frequently the mail system attempts to send messages

Communication carried out between SMTP-client and SMTP-server for mail transfer is in the form of readable text.

5.6.2.1 SMTP Commands and Semantics

The SMTP commands define the mail transfer or the mail system function requested by the user. SMTP commands are character strings terminated by ⟨CRLF⟩. The command codes themselves are alphabetic characters terminated by ⟨SP⟩ if parameters follow and ⟨CRLF⟩ otherwise. In the following list, the commands are given in parentheses.

Hello (`HELO`): This command is used to identify the sender-SMTP to the receiver-SMTP. The argument field contains the host name of the sender-SMTP. The receiver-SMTP identifies itself to the sender-SMTP in the connection greeting reply, and in response to this command. This command and an OK reply to it confirm that both the sender-SMTP and the receiver-SMTP are in the initial state; that is, there is no transaction in progress and all state tables and buffers are cleared.

Mail (`MAIL`): This command is used to initiate a mail transaction in which mail data are delivered to one or more mailboxes. The argument field contains a reverse path.

Recipient (`RCPT`): This command is used to identify an individual recipient of the mail data; multiple recipients are specified by multiple use of this command.

Data (`DATA`): The receiver treats the lines following the command as mail data from the sender. This command causes the mail data to be appended to the buffer. The mail data may contain any of the 128 ASCII character codes. To indicate the end of mail data, SMTP uses a line containing only a period, that is, the character sequence ⟨CRLF⟩.⟨CRLF⟩ appears at the end of mail data.

Send (`SEND`): This command is used to initiate a mail transaction in which the mail data are to be delivered to one or more terminals. The argument field contains a reverse path. This command is successful if the message is delivered to a terminal.

Send or mail (`SOML`): This command is used to initiate a mail transaction serving to deliver the mail data to one or more terminals or mailboxes. The mail data are delivered to a recipient's terminal if the recipient is active on the host (and accepting terminal messages); otherwise delivery is to the recipient's mailbox. The argument field contains a reverse path. This command is successful if the message is delivered to a terminal or the mailbox.

Send and mail (SAML): This command is used to initiate a mail transaction in which the mail data are delivered to one or more terminals and mailboxes. The mail data are delivered to a recipient's terminal if the recipient is active on the host (and accepting terminal messages) and to all recipients in their respective mailboxes, including those that have received the message at their terminals. The argument field contains a reverse path. This command is successful if the message is delivered to the mailbox.

Reset (RSET): This command specifies that the current mail transaction is to be aborted. Any stored sender, recipients, and mail data must be discarded, and all buffers and state tables cleared. The receiver must send an OK reply.

Verify (VRFY): This command asks the receiver to confirm that the argument identifies a user. If it is a user name, the full name of the user (if known) and the fully specified mailbox are returned. This command has no effect on the reverse path buffer, the forward-path buffer, or the mail data buffer.

Expand (EXPN): This command asks the receiver to confirm that the argument identifies a mailing list, and if so, to return the membership of that list. The full name of the users (if known) and the fully specified mailboxes are returned in a multiline reply. This command has no effect on the reverse path buffer, the forward path buffer, or the mail data buffer.

Help (HELP): This command causes the receiver to send helpful information to the sender. The command may take an argument (e.g., any command name) and return more specific information as a response. This command has no effect on the reverse path buffer, the forward path buffer, or the mail data buffer.

Noop (NOOP): This command does not affect any parameters or previously entered commands. It specifies no action other than that the receiver send an OK reply. This command has no effect on the reverse path buffer, the forward path buffer, or the mail data buffer.

Quit (QUIT): This command specifies that the receiver must send an OK reply and then close the transmission channel.

Turn (TURN): This command specifies that the receiver must either (1) send an OK reply and then take on the role of sender-SMTP or (2) send a refusal reply and retain the role of receiver-SMTP.

5.6.2.2 SMTP Command Syntax

```
HELO ⟨SP⟩ ⟨domain⟩ ⟨CRLF⟩
MAIL ⟨SP⟩ FROM:⟨reverse-path⟩ ⟨CRLF⟩
RCPT ⟨SP⟩ TO:⟨forward-path⟩ ⟨CRLF⟩
```

```
DATA ⟨CRLF⟩
RSET ⟨CRLF⟩
SEND ⟨SP⟩ FROM:⟨reverse-path⟩ ⟨CRLF⟩
SOML ⟨SP⟩ FROM:⟨reverse-path⟩ ⟨CRLF⟩
SAML ⟨SP⟩ FROM:⟨reverse-path⟩ ⟨CRLF⟩
VRFY ⟨SP⟩ ⟨string⟩ ⟨CRLF⟩
EXPN ⟨SP⟩ ⟨string⟩ ⟨CRLF⟩
HELP [⟨SP⟩ ⟨string⟩] ⟨CRLF⟩
NOOP ⟨CRLF⟩
QUIT ⟨CRLF⟩
TURN ⟨CRLF⟩
```

5.6.2.3 Replies to SMTP Commands

SMTP command replies are devised to ensure the synchronization of requests and actions in the process of mail transfer, and to guarantee that the sender-SMTP always knows the state of the receiver-SMTP. Every command must generate exactly one reply. An SMTP reply consists of a three-digit number (transmitted as three alphanumeric characters) followed by some text. The number is intended for use by automata to determine what state to enter next; the text is meant for the user. Table 5.8 lists some well-known reply codes by function group.

TABLE 5.8
Reply Codes by Function Group

Error code	Description
500	Syntax error, command unrecognized [This may include errors such as command line too long.]
501	Syntax error in parameters or arguments
502	Command not implemented
503	Bad sequence of commands
504	Command parameter not implemented
211	System status, or system help reply
214	Help message [Information on how to use the receiver or the meaning of a particular nonstandard command; this reply is useful only to the human user.]
220	⟨domain⟩ Service ready
221	⟨domain⟩ Service closing transmission channel
421	⟨domain⟩ Service not available, closing transmission channel [This may be a reply to any command if the service knows it must shut down.]
250	Requested mail action OK, completed
251	User not local; will forward to ⟨forward-path⟩
450	Requested mail action not taken: mailbox unavailable (e.g., mailbox busy)
550	Requested action not taken: mailbox unavailable (e.g., mailbox not found, no access)
451	Requested action aborted: error in processing
551	User not local; please try ⟨forward-path⟩
452	Requested action not taken: insufficient system storage
552	Requested mail action aborted: exceeded storage allocation
553	Requested action not taken: mailbox name not allowed (e.g., mailbox syntax incorrect)
354	Start mail input; end with ⟨CRLF⟩.⟨CRLF⟩
554	Transaction failed

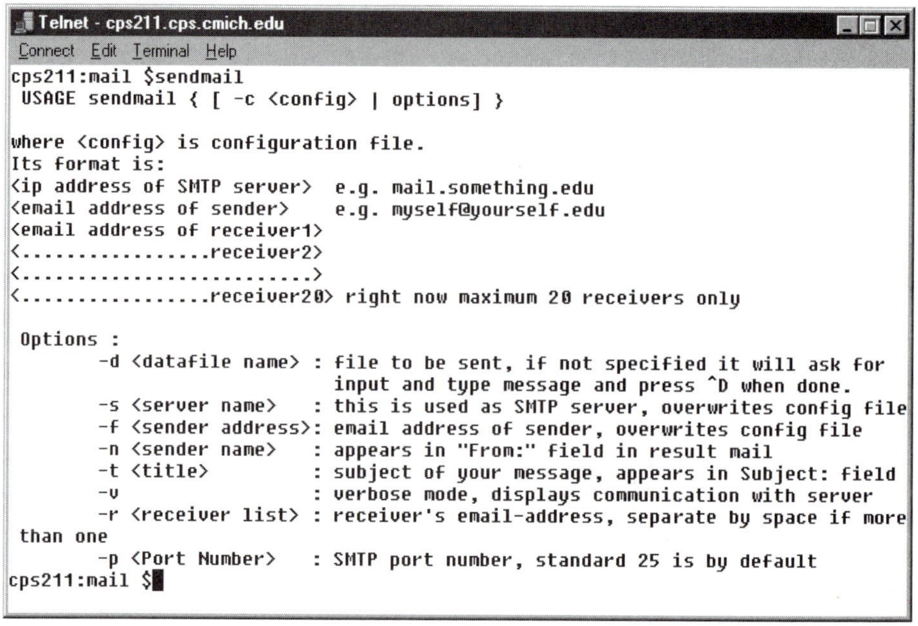

Telnet - cps211.cps.cmich.edu
Connect Edit Terminal Help
cps211:mail $sendmail
USAGE sendmail { [-c <config> | options] }
where <config> is configuration file.
Its format is:
<ip address of SMTP server> e.g. mail.something.edu
<email address of sender> e.g. myself@yourself.edu
<email address of receiver1>
<.................receiver2>
<.........................>
<.................receiver20> right now maximum 20 receivers only
Options :
-d <datafile name> : file to be sent, if not specified it will ask for input and type message and press ^D when done.
-s <server name> : this is used as SMTP server, overwrites config file
-f <sender address>: email address of sender, overwrites config file
-n <sender name> : appears in "From:" field in result mail
-t <title> : subject of your message, appears in Subject: field
-v : verbose mode, displays communication with server
-r <receiver list> : receiver's email-address, separate by space if more than one
-p <Port Number> : SMTP port number, standard 25 is by default
cps211:mail $

Figure 5.31 Screen snapshot of sendmail.

5.6.2.4 A Nonstandard "Sendmail" Program

Now we illustrate a nonstandard "sendmail" program that is Unix based and developed in C using BSD 4 sockets. It can be used to send text-mails to multiple recipients. The program requires the SMTP server name, the sender's e-mail address, and receiver's e-mail address as basic input. It connects to the server's SMTP port (which is default 25 if not otherwise specified) and sends standard commands to send the message. The user can specify the file name to be sent; otherwise, the program will prompt the user to type a message. The user can end the message with Ctrl-D. The program uses HELO, MAIL, RCPT, DATA, dot (.), and QUIT commands only for communication with the SMTP server. It is not capable of sending attachments. Various command line options available with "sendmail" are displayed when it is invoked with wrong syntax or without any arguments. Figure 5.31 shows the output of program when invoked without any command line arguments.

A configuration file that contains the SMTP server name, from-address, and to-addresses can be specified in sendmail. If the program is invoked with the '`-c`' option and command line parameters like '`-s`' and '`-f`', then command line parameters will override the values in configuration file. But in the case of recipients, the mail will be sent to both addresses specified in the file and addresses specified in the command line with the '`-r`' option. Figure 5.32 shows a list of "verbose" commands invoked by sendmail during communication with an SMTP server.

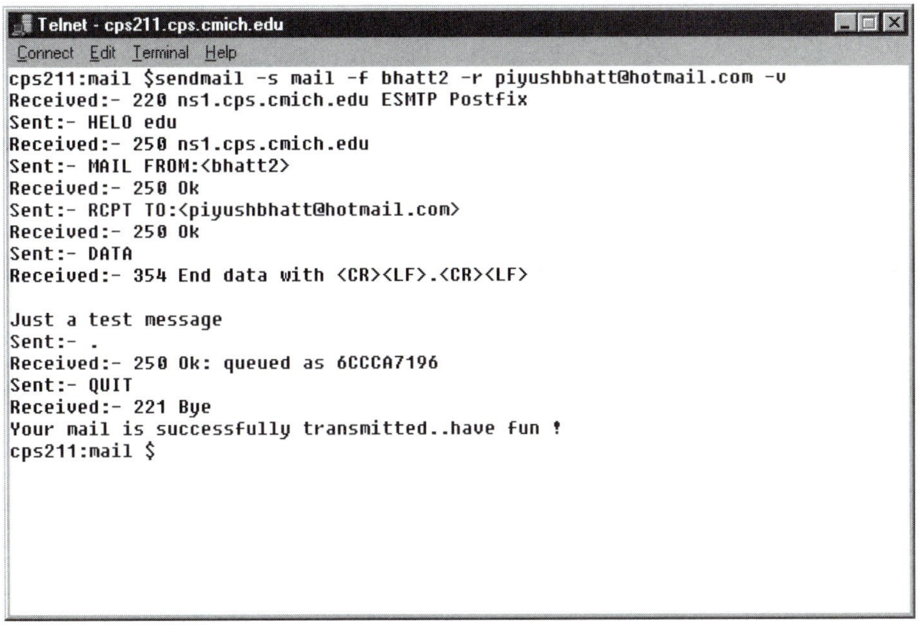

Telnet - cps211.cps.cmich.edu
Connect Edit Terminal Help
cps211:mail $sendmail -s mail -f bhatt2 -r piyushbhatt@hotmail.com -v
Received:- 220 ns1.cps.cmich.edu ESMTP Postfix
Sent:- HELO edu
Received:- 250 ns1.cps.cmich.edu
Sent:- MAIL FROM:<bhatt2>
Received:- 250 Ok
Sent:- RCPT TO:<piyushbhatt@hotmail.com>
Received:- 250 Ok
Sent:- DATA
Received:- 354 End data with <CR><LF>.<CR><LF>
Just a test message
Sent:- .
Received:- 250 Ok: queued as 6CCCA7196
Sent:- QUIT
Received:- 221 Bye
Your mail is successfully transmitted..have fun !
cps211:mail $

Figure 5.32 Use of "sendmail" during communication with an SMTP server.

5.6.3 Post Office Protocol (POP3)

The POP3 protocol is used to pop mail messages from the POP3 server. This protocol is described in RFC 1725 and is very similar to SMTP inasmuch as it uses text commands for popping mail messages.

Initially, the server host starts the POP3 service by listening on TCP port 110. When a client host wishes to make use of the service, it establishes a TCP connection with the server host. When the connection is established, the POP3 server sends a greeting. The client and the POP3 server then exchange commands and responses, respectively, until the connection is closed or aborted.

Commands in POP3 consist of a keyword, possibly followed by one or more arguments. A CRLF pair terminates all commands. Keywords and arguments consist of printable ASCII characters. Keywords and arguments are each separated by a single space character. Keywords are 3 or 4 characters long. An argument may be up to 40 characters long.

Responses in the POP3 format consist of a status indicator and a keyword possibly followed by additional information. A CRLF pair terminates all responses just like the commands. There are currently two status indicators: positive ("`+OK`") and negative ("`-ERR`").

5.6.3.1 The popmail Program

The popmail program is developed in C using BSD-4 sockets to run on a Unix platform. It is used to pop the mail messages of a user from a single POP3 server. The program

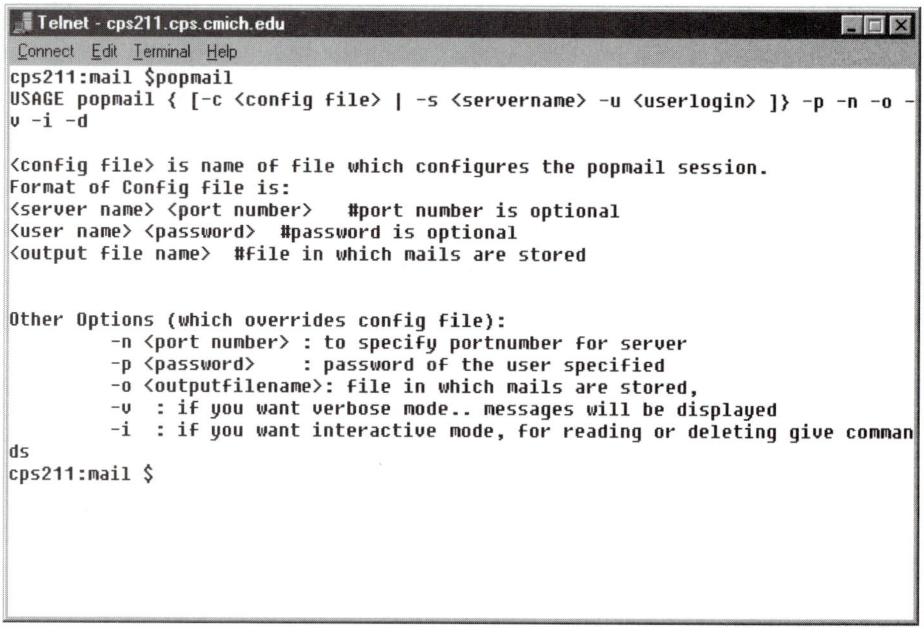

Figure 5.33 Use of popmail.

needs the server name, plus the user name as argument and it asks for a password if one is not specified in command line. In demo versions, one password is displayed on the screen when typed, just to be sure. If the program is invoked without arguments or with wrong arguments, it gives the output as shown in Figure 5.33, which helps the user to give correct parameters.

The popmail program checks the number of messages in a user's mailbox. If there are no messages, it displays the new message and exits. If user has mail in his mailbox and if popmail is not invoked for interactive operation, with the '`-i`' option, all the mails will be saved in the readmail file by default. The output filename can be specified with the '`-o`' option. If invoked with the '`-i`' option, it prompts the user with "`(r, s, d, q)?`" (where '`r`' means read, '`s`' means 'save', '`d`' means delete, and '`q`' means quit).

To read message number 1, the user can type '`r 1`'. And same for '`d`'. If '`r`' is used without a parameter, it reads the last message (by default 0) plus 1 and rotates from 1 to last number mail. If '`d`' is used without parameter, it deletes the last message read. '`s`' needs an argument for filename to store the last read message. So, to save message number 2 in file '`m2`', the sequence of command will be to read 2 (`r 2`) and then save to file (`s m 2`).

An option not listed earlier, is '`-b`'. When followed by a numerical parameter, '`-b`' is used to invoke the program as background process to keep checking the user mailbox periodically. The period is specified as number of minutes in argument of '`-b`'. To run popmail as a background process and keep checking for mail every 5 minutes, the syntax is:

```
$popmail -s server -u user -p password -b 5 &
```

After invoking this, popmail will check user's mailbox every 5 minutes and, if new mail messages have arrived, it will display the number of messages now in the user's mailbox. If all messages are deleted, the display will show that the user now has 0 messages.

If any other program checks the mail at the same time that the background process is being invoked, popmail may declare a "mailbox locked" error and exit. It has no functionality to check for semaphores and locking of mailboxes.

Typical use of popmail is displayed below in "verbose" mode.

```
cps227:mail $popmail -c cps -i -v

Configuration File:cps
Server Name: mail.cps.cmich.edu Port Number: 110
User Name : bhatt2 Password : XXXXXXXX
Out File Name: readmail
VerBoseMode: ON
You have 4 new messages
(r, s, d, q)?r 4
Delivered-To: bhatt2@cps.cmich.edu
Received: from cps.cmich.edu (cps211 [141.209.131.211])
        by ns1.cps.cmich.edu (Postfix) with SMTP
        id 708A87198; Sat, 30 Jan 1999 18:29:31 -0500 (EST)
Date: Sat Jan 30 18:28:45 1999 EST
X-Mailer: ⟨ Why Do You Want To Know ⟩
Content-Type: TEXT/PLAIN; charset=US-ASCII
Message-Id: ⟨19990130232931.708A87198@ns1.cps.cmich.edu⟩
From: bhatt2@cps.cmich.edu
X-UIDL: 8e77e96ddc2b8d890340e4f1988a2564
Status: U

test message 2

(r, s, d, q)?d 4
 Mail # 4 successfully deleted..
(r, s, d, q)?r 3
Delivered-To: bhatt2@cps.cmich.edu
Received: from cps.cmich.edu (cps211 [141.209.131.211])
        by ns1.cps.cmich.edu (Postfix) with SMTP
        id 5209D7196; Sat, 30 Jan 1999 17:16:41 -0500 (EST)
Date: Sat Jan 30 17:15:55 1999 EST
X-Mailer: ⟨ Why Do You Want To Know ⟩
Content-Type: TEXT/PLAIN; charset=US-ASCII
Message-Id: ⟨19990130221641.5209D7196@ns1.cps.cmich.edu⟩
From: bhatt2@cps.cmich.edu
X-UIDL: 0bac6b1db0a7310a40ea93bc30a77141
Status: RO

hello, how are you?
Just test message..

(r, s, d, q)?q
cps227:mail $
```

5.6.3.2 Basic Operation of POP3

A POP3 session passes through three major states, namely, authorization, transaction, and update.

Authorization: When TCP connection is made with the POP3 server, it enters the authorization state. The client must identify itself with username and password. Once identified, it enters the transaction state.

Transaction: During this state, client can invoke command like `LIST`, `STAT`, `RETR`, and `DELE`. When `QUIT` is invoked in this state, the session enters the update state.

Update: The server releases all the resources it acquired during the POP3 session and sends '`Bye`' to the client. The server closes the TCP connection, and thus the POP3 session ends.

5.6.3.3 Some POP3 Commands

POP3 commands that are valid in the authorization state include the following:

`USER name:` Specifies the name of the user whose mail messages are to be popped

`PASS string:` Gives the ASCII password for the user name

`QUIT:` Closes the connection with the POP3 server and quits the POP3 session.

The following commands are valid in the transaction state:

`STAT:` Gives the total number of mail messages user currently has.

`LIST [msg]:` Lists all the messages and their size. It does not list the data. If a message number is specified, it gives the size of any existing messages.

`RETR msg:` Retrieves the message content, if any, specified in the argument. The message is sent to the client through a TCP connection, the client receives all the data and either stores them or displays them to the user.

`DELE msg:` Deletes from the user's mailbox the message specified as the argument.

`NOOP:` No operation is invoked to keep the connection up with the POP3 server. The POP3 server closes the connection if the client remains idle for a specific time.

`RSET:` Resets the session.

`QUIT:` Closes the connection and quits the session.

5.6.3.4 Optional POP3 Commands:

The following optional POP3 commands are valid in the transaction state:

`TOP msg n:` Retrieves the header information of the message specified in argument; '`n`' specifies how many lines of the body of the message are to be read.

`UIDL [msg]:` This is the message identification number's list.

5.6.3.5 POP3 Replies

```
+OK
-ERR
```

5.6.4 Remote Login and Telnet

Telnet is intended to provide access, in the form of a terminal session, to a computer connected to the network. The telnet service is attached to TCP port 23. Unix processors currently incorporate the command rlogin, which offers almost the same functionality as telnet but provides better support for the unix environment. Both telnet and rlogin run on the prototype networks smoothly. Telnet (and FTP) are configured in the file `/etc/services and /etc/passwd`

Remote login allows a user at one site to gain access to a computer at another site. For reasons of security, however, telnet offers a solution in which users are required to provide login information. To use telnet, one invokes the client at a local machine and then establishes a connection with a telnet server at the remote site. To provide homogeneity between the user and the remote server, telnet provides translation services between the two machines. The client transforms the output from the actual terminal to standard code, which is then transferred to the server, and the server transforms the information into characters acceptable by the remote host. This type of terminal connection is called a virtual terminal (VT) connection because the remote host assumes a local terminal that is nonlocal and connected via TCP/IP.

5.7 DOMAIN NAME SYSTEM (DNS)

Machines understand numeric addresses, but it is easier for users to remember names. The domain name system identifies hosts on the Internet with unique names that may be translated unambiguously to the corresponding IP addresses. With the domain name known, a program can obtain the associated IP address by engaging with a name server in a client–server session.

To maintain the individual identities of hosts on the Internet, which contains millions of hosts, the designers have come up with a hierarchical system comprising a number of labels separated by dots. For example, a host at a company called ABC Systems might have the domain name anyhost.abcsys.com. A host can have any number of identifying labels to make its name unique as long as the label length does not exceed 63 characters.

5.7.1 Mapping Domain Names to IP Addresses

The DNS implementation is just like a tree in which each node represents one possible label. The right-most label corresponds to the node closes to the root, whereas the left-most label corresponds to the host name and it is the farthest node from the root. The Internet's organization domain consists of labels describing the types of client organization. The following labels are used to describe organization type:

.com: Commercial organization

.edu: Educational institution

.gov: Government organization

.int: International organization

.mil: Military

.net: Network-related support center

.org: Organization not included in any other category

In our example hostname (anyhost.abcsys.com), anyhost is the host name in the commercial organization registered as abcsys. It is noticed that the number of levels in the hierarchy is not limited to three. The country name is used as a suffix after the organization type: for example, .us for the United States, .fr for France, .jp for Japan.

The resolution of a domain name to its IP address occurs as follows. The host looking for the IP address for a domain name searches its local domain name server to locate the IP address. If address resolution is not reached, a message is sent to the name server of the next level, that is, the level next closest to the root. The name resolution process continues until the root is reached; then further resolution is achieved by moving down the tree toward the destination name server, which provides the correct IP address to the source host.

5.7.2 DNS Messages

DNS messages are exchanged between stations to perform certain tasks during address resolution. The DNS message format is shown in Figure 5.34. The identification field is

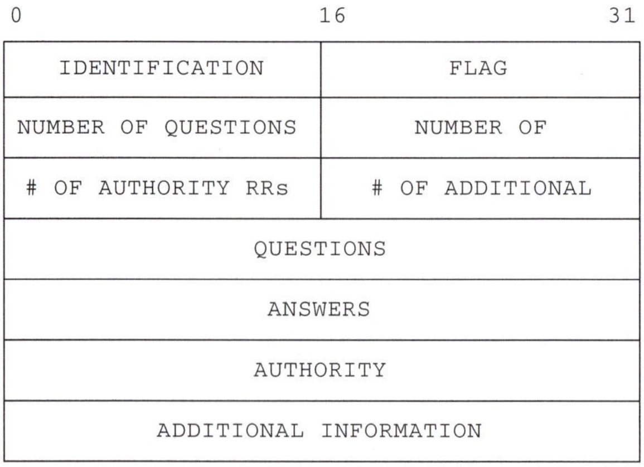

Figure 5.34 DNS message format.

set by the client and returned by the server. The 16-bit field flag consists of the following:

0th-bit field: Contains a 0, meaning that the message is a query, or a 1, meaning that it is a response.

1–4 bit fields: Opcode 0 indicates a normal value (standard query); 1 indicates an inverse query; 2 is the server status request.

5th bit field: Authoritative answer. The name server is authoritative for the domain in the question section.

6th bit field: This is set if the message is truncated. With UDP this means that the total size of the reply exceeded 512 bytes, and only the first 512 bytes of the reply were returned.

7th bit field: Recursion desired. This bit can be set in a query and is then returned in the response.

8th bit field: Recursion available.

9–11 bit field: Must be 0.

12–15 bit field: Return code: 0, no error; 3, name error.

Each field labeled "number of . . ." gives a count of the entries in the corresponding section in the message. The question section field contains queries for which answers are desired. The client fills in only the question section; the server returns the question and answers with its response. Each question has a query domain name field followed by query type and query class fields (as shown in Figure 5.35). The answers, authority, and additional information sections (Figure 5.34) consist of a set of resource records that describe domain names and mappings. Each resource record describes one name.

5.7.3 Recent Advances in Internet Domain Name Hierarchy

Industry representatives, Internet users, and academics around the world have essentially agreed on the structure for a new global organization for the assigning and management of top-level domain names. The agreement followed months of discussions, meetings, and white papers that had as their common goal the formation of a new entity to replace the Internet Assigned Numbers Authority (IANA). The IANA is the U.S. government–funded organization that handles the back-end administration of so-called top-level domain names, such as .com, .net, and .org. The international Internet community called for this new, global nonprofit organization to be put into place to allocate and manage top-level do-

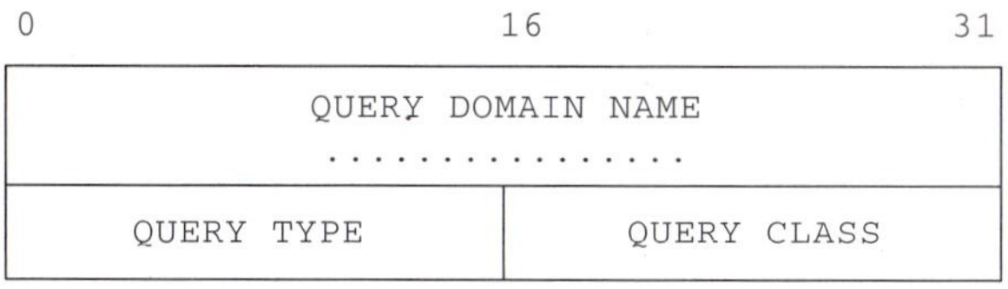

Figure 5.35 Question-type DNS message.

main names. The way top-level domain names are allocated, registered, and managed caused heated debate when the existing U.S.-controlled system came under fire for its noninternational approach.

5.8 TCP/IP FOR PCS

5.8.1 Serial Line Internet Protocol (SLIP)

RFC 1055 described SLIP as a "non-standard" back in 1988 because it was not a defined standard but was a de facto standard for point-to-point serial connections running TCP/IP. It is a simple protocol that provides no addressing, packet-type identification, error detection/correction, or compression mechanisms. Because the protocol does so little, though, it is usually very easy to implement.

The SLIP protocol defines two special characters: END and ESC. END is octal 300 (decimal 192) and ESC is octal 333 (decimal 219), not to be confused with the ASCII ESCape character. To send a packet, a SLIP host simply starts sending the data in the packet. If a data byte is the same code as END character, a 2-byte sequence of ESC and octal 334 (decimal 220) is sent instead. If it is the same as an ESC character, a 2-byte sequence of ESC and octal 335 (decimal 221) is sent instead. When the last byte in the packet has been sent, an END character is transmitted.

The following are commonly perceived shortcomings in the SLIP protocol:

Addressing: The computers in a SLIP link need to know each other's IP address for routing purposes. Also, when using SLIP for hosts to dial up a router, the addressing scheme may be quite dynamic and the router may need to inform the dialing host of the host's IP address.

Error detection/correction: Noisy phone lines corrupt packets in transit, since the line speed is quite low in the standard 56 Kbps environment.

5.8.2 Point-to-Point (PPP)

As the number of hosts supporting IP started to grow in the late 1980s, there was a need to standardize Internet encapsulation of IP over point-to-point links. Point-to-point links, also called serial links, are among the oldest methods of data communications, and almost every host supports point-to-point connections. The Point-to-Point Protocol (PPP) is designed to standardize communications over such links. In addition to standardizing the IP encapsulation over point-to-point links, PPP handles other issues. Some of these issues are assignment and management of IP addresses, asynchronous (start/stop) and bit-oriented synchronous encapsulation, network protocol multiplexing, link configuration, link quality testing, error detection, and option negotiation for such capabilities as network layer address negotiation and data compression negotiation. PPP addresses these issues by providing an extensible link control protocol (LCP) and a family of network control protocols (NCPs) to negotiate optional configuration parameters and facilities. Today, PPP supports other protocols besides IP, including Internetwork Packet Exchange (IPX) and DECnet.

PPP provides a method for transmitting datagrams over serial point-to-point links. It has three main components:

- A method for encapsulating datagrams over serial links. PPP uses the HDLC protocol, described in Chapter 3, as a basis for encapsulating datagrams over point-to-point links.
- An extensible LCP to establish, configure, and test the data link connection.
- A family of NCPs for establishing and configuring different network layer protocols. PPP is designed to allow the simultaneous use of multiple network layer protocols.

To establish communications over a point-to-point link, the originating PPP first sends LCP frames to configure and (optionally) test the data link. After the link has been established and optional facilities have been negotiated as needed by the LCP, the originating PPP sends NCP frames to choose and configure one or more network layer protocols. When each of the chosen network layer protocols has been configured, packets from each one can be sent over the link. The link will remain configured for communications until explicit LCP or NCP frames have closed the link, or until some external event occurs (e.g., an inactivity timer expires or a user intervenes).

SLIP and PPP are similar protocols in several aspects. SLIP/PPP provides the ability to transport TCP/IP traffic over serial lines, such as dial-up telephone lines, between two computers running TCP/IP-based network software. This allows a home user to get direct Internet access from his own PC with just a simple modem and a telephone line. With SLIP/PPP, GUI-based web browser ftp clients, and so on may be run from the home PC. SLIP/PPP is really a form of direct Internet connection in the following senses:

- The home computer has a communications link to the Internet, even if it is via a service provider.
- The home computer has the networking software that can speak TCP/IP with other computers on the Internet.
- The home computer has an identifying address (IP address) at which it can be contacted by other computers on the Internet.

While SLIP and PPP are largely similar, there are some key differences. PPP is a newer protocol, better designed, and more acceptable to the sort of people who like to standardize protocol specifications. PPP has some additional benefits. Unlike SLIP (which can transport only TCP/IP traffic), PPP is a multiprotocol transport mechanism. This means that PPP not only transports TCP/IP traffic, but can also transport IPX and Appletalk traffic, to name just a few. Better yet, PPP lets the user transport all these protocols at the same time, on the same connection.

Such multitasking is often not a concern for most users, since their purpose of using either SLIP or PPP is to connect to the Internet, which uses TCP/IP only. Therefore, there is no need to transport other protocols. Another key difference is the configuration and management details. With SLIP, you must know the IP address assigned to you by your service provider. You also need to know the IP address of the remote system you will be dialing into. If IP addresses are dynamically assigned, your SLIP software needs to be able to pick up the IP assignments automatically; otherwise, you will have to set them up manually. You may also need to configure such details as MTU (maximum transmission unit) and MRU (maximum receive unit). PPP addresses this problem by negotiating configuration parameters at the start of the connection. This can greatly simplify the configuration of a PPP connection. PPP provides two methods with which logins can be automated: PAP (Password Authentication Protocol) and CHAP (Challenge–Handshake Authentication Protocol). Both provide the means for a system to automatically send the login userid/password information to the remote system.

5.8.3 Winsock

The Windows Sockets standard (WinSock) provides developers of Microsoft Windows applications with the ability to write one piece of source code to the WinSock application programming interface (API) and have that application work over any Winsock-compliant TCP/IP stack. The API is derived from JSB's Virtual Socket Library and is based on Berkeley Sockets BSD 4.3 with asynchronous extensions. Version 1.1 of Winsock addresses only connectivity over TCP/IP. Version 2 of the Winsock standard also allows communication over protocols such as IPX/SPX, DECNet, NetBUIE, and Vines IP.

The benefits of Windows Sockets are clear. No matter what the communication protocol, developers need learn only one API. As a result, development time is drastically reduced.

5.9 INTERNET APPLICATIONS

5.9.1 World Wide Web (WWW)

The growth of the Internet provided people with a means of sharing and distributing information. The World Wide Web, which developed as a part of the Internet, allows people to easily create "web sites" where other users can access information. The opportunities the World Wide Web offers have in many ways fired people's imaginations and opened corporate eyes to potential commercial opportunities. A host of tools can be used to access information on the web, the most commonly used are HTTP, FTP, gopher and telnet, which are supported in most web browsers.

5.9.1.1 Web Browser

A web browser is a tool that translates HTML (the code that defines the content of a web document) for display on a client, which may be a character terminal, an X workstation, or a PC. A browser has the ability to display the vast library of information that is available on the World Wide Web. The first web browsers were introduced as a means of interpreting HTML code and displaying the text-based information. However as web browser technology progressed, new functionalities started to be built into browser products, allowing information other than text to be delivered to the desktop.

Web browsers can also handle multimedia files that contain audio and visual information through the use of "built-in viewers." These are internal and sometimes external applications that understand how to interpret specific file types such as GIF or JPEG. Additional "helper" applications can be configured to handle other file types. Web browsers also provide a simple method of organizing information through the use of "hotlists" to quickly move to favorite or frequently visited web sites. Key areas in web browser technology are performance, security, and ease of use.

5.9.1.2 HTML

HyperText Markup Language (HTML) is the language in which most web documents are created. HTML is essentially a simple set of codes or "tags" that define the appearance of text in a document. When a document is retrieved from the WWW, the user's web browser reads the codes and displays the HTML-formatted text. The codes define items such as text alignment, text size, graphical image locations, document backgrounds, and hypertext links to related documents. HTML supports facilities that can create tables or align text. To take advantage of the HTML features, however, the client (i.e., the web browser) must be able to support them.

When it comes to creating HTML documents for inclusion on a web server, there are three possibilities. You can directly write the source code, use an HTML editor, or convert existing documents into the HTML format.

5.9.1.3 Web Servers

Software that is used on the World Wide Web is designed around a distributed client–server architecture. Web clients such as browsers send requests for information to web servers, which upon receipt send the requested information back to the client. In this environment a web server is basically a transaction processor. A web browser uses a TCP/IP connection to pass a request to a web server running an HTTP service (under Windows NT) or as a daemon (under Unix). The HTTP service then responds to the request, typically by sending an HTML document to the web browser, which then displays the web page using its own resources (fonts, colors, etc.).

The main functionality provided by standard Unix Web server includes the following:

Access authorization: Allows the server to restrict access to locations or files—for example, to anyone who fails to provide the correct username and password.

Proxy services: Allows the server to be customized to assist in securing a given Internet site.

Server management: Allows the administrator to perform certain management tasks, such as log server requests and statistics.

Imagemap support: Allows regions within an image in an HTML document to link to different HTML pages.

5.9.1.4 CGI

Web servers usually expand on the basic transaction in a number of ways. For example, many web servers support the common gateway interface (CGI), which allows a program or script to be run on the server as part of the original HTTP request for a web page. The content of the outgoing web page is then altered to incorporate the output from the program or the script. CGI is the standard for interfacing external applications with web servers. A CGI program resides on the web server, has the ability to run in real time, and delivers dynamic information, as opposed to the static information contained in most HTML documents. CGI programs can be written with Unix Shell Scripting or with development languages such as Visual Basic, C, or C++. With this technology it is possible, for example, to create documents that give web users the ability to query a corporate database.

5.9.1.5 VRML

A newer web language called VRML (pronounced "vermel") was conceived in early 1994 as a means of bringing three dimensional graphics onto the World Wide Web. Originally VRML stood for Virtual Reality Markup Language, although later, in recognition of the graphical nature of the tool, it was renamed Virtual Reality Modeling Language. The technology allows three-dimensional objects to be viewed and manipulated within a VRML browser. This capability has many potential commercial and academic applications, and there are many sample VRML documents available on the World Wide Web to illustrate its potential. A VRML document could, for example, let the user walk through a three-dimensional room of objects or furniture, or alternatively, view an atom's molecular structure in three dimensions.

5.9.2 Recent Developments

5.9.2.1 Intranet

An intranets provides a highly effective communications platform that is suitable in a corporate environment. A basic intranet should be extensible and able to provide timely information. It can serve as an "information hub" for an entire company, its remote offices, partners, suppliers, and customers. The application of Internet technologies in an intranet setting can dramatically increase the flow and value of information in an organization. Users can gain quick and timely access to a much wider variety of existing information residing in a variety of original forms and sources, ranging from word processing files to databases, Lotus Notes, and other resources. In addition, an intranet can displace traditional paper-based information distribution applications, lowering costs and increasing the timeliness of information flow.

The main components of the Internet are as follows.

Communications protocol: The ability to connect and communicate between networks and individual desktop devices.

File transfer: The ability to transfer files between point-to-point locations.

Mail: The ability to provide direct point-to-point communication between individuals or groups.

Web browsing: The ability to provide access to information on a one-to-many basis, on demand.

Terminal emulation: The ability to access existing infrastructure applications.

User interfaces: The ability to deliver information of increasing technical complexity to the desktop in a transparent, seamless, and intuitive manner.

Intranet technologies provide the tools, standards, and new approaches for meeting the problems of today's business world. The first organizations to use Internet technologies on corporate networks generally moved traditional paper-based information distribution online. These organizations have focused on a core group of supporting or mission-critical information sets, including the following:

Competitive sales information
Human resources/employee benefits statements
Technical support/help desk applications
Financial
Company newsletters
Project management
ISO 9000 documentation

These companies typically provide a "corporate home page" as a launch pad for employees to find their way around the corporate intranet site. This page may have links to internal financial information, marketing, manufacturing, human resources, and even nonbusiness announcements.

5.9.2.2 Extranet

An extranet securely connects a virtual business community of partners, customers, suppliers, and other parties over the Internet. Because an extranet gives various levels of access to a diverse user base, it requires a high level of security and access control. Extranets are changing the way corporations do business. Businesses are discovering that a sophisticated extranet can improve corporate productivity, increase customer satisfaction, reduce costs, and improve efficiency while potentially increasing revenue.

Companies are deploying extranets to accomplish the following:

- Publish and exchange proprietary, confidential, and/or timely information.
- Exchange CAD files in real time to speed design turnaround.
- Improve communication on collaborative projects.

- Share product catalogs exclusively with wholesalers.
- Train resellers and employees online.
- Extend applications, such as online banking applications, to customers.
- View private data online.

Though all the current Internet technologies, such as extranet, intranet, virtual private network (VPN), and firewalls are technologies that are used for security, they differ in purpose and in structure. Firewalls are best used on the perimeter of a network to block unwanted traffic. Virtual private networks are best used as encrypted tunnels across the Internet to connect branch offices and remote employees. Intranets and extranets closely resemble each other, though intranets reside behind a firewall and are accessible only to members of the same organization. Extranets securely connect outside, authorized individuals to resources on a protected network.

Extranets are enabling companies to share information formerly available only to known and trusted users, such as business employees. However, security is more important than ever as companies extend private resources to outside individuals, such as business partners and customers.

Virtual private networks are considered to be one of the hot commodities of the networking industry. They hold the promise of enabling corporations to finally conduct business online in a secure fashion. It is expected that VPNs will redefine electronic commerce from the narrow connotation of credit card transactions to include full-scale business negotiations. In other words, VPNs will enable simple e-commerce applications to mature into fully functional e-business applications. VPNs will provide the framework for initial contact with a client, sales negotiation, order fulfillment, and ongoing support. VPNs will also automate the supply chain, facilitate collaborative projects with partners, and improve productivity for both in-house and remote employees through streamlined, secure access to critical information.

5.10 CHAPTER SUMMARY

This chapter started with a general overview of Internet architecture that used the dot decimal notation to discuss Internet addressing and ARP. IP datagram format was discussed, with explanation of packet fragmentation and reassembly. It was shown how the ICMP is used on the Internet to facilitate error control and reporting. Next, the new IPv6 protocol was discussed.

Routing protocols play an important role in forwarding the data appropriately in networks. Problems such as congestion and errors may be due to inefficient routing used in networks. Congestion obviously leads to excessive latencies, degrading the overall network's performance. We discussed how routing can be used effectively in subnets and on the Internet.

Two transport layer packet services, UDP and TCP, were discussed. We discussed the UDP and TCP headers and services and illustrated the reliability of TCP by going over its different phases (e.g., connection establishment, termination). Different Internet services (applications) such as telnet, ftp, POP, SMTP, and rlogin were discussed and explained. Next, the DNS for mapping the domain names to IP was considered.

Use of PCs has increased dramatically over the last years. We discussed how TCP/IP is implemented and used in PCs via SLIP and PPP, and how Winsock is playing an important role in providing this service. Finally we talked about recent advances in Internet services, which include web development.

5.11 PROBLEMS

5.1 Categorize the major differences between a WAN and a LAN.

5.2 Describe the connectionless and connection-oriented service paradigms for network nodes. Specifically, indicate the type(s) of service(s) used in LAN and WAN.

5.3 Use the Bellman–Ford and Dijkstra algorithms to find the shortest-path tree from each node to node 1 for the graph in Figure P5.3.

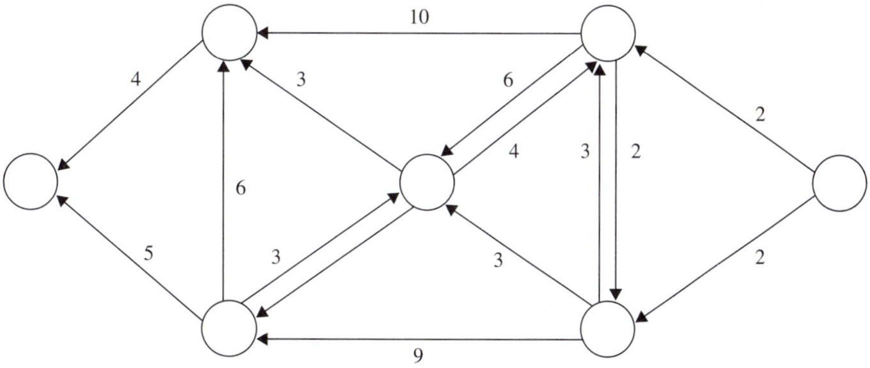

Figure P5.3

5.4 Use the Prim–Dijkstra and Kruskal algorithms to find a minimum weight spanning tree of the graph in Figure P5.4.

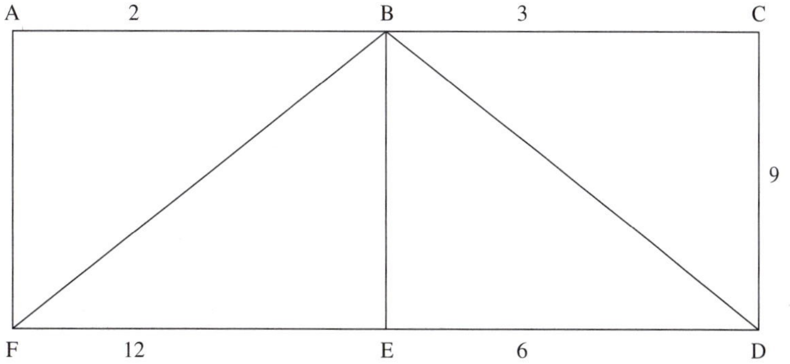

Figure P5.4

5.5 Consider that the weights given in Problem 5.4 are the distances. Apply Dijkstra's algorithm to find the shortest distance from node A to node E. Try to show the values (in each iteration) of all variables by using graph theoretical notation used to explain the algorithm.

5.6 Show (sketch) the packet distribution in the network (of Figure P5.6), using the pure flooding algorithm with the maximum lifetime field set at the hop count of 3. Source is 1 and destination is 6. Packets that have reached the destination are not duplicated. Use packet identifiers 1, 2, and 3 to indicate the duplication of original frame created for three outgoing links from 1. How many frames would be generated on the network for the third hop travel?

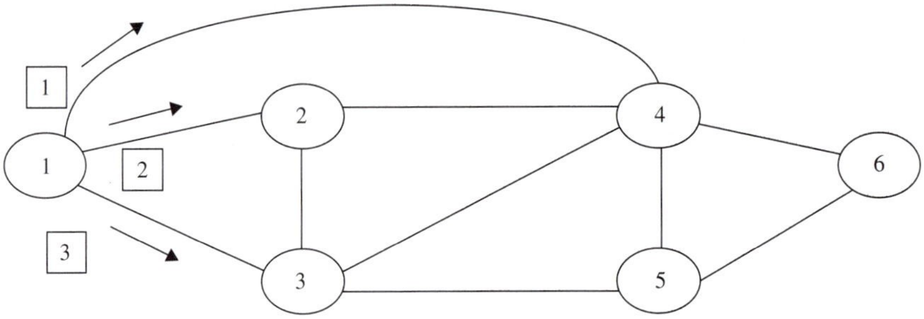

Figure P5.6

5.7 Consider the network in Figure P5.7. Suppose that A establishes a virtual circuit to D, and B establishes a virtual circuit to C. If both virtual circuits go through X and Y, what would X's and Y's routing tables look like?

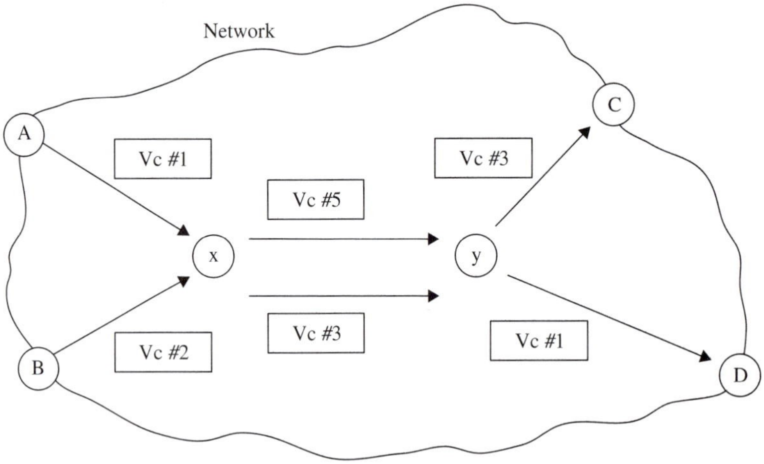

Figure P5.7

5.8 Consider a campus network that runs RIP, where router K has the following routing table:

Destination	Distance	Route
Net 1	0	Direct
Net 2	0	Direct
Net 4	8	Router L
Net 17	5	Router M
Net 24	6	Router J
Net 30	2	Router Q
Net 42	2	Router J

Suppose router K receives the following routing update from router J:

Destination	Distance
Net 1	2
Net 4	3
Net 17	6
Net 21	4
Net 24	5
Net 30	10
Net 42	3

Give router K's routing table after it incorporates this update from router J.

5.9 Consider the situation of routing instability (assuming distributed adaptive routing) illustrated in Figure P5.9. Explain the instability that may be caused in the network if the link cost is equal to link flow plus x, where x is between 0 and 1 (inclusive) and ϵ is very small. Specifically, demonstrate that the instability vanishes for $x = 1$, with the only oscillation in routing for node 4.

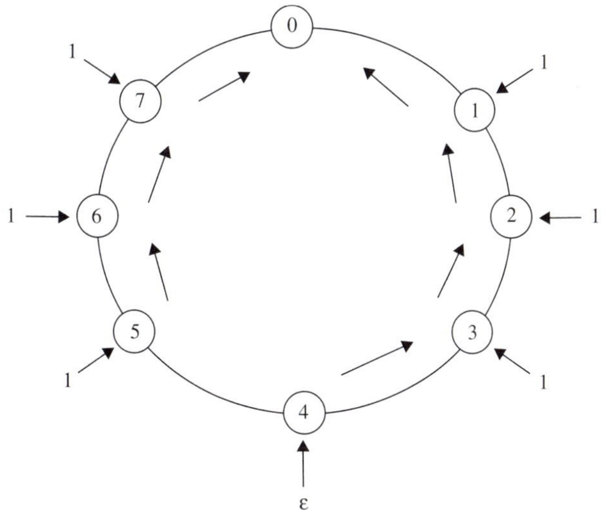

Figure P5.9

5.10 Write a program to generate the first three choices of a routing matrix for the given 10-node network of (Figure P5.10), using the path length as performance measure. The output should contain the first three choices of paths between each source–destination pair, with the calculated path lengths and shortest paths for each case. For example, for 1–3 the best paths would be 1-5-3 (first choice), 1-7-3 (second choice), and 1-7-9-10-6-3 or 1-7-9-5-3 (as third choice), with total path lengths of 1.3, 1.4, and 3.2 km, respectively.

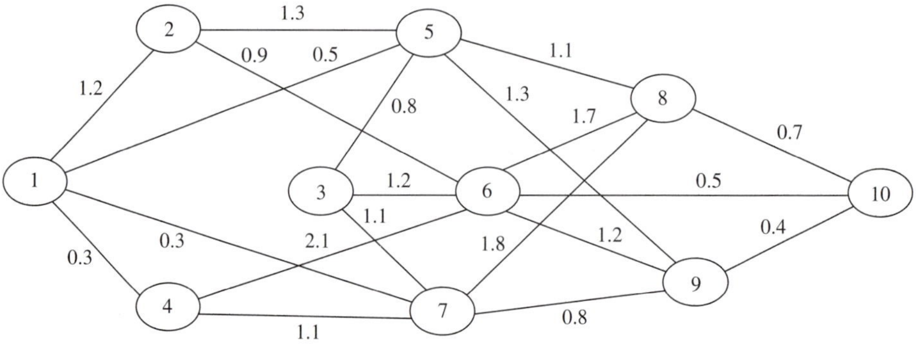

Figure P5.10

5.11 Rewrite the following hexadecimal IP addresses in dotted decimal notation. Identify the address class, host id, and net id (in hex 8-bit groups) in each case.

C0290614
C2124679

5.12 A class B Internet address has a subnet mask of 255.255.240.0. What is the maximum number of hosts per subnet?

5.13 Draw an internetworking connection with the following specifications. Show all the hosts, their addresses, and connections. The address selection is up to you.
(a) One token ring network with four hosts (class A)
(b) One Ethernet with four hosts (class C)
(c) The token ring connected to the Ethernet by a gateway.

5.14 What are the network number, broadcast address, class, and number of usable host addresses for the following IP/Subnet Mask.

IP: 10.1.0.100 Mask: 255.255.254.0

5.15 Which of the following characteristics accurately reflect the given IP/Subnet mask?

IP: 194.122.14.5 Mask: 255.255.255.0

(a) Host ID is 14.5
(b) Class C address
(c) Network ID of 194.122.14
(d) Class B address
(e) Network ID is 194.122

5.16 The most typical loopback address used to test network applications is 128.0.0.1? (T/F)

5.17 A computer that receives an ARP request will broadcast its response? (T/F)

5.18 The Ethernet standard specifies that the type field in an Ethernet frame carrying an ARP message must contain the Hex value 0x806? (T/F)

5.19 ARP permits address resolution to occur: (Choose one)
(a) on a single network
(b) on a single domain level
(c) on a small linked internet
(d) anywhere on the Internet.

5.20 The Address Mask field in a routing table is:
(a) the destination
(b) the next hop to go to reach the destination
(c) defined as the current routers address
(d) used to extract the network part of an address

5.21 Which of the following is not a ICMP informational message:
(a) echo Request
(b) address Mask Reply
(c) time Exceeded

5.22 An ICMP message has arrived with the following header given in hex:

```
05 00 11 12 11 0B 03 02
```

What are the message type and code? Explain the purpose of this message. Write the last four bytes in IP address format. What do these four bytes signify?

5.23 How many responses does a computer expect to receive when it broadcasts an ARP request? Why?

5.24 ARP only permits address resolution to occur on a single network. Could ARP send a request to and get a reply from a remote server using an IP datagram?

5.25 Write a short program and test it to print the host information including X, Y, Z, and W parts of the Internet address X.Y.Z.W (both in binary and in decimal) for a host typed in as a command argument. The program could be run in one of two modes decided by the user interactively. In the first mode of operation, the program should use an appropriate call to get the host name of the machine on which the program is running. In the second mode of operation, any host name can be a part of the command argument. For example, the second mode command could be "`host_test cps215.cps.cmich.edu`". In both modes, it must respond with the four address parts and the address class as the output.

5.26 Write down the name, IP address, and Ethernet address of your current machine. What commands did you use to find them?

5.27 A router with IP address `125.43.23.8` and Ethernet physical address hex `2345AB4F67CA` has received a packet for a destination host with IP address `125.11.78.10` and Ethernet physical address hex `AABBA24F67CC`. Assuming no subnetting, show the entries in the ARP request packet sent by the router.

5.28 Will a computer that receives an ARP request broadcast its response? Explain.

5.29 The Ethernet standard specifies that the type field in an Ethernet frame carrying an ARP message must contain a certain hex a decimal value. What is it?

5.30 On which of the following does ARP permit address resolution to occur?
(a) A single network
(b) A single domain level
(c) A small linked internet
(d) Anywhere on the Internet.

5.31 In ARP, the address mask field in a routing table is:
(a) The destination address
(b) The next hop to go to reach the destination
(c) Defined as the current router's address
(d) Used to extract the network part of an address

5.32 Is "time exceeded" an ICMP informational message? Explain.

5.33 How does a computer know if an arriving frame contains an IP datagram or an ARP message? Explain.

5.34 How many responses does a computer expect to receive when it broadcasts an ARP request? Why?

5.35 ARP permits address resolution to occur on a single network only. Could ARP send a request to and get a reply from a remote server using an IP datagram? Explain?

5.36 Describe the three main methods of address resolution and tell where they are most often used.

5.37 Explain with an example what function masks perform in a routing table. How are masks determined?

5.38 IP is a best-effort delivery system. Is there any datagram transmission problem it does not handle? Explain.

5.39 Packets are either lost or delivered at the data link layer. At the transport layer, however, a packet may also be delayed inside the network. What problems this may cause?

5.40 Suppose a TCP message that contains 4096 bytes of data and 20 bytes of header is passed to IP for delivery across two networks in the Internet (i.e., from the source host to a gateway and then to the destination host). Also, suppose the first network uses 18-byte headers, the second network uses 6 byte headers, the IP header is 20 bytes long, the MTU of the first network is 1024 bytes, and the MTU of the second is 512 bytes. (Recall that each network's MTU gives the total packet size that may be sent, including the network header.) Schematically show the packets that are delivered to the network layer at the destination host. Be sure to show how many bytes are in each packet, how those bytes correspond to various headers, and the offset and length fields of the IP header (in decimal format). Assume that datagrams are fragmented into $n - 1$ full packets and 1 partial packet. Don't forget IP's rule about breaking datagrams into fragments on 64-bit boundaries.

5.41 A 1000-byte chunk of data (including the transport layer header) is to be transmitted over the Internet by means of an IP datagram that supports a maximum of 256 byte frames (i.e., MTU is 256 bytes). Assuming the minimum IP header in each IP datagram, derive the number of fragments required and the contents of the following fields in each datagram header:

- A valid datagram identification
- Total length field
- Fragment offset (in decimal)
- More fragments flag (bit)

5.12 REFERENCES

BOOKS

Albitz, P., *DNS and BIND*. Sebastapol, CA: O'Reilly & Associates, 1997.

Baker, R. H. *Extranets: The Complete Sourcebook*. New York: McGraw-Hill, 1997.

Bayles, D. L., *Extranets—Building the Business-to-Business Web*. Englewood Cliffs, NJ: Prentice Hall, 1998.

Bertsekas, D., and R. Gallager, *Data Networks*, 2nd ed. Englewood Cliffs, NJ: Prentice Hall, 1992.

Black, U. D. *TCP/IP and Related Protocols*. New York: McGraw-Hill, 1992.

Black, U. D. *Advanced Internet Technologies*. Englewood Cliffs, NJ: Prentice Hall, 1999.

Bradner, S. O., and A. Mankin, eds., *IPng: Internet Protocol Next Generation*. Reading, MA: Addison-Wesley, 1995.

Comer, D. E., *Internetworking with TCP/IP*, vol. I: *Principles, Protocols, and Architecture*, 3rd ed. Englewood Cliffs, NJ: Prentice-Hall, 1995.

Comer, D. E., *The Internet Book: Everything You Need to Know About Computer Networking and How the Internet Works*, 2nd ed. Englewood Cliffs, NJ: Prentice Hall, 1997.

Comer, D. E., and R. E. Droms, *Computer Networks and Internets*, 2nd ed. Englewood Cliffs, NJ: Prentice Hall, 1999.

Comer, D. E., and D. L. Stevens, *Internetworking with TCP/IP*, vol. II: *Design, Implementation, and Internals*, 3rd ed. Englewood Cliffs, NJ: Prentice-Hall, 1999.

Evans, T., *Building an Intranet*. Indianapolis: SAMS Publishing, 1996.

Forouzan, B. A., *TCP/IP Protocol Suite*. New York: McGraw-Hill, 2000.

Gralla, P., *How Intranets Work*. Emeryville, CA: Ziff-Davis Press, 1997.

Gralla, P., *How the Internet Works*. Emeryville, CA: Ziff-Davis Press, 1998.

Heinle, N., *Designing with JavaScript—Creating Dynamic Web Pages*. Sebastapol, CA: O'Reilly & Associates, 1997.

Held, G., *Voice-Over Data Networks*. New York: McGraw-Hill, 1998.
Huitema, C., *IPv6: The New Internet Protocol*. Englewood Cliffs, NJ: Prentice Hall, 1996.
Hunt, C., *Networking Personal Computers with TCP/IP*. Sebastapol, CA: O'Reilly & Associates, 1995.
Kercheval, B., *DHCP: A Guide to Dynamic TCP/IP Network Configuration*. Englewood Cliffs, NJ: Prentice Hall, 1999.
Miller, P., *TCP/IP Explained*. Oxford: Digital Press, 1997.
Minoli, D., and E. Minoli, *Delivering Voice Over IP Networks*. New York: Wiley, 1998.
Moy, J., *OSPF*. Reading, MA: Addison-Wesley, 1998.
Perlman, R., *Interconnections: Bridges and Routers*. Reading, MA: Addison-Wesley, 1992.
Shay, W. A., *Understanding Data Communications and Networks*. Boston, MA: PWS Publishing, 1995.
Siyan, K. S., *Inside TCP/IP*, 3rd ed. Indianapolis, IN: New Riders, 1997.
Stevens, W. R., *TCP/IP Illustrated*, vol. 1: *The Protocols*. Reading, MA: Addison-Wesley, 1994.
Stevens, W. R., *TCP/IP Illustrated*, Vol. 3: *TCP for Transactions, HTTP, NNTP, and the UNIX Domain Protocols*. Reading, MA: Addison-Wesley, 1996.
Stewart, J. W., III, *BGP4: Inter-Domain Routing in the Internet*. Reading, MA: Addison Wesley, 1999.
Thomas, S. A., *IPng and the TCP/IP Protocols*. New York: Wiley, 1996.
Wright, G. R., and W. R. Stevens, *TCP/IP Illustrated*, vol. 2: *The Implementation*. Reading, MA: Addison-Wesley, 1995.

ARTICLES

Allman, M., V. Paxson, and W. R. Stevens, TCP Congestion Control. RFC 2581, 1999.
Gilligan, R. E., S. Thomson, J. Bound, and W. R. Stevens, Basic Socket Interface Extensions for IPv6. RFC 2553, 1999.
Stevens, W. R., and M. Thomas, Advanced Sockets API for IPv6. RFC 2292, 1999.
The World Wide Web Consortium, *Proceedings of the Fourth International World Wide Web Conference*, World Wide Web Journal. Sebastapol, CA: O'Reilly & Associates, 1995.
The World Wide Web Consortium, "Building an Industrial Strength Web." *World Wide Web Journal*, Sebastapol, CA: O'Reilly & Associates, 1996.
The World Wide Web Consortium, "Key Specifications of the World Wide Web." World Wide Web Journal. Sebastapol, CA: O'Reilly & Associates, 1996.
The World Wide Web Consortium, "The Web After Five Years." *World Wide Web Journal*. Sebastapol, CA: O'Reilly & Associates, 1996.

WORLD WIDE WEB SITES

Yahoo's links on TCP/IP
http://dir.yahoo.com/Computers_and_Internet/Communications_and_Networking/Protocols/TCP_IP/
Yahoo's links on TCP/IP
http://dir.yahoo.com/Business_and_Economy/Companies/Computers/Communications_and_Networking/Software/TCP_IP/
Network Research Group (NRG) of the Information and Computing Sciences Division (ICSD) at Lawrence Berkeley National Laboratory (LBNL)
http://www-nrg.ee.lbl.gov/
H. M. Kriz, "Windows and TCP/IP for Internet Access," University Libraries
Virginia Polytechnic Institute & State University

http://learning.lib.vt.edu/wintcpip/wintcpip.html
TCP/IP information at UC Davis
gopher://gopher-chem.ucdavis.edu/11/Index/Internet_aw/Intro_the_Internet/intro.to.ip/
RFC-768; The User Datagram Protocol (UDP)
RFCs may be obtained via FTP from one of the following primary repositories:
ftp://ftp.isi.edu
ftp://ftp.wuarchive.wustl.edu
ftp://ftp.src.doc.ic.ac.uk
ftp://ftp.ncren.net
ftp://ftp.sesqui.net
ftp://ftp.nic.it
ftp://ftp.imag.fr
http://www.normos.org
http://www.faqs.org/rfcs

RFCS

RFC-791, The Internet Protocol (IP), 1981
RFC-792, The Internet Control Message Protocol, 1981
RFC-793, The Transmission Control Protocol (TCP), 1981
RFC-826, The Address Resolution Protocol (ARP), 1982
RFC-854 and RFC-855, The Telnet Protocol, 1983
RFC-951, The Bootstrap Protocol (BOOTP), 1985
RFC-1034 and RFC-1035, The Domain Name Service (DNS) Protocols, 1987
RFC-1350, The Trivial File Transfer Protocol (TFTP), 1992
RFC-2068, The Hyper Text Transfer Protocol (HTTP), 1997
RFC-2292, Advanced Sockets API for IPv6, 1998.
RFC-2553, Basic Socket Interface Extensions for IPv6, 1999.
RFC-2581, TCP Congestion Control, 1999.

6

ACCESS AND HIGH-SPEED NETWORKING TECHNOLOGIES

Until recently, remote access and integrated services were useful mainly for surfing the web for fun. Now, however, these technologies are very much related to productivity. Millions of people currently telecommute. In many circumstances, the 56 Kbps speed of telephone modems does not provide an adequate data rate for the resources people need for their jobs. Dial-up modems cannot deliver the media-rich content that pervades the Internet web sites. With the three-dimensional graphics, multimedia, animation, and video becoming a part of every database, even ISDN running at 128 Kbps may not be fast enough for these new applications.

With any remote access connection technology, a well-designed and well-implemented remote node system—sometimes called a remote LAN access system—can link remote users to the LAN and give the users the same capabilities they have back at the local site of LAN. In such a system, a remote LAN access system server is directly attached to the LAN running Ethernet, token ring, or FDDI. The server then connects to multiple remote clients through modems or other appropriate adapters for ISDN, digital 56 Kbps, frame relay, HDSL, and so on, running PPP (Point-to-Point Protocol) or SLIP (Serial Line IP). The remote workstation can see all the file servers and other network resources that local workstations see. Except for the inevitable slowdown resulting from the asynchronous connection, a good remote LAN access system should look, feel, and function like a local network node.

It is generally believed that the next step for private uses after a 56 Kbps modem line is a T1 line to the home. The T1 connection, which is commonly used to connect various enterprise network sites in a corporate environment, offers a bandwidth of 1.544 Mbps. However, strict wiring conditions and use of repeaters make T1 services expensive and add considerably to installation time. Thus T1 setups are rarely suitable for home offices and small businesses.

While the telephone companies have made a commitment to ISDN, several other high-speed communication vehicles are being explored, as well. The most excitement among these alternative approaches is being generated by cable modems, which use cable TV coaxial cable, and digital subscriber line (DSL) technology. DSL, in turn, uses the same copper twisted-pair cables as voice grade telephone wiring, but delivers a data rate tens of times higher than analog modems. We mentioned cable modem technology in Chapter 4. In this chapter we discuss the common access technologies available today, including frame relay, ISDN, HDSL (high-bit-rate digital subscriber line) and ADSL (asynchronous digital subscriber line).

ISDN is considered to be the mainstay of residential and business telecommunications services. There are two main types of ISDN: basic rate interface (BRI) offers two 64 Kbps B-barrier channels and one 16 Kbps D-data channel, while the primary rate interface (PRI) has 23 B channels (64 Kbps each) and one D channel. Each B channel can support simultaneous and independent voice, video, or data connections. The carriers believe that ISDN can meet the needs of almost every user.

Frame relay, which was initially designed as a barrier service for ISDN, now has become an independent technology owing to its popularity and ease of migrating to ATM services. Assuming operation over reliable digital links, frame relay, which originated from the X.25 standard, provides bandwidth sharing, voice-over data capability, and much greater efficiency and speed than X.25.

HDSL and ADSL use special 2B1Q (two binary and one quaternary) coding. Because these are end-to-end digital technologies like ISDN, the total bandwidth may be divided into many parts. HDSL in its original configuration uses two sets of twisted-pair wire to provide 1.5 Mbps (T1 speed) in each direction for a distance of 12,000 feet without repeaters. In addition, because HDSL does not require special line conditioning and allows wires with different diameters, HDSL can be installed in hours, in contrast to T1 service, which takes weeks to install and set up. The benefit of HDSL is that it provides T1 speed in both directions with much lower cost than T1. Thus HDSL is a low-cost alternative to running fiber between buildings. ADSL provides either 1.5 or 6 Mbps from the network to the user and 64 or 640 Kbps in the reverse direction, depending on the distance (12,000 or 18,000 feet). ADSL is designed to take advantage of the fact that video-on-demand, telecommuting, and Internet access traffic are inherently asymmetrical: that is, a user sends a short message up to the network and receives massive data in return. Indeed the "asymmetric" part of ASDL indicates that different speeds are used in the forward and reverse directions.

There is an increased risk of unauthorized activity any time network access is given to more people. This problem is particularly acute with dial-up connections. Unless appropriate controls are in place, anyone can dial into such networks.

Remote network devices typically provide some form of security. We postpone a discussion of general security issues until Chapter 10. However, there are special security considerations for a remote access system. The most common methods are ICLID (**in**coming **c**aller **i**dentification) and authentication. Authentication is the mechanism that makes sure the user dialing in to the remote access system is the approved user for the particular service. User ID and password are the simplest but not very secure.

More advanced options include callback to authorized telephone numbers and the smart card approach. Digital services such as ISDN can automatically supply the telephone number of the calling party. Network devices can check the ICLID before accepting the call to determine whether the call is coming from a legitimate user. If the number is not recognized, the call is rejected. This simple security procedure is not foolproof, however. It will not work if the call originates at a location that does not issue the calling telephone number, and it can be fooled by using call forwarding to route the connection through a second number. Besides authentication, other main issues are access control, privacy, and detection.

- **Access control** involves physically securing the server of the remote access system. The input/output devices, such as keyboard, mouse, and display, can be locked by the control software. The better control software can restrict a specific user from accessing unauthorized resource in the system.
- **Privacy** issues usually involve data encryption by either the private key or public key method. The use of an encryption technique keeps unauthorized persons from eavesdropping on remote sessions.
- **Detection** mechanisms can inform the network administrator of outside intruders and identify their point of entry. Any remote access system should keep security logs as the minimum measure for detecting when the security system has been compromised.

6.1 INTEGRATED SERVICE DIGITAL NETWORK (ISDN)

ISDN has been available for more than a decade, and its current popularity as an access method owes much to the evolution of the Internet. Many RBOCs (Regional Bell Operating Companies) offer ISDN-based Internet access. Part of the reason ISDN is not very popular is its varying availability of the service. ISDN is offered in many areas, but not all local switches are configured for it. Even if the local telephone company offers ISDN, not all ISPs (Internet service providers) offer ISDN access, and those that do may not support all types of ISDN.

The initial ISDN standard created by leading industries in the United States is called National ISDN 1 or NI-1. The purpose was to make the purchase of equipment and software easier for end users. With the standard, end users do not have to know switching details. However, there were problems agreeing on this standard. Ultimately, all the RBOCs did support NI-1. A more comprehensive standardization initiative, National ISDN 2 or NI-2, was later adopted. Some manufacturers of ISDN communications devices have worked with the RBOCs to develop their own configuration standards for their own equipment. Such initiatives, along with more competitive pricing, inexpensive ISDN connection equipment, and the desire of people to have relatively low-cost, high-bandwidth Internet access have made ISDN more popular in recent years.

6.1.1 ISDN Signaling and Architecture

ISDN carries three separate digital signals over a standard, two-wire phone line. Together these constitute the basic rate interface. The telephone company uses one of these channels to handle control and signaling information; the remaining two, the B channels, can carry voice, data, or both. B channels run at either 56 or 64 Kbps, depending on the equipment at the local telephone company switch or central office.

The B channels also can be aggregated to act as a single channel running at 128 Kbps. With some ISDN equipment, the channels can be aggregated dynamically, with one result that the user pays for bandwidth only when needed. B channels are logical "pipes" in a single ISDN line. Two B channels can be combined to download data at very high speeds to a PC. One of the B channels can also be freed up for another device, such as a telephone or fax machine, with the remaining B channel used for data transfer. Each provides a 64 Kbps clear channel whose entire bandwidth is available for data, since call setup and other signaling is done through a separate D channel.

B channels form circuit-switched connections that resemble analog telephone connections. There are end-to-end physical circuits temporarily dedicated to transfer between two devices. The circuit-switched nature of B-channel connections, combined with their reliability and high bandwidth, makes ISDN suitable for voice, video, fax, and data applications. B channels can be used to transfer any layer 2 or higher protocols across a link within the OSI model. Although B channels are normally used for on-demand connections, they can also be configured as semipermanent connections that are always "up," much like a leased line. This is possible because the advantage of the circuit-switched networks on which the B channels are based. This connection can cause a capacity problem for carriers and ISPs, but an emerging standard called always on/dynamic ISDN (AO/DI) provides a solution.

The ISDN D channel is used mostly for administrative signaling (e.g., to instruct the carrier to set up or terminate a B channel call, to ensure that a B channel is available to receive a call, or to provide signaling information for such features as caller identification). The D channel uses packet-switched connections, which are best adapted to the intermittent but latency-sensitive nature of signaling traffic. This results in a vastly reduced call setup time of 1–2 seconds on ISDN calls (vs 10–40 seconds with an analog modem). The D channel transmits at either 16 Kbps (for BRI service) or 64 Kbps (for PRI service, explained shortly).

Unlike the B channels, which function as "pipes," the D channel is associated with the higher level protocols at layers 2 and 3 of the OSI model that form the packet-switched connections. Within the ITU layer 3 protocol specifications for use on the D channel, Q.931 is the call control protocol component. This layer 3 signaling protocol is transferred on the D channel by means of the link access procedure—D channel (LAP-D), which is a layer 2 HDLC-like protocol.

Another variety of ISDN is the primary-rate interface. PRI is mainly used for connecting large networks or by ISPs for linking their backbones. PRI is intended for users with greater capacity requirements. Typically the channel structure is 23 B channels plus one 64 Kbps D channel for a total of 1536 Kbps. In Europe, PRI consists of 30 B channels plus one 64 Kbps D channel for a total of 1984 Kbps. It is also possible to support multiple PRI lines with one 64 Kbps D channel by using nonfacility associated signaling (NFAS). BRI, on the other hand, makes sense for small branch offices or residential users.

H channels in ISDN provide a way to aggregate B channels. They are implemented as follows:

- H0 = 384 Kbps (6 B channels)
- H10 = 1472 Kbps (23 B channels)
- H11 = 1536 Kbps (24 B channels)
- H12 = 1920 Kbps (30 B channels)—international (E1) only

To access BRI service, it is necessary to subscribe to an ISDN phone line. The customer must be within 18,000 feet (about 3.4 miles, or 5.5 km) of the telephone company central office for BRI service; beyond that, expensive repeater devices are required, or ISDN service may not be available at all. Customers also need special equipment to communicate with the phone company switch and with other ISDN devices. These devices include ISDN terminal adapters (sometimes incorrectly called "ISDN modems") and ISDN routers.

6.1.2 ISDN Protocols

The ITU I- and G-series documents specify the ISDN physical layer. The U interface for BRI is a two-wire, 160 Kbps digital connection. Echo cancellation is used to reduce noise, and data encoding schemes (2B1Q in North America, 4B3T in Europe) permit this relatively high data rate over ordinary single-pair local loops.

On U interfaces, 2B1Q is the most common signaling method providing 2 bits per baud, 80 Kbaud/s and a transfer rate of 160 Kbps. Each U interface frame is 240 bits long. At the prescribed data rate of 160 Kbps, each frame is 1.5 ms long.

The ISDN frame consists of 9 quaternaries (2 bits each), 12 repetitions of 18 data bits (8 for each B channel, 2 for the D channel) and the 6-bit maintenance field that consists of error-checking information. Data are transmitted in a superframe consisting of eight 240-bit frames for a total of 1920 bits (240 octets). The inverted sync field in the first frame identifies the start of a superframe.

The ISDN data link layer is specified by the ITU Q-series documents Q.920 through Q.923. All the signaling on the D channel is defined in the Q.921 specification.

The layer 2 protocol used by ISDN is link access protocol—D channel (LAP-D), which is almost identical to the X.25 LAP-B protocol. Figure 6.1 shows the structure of a LAP-D frame.

The 2-byte address field of the LAP-D frame consists of a 6-bit service access point identifier (SAPI) that indicates the intended use of the D channel, such as call control for network and upper layers, indication of data use of the D channel, and management. Following by SAPI, a 1-bit C/R field indicates whether the frame is a command or a response. A 1-bit EA0 field (usually set to 0) indicates whether this is the final byte of the address. Next comes the 7-bit terminal endpoint identifier (TEI) field; it is the unique address of the ISDN device that is capable of supporting ISDN standards (commonly called as terminal equipment, or DTE). The last bit in the address field is another address field, EA1, usually set to 1 for the second byte.

The 2-byte control field in a LAP-D frame indicates the frame type (information, supervisory, or unnumbered) and sequence numbers as required. The information field contains the layer 3 protocol information and user data. The frame ends with a 2-byte CRC checksum and the end flag.

Flag (01111110)	Address	Control	Information	CRC	Flag (01111110)

Figure 6.1 LAP-D frame format.

The layer 2 connection establishment process is very similar to the X.25 LAP-B setup.

1. The terminal endpoint (TE) and the network initially exchange receive ready (RR) frames, listening for someone to initiate a connection.
2. The TE sends an unnumbered information (UI) frame with a SAPI of 63 (management procedure, query network) and TEI of 127 (broadcast).
3. The network assigns an available TEI (in the range of 64–126).
4. The TE sends a set asynchronous balanced mode (SABME) frame with a SAPI of 0 (call control, used to initiate a setup) and a TEI having the value assigned by the network.
5. The network responds with an unnumbered acknowledgment (UA), SAPI = 0, TEI = assigned.

At this point, the connection is ready for a layer 3 setup.

After a connection has been established, the B channel sends data. The network layer functions of the D channel are defined by the ITU-T Q.931 standard. The frame at this layer consists of the following fields:

Protocol discriminator (8 bits): Identifies the protocol in use; for Q.931, it is 0000 1000.

Call reference (2 or 3 bytes): Call sequence number.

Message type (1 byte): Purpose of the message.

Information elements (variable): Connection details such as sender/receiver addresses, routing information, and the desired network type for B-channel exchange.

Although the layer 3 protocol takes care of most ISDN signaling requirements, LAP-D also plays a very important role in terms of low-level signaling to ISDN devices, which often share a single D channel (in contrast to B channels, which are temporarily dedicated to specific ISDN devices). LAP-D frames contain the information that ensures that incoming calls are routed to the appropriate ISDN device, and they pass along the addressing information that distinguishes ISDN devices on a single line from each other. This is accomplished with the terminal endpoint identifier (TEI) and service access point identifier (SAPI) fields within the LAP-D frame.

6.1.3 ISDN Advantages

The modem was a big breakthrough in computer communications. It allowed computers to communicate by converting their digital information into an analog signal to travel through the public phone network. There is an upper limit to the amount of information that an analog telephone line can hold. Currently, it is about 56 Kbps. Commonly available modems have a maximum speed of 56 Kbps but are limited by the quality of the analog connection and routinely go no faster than 45 Kbps.

ISDN allows multiple digital channels to be operated simultaneously through regular phone wiring. The change comes about when the telephone company's switches can support digital connections. Therefore, the same physical wiring can be used, but a digital signal, instead of an analog signal, is transmitted across the line. This scheme permits a much higher data transfer rate than is possible with analog lines. BRI ISDN, using a channel aggregation protocol, supports an uncompressed data transfer speed of 128 Kbps. In addition, the latency (i.e., the amount of time it takes for a communication to begin) on an ISDN line is typically much smaller than that on an analog line.

Previously, it was necessary to have a phone line for every device that was intended to serve simultaneously with others. For example, one line each was required for a telephone, fax, computer, bridge/router, and live videoconferencing system. Transferring a file to someone while talking on the phone or seeing a live image on a video screen would require several expensive phone lines. It is possible to combine many different digital data sources and have the information routed to the proper destination. Since the line is digital, it is easier to keep the noise and interference out while combining these signals. ISDN technically refers to a specific set of digital services provided through a single, standard interface. Without ISDN, distinct interfaces are required instead.

Another ISDN advantage is in terms of signaling. Instead of the phone company sending a ring voltage signal to ring the bell in the home phone ("in-band signal"), it sends a digital packet on a separate channel ("out-of-band signal"). The out-of-band signal does not disturb established connections, and call setup time is very fast. For example, a v.34 modem typically takes 30–60 seconds to establish a connection; an ISDN call usually takes less than 2 seconds. The signaling also indicates who is calling, what type of call it is (data/voice), and what number was dialed. Available ISDN phone equipment is then capable of making intelligent decisions on how to direct the call.

6.1.4 Broadband ISDN

The bit rate demands of many real-time applications expected in the future, such as videoconferencing (1–100 Mbps) and HDTV ($\geq$1Gbps) are beyond the capabilities of BRI and PRI. However, an extension of ISDN called broadband -ISDN, or B-ISDN, has the capability of supporting data transmissions of the order of hundreds of megabits per second.

The services that can be provided across broadband ISDN cables are defined by ITU-T as interactive services and distribution services. Some examples of interactive services are video telephony, high-speed data and information transfer, and message and retrieval services. Video telephony is the transfer of voice, moving pictures, and scanned images and documents between two points. Areas potentially utilizing such technology include sales, consulting, teaching, and legal services. The problem with gaining widespread use of video telephony lies in the prohibitive costs of terminal equipment. In the future, as demand and competition increase, the cost of such equipment will fall and the service will become more widespread. Video conferencing and video surveillance are among the other services that are expected to be implemented via video telephony.

High-speed data transfer services include LANs and WANs, as well as the Internet and other computer networking. Other applications include document transfers, facsimile, and multimedia documents including text, graphics, voice, and audiovisual information. Messaging services transmit information on a user-to-user basis, but without requiring the availability of both users at once. Since this service consists mainly of text transfers, it asks little of the available resources. The primary application is e-mail, but paging services and more are possible expansion activities.

Retrieval services involve retrieval of information stored at remote sites. The information is made available at public sites and supplied to the user on a demand basis. Items transferred in this way could include medical information, stock market information, and audio and video files.

Distribution services are provided with and without presentation control. Distribution services ensure a continuous flow of information from a central source to any number of users. All the users have access to the information but no control over it. This type of service includes TV program distribution and document distribution. TV program distribution is the most common application within distribution services. With the capacities of broadband ISDN, higher quality, higher resolution, interference-free television can be provided. This quality should be equal to that provided in theatres, but will require transfer rates around 1 Gbps. ITU-T has suggested that data compression be used to reduce the bit rate requirements and enhance coexistence with multiple broadband ISDN services.

Distribution with presentation control is designed to be done with a central control, but information is transmitted in cycles. The user has individual access to the cyclical distributed information, but unlike TV program distribution, the user has control over the start and order of presentation. Applications of such systems would be in educational training and other services for which it is essential that users be able to control the time of access.

B-ISDN defines three access methods designed to support three levels of user needs: symmetrical, at 155.52 Mbps (same for inbound and outbound), asymmetrical, at 155.52 Mbps (outbound) and 622.08 Mbps (inbound), and symmetrical, at 622.08 Mbps (same for inbound/outbound). The functional groupings of equipment in the B-ISDN model are the same as those for regular or narrowband ISDN (N-ISDN).

The protocol for B-ISDN adopts the following layered approach:

- Physical layer
- ATM layer
- ATM adaptation layer (AAL)
- Higher layers

The higher layers in the B-ISDN protocol are service layers for video, SMDS, frame relay, access, and network signaling. ATM (asynchronous transfer mode) is often referred to as fast packet switching. Thus B-ISDN is often taken as a packet-based network. However ITU-T recommendations state that B-ISDN needs to be able to handle both packet- and circuit-switched applications. Thus the use of an ATM adaptation layer (AAL) is required. The AAL is required to handle non-ATM protocols, such as LAP-D.

The ATM layer provides the packet transfer capabilities, while the physical layer provides the base network functions. There are line control protocols for both an electrical and an optical interface. The transmission convergence sublayer handles the way the ATM cells are multiplexed. There are two implementation options in relation to the transmission structure. The first is to incorporate no multiplexing frame structure. The second option utilizes the SONET (**s**ynchronous **o**ptical **net**work) protocol developed by BellCore. This protocol is based on a synchronous digital hierarchy (SDH) frame.

6.2 CABLE MODEM SYSTEMS

Several new and improved cable network structures and technologies that make use of the existing cable systems were developed and facilitated to enhance network performance and simplify installation.

One new technology, called cable modem, provides multimegabit-per-second speeds without the hassle of dial-up connections. It is faster than ISDN and cheaper, too. A cable modem is a device that enables you to hook up your PC to a local cable TV line and receive data at about 1.5 Mbps. This data rate far exceeds the prevalent data rates for telephone modems (28.8 and 56 Kbps) and ISDN ($\leq$128 Kpbs) and is close to the data rate available to users of digital subscriber line (DSL) telephone service. A cable modem can be added to or integrated with a set-top box that turns a TV set into an Internet channel. For PC attachment, the cable line must be split so that part of the line goes to the TV set and the other part goes to the cable modem and the PC. Figure 6.2 shows such a setup.

A cable modem is really more like a NIC than a computer modem. All the cable modems attached to a cable TV company coaxial cable line communicate with a cable modem termination system (CMTS) at the local cable TV company office. All cable modems can receive from and send to the CMTS only; they cannot communicate with other cable modems on the line.

The actual bandwidth for Internet service over a cable TV line is up to 27 Mbps on the download path to the subscriber and about 2.5 Mbps of bandwidth for interactive responses in the other direction. However, since the local provider may not be connected to the Internet on a line faster than a T1 at 1.5 Mpbs, a more likely data rate will be close to 1.5 Mpbs. Among companies using cable TV to bring the Internet to homes and businesses are @Home, a service of TCI, and Time-Warner.

In addition to the faster data rate, that is one advantage of cable over Internet access by means of telephone lines, cable is a continuous connection. The networks are being set up for asymmetrical communication; links to a user from the network (called the down-

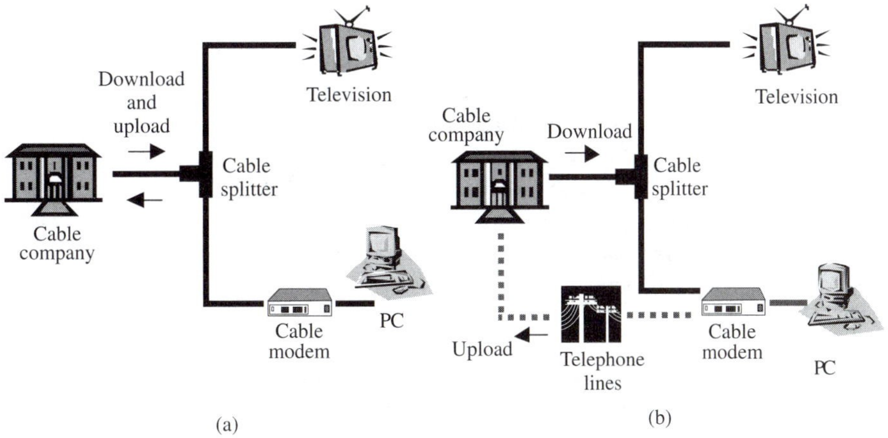

Figure 6.2 Cable modem with (a) RF and (b) telephone return services.

stream connection) are expected to run significantly faster than links from the user to the network) (upstream connections). All head ends are connected by fiber. They all run at 10 Mbps, but 27 Mbps is possible if the cable plant can handle it. It is hoped that asymmetrical rates as high as 20 Mbps downstream and 128 Mbps upstream will be offered, a remarkable improvement in comparison to 64 or 128 Kbps from ISDN.

The idea behind asymmetry in cable modems is that a home user surfing the net, who needs a high-speed connection to receive images and large files, is expected to send only mouse clicks and e-mail back into the system. Therefore, 20 or 27 Mbps downstream versus 128 Kbps upstream should be more than adequate for such an application. In addition, separate channels are used for downstream and upstream traffic. Typically, the downstream connection would use a channel higher in the frequency—channel 60 or above, perhaps. The upstream link would use one lower in the frequency—channel 40 or below. Note that cable companies seldom use these channels at the edge of the spectrum.

Security is a major issue for cable connection because cable service is modeled for consumer rather than business use. Another major concern with this business model is its failure to address interoperability between cable modem service providers.

One of the technologies in use is a hybrid fiber coaxial network, which consists of the cable modem and the cable network itself. This type of cable-based network is actually made up of a combination of fiber optics and coaxial known as "hybrid fiber coax (HFC)." A hybrid fiber coax network uses fiber optic cables to build backbone and the coaxial cables to build branches. The connections from a service center to neighbor nodes are fiber optics, and the connections from the neighbor nodes to the users are coaxial cables, of which the original cable network is formed. The signal in fiber-optic cables runs from the head-end to a location near the subscriber. At that point, the signal is converted to electronic form and transmitted to the subscriber premises using coaxial cables.

A technology similar to that of hybrid fiber coax networks is the "fiber to the curb" network, in which fiber optic cables are used from the backbone to the end users. Because the fiber optic cable connections need to be built from scratch, this service is extremely expensive.

The first-generation cable modem systems were proprietary. Cable modems from different vendors did not work on the same head-end or Cable Modem Termination System (CMTS). Many old cable modem systems use telephone line for "return" or upstream connection. The second-generation system does two-way communications on the cable and is based on Data Over Cable Systems Interface Specifications (DOCSIS). Modems from different vendors work together. DOCSIS 1.0, which was ratified by the ITU in March of 1998, defines interface requirements for cable modems and the CMTS.

Through the newer DOCSIS 1.1 cable modem system with Voice Over Internet Protocol (VoIP) capability through Media Gateway Control Protocol (MGCP), a user can make telephone call along with data access to an IP backbone and Internet through the HFC network.

Cable modem system is "always on." It uses one of the unused cable television channels and is a type of shared media systems like Ethernet. Single CMTS normally drives about 1-2000 simultaneous cable modem users on a single television channel. If more cable modems subscribers are added, additional channels may be allocated to the CMTS.

Cable-Point, a technology developed by Intel to utilize the existing cable system, adds an extra internal card to be installed on a PC and an external box connected to the PC through the card. The box performs an RF signal conversion between analog and digital signals to allow communication to the connected computer. Data are transmitted through the high part of the frequency spectrum of a cable television channel.

6.3 DIGITAL SUBSCRIBER LINE (DSL) TECHNOLOGY

Digital Subscriber Line (DSL) technology is based on the assumption that digital data may not have to change to analog form before transmission. Digital data could be transmitted to a user's computer directly by using a much wider bandwidth. Meanwhile, the bandwidth can be divided so that some of the bandwidth is used to transmit analog signals so that both telephone conversation and computer data exchange can be simultaneously carried on at the same line. In particular, the data part of the line is continuously connected. This technology brings broadband communications and rich-media information to homes and small businesses over ordinary copper telephone lines.

There are two major types of DSL: symmetric and asymmetric. A symmetric transmission system provides data to be transmitted at the same bandwidth (speed) in both directions—from the user to the provider and vice versa. Asymmetric communications allow faster data transmission in one direction, usually from the provider to the user, and a slower lane from the user to the provider. This arrangement is significant particularly because it causes a large difference in the top speed of the system and because it follows user behavior more closely.

Each major type of DSL has some subtypes that use a different signaling method or require different wiring. Symmetric DSL types come in several forms:

- Symmetric DSL (SDSL)
- High-bit-rate DSL (HDSL)
- High-bit-rate DSL type 2 (HDSL2)
- Unidirectional high-bit rate DSL (UDSL)
- ISDN DSL (IDSL)

Asymmetric DSL types are as follows:

- Asymmetric DSL (ADSL)
- Rate-adaptive DSL (RADSL)
- DSL Lite/G.Lite/Universal DSL
- Very-high-bit-rate DSL (VDSL)
- Broadband DSL (BDSL)

The different DSL-based technologies are often combined and referred to as xDSL. Of these major types we will concentration on the high-bit-rate digital subscriber line (HDSL) and the asymmetric digital subscriber line (ADSL).

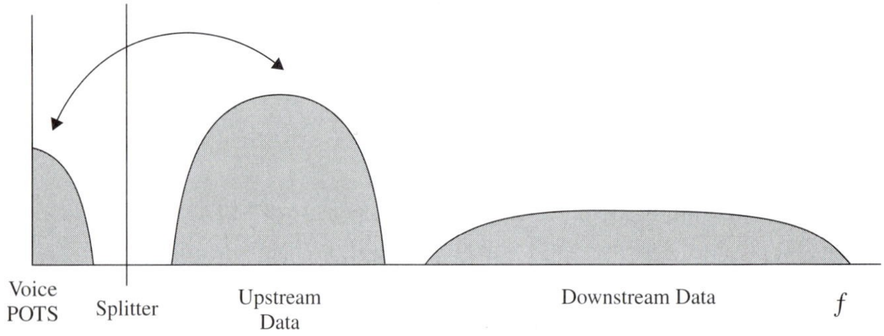

Figure 6.3 ADSL frequency spectrum allocation.

For ADSL, the frequency spectrum is separated and allocated for voice, upstream and downstream as shown in Figure 6.3.

The voice part uses the bottom 4 Khz frequency spectrum as POTS, and the line current remains unaffected thus meeting the "lifeline" regulatory requirement. The upstream utilizes the lower frequency spectrum to take advantage of the reduced Near-End Cross Talk (NEXT) while the downstream uses the remaining frequency spectrum to maximize bandwidth topping out at about 1 Mhz.

At the service provider side, a Digital Subscriber Line Access Multiplexer (DSLAM) interconnects multiple DSL users to a high-speed backbone network—typically connecting to an ATM network that can aggregate data transmission at gigabits per second rates. At the customer premises, a DSL modem with a splitter is installed. However, it is possible to manage the splitting voice and data remotely from the central office. This is known as splitterless DSL, DSL Lite, G.Lite, or Universal ADSL, which is the most widely installed DSL system for residential Internet access.

In the telephone industry, the wiring from a telephone company's local central office to the end user's residence has traditionally been the area of greatest concern for telecommunication and network managers. Typically when copper wiring is used, this "last mile" is the weakest link of any network in terms of bandwidth, quality of performance, and downtime. The xDSL technology enables telephone companies to make full use of existing copper cable to meet the demand for updated services and to extend their services in developing countries beyond the reach of existing technology while reducing operating costs.

In DSL, analog voice signals are generated in each local subscriber loop. These signals are received by the local exchange, where they are quantized and coded into digital information channels. To carry multiple channels to a destination exchange economically, information from each channel can be time-division-multiplexed onto a single transmission medium. In the United States, T1 is used for basic digital multiplexing, and 24 digitized voice channels are multiplexed together over a four-wire cable: two wires for transmission and two for receiving. The counterpart of T1 for European and many Asian countries such as China is E1, which contains 32 voice/data channels in two pairs of wires.

DSL is a broadband access technology that enables fast-moving service providers to add incremental sources of revenue and increase margins over time. With DSL, service providers can gain competitive advantage by offering differentiating small office/home office (SOHO), telecommuting, and branch-to-branch high-speed Internet access, intranet access, and new-world voice services that link users to a new and exciting public packet infrastructure.

xDSL is considered to be a high-speed rival to ISDN because it offers several times the bandwidth available from existing copper wire. The technology was originally meant to be used by telephone companies for delivering video over telephone lines, making them more competitive with cable operators. But as interest in the Internet has grown, xDSL is being seen as a viable access method—even if it has yet to establish itself.

Unlike ISDN, xDSL does not require an upgrade of central office switches; instead, separate xDSL modems are placed at the central office. The system also works over an ordinary pair of phone wires, which means that analog phones can be used while a high-speed data connection is maintained at the same time. HDSL functions are performed by an HDSL transmission unit (HTU) installed at each end of the T1 or E1 link. At the subscriber end, the remote HTU (HTU-R) is installed on the subscriber's premises. At the central office, the line is terminated by a central HTU (HTU-C).

Each HTU type contains a data pump circuit for each pair of wires. The data pump extracts data and timing information from the incoming signal and produces a stream of binary data using a 2B1Q (two-binary, one-quaternary) decoding. Data are passed on to the framer/multiplexer chip, which takes the signals from two data pumps, interlinks them as appropriate, inserts the T1 or E1 framing bit that identifies each signal, and sends it on to the T1 or E1 interface chip. All HDSL circuits are controlled by an advanced embedded microprocessor and corresponding software (firmware). The technique incorporating both software and an embedded system plays a critical role not only in the HDSL implementation itself, but also for performance monitoring, diagnosis, and management of the HDSL system, via integrated SNMP (simple network management protocol).

DSL delivers 378 Kbps to 2 Mbps with HDSL services, and up to 8 Mbps downlink speed with ADSL services, dwarfing ISDN's 128 Kbps data rate—about four times the speed of top-of-the-line analog modems. Furthermore, DSL uses frequency division multiplexing to carry voice and data over a single physical connection. This eliminates the need to run a second line into residential or business buildings. The ADSL version of xDSL supplies three separate frequency channels over the same line. One channel carries telephone conversations, another carries a 16 to 640 Kbps data signal upstream from a user to the Internet and the third channel is a high-speed downstream connection running anywhere from T1 (1.544 Mbps) to 9 Mbps. The splitterless G.Lite, based upon ITU-T standard G-992.2, provides support for data rate from 1.544 Mbps to 6 Mbps downstream, and from 128 Kbps to 384 Kbps upstream.

Despite the bandwidth advantages, ADSL faces some technical constraints. Range is quite limited—in fact, the higher the data rate, the closer the subscriber and central office must be. For instance, a 9 Mbps link requires the two facilities to be within 9000 feet of each other. (In contrast, the maximum distance for T1 lines is 5000 feet, and for ISDN 18,000 feet.) To reach subscribers outside this range, operators must install upgrades or repeaters. ADSL involves a point-to-point connection, which means that the service can be brought to one customer at a time. Cable operators, on the other hand, have to upgrade the entire delivery system for every neighborhood they target.

To upgrade to ADSL, telecom operators need to install ADSL equipment at both ends of the loop connecting subscribers and the central offices. A potential drawback for customers, however, is that equipment from different vendors is likely to be incompatible. Cost also could be a problem: estimates put the price of ADSL modems to be higher than $1000 each at this time. As volume increases, prices should come down, particularly if industry standard chip sets become available.

Unlike ISDN, xDSL is not a protocol. Instead, it is a physical layer transport technology. Whereas the users of ISDN have been plagued with the complexity of the protocol, xDSL users will be employing protocols that are well defined and implemented, such as IP, frame relay, or ATM.

6.4 SWITCHED MULTIMEGABIT DATA SERVICE (SMDS)

SMDS is a WAN service designed for LAN interconnection through the public telephone network. It is a connectionless, cell-switched data transport service that offers total end-to-end applications solutions. With SMDS, organizations have the flexibility they need for distributed computing and bandwidth-intensive applications. At the same time, because SMDS supports both existing and emerging technologies, it provides the scalability organizations need to support the applications of the future.

Being a connectionless service, SMDS is different from other similar data services like frame relay and ATM. SMDS also differs from these services in the sense that it is a true service and is not tied to any particular data transmission technology. SMDS services can be implemented transparently over any type of network. Unlike frame relay technology, the SMDS circuit is committed at its specified full-time speed. Upgrading to the next speed is technically easy, requiring no additional in-house hardware or system software.

Like the frame relay service, SMDS is a relatively new service being offered by telecommunication companies. It aims to provide high-speed, high-quality data communications over a wide area. SMDS was originally proposed by BellCore in the United States and later adapted for use in Europe by the European SMDS Interest Group (ESIG). Efficiently interconnecting connectionless LANs such as Ethernet with connection-oriented WANs has proved to be a challenge, and many of the solutions put forward were not ideal. SMDS sought to provide a connectionless solution in which LANs may exchange frames without setting any kind of connection. Datagrams are simply transmitted onto the SMDS network. These datagrams are of variable length, and a maximum size of 9188 bytes allows SDMS to encapsulate an entire frame from most LANs.

6.4.1 SMDS Features

SMDS was designed to support the interconnection of connectionless LANs and does not provide explicit support for the transmission of audio or video. For example, it does not support data stream synchronization, and thus multimedia applications must perform synchronization by multiplexing data, audio, and video streams and present these combined streams to the SDMS network for delivery. Even though the SMDS protocol has a high overhead, which consumes as much as 25% of the available bandwidth, the maximum speed of around 45 Mbps (eventually 155 Mbps) is suitable for most multimedia applications. The ability of SMDS to support multicasting also allows broadcasting and near video-on-demand services.

One of the SMDS features is group addressing, by means of which files or information can be sent to multiple users at one time. This means that files can be sent to all members of a group and one member can make last-minute changes. That member can then send all the members of the group the updated copy in a couple of minutes. Another feature of SMDS, called multiple addressing per subscriber network interface (SNI, discussed shortly), allows subscribers to have a variety of unique addresses, which make it possible to reach individual users in a business. Multiple addressing means that a separate number can be kept for each important customer. This permits the use of a single SMDS connection to provide dedicated access for specific customers.

Address validation in SMDS is a security feature that prevents unauthorized users from gaining access to the network. Along with the address validation, address screening may be used to provide security from unauthorized network users who may try to access files. Thus a company from another group that knows a competitor's group address may try to gain access to the competitor's files, but screening will ensure that access is denied. Another feature that SMDS offers is customer network management (CNM). This option allows administrators to see their portion of the network as though it were a private network, enabling them to capture traffic statistics for trend analysis and alerting them to potential errors.

Another feature of SMDS is the ability to screen the addresses of incoming and outgoing packets. An address screen can either limit packets to a set of allowed hosts or list a series of forbidden hosts. By using this feature, subscribers can construct virtual private networks, preventing the loss of confidential data over the network. Similar in operation to address screening, source address validation is a feature used by service providers to ensure that frames injected into the network originate from valid address assigned to the appropriate subscriber. This feature allows accurate billing information to be collected and prevents "address spoofing," which would compromise the security provided by address screening.

SMDS provides a range of scalable, high-speed 1.17–34 Mbps connections to link an organization's networks directly to a DS-3 (45 Mbps) network. It is particularly beneficial for high-traffic networks to support the access of large databases and the quick transfer of multimegabit files, including high-resolution images and video. The moderate bandwidth connections offered by SMDS suit the LAN interconnection requirement well, since these numbers are within the range of most popular LAN technologies. Typical SMDS services are 1.17, 4, 10, 16, 25, and 34 Mbps. Figure 6.4 shows a LAN connection to the Internet via SMDS.

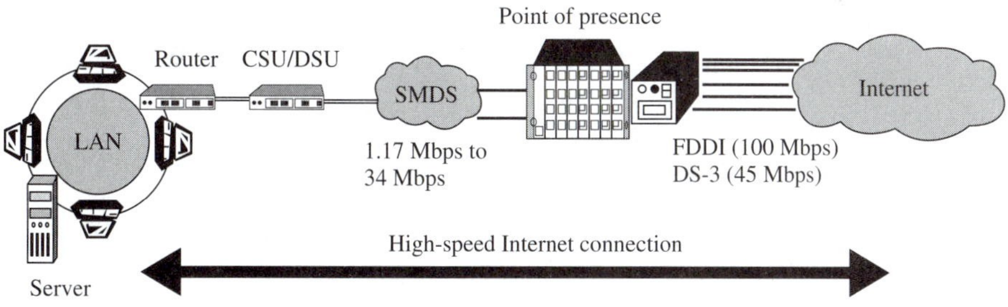

Figure 6.4 LAN connection to the Internet via SMDS.

The typical system requirements for the connection are SMDS with a channel service unit/data service unit (CSU/DSU), a high-speed Internet router, and TCP/IP application software. The CSU/DSU connects the subscriber via a router or a gateway to the service provider network. The point-of-presence system is typically a series of SMDS switches in the provider's network.

The interface protocols for SMDS are defined in terms of various system components. The first interface protocol is the SMDS Interface Protocol (SIP), which connects the CSU/DSU to the public network, thereby defining the subscriber network interface (SNI). This interface takes much of its design from the IEEE 802.6 Distributed Queue Dual Bus Protocol, making SIP a very robust protocol. The SIP interface is fundamental to SMDS, since almost all the other protocols are based to some extent on SIP.

Another interface is called the data exchange interface (DXI), which connects the customer's equipment to the CSU/DSU. This protocol is based on HDLC. DXI excludes some features of the IEEE 802.6 DQDB standard that are not important for SMDS. This results in reducing the cost of DXI, and consequently many SMDS service providers now offer DXI as the primary interface between the customer's equipment and the public network.

An interface called SIP relay allows users of frame relay equipment to access SMDS networks. This interface encapsulates SIP frames within frame relay's LAP-F protocol (discussed later). These frames are delivered to an interface within the service provider's network, where they are placed on the SMDS network. Finally an ATM interface allows SMDS service to be provided over ATM networks. This is accomplished by replacing the lower layers of SIP with ATM layers.

A variety of transmission technologies can be used with SMDS, including DS-1 (E1 in Europe), DS-3 (E3 in Europe), DS-3-based ATM, ISDN, and frame relay. Since, in its initial offering, the physical medium used by SMDS was limited to either DS-1 or DS-3 (E1 and E3 in Europe), a mechanism was required to provide users with intermediate data rates. This mechanism prevents users from having to pay for bandwidth they cannot utilize, making SMDS more economical. The intermediate rates are called access classes. Access classes operate by limiting the sustained information rate available to the user. This limit affects only traffic injected into the network, since limiting traffic leaving the network could lead to an excessive use of buffer space in the public network. The sustained information rate is limited by means of the so-called credit manager algorithm. The algorithm works by requiring that outgoing packets have a certain amount of "credit" to be transmitted. If a packet has sufficient credit, it is transmitted, and the required amount of credit is deducted from the total. If a packet does not have sufficient credit, it is discarded. Additional credit, up to a defined maximum, is allocated at the sustained information rate. The credit manager algorithm therefore allows short bursts to use the full bandwidth of the link, while limiting the sustained rate on the basis of the rate at which credit is accrued.

There are five access classes defined for DS-3, and four for E3. No access classes are defined for DS-1 or E1, since smaller bandwidths were not deemed to be desirable (although slower connections to SMDS were later added). The access classes for DS-3 are defined for sustained information rates of 4, 10, 16, 25, and 34 Mbps (full DS-3). Similar access classes are defined for E3 also. It may be noted that some of these classes match the rates of existing LAN technologies. Having these access classes will provide a response to the growing demand for high-performance interconnection of connectionless LANs.

SMDS is designed to be a reliable service, operating 24 hours a day, 365 days a year, with no less that 3500 hours between service outages. In the event of an outage, service should be restored within 3.5 hours. This objective is defined for service through a single carrier, from one SNI to another. The defined goals imply that SMDS should be available 99.9% of the time. Thus the access reliability should be better than even that of leased-line technologies. It is possible to achieve this goal because the use of switching technology allows SMDS packets to be routed around network failures.

Another set of objectives relates to the accuracy of the delivered information in terms of delivery, bit error, duplicates, and missequenced packet probabilities. The objectives range from a probability of 0.0001 of a packet not being delivered to a probability of 5×10^{-13} of a packet containing a bit error.

The final set of performance objectives addresses the delays through the transmission network. The delays depend on a number of factors, including packet length and the type of SNI at the source and destination. For medium-sized packets (1600 bytes), delays range from 472 ms for 56–56 Kbps transmissions to 20 ms for DS-3–DS-3 transmissions. These times are expressed as 95th-percentile delays (95% of packets will have a delay less than or equal to these figures). These delays are defined for a local connection only. If multiple carriers are used (i.e., long distance), the delays may be higher.

6.4.2 SMDS Addressing and Protocols

Within the public network, the **inter**switching **s**ystem **i**nterface (ISSI) connects individual SMDS switches. ISSI includes other functions relevant to the provider, including routing and maintenance functions. In addition to ISSI, the **interc**arrier **i**nterface (ICI) is used in the public network. ICI is used to make connections between different SMDS service providers, such as between a local SMDS provider and a long distance provider.

Each SMDS message contains a source address and a destination address. The address is constructed of two fields: a 4-bit address type and a variable length E.164 address. The address type field indicates one of two address types: individual or group. Individual addressing implies a single-source, single-destination message. Group addressing allows a single source to send identical packets to a number of destinations. Group addressing is achieved by allocating a dedicated address to represent a group of machines. These addresses are semipermanent, since the service provider must allocate them. The E.164 address consists of two fields: a country code and a national significant number. These fields together create the address of the actual subscriber network interface (or group, in the case of a group address). Multiple addresses per SNI are allowed. For a high-speed connection (DS-1 or higher), up to 16 individual and 48 group addresses are allowed per SNI. For low-speed connections, up to 2 individual and 3 group addresses per SNI are permitted.

The SIP protocol is designed for the SNI. Based on the IEEE 802.6 DQDB, SIP is defined in three layers. SIP level 3 processes data frames from upper layers, which can be up to 9188 bytes in length. SIP level 2 splits the level 3 frames into a series of 53-byte packets. SIP level 1 is composed of two sublayers, governed by the Physical Layer Convergence Protocol (PLCP) and the Physical Medium–Dependent Protocol (PMDP). The

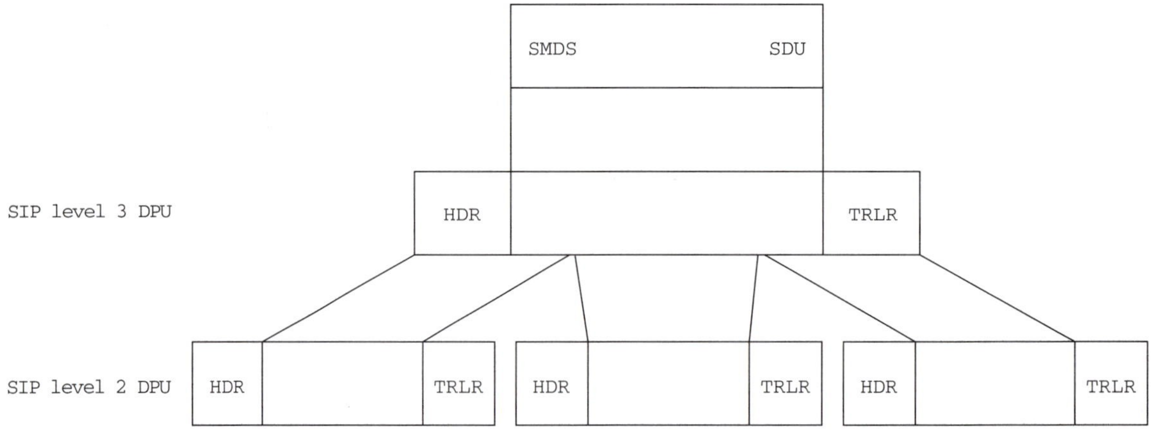

SDU = Service data unit
PDU = Protocol data unit
HDR = Header
TRLR = Trailer

Figure 6.5 Encapsulation of user information by SIP layers.

PLCP defines how cells are mapped onto the physical layer, while the PMDP defines the actual physical medium. The relation between the different SIP layers is shown in Figure 6.5.

SIP level 3 accepts variable length data from higher layers, adding the appropriate header, frame check sequence, and trailer fields. The information field can be up to 9188 bytes in length. A pad field is used to ensure that the data field ends on a 4-byte (32-bit) boundary. An optional CRC-32 field can be included to provide error checking. The header field is 36 bytes long and includes source and destination addressing, length, carrier selection, and higher layer protocol information (Figure 6.6).

SIP level 2 segments the level 3 frame into a series of short, fixed length segments designed for transmission over the telephone network (Figure 6.7). SIP level 2 conforms to IEEE 802.6 DQDB, so hardware designed for DQDB can be used with SMDS. Each of the small level 2 segments is 53 bytes in length and contains 44 bytes of data. The 44-byte data segment is preceded by 7 bytes of header and followed by a 2-byte trailer. The header contains fields for access control, network control (unused in SMDS), segment type (beginning of message, continuation of message, end of message), a sequence number, and a message identifier. The trailer contains a payload length field, indicating how much of the data in the segment is meaningful. The remainder of the segment holds a 10-bit cyclic redundancy check field protecting the last 47 bytes of the segment (the first 5 bytes contain another CRC).

SIP level 1 is responsible for placing the 53-byte segments created by SIP level 2 onto the physical medium. A variety of different physical media are supported. In North America, the most common physical layer is either DS-1 (1.544 Mbps) or DS-3 (44.736

Field length (bytes)	1	1	2	8	8	1	4 bits	4 bits	2	12	9188	0, 4	1	1	2
	RSVD	BEtag	BA size	DA	SA	X+ HLPI	X+	HEL	X+	HE	Info+ Pad	CRC	RSVD	BEtag	Length

RSVD = reserved
BEtag = beginning-end tag
BAsize = buffer allocation size
DA = destination address
SA = source address
X+ = carried across network unchanged
HLPI = higher-layer protocol identifier
HEL = header extension length
HE = header extension
Info + Pad = information + padding (to ensure that this field ends on 32-bit boundary)
CRC = cyclic redundancy check

Figure 6.6 SIP level 3 protocol data unit.

Mbps). However, SIP is in the process of being extended to support faster links such as OC3.

As mentioned earlier, the protocol for the data exchange interface (DXI) was developed to help control the costs of the SIP protocol by communicating between a customer's router and the CSU/DSU. Since the SIP protocol is based on IEEE 802.6, it contains many features (e.g., the ability to have many hosts on a single link) that are infrequently used in SMDS. These features predominate in SIP levels 2 and 1, where the DQDB protocols are used. Therefore, only the CSU/DSU is required to implement the full SIP protocol. Routers and terminals require only a small software upgrade to speak SIP level 3. The DXI protocol is designed to operate over a single link and is based on high-level data link control. HDLC frames are used to encapsulate SIP level 3 frames for transmission over the link. DXI is a two-layer protocol, with a DXI link layer and a DXI physical interface.

As discussed earlier, HDLC uses the binary flag "01111110" to delimit individual frames. Bit stuffing is used to prevent such a sequence from occurring within the data

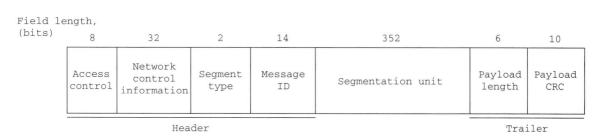

Field length, (bits)	8	32	2	14	352	6	10
	Access control	Network control information	Segment type	Message ID	Segmentation unit	Payload length	Payload CRC
	Header					Trailer	

Figure 6.7 SIP level 2 protocol data unit.

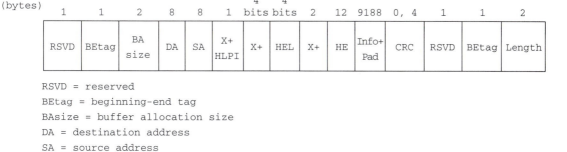

frame. In the DXI protocol, the address bits are used to divide the channel into a number of logical links, allowing individualized flow control for different messages. Other address bits are used to indicate the destination (either the router or the CSU/DSU) and to differentiate between command and data frames.

SMDS service can also be provided over ATM networks by using the ATM interface. This is accomplished by replacing the services defined in SIP to ATM, using ATM adaptation layer 3/4. SIP level 1 is replaced by the ATM physical layer, SIP level 2 by the ATM layer and the segmentation and reassembly (SAR) sublayer of AAL 3/4, and SIP level 3 by the CPCS (common part convergence sublayer), and SSCS (service-specific convergence sublayer) of AAL 3/4. Finally the SIP_CLS (SIP connectionless service layer) provides the appropriate interface to the upper layers, hiding the ATM implementation from the user.

6.5 FRAME RELAY

Frame relay is a protocol standardized by ANSI and the CCITT. Frame relay provides an interface between customer premise equipment (CPE) and a wide-area network that delivers data to remote CPE. Compared with traditional packet switching technology such as X.25, frame relay is a less sophisticated, less expensive, and more efficient wide-area communications technology. On the other hand, in comparison to circuit switching, frame relay can provide substantial cost and performance benefits for the bursty traffic of Internet access and other LAN interconnections. Since 1991, frame relay's ability to efficiently support bursty communications needs has led to rapid growth in market activity and user acceptance. Public frame relay services are quickly becoming available throughout most of the world. Businesses and organizations are upgrading their private networks to support the new interface protocol. Today, dedicated connection and access to the Internet drive the demand for frame relay. Frame relay is becoming an increasingly popular way to connect LANs in multiple locations. Frame relay emerged from the need for the following:

- Higher speed packet technology driven by LAN internetworking and wide-area communications
- Integrating traffic from legacy and client–server applications over the same links
- A definition for a streamlined packet technique from ISDN

A compelling reason for using frame relay is its cost-effectiveness versus leased lines. More importantly, frame relay's interworking capability with ATM makes it a low-cost alternative and transitional technology to ATM. The improved performance, as well as new features and services, enable users to build a backbone network capable of supporting higher bandwidth WAN for multiprotocol data, voice, and video. Along with ATM, frame relay represents the major improvement in packet switching technology of the past quarter-century.

6.5.1 Protocol and Architecture

Frame relay is an interface protocol providing OSI layer 2 and layer 3 services, which evolved from X.25 and ISDN packet switching technique and standards. In Chapter 3 (Figure 3.17), we showed that in the spectrum of switching techniques, frame relay lies between the two traditional switching mechanisms—circuit switching and packet switching.

As we recall, layer 2 is responsible for establishing and maintaining a reliable connection across a link, guaranteeing that frames of data are successfully transferred across a link. While multiplexing and switching of logical connections take place at layer 2 in frame relay, virtual channels can be established across a frame relay network characteristic of both levels: the ability to establish a point-to-point connection between two devices and the ability to multiplex calls over a link.

Figure 6.8 compares the frame relay with X.25 protocols and the OSI layer model. For the frame relay protocols, the control plane of operation is involved in the establishment and termination of logical connections between a subscriber and the network, and the user plane of operation is responsible for the transfer of actual data between the end users. The Q.922 used for both control plane and user data plane is an enhanced version of LAP-D. The core functions of Q.922 in the user data plane comprise a sublayer of the data link layer to provide the minimal service of transferring user data frames between end users without error handling and flow control. In the control plane, Q.922 provides a reliable data link control service for delivery of I.451/Q.931 messages.

The frame relay format is similar to X.25 LAP-B and LAP-D frames, which include both user data and the address information used to route the frame (see Figure 6.9). A typical frame has a 3-byte (default) header that includes a 1-byte flag (01111110) and a 2-byte address field, including 10 bits used for the data link connection identifier (DLCI). These 10 bits permit over a thousand virtual circuit addresses on every interface. The DLCI field can be increased up to 24 bits, resulting in a 4-byte address field or a 5-byte header, to offer 16 million possible virtual circuits. The length of the address field, including the DLCI, is determined by the address field extension (EA) bits.

The remaining bits in the header are used for congestion indication and other control functions in the network. Notification of congestion conditions may be sent by the net-

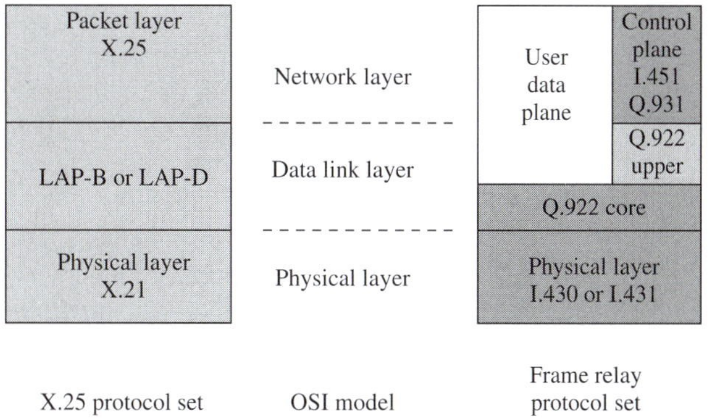

Figure 6.8 Protocol stacks for frame relay protocol and X.25, and the OSI layer model.

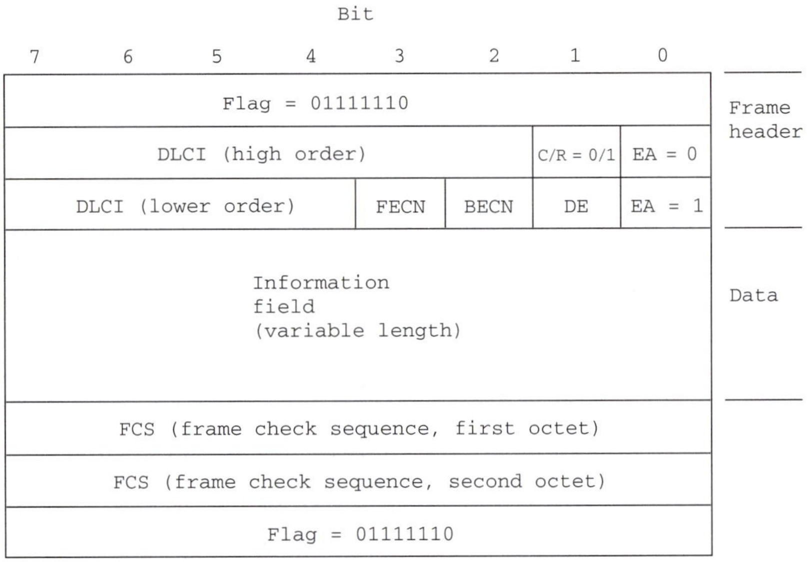

C/R = Command/response
EA = Address extension
DE = Discard eligibility indicator
BECN = Backward explicit congestion notification
FECN = Forward explicit congestion notification
DLCI = Data link connection identifier

Figure 6.9 Frame relay format.

work to the access devices through the use of forward and backward explicit congestion notification bits (FECN and BECN). Access devices are responsible for restricting data flow under congestion conditions. To manage congestion and fairness, frames may be selectively tagged for discard with the discard eligibility (DE) bit. The command/response (C/R) bit is application specific.

The frame relay standard provides mechanisms for congestion control techniques, but does not guarantee their implementation. These vendor-specific issues may result in product performance differences, but they do not normally interfere with basic frame relay interoperability.

There is only one frame type for user data. There are no control frames, no in-band signals, and no sequence numbers. Thus frame relay cannot be used for flow control. The I field (information field) of the frame that carries higher layer data is of variable length. The actual maxima are vendor specific, while 4096 bytes is the theoretical maximum of length. The I field contains the user data that are passed between devices over a frame relay network. The user data may contain various types of protocol (protocol data units) used by the access devices. The I field may also include multiprotocol encapsulation according to IETF RFC-1294, the industry standard, which is in the current I field. With or without multiprotocol encapsulation, the protocol information sent in the information field is transparent to the frame relay network.

At the end of the frame, preceding the frame delimiter flag (01111110), a 2-byte-long frame check sequence (FCS) is used between the access device and the network to ensure the bit integrity of the frame. Frames with unrecognized DLCI or incorrect FCS are simply discarded. It is the responsibility of upper layers in the end point devices to recognize the dropped frames and recover them by reinitiating transmission.

Frame relay is more efficient than traditional layer 2 and layer 3 protocols such as LAP-B and X.25 in terms of speed of operation and ease of implementation because the error-handling overhead is stripped out. Obviously, this increased efficiency requires two important assumptions:

- It requires that the physical/electrical connections (the physical layer) be clean, that is, have a very low error rate, as is seen with the much improved physical channels and fiber optic connections used increasingly by public carriers in recent years. When errors do occur, frame relay can identify the errors but does not retransmit any damaged frames. Rather, damaged frames are simply dropped, with the expectation that the end devices will detect the missing frames and retransmit the data.
- It assumes that the end devices with functions implemented in higher level protocols are intelligent and able to guarantee the integrity of communications between end devices.

With such redundant error handling and send/acknowledgment capabilities, the result is a streamlined communications process with reduced delay and much higher throughput. Figure 6.10 shows the data flows in traditional packet switching and frame relay.

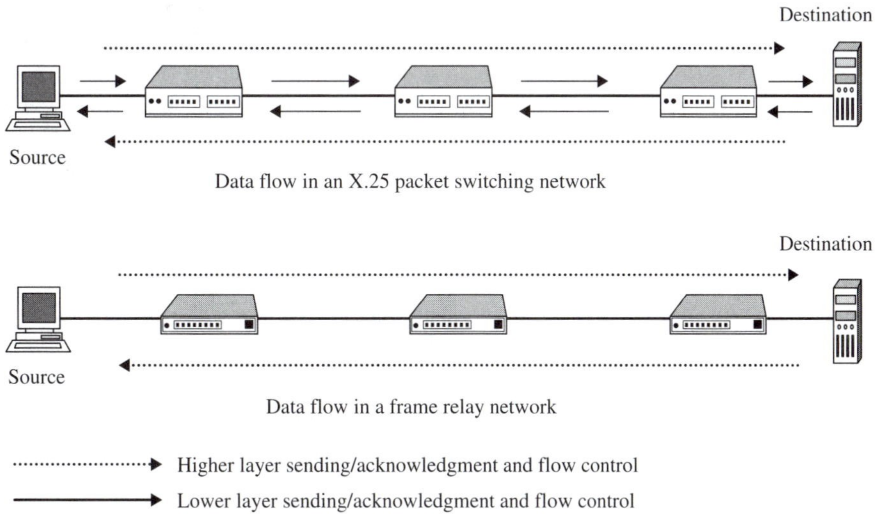

Figure 6.10 Error and flow controls in traditional packet switching and in frame relay.

In summary, frame relay is designed to eliminate all the unnecessary overhead associated with X.25 for error handling and flow control functions. The frame relay protocol differs from X.25 in the following respects:

- Multiplexing and switching of virtual circuit connections are implemented at layer 2 rather than layer 3. Layer 3 processing for user data frames is completely removed.
- Call control signaling is carried through the control plane, which is separate from the user data plane, making it unnecessary for intermediate nodes to maintain state tables on an individual virtual circuit.
- No error handling and flow control functions are provided. Since there is no sequence number in the frame header, correct frame order is not guaranteed either. All these functions are expected to be implemented in the transport layer and higher layers of networks. To alleviate the impact of errors, congestion control functions are added to frame relay protocols.

These features apparently complement and fit well with TCP/IP conventions. TCP establishes reliable and robust transport connections over a network and IP carries data packets across a network. This combination handles retransmission of data, should errors occur.

Since most of the error handling and retransmission functions in X.25 and LAP-B are eliminated from frame relay implementation, frame relay is less process intensive than X.25. The resulting high performance and low cost of frame relay for Internet communications have helped fuel a market for frame relay that exceeds initial expectations.

6.5.2 Implementation

Frame relay uses variable length frames to transport the user traffic across the interface. Frame relay can carry user data regardless of the higher level format. Frame relay reduces the complexity of the physical network without disrupting higher level network functions. In fact, as discussed earlier, it actually relies on and utilizes these higher layer protocols to its advantage. It provides a common network transport for multiple traffic types while maintaining transparency to higher levels. The frames contain addressing information that enables the network to route them to the proper destination.

Frame relay's simplified link layer protocol can be implemented over existing technology. Access devices often require only software changes or simple hardware modifications to support the interface standard. Besides specially designed frame switches and cell switches (ATM), existing packet switching equipment and T1/E1 multiplexers often can be upgraded to support frame relay over existing backbone networks. Also, many types of existing internetworking equipment (e.g., routers) can be upgraded to frame relay without hardware changes.

Frame relay provides a user with multiple independent data links to one or more destinations. Traffic on these data links is statistically multiplexed to provide efficient use of access lines and network resources. Since the multiplexing is at the link layer, end-to-end delay is minimized. Frame relay networks transfer the user data in a frame without regard to its contents, thereby providing service effectively as transparent as that available from

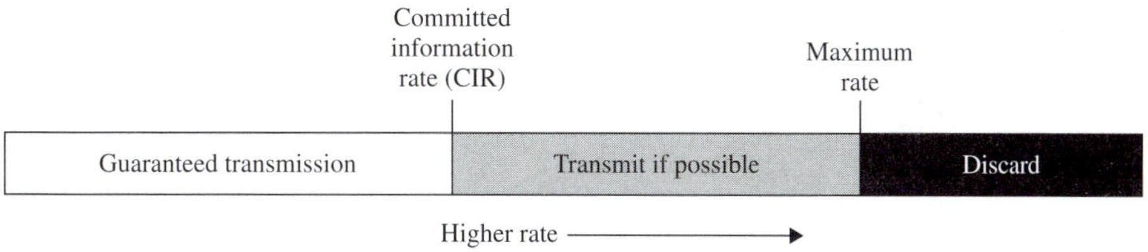

Figure 6.11 Access rate, committed information rate (CIR), and maximum rate associated with frame relay networks and services.

a leased line. The prevalent speed of frame relay service is fractional T1/E1 and full T1/E1 ranging from 56/64 Kbps through 2 Mbps, while high-speed services with rates up to T3 (45 Mbps) are gradually becoming available. When a fractional T1/E1 link is desired, a full T1/E1 may still be used for access, depending on the carrier offering, with the unused portion of the T1/E1 for transporting other traffic, such as voice. A standard CSU/DSU is used in conjunction with the 56/64 Kbps or T1/E1 service.

Three important speeds in the implementation of a frame relay service are the access rate, the committed information rate (CIR), and the maximum rate. The access rate is the physical speed at which the user receives information, and delivers it to the entrance of the frame relay network via the user-to-network interface (UNI). The minimum level of service for a particular virtual circuit is defined as the CIR. During a file transfer, for example, the network is committed to deliver only the contracted CIR. However, if spare capacity exists because other virtual circuits are not using their assigned bandwidth, this virtual circuit could burst in excess of the CIR up to the access rate. When the access rate surpasses the maximum rate, frames are dropped (Figure 6.11).

Frame relay networks are made up of three elements:

- **Frame relay access equipment** is the customer premises equipment (CPE) that sends information across the frame relay wide-area network. In general, the same frame relay access equipment may be used for either private frame relay switching network equipment or public frame relay services.

- **Frame relay switching equipment** comprises devices responsible for transporting the frame-relay-compliant information received from and sent to the access equipment. Either variable length information units (frames) or fixed-length information units (cells) can be used by the frame relay switching equipment based on different implementations.

- **Public frame relay services** are offered by public service providers (carriers), which deploy frame relay switching equipment for the network and maintain access to the network via the standard frame relay interface. Both frame relay access equipment and private frame relay switching equipment may be connected to services provided by a carrier.

A typical setup for frame relay access is the frame relay assembler/disassembler (FRAD), which is the frame relay counterpart to the X.25 packet assembler/disassembler (PAD) in that it aggregates packets (from TCP, SNA, IPX, etc.) into frames and converts them into frame relay format. A typical frame relay network uses FRADs or routers in each of an organization's branch offices or remote sites. The sites can be linked over a private frame relay network or over a carrier's public frame relay network. Access equipment may be hosts, bridges, routers, packet switches, specialized frame relay PADs, or any other similar devices.

Since frame relay is only an interface specification, the network may route the frames in different ways. Frame relay switching equipment may be frame switches, cell switches, T1/E1 multiplexers, modified packet switches, or any specialized frame relay switching equipment that implements the standard interface and is capable of switching and routing information received in frame relay format.

The common architectures used to implement a frame relay service are frame switching and cell switching. With frame switching, the variable length frames are kept intact as they traverse the network via routing by server modules located at each node. Examples of frame switches include routers and modified X.25 packet switches. With a frame switch, frames from all ports at a node are received and queued by a server module. Frames in excess of the CIR for each virtual circuit are marked as discard eligible (DE) by the network. The server also processes and queues frames from other nodes via trunks. All these frames are queued either in shared buffers or in virtual circuit buffers, depending on the specific frame relay network structure.

With cell switching (also called cell relay, or ATM), the frames are broken into smaller, fixed-length information units, or cells, and hardware-based techniques are used to switch them across the network in accordance with the information in the cell header. The cell switching network passes the information along in small entities, then reassembles the cells into frames at the destination. Both frame switching and cell switching networks are common network architectures for frame relay today. While each architecture has its own set of advantages, and both are fully compliant with frame relay, the choice of platform will determine the quality of service (QoS) received by the users. For frame relay, QoS is defined in terms of maximum application performance, lower cost of networking, predictability, fairness, scalability, and expandability.

Some frame relay implementations use time division multiplexing to divide the bandwidth of the network trunk in an effort to support a mixture of traffic types. Voice and data will use part of the trunk in TDM mode, while frame relay traffic uses the remaining of the bandwidth—typically 256 Kbps in a T1 connection. These TDM hybrids allocate trunk bandwidth in a static manner. That is, each voice or data connection has dedicated bandwidth, and the frame relay traffic is confined to dedicated channels. Unlike ATM, as explained in Chapter 7, frame relay/TDM hybrids cannot reduce bandwidth costs by dynamically reallocating unused bandwidth to other channels. With the circuit-switching-based frame relay/TDM user-to-network interfaces, the user must preallocate some or all bandwidth to the frame relay service, regardless of whether that bandwidth will be used. This hybrid approach does not deliver the intrinsic benefits of frame relay.

Note that frame relay is an interface and a service, which may be implemented by using either frame switches or cell switches (cell relay, or ATM), while ATM refers to both switches and a service.

At the heart of a frame relay connection is the type of virtual circuit called a data link connection (DLC). Just as a typical telephone cable may carry multiple conversations through multiplexing, a single physical frame relay interface may contain many individ-

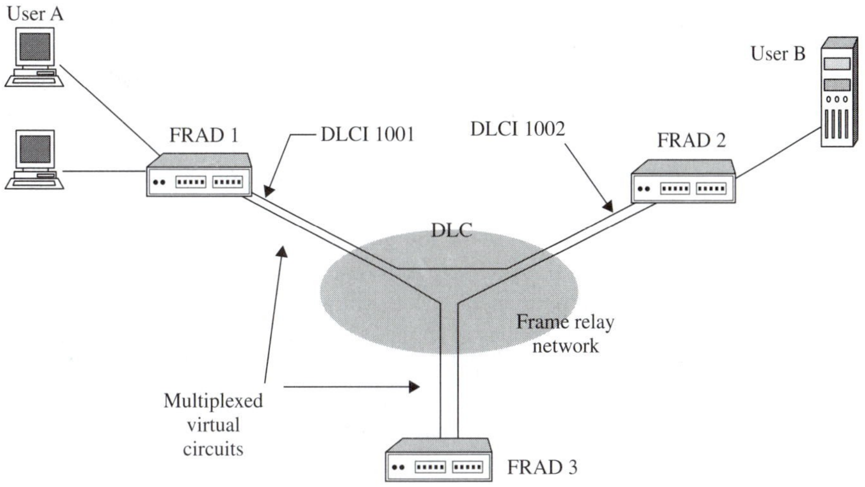

Figure 6.12 DLCI numbers are used to route frame relay traffic and have only local significance.

ual communications channels, rather than requiring separate physical facilities for each channel. As described earlier, each frame contains a circuit number called a DLCI, or data link connection identifier, which denotes the virtual circuit that "owns" that particular frame of information. The frames are routed by the network to their destination based on the circuit numbers in the frames.

Unlike IP addresses and the port numbers for TCP/IP protocols, the DLCI has only local significance and may be dynamically assigned to a virtual circuit. The DLCI numbers of a frame relay connection often are different at the local and remote ends. In Figure 6.12, for example, DLCI 1001 at FRAD 1 is routed by the frame relay network to FRAD 2, where it appears as DLCI 1002. All data sent by user A on DLCI 1001 from FRAD 1 will appear on DLCI 1002 at FRAD 2. All data sent via DLCI 1002 from FRAD 2 will appear on DLCI 1001 at FRAD 1. If user A at FRAD 1 tries to sends data on DLCI 1002, the data will not appear at FRAD 2 on DLCI 1002. Instead, the frames will be dropped because of the unknown DLCI number. The presence of incorrect DLCI numbers, due either to installer error or to misinformation from the carrier, is a common installation mistake.

Since DLCI numbers operate over point-to-point links between equipment and there may be a number of links or hops over which a DLC's data may travel before reaching the last hop and being delivered to the destination, both the local and remote DLCIs for any DLC across a frame relay network must be given before the virtual circuit is established. The routing tables in each intervening frame relay switch in the carrier's network, or a private network, take care of routing the frames to the proper destination. The entries in the connection routing table are responsible in mapping incoming frames from one channel to an outgoing channel and translating the DLCI in the frame before transmission based on the so-called chained-link path routing algorithm. When it is time for a particular frame to be transmitted, it travels across the trunk to the next node, which reads the destination address, queues the frame, and then routes it to the next node. This process continues until the destination node is reached, whereupon the frame is routed to a port that interfaces with the destination device.

Special management frames with special DLCI addresses—DLCI 0 and DLCI 8191—may be passed between the network and the access device. These frames monitor the status of the link and indicate whether the link is active or inactive. They can also pass information regarding stands of permanent virtual circuits (PVC) and DLCI changes. This frame relay management protocol is sometimes referred to as the local management interface (LMI). Its function is to provide information about PVC status.

Frame relay service can be based on switched virtual circuits (SVCs), or "switched access" to permanent virtual circuits (PVCs).

For a PVC-based frame relay service, the DLCs are predefined by both sides of the connection. The network operator—be it a private network or a carrier—assigns the end points of the circuits. While the actual path taken through the network may vary from time to time when automatic alternate routing is taking place, the source and destination of the circuit will remain the same. This type of circuit behaves like a dedicated point-to-point circuit. The majority of current frame relay services are PVC-based because the original service offerings for frame relay were PVC-based—a form carried on to the present day. With a PVC frame relay service installed, the user can send traffic at the maximum speed all day long at a flat price.

PVCs have served the needs of most existing telecommunication applications and continue to efficiently serve these users. There is, however, growing support for SVC capabilities to meet emerging applications and stimulate intercorporate communications.

Like a typical telephone call, there is a call setup procedure—dialing for each conversation—that must take place to establish a connection. Also similar to its X.25 SVC counterpart, SVC-based frame relay is initiated by the actual user of the circuit (caller), and the user specifies the destination of the desired connection. A virtual circuit—using a virtual circuit number (DLCI)—is established for the duration of the call. The SVC will be torn down after completion of the communications. Users establish calls over an SVC frame relay network by requesting a destination based on the internationally recognized X.121 or E.164 numbering plan in a manner similar to X.25, ISDN, and Internet services.

With the SVC frame relay service, bursty data may be sent at very high speed while users still save money if the average usage is low. In addition, network administration will be simplified, since carriers and customer network administrators are relieved of the burden of preconfiguring and managing changes of a PVC network topology. This can be particularly beneficial to large and dynamic networks with tens of thousands of sites to connect. SVC provides a true bandwidth-on-demand service, with throughput parameters requested at call setup time. The network will set up an SVC with the requested CIR. If the requested bandwidth is unavailable, the network will negotiate the parameters with the FRAD or router at the customer site. This provides the customer with customized services based on the application in use. For example, a short interactive session might use an SVC with a low or zero CIR, while a large file transfer of time-critical data might require an SVC at a high CIR value. One special application of SVC is associated with an inexpensive solution for disaster recovery, where a mirror image of the database residing on a disaster recovery server can be regularly updated through an SVC frame relay service.

PVC and SVC can share a single physical circuit. Unlike the case of a traditional telephone call, no network resources except the DLCIs are used when there is "silence"

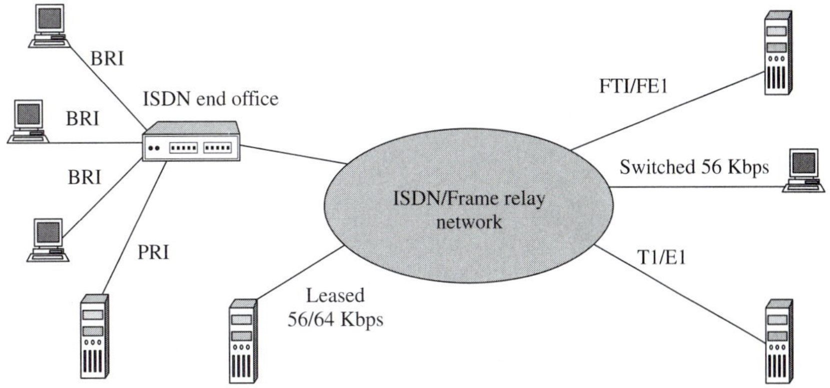

Figure 6.13 Frame relay local loops using T1/E1, FT1/FE1, leased/switched 56/64 Kbps, and BRI/PRI ISDN.

in the circuit. With the standards in place and the products and services beginning to roll out, SVC services will provide the on-demand any-to-any connectivity that is available from current telephone networks.

A local loop connection is always needed from user premises to the carrier's nearest service access point (SAP) to access a frame relay service. A traditional local loop is provisioned over leased circuits such as a 56/64 Kbps DDS service, a fractional T1/E1 (FT1/FE1), or T1/E1 service, although there is no requirement that a leased circuit be used. Actually, switched 56 Kbps and, in particular, basic rate ISDN (BRI) may soon become the local loop of choice owing to the popularity and dropping cost of ISDN and the strong demand for access to Internet and online information services as well as the need for telecommuting. As we noted earlier, a primary rate ISDN (PRI) connection can support 25–30 simultaneous connections, while a full T1/E1 frame relay link can accommodate many more virtual circuits, depending on traffic loading, at a much higher price than ISDN (Figure 6.13).

6.5.3 Frame Relay Performance Analysis and Congestion Control

The improved efficiency due to simplified operation and reduced overhead are the major advantages of frame relay over X.25, the conventional packet switching technology. Bandwidth sharing is frame relay's advantage over TDM circuit switching. On the other hand, owing to protocol simplification and bandwidth sharing, frame relay needs special treatment regarding performance and congestion control.

Frame relay standards define only the interface between the CPE and the frame relay wide-area network. Because the interface is standardized, any wide-area networking platform, frame switch, or cell relay (even TDM) with a frame relay interface will provide the basic benefits just described. Within the wide-area network, each vendor is free to implement the frame relay service based on any switching fabric. In evaluating platforms that support frame relay, it is important to understand that the implementation of each will yield different characteristics that determine quality of service.

Generally speaking, a frame relay service has the following goals:

- To maximize application performance and lower the cost of networking.
- To offer predictable performance and fair allocation of resources.
- To be scalable, expandable, and able to support future needs.

The reliability of the communications channels plays a particularly important role for a frame relay network because of the lack of error-handling functions in the lower layer. Because frames with errors are simply dropped, frame discards are very disruptive to protocols and applications of all kinds. Applications such as Novell IPX/SPX, SNA, and TCP/IP that individually acknowledge each frame or frame burst will wait for a time-out and then retransmit, thus reducing throughput and response time. Particularly with a sliding window acknowledgment mechanism, such as with TCP/IP and SNA, the discarding of a single frame will cause retransmission of the entire window and the reduction of the window size to one. Fortunately, the use of relatively error-free digital lines means that network-induced bit errors will have virtually no impact on throughput. However, if congestion is allowed to form, the resultant discards will be disastrous for frame relay performance.

Throughput also varies according to the delay caused by processing, queuing, and serialization within a network.

- **Processing delay** occurs in customer premises equipment, frame relay switches, and physical transmission facilities. A CPE adds about 3 ms of delay. A frame switch needs more time, typically about 12 ms, to process and determine where to route the frame based on addressing by relatively slow software. However, the processing of fixed-length cells by hardware in an ATM switch results in switching decisions made in very little time. The delay due to the transmission of the signal is estimated to be 1 ms per 100 circuit miles.

- **Queuing delay** occurs when trunk buffers or buffers in a server module of a frame switch fill, and a newly arriving frame must wait for other frames to be served before it can be transmitted. These buffers are often shared by all virtual circuits active in the network. Classical queuing theory and simulation indicate that for any given network load, a fixed-length short cell switch has a queue depth significantly less than that of variable length frame switch.

- **Serialization delay** is the amount of time necessary for a frame or cell to be fully received by a node before it can be processed. This delay is function of the length of the frame or cell, the number of hops the virtual circuit must traverse, and the trunk speed available to the frame relay traffic.

With a frame switch that has a 1024 Kbps processing speed, the time needed for an entire 1500-octet frame to be processed at each node is $(1500 \times 8)/1024 = 12$ ms. At each intermediate node, more delay is added. As the network size increases, so does the amount of processing delay. Moreover, because of the store-and-forward nature of frame switches, the length of the frame can have serious consequences.

If low delay and high reliability are not present, raising the CIR will not be able to significantly affect performance. When the network delay increases, the performance will degrade, especially for delay-sensitive protocols such as SNA, DECnet, and any protocol that apply the stop-and-wait protocol, which waits for an acknowledgment after sending each frame or frame burst.

Generally speaking, the higher the CIR, the more delay sensitive the network will be. Simple arithmetic reveals that an IPX/SPX network with a 100 ms delay (round trip needs 2×100 ms) is delay-limited to about 40 Kbps, assuming an average frame length of 1000 octets:

$$\frac{1000 \times 8}{0.1 \times 2} = 40{,}000 \text{ bps}$$

In fact, assuming a 200 ms round-trip delay, many Novell IPX/SPX installations will be limited to approximately 23 Kbps even if a higher CIR is configured. If the CIR subscribed is 128 Kbps, the money is essentially wasted. The shorter the average frame length, the worse the degradation will be due to delay as calculable from the foregoing formulas. A higher CIR offers no advantage, since delay may reduce the throughput.

Fortunately, cell relay architectures are able to overcome this problem. Cells are transmitted across the network without the need to rebuild the frame at each hop. When one assumes a trunk speed of 1024 Kbps, each cell takes only 0.2 ms to be received. Only at the destination node does the entire frame need to be rebuilt; thus only one store-and-forward delay is present with a true cell relay platform, regardless of the size of the network.

Some frame switches try to use proprietary fragmentation algorithms to break very large frames into smaller ones, to reduce serialization and queuing delay problems. Unfortunately, fragmentation performed without proactive congestion avoidance does more harm than good. Frame drops still happen because of congestion. When fragments are shorter but more numerous, discarding fragments in many scattered frames is worse than discarding an entire single frame: now all the affected frames will be retransmitted by the end devices, thus creating exponentially growing congestion.

Fairness is an important factor for a shared network. The ideal network would guarantee every virtual circuit the opportunity to transmit at least at its committed information rate. Bandwidth unused at any given instant would be fairly reallocated to active virtual circuits in proportion to assigned CIRs. In the meantime, the throughput of one virtual circuit should not degrade below its CIR or suffer increased delay because another virtual circuit is transmitting excessively. If the router is bursting above its CIR, the network will mark frames in excess of its CIR as DE. Still, this burst can cause the level of the shared server module buffers across the network to increase toward a critical level, causing delay and the setting of FECN/BECN for all users.

It is well known that bandwidth is more expensive than telecommunication equipment. To lower networking costs, one must minimize wasted bandwidth and heavily utilize trunk bandwidth. Yet owing to the highly bursty nature of today's communications, this strategy will result in discards or protocol time-outs. Knowing this, network designers specify only 50% average trunk utilization for a frame relay network that uses frame switches, to ensure reliable operation. This is still a substantial improvement, considering that in a voice call 60% of the bandwidth is wasted because it carries voice silence, and 100% of the bandwidth is wasted when the communication is terminated for a TDM channel. Simulations and experience show that when a frame relay service is provided by ATM switches with proactive and effective congestion avoidance, trunks can be highly loaded up to 95% average utilization.

Congestion control and avoidance mechanisms play vital roles in providing good frame relay services. Besides the DE field, there are two fields in the frame header, FECN and BECN, that are specially designed for congestion control.

- **Forward explicit congestion notification (FECN)** bits indicate that the transmitting frame may encounter congesting channel conditions, and that congestion avoidance procedures should be initiated for traffic in the forward direction.

- **Backward explicit congestion notification (BECN)** bits indicate that a received frame has encountered congested channel condition, and congestion avoidance procedures should be initiated for traffic in the opposite direction.

If a frame relay network and its users indiscriminately admit all traffic into the network regardless of available bandwidth, the switching buffers along the channel will fill and the delay will increase. Increased delay lowers throughput. Once the delay experienced by users reaches a critical level, the switching server module starts FECN/BECN on all frames to notify end devices. Congestion occurs. The router typically does not pass the congestion notification to the higher level protocol, so the offending user does not reduce its information rate. Most end devices also ignore this notification. If the congestion continues to grow, and buffers begin to overflow, the DE bit forces the frames to be discarded. When frames are discarded, throughput is lowered even more as frames are retransmitted and sliding windows reduce in size.

The congestion situation in a frame relay network is much more serious than in a traditional packet switching network like X.25 because of lack of low-level flow control mechanisms. Making the matter even worse, the extremely high delay prevents the FECN/BECN feedback mechanism from operating effectively at the frame level. By the time an adjustment message was finally received at the source of traffic, the network would already be congested.

A much better solution is a proactive approach that prevents congestion whenever possible. If there is no spare capacity in the network, traffic should not be fed into it at the access rate, only to cause additional discards and worsen the congestion. More advanced and complicated algorithms would lower the transmission rate of the virtual circuit down to the CIR if necessary, and buffer the excess burst at the edge of the network. The buffer would be large enough (typically 64 KB) to hold the very large window size of standard protocols. The algorithm would closely monitor the network traffic load, continuously determine whether spare capacity exists across the entire path of each individual virtual circuit, and then independently adjust each information rate. If an individual buffer filled, the algorithm would set congestion notification bits on only the overactive virtual circuit. If the user device still continued to transmit excessively, putting the buffer in danger of overflowing, discards would occur on the offending virtual circuit only. As the network load increased and decreased, such an algorithm would dynamically and fairly adjust the rate of transmission for each virtual circuit, to simultaneously maximize performance and prevent trunk buffers from becoming congested.

6.5.4 Voice over Frame Relay

Bandwidth-on-demand services such as frame relay and ATM are recognized as methods to provide cost and performance benefits for data traffic. What is less well known is that frame relay can be applied to voice traffic as well. However, frame relay networks were not originally designed to differentiate between different applications. Early frame relay services were almost exclusively used for interconnecting LANs carrying data bursts. At that time, only one type of CPE—the router—could support frame relay traffic. Routers are good at handling variable lengths of frame with data but are not geared for handling time-sensitive traffic such as voice. Routers cannot prioritize voice traffic over data in a way that ensures that voice frames arrive at the destination at a constant rate, to allow natural speech flow.

Voice conversation consists of speech bursts separated by short or long periods of silence, such as the pauses between sentences. Digitized voice traffic turns out to be similar in many respects to data traffic—both can be described as bursty and intermittent. Older TDM can keep constant voice data to be sent at constant rate by dedicating individual bandwidth channels for each application or user. Because of the bursty nature of communications, however, most of the expensive bandwidth was wasted carrying idle data and silence.

Thanks to advances in technology, it is now realistic to try to carry time-sensitive voice and SNA over frame relay with relatively minor adjustments to network infrastructure. Long distance calls are expensive. It is easy to cost-justify voice over frame relay, particularly for international long distance calls over international connections. However, because of the nature of frame relay networks and interoperation issues, voice over frame relay is best at *on-net* communications (i.e., connecting sites within the frame relay network). It is not now suited for calling *off-net*, although future changes in standards and new frame relay technologies may make that possible.

Voice traffic is not as tolerant as data traffice of delay, in particular the *changing* delay. When delay reaches 100 ms, voice distortion becomes noticeable. One of the main causes of changing delay in frame switch networks is variable frame sizes. Architectures that use frame switching internally are vulnerable because a voice frame could become delayed behind a long frame used for file transfer. The important issue that must be resolved to enable good quality voice traffic to be sent over a frame relay network is how to control the amount of delay that a frame experiences when traveling across the network. Supporting voice traffic on a frame relay network requires the premises equipment to do at least three things:

- Segment long data frames to allow voice traffic a better chance to transmit.
- Prioritize voice frames and other time-sensitive data such as SNA session before regular data traffic.
- Set the DE bit for noncritical data traffic.

The frame relay CPE or FRAD not only must be able to segment large data frames into smaller frames, it also should be intelligent enough to know to make segmentation only when voice frames are waiting to be transmitted. In that way, a higher priority voice frame always gets priority and is transmitted even if the system is in the process of transmitting a large data file. On the other hand, voice has more tolerance for lost frames than data.

In other words, if a single frame is lost in a data transmission, it must be retransmitted, but a few missing voice frames might not be noticeable in a conversation. This leads to an expedient approach that selectively marks voice frames as DE frames and always sends voice as high priority so that a basic level of voice information can be delivered across the network.

Successfully supporting voice traffic on frame relay also requires equipment with correct interfaces and support for voice-related functions, such as echo cancellation, silence detection and suppression, and fax detection and demodulation.

With frame relay/TDM hybrids, a portion of the trunk bandwidth is set up as a dedicated channel used for frame switching between nodes. In terms of frame relay, these frame relay/TDM hybrids can be thought of as frame switches with lower speed trunks. With circuit switching systems, the system must preallocate some bandwidth to the frame relay service, regardless of whether that bandwidth will be used. Bandwidth allocation for a frame relay/TDM hybrids faces the same problems associated with frame switching, but its use also further magnifies serialization delay and minimizes effective bursting. There is failure to deliver the intrinsic benefits of frame relay, resulting in poor support for frame relay as well as voice and circuit data traffic.

Adding voice to frame relay networks blurs the line between frame relay and ATM. ATM's overhead makes it less efficient than frame relay for many data applications, but frame relay's variable length frames are not as conducive to voice applications because they are sensitive to delay variation. ATM's solution is to package all traffic into small, fixed-length cells. Frame relay's solution is to package the voice traffic into fixed-length cells, while keeping the data frames with variable length to reduce overhead. Only when the data frames become excessively long are they chopped into smaller segments to allow more uniform voice traffic. Combined with an effective proactive feedback mechanism for congestion avoidance, frame relay can guarantee a minimum level of voice service. When the feedback loop operates at the cell level, the delay is very low; hence good voice quality is guaranteed. By using ATM architecture with effective congestion avoidance algorithms, high quality voice is provided.

6.5.5 Migration Toward ATM

The majority of today's data communications and network equipment is frame-based. Frame relay is an effective means of providing high performance to these legacy devices.

Broadband ATM, in both the LAN and WAN environments, has been recognized as the best networking architecture to support very high bandwidth applications. To avoid a throwaway investment, one should make sure that the current platform and existing equipment are able to support new advanced applications when these become necessary. Building a frame relay service on ATM switches is a good way to get cost and performance benefits with today's narrowband traffic, without being forced to deploy a new architecture in the future to support these emerging applications.

The philosophy behind frame relay and ATM is the same—keep things simple, and take advantage of the more reliable physical layer of modern communication channels. The basic difference between frame relay and ATM is that ATM uses fixed-length cells (5 + 48 octets), while frame relay has a variable frame length. The ATM Forum's data exchange interface (DXI), and frame-based user-to-network interface (FUNI) specify a

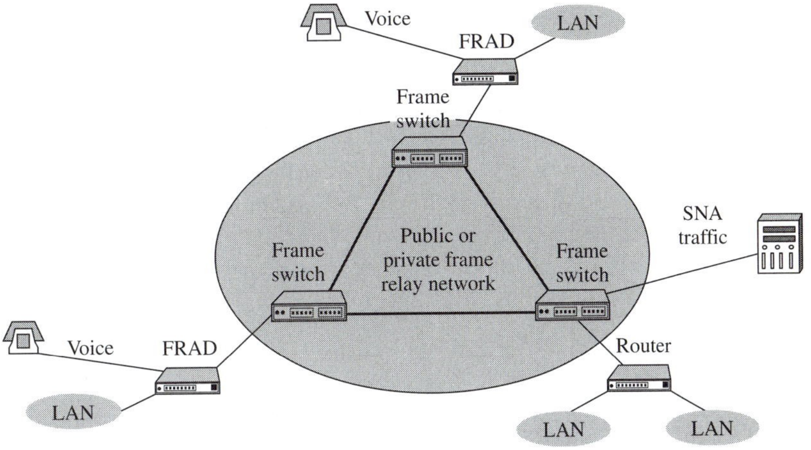

Figure 6.14 A public or private frame relay network can provide different data communications and voice services.

ATM access protocols that take advantage of existing low-cost, frame-based user equipment. Thus, the DXI and FUNI allow the user equipment to send traffic in frames, as opposed to cells. With another piece of equipment to perform the function of segmenting the frame traffic into ATM cells, DXI and FUNI software can run on the same hardware that supports frame relay and X.25. This topic is further discussed in Chapter 7.

In summary, frame relay has attracted high interest because it is a technology that has been developed in response to the following trends:

- Improved telecommunication conditions with much lower error rates
- The proliferation of powerful end point devices, such as PCs and workstations
- The use of intelligent protocols, such TCP/IP, XNS, and DECnet

Frame relay is a very effective way to support today's LAN interconnect, SNA applications, and even voice traffic (Figure 6.14). Frame relay can be implemented on either frame switch or cell switch, as well as traditional circuit switching TDM. Frame relay provides the following advantages over conventional packet switching and circuit switching technologies.

1. Higher efficiency: Simplified architecture and operations dramatically reduce overhead, thus streamline the traffic along the network.

2. Bandwidth sharing: The statistical sharing of bandwidth by multiple virtual circuits increases bandwidth utilization and allow higher traffic load; voice traffic can ride over a frame relay network inexpensively, and each virtual circuit may burst to a speed higher than could be afforded with leased lines or TDM channels.

3. Single physical interface: Independent of the WAN architecture, frame relay offers a single physical interface that can support multiple virtual circuits, thus supporting a full range of connectivity without the expense of multiple line interface cards and leased lines.

4. Better security: Since frame relay operations are limited to the physical and data link layers, frame relay networks tend to be more secure than traditional packet switching protocols that extend to the network layer, which makes a frame easier to be manipulated adversely.

5. Easy to upgrade: Because most user devices such as routers and intelligent hubs are upgradable via software changes to support frame relay, the investment in existing CPE can be preserved. More importantly, because of frame relay's internetworking capability with ATM, frame relay becomes a low-cost alternative to ATM.

Even in the endless pursuit of more and more bandwidth, it is apparent that broadband ATM is not needed, and cannot be afforded, at every location in the network today. With the standard for frame relay to ATM internetworking in place, when wide-area trunks need to be upgraded to support future very-high-bandwidth applications, it will be possible to upgrade to broadband on a trunk-by-trunk basis. Frame relay technology offers a feasible way to migrate toward ATM.

6.6 FAST ETHERNET, 100VG-ANYLAN, AND GIGABIT ETHERNET

The dramatic improvements in the ability to transmit and switch high-speed digital information make it possible for today's networks to carry voice, data, images, video, and multimedia information for a large number of users and various applications. The development of Ethernet technology has played an important role in LAN and backbone network applications. We present three Ethernet-based techniques that are likely to gain popularity in the industry.

6.6.1 100VG-AnyLAN

Originally called 100Base-VG, the new high-speed network design known as 100VG-AnyLAN was accepted by the IEEE as a separate specification because it is not based on the existing Ethernet architecture. This new specification, also known as IEEE Standard 802.12, is a 100 Mbps high-speed networking standard originally developed to transmit Ethernet or IBM Token Ring packets on existing wiring. The VG "voice grade," meaning that the 100VG technology runs standard Ethernet and Token Ring frames at 100 Mbps utilizing voice grade category 3, 4, and 5 unshielded twisted-pair (UTP) wire, shielded twisted pair (STP) wire, and optical fiber. Data packets are transferred from node to node by a hub based on the address of the data packet. This ensures orderly transmission and eliminates collisions. Because of this managed transfer of data, 100VG-AnyLAN can transmit data at peak speeds of 96 Mbps, in comparison to 4–6 Mbps on Ethernet. In addition, 100VG-AnyLAN can provide guaranteed bandwidth for emerging time-sensitive applications such as multimedia. Low costs and an easy migration path from existing 10BaseT and Token Ring networks promise to make 100VG-AnyLAN the best alternative for upgrading 10BaseT as well as token ring users to 100 Mbps speeds.

Increased network performance is not the only requirement of today's networks. To attract customers from the mainstream marketplace, network technologies must also address other needs, primarily related to costs. Costs encompass both the purchase of new components and the price of migrating from an organization's existing network. Building or upgrading a network can require the inclusion of network adapters for each network client, concentrators or hubs in the building wiring closets, and other new components. The costs of migration must also be considered: monetary costs as well as the organizational costs of disruptions and training. With today's costs of labor, a significant percentage of the replacement value of an existing local-area network goes for installed cabling. Obviously, any networking technology that lets organizations use their installed cabling is preferable to one that requires new cabling.

100VG-AnyLAN brings together the best characteristics of both Ethernet and Token Ring, combining the simple, fast network access familiar to Ethernet users with the strong control and deterministic delay characteristics of token ring networks. In addition, by providing a single high-speed network hardware infrastructure capable of supporting Ethernet and token ring frames, 100VG-AnyLAN is considered to be the logical successor to both Ethernet and Token Ring technologies.

The key to 100VG-AnyLAN's powerful capabilities is its leverage of the physical star topology used in most modern networks. Ethernet and Token Ring, while initially conceived and implemented as shared-media bus and ring topologies, respectively, have come to be physically implemented via a star topology, with a central hub and individual connections to each connected node. Taking advantage of this star topology, 100VG-AnyLAN uses intelligence in the hub to better manage network usage and improve network control. This central intelligence implements a powerful frame switching technique called demand priority. Demand priority hubs arbitrate requests from connected nodes for access to the network, building in a natural flow control that allows 100VG-AnyLAN to minimize network latency, maximize network throughput, and enable support for time-sensitive applications such as multimedia.

The demand priority system gives higher priority stations faster and more frequent access to the network. A node using demand priority, can transmit a packet by signaling its request to the hub. If the network is idle, the hub will immediately acknowledge the request and the node will begin transmitting its packet to the hub. As the packet arrives at the hub, the hub decodes the destination address contained in the packet and automatically switches the incoming packet to the outbound destination port. If more than one request is received at the same time, the hub uses a round-robin arbitration scheme to acknowledge each request in turn, until all requests have been serviced.

In a demand priority network, data packets are directed only to their respective intended destination ports. Since no other station on the network sees the data packet, its source, or its destination, this frame switching technique provides a level of link privacy or security that is not available with existing Ethernet, token ring, or FDDI networks. For network diagnostic purposes, network administrators can activate individual hub ports to monitor all traffic passing through the hub.

Since stations do not transmit their packets until they receive an acknowledgment from the hub, demand priority networks have a natural flow control that avoids packet collisions and allows prioritization of network traffic. By avoiding packet collisions, demand priority eliminates the network overhead consumed by packet collisions and recovery and substantially increases usable network throughput. In doing so, demand priority simplifies network operation and improves network characteristics such as latency or network access delay. Because the round-robin arbitration scheme is completely deterministic, the maximum latency seen by a packet of information is deterministic as well.

Compared to Token Ring–based networks, the round-robin arbitration scheme of demand priority hubs effectively collapses the token-passing process into the operation of the hub itself, eliminating delays due to token rotation and reducing latency for stations on an otherwise idle network. In addition, demand priority relaxes the limits on the number of stations in a single ring or subnet. Unlike traditional token ring environments, the latency experienced by individual stations on a demand priority network is unaffected by the number of idle stations connected to the network.

The ability to guarantee continuous, uninterrupted bandwidth is one of the critical network requirements for efficient support of time-sensitive applications such as multimedia. By prioritizing network traffic and taking advantage of the natural flow control inherent in demand priority networks, 100VG-AnyLAN is able to guarantee bandwidth to specific applications regardless of other traffic on the network.

To minimize delays experienced by time-sensitive applications, demand priority networks recognize and acknowledge high-priority transmission requests before normal-priority requests. In this way, 100VG-AnyLAN can effectively guarantee bandwidth to high-priority applications without regard to the level of normal-priority requests. To ensure that normal-priority data traffic is not blocked by high-priority requests, demand priority incorporates natural safeguards to guarantee transmission of all traffic, regardless of priority.

Given the multiple advantages of 100VG-AnyLAN, however, it is unclear whether the demand priority scheme scales effectively across large networks with multiple routers and hubs. This new technology design does not integrate easily into Ethernet networks, requiring compatible bridges, routers, or switches, which are not yet widely available.

Another problem is that demand priority technology is backed by a very small number of vendors and has yet to prove itself as a viable, high-bandwidth option. Apparently, Fast Ethernet and ATM provide greater compatibility with existing networks as well as better performance across large LANs. Fast Ethernet and ATM are also more widely supported by vendors and demanded by end users.

6.6.2 Fast Ethernet

With more than several million installed nodes, Ethernet is the most widely used networking technology today. Its advantages include a flexible protocol that is speed and media independent, enabling users to deploy the network in a variety of topologies using numerous wiring types. To preserve this advantage, the specification for fast Ethernet parallels the structure of 10 Mbps Ethernet, adding extensions to support 100 Mbps performance.

Fast Ethernet, or 100BaseT Ethernet, is the result of an industry-wide collaboration called the Fast Ethernet Alliance (FEA) aimed at extending the Ethernet specification for higher performance while preserving its core protocols and compatibility with applications, media, and wiring types. In 1995 the FEA succeeded in drafting, submitting, and winning approval for the new Fast Ethernet IEEE standard, known as IEEE 802.3u. In specifying the Fast Ethernet extensions, two very important goals were kept in mind: preserving network investments (including cabling, training, and user-written application code) and maintaining interoperability between Fast Ethernet products to allow customers flexibility in the network infrastructure. As a result, the new standard enables users to incrementally add Fast Ethernet to existing networks, minimizing migration effort and costs.

The FEA reached its goal of ensuring open, cost-effective, interoperable 100 Mbps Ethernet solutions. The June 1995 ratification of the 100BaseT IEEE 802.3u standard coupled with the over a hundred products announced by vendors fulfills the alliance's original charter. Having achieved these goals, the alliance disbanded with the closing of the Networld+Interop trade show in Atlanta in September 1995.

Fast Ethernet has the ability to automatically adjust to 10 Mbps or 100 Mbps operation and can be easily added to existing networks to improve performance at the desktop, server, and backbone levels. As a natural evolution of Ethernet, Fast Ethernet offers many benefits, including but not limited to the following:

- It operates with the existing network infrastructure, including cabling, protocol stacks, and user-written application code.
- It retains the knowledge base of network administrators and users.
- It offers 10 times the speed of 10 Mbps Ethernet at less than twice the price.
- It supports new and existing bandwidth-intensive applications.
- It ensures a virtually risk-free upgrade path to support higher performance workstations and servers.

Like Ethernet, Fast Ethernet is implemented as a star topology. With its higher bandwidth, Fast Ethernet supports many more users than Ethernet within the same physical distance. For example, a typical Ethernet configuration using a single 10 Mbps hub may adequately support a dozen users, while Fast Ethernet may provide enough bandwidth for a hundred or more users.

One of the only constraints for implementing Fast Ethernet is that the network diameter is limited to approximately 200 m. However, by using routers, bridges, or switches, this constraint can be overcome. In practice, LAN size is more likely to be determined by the hundred-meter limit of UTP links and the physical characteristics of building size and user locations.

Fast Ethernet consists of five component specifications: the media access control (MAC) layer, the media-independent interface (MII), and three physical layers supporting the most widely used cabling types. Each of these components was designed to preserve compatibility with 10 Mbps Ethernet and existing installations.

The media access control (MAC) layer is fundamental to the structure of both Ethernet and Fast Ethernet networks. As its name implies, this layer arbitrates transmissions between all nodes attached to the network, using the CSMA/CD algorithm. This arbitration algorithm ensures that one and only one station transmits data into the network at a time. Fast Ethernet simply increases the speed of data transmission by a factor of 10. Access method, packet format, packet length, error control, backoff algorithm, and management information remain unchanged. This consistency ensures that all network services and protocols written for 10 Mbps networks work unchanged over 100 Mbps Ethernet.

In the original 10 Mbps Ethernet specification, the attachment unit interface (AUI) enables use of a wide variety of cabling types. However, the reliance of this interface on certain mechanisms for data encoding, synchronization, and network management limits its use at higher network speeds. In Fast Ethernet a new interface is designed to replace the AUI, allowing more flexible support for various media types. The interface unit is called the media-independent interface (MII). It is a synchronous digital interface carrying unencoded data over separate transmit and receive paths. Internal and external media transceivers connect the MII to the three physical layers that support the network wiring. The MII also supports interconnection of two integrated circuits located on the same printed circuit assembly: interconnection of two circuit boards and plug-in physical modules.

The MII offers several features for high-bandwidth networking, including the following:

- Independent paths for transmitting and receiving data
- Support of a future IEEE specification for full-duplex operation
- Multivendor interoperability
- Compatibility between networking application-specific integrated circuits (ASICs) and physical modules supplied by different vendors
- Simple resource management
- Two-wire serial management interface for query and control of physical resources

The Fast Ethernet physical layers are similar to the 10BaseF (fiber) and 10BaseT (twisted pair) physical layers in 10 Mbps Ethernet. By maintaining this close compatibility with installed networks, Fast Ethernet provides an immediate performance improvement and allows users to retain much of their existing cabling investment.

Ethernet is deployed across several types of network cabling, including coaxial cables, fiber optics, and various categories of UTP and STP wiring. In the Fast Ethernet specification, also known as IEEE 802.34, media-dependent functions are grouped into three physical layers: 100BaseTX, 100BaseT4, and 100BaseFX. These three physical layers support an estimated 80% of currently installed 10 Mbps Ethernet cable structures and can be mixed in the same LAN for maximum flexibility in the network infrastructure.

The Fast Ethernet physical layers support the following media options:

- Half-duplex operation on four pairs of category 3, 4, or 5 UTP (100BaseT4)
- Half- or full-duplex operation on two pairs of category 5 UTP or STP (100BaseTX)
- Half- or full-duplex transmission over multimode fiber optic cable (100BaseFX)

6.6.3 Gigabit and 10-Gigabit Ethernet

The higher the speed, the better the expected performance. With the growing multimedia traffic on the Internet, however, speed is simply never adequate. Gigabit Ethernet (1000 Base-X) designed to deliver data at 1 Gbps (i.e., at one billion bits per second, or 1000 Mbps, is an extension of the highly successful 10 Mbps and 100 Mbps IEEE 802.3 standards. The most important feature of Gigabit Ethernet is that while offering a raw data bandwidth of 1000 Mbps, it maintains full compatibility with the huge installed base of Ethernet nodes. In 1998 IEEE approved the standard over fiber cables based on the recommendations of Working Group 802.3z. The most recent logical development in the Ethernet family is the 10-Gigabit Ethernet expected to operate at 10 Gbps, that is 1000 times faster than the original Ethernet.

The 802.3z taskforce identified three objectives for link distances: a multimode fiber optic link with a maximum length of 500 m; a single-mode fiber optic link with a maximum length of 3 km; and a copper-based link with a length of at least 100 m. In addition, the task force investigated technology that would support link distances of at least 100 m over category 5 UTP (1000BaseT). Because of the complexity of implementing gigabit technology over category 5 UTP, the task was undertaken by another taskforce, IEEE 802.3ab (1000BaseT or Gigabit Ethernet over copper), whose recommendations were approved in 1999. The new 1000BaseT specifications outline operation, testing, and usage requirements of Gigabit Ethernet for the installed base of category 5 copper wiring, which includes most of the cabling inside buildings.

The IEEE 802.3ae task force is responsible for defining the specification for 10 Gigabit Ethernet. The standard is scheduled to be ratified by early 2002. The key goals of the task force are to preserve the Ethernet frame format, including min/max frame size, define two families of physical interfaces LAN-PHY at 10.000 Gbps and an optional PHY for attachment to the wide area network (WAN-PHY) at a data rate compatible with OC-192c. In addition, it is expected to support full-duplex operation only with physical layer specifications that support link distances of at least 650 m over multimode fiber (MMF), 300 m over installed MMF and 2-40 km over single mode fiber (SMF). The main difference from the previous Ethernet specifications is the adoption of a optical transceivers.

In addition to the taskforce, a 10-Gigabit Ethernet Alliance (originally called Gigabit Ethernet Alliance) was organized around common objectives that support the activities of IEEE 802.3. This multivendor effort is committed to providing open, cost-effective, and interoperable Gigabit Ethernet solutions for both fiber and copper cabling environments. Representatives from approximately 120 networking, computer, component, and test equipment companies participate in the alliance.

Gigabit Ethernet supports the following topologies:

- 1000BaseSX (short wavelength fiber)—802.3z task force
- 1000BaseLX (long wavelength fiber)—802.3z task force
- 1000BaseCX (short run copper cable)—802.3z task force
- 1000BaseT (100 m four-pair category 5 UTP cable)—802.3ab task force

In general, Gigabit Ethernet has several appealing features. It supports a raw data bandwidth of 1000 Mbps over fiber and copper. Half- and full-duplex operating modes are included in the specification. Run lengths of 20 km, using fiber at full duplex, are possible. Single-mode fiber supports gigabit data rates up to 500 m and with multimode fibers the support is up to 2 km. Gigabit Ethernet is also compatible with the millions of existing Ethernet nodes, since it uses standard Ethernet frame and CSMA/CD access methods with the Ethernet slot time extended from a minimum of 512 bits to a minimum of 512 bytes (4096 bits) and support for one repeater per collision domain.

Gigabit Ethernet is backward compatible with 10BaseT and 100BaseT technologies. From the backbone point of view, Gigabit Ethernet competes with FDDI and ATM. Gigabit Ethernet uses 8B/10B encoding and short wavelength (780 nm) fiber optic component technology borrowed from the fiber channel and increased in speed by 20% to achieve 1 Gbps throughput.

The most acknowledged usage of the Gigabit Ethernet is its backbone use as a link between switches, although it could also be used as a link for servers. Gigabit Ethernet has advantage of being more compatible with the existing Ethernet structure. However Gigabit Ethernet probably will not replace ATM as some predict. ATM is unique in providing quality of services and in its ability to handle traffic with different requirements in terms of bandwidth and latency. Gigabit Ethernet's technologies do not currently support the same services and provide a permanent or strategic solution, although its huge bandwidth does alleviate the issues. However, the 10 Gigabit Ethernet is intended to link backbone switch routers and provide connectivity for MAN applications.

6.7 FDDI AND CDDI

Fiber distributed data interface (FDDI) is an ANSI X3T9.5 standard developed in the mid-1980s for 100 Mbps fiber optic networks. After completing the FDDI specification, ANSI submitted FDDI to the International Standards Organization. ISO has created an international version of FDDI that is completely compatible with the ANSI standard version. Today, although FDDI implementations are not as common as Ethernet or token ring, FDDI has gained the popularity that continues to increase as the cost of FDDI interfaces diminishes.

FDDI is frequently used as a backbone technology as well as a means of connecting high-speed computers in a local area. Its sophisticated feature set and built-in network management facilities make it an appealing solution for high-end workgroups. As one of the highly mature high-speed networking technologies, it is supported by a large number of vendors and products and offers a high level of interoperability with existing networks using compatible bridges, routers, and switches.

FDDI supports media access control, data encoding/decoding, and fault-tolerant transmissions. The FDDI specification also defines drivers for fiber optic components, allowing development of a wide array of compatible network devices. Because of its high cost per node, FDDI is most often used as a backbone interconnect between lower speed subnets, or for small workgroups using data-intensive applications. FDDI's high cost and limited upper range of speed leave it open to challenge by Fast Ethernet for high-bandwidth desktop networks and by ATM and Gigabit Ethernet as a high-speed backbone solution.

The fiber optic transmission used in FDDI allows vast amounts of information transfers in a flash. The optical fiber cables enable ultrahigh-speed transmission and reception of massive volumes of data and hi-fi voice and video information.

FDDI defines use of two types of fiber: single mode and multimode (explained in Chapter 2). The characteristics of single-mode fiber render it most appropriate for interbuilding connectivity (localized setup), while multimode fiber is often used for intrabuilding connectivity (backbone applications). Multimode fiber uses light-emitting diodes (LEDs) as the light-generating devices, while single-mode fiber generally uses lasers.

The FDDI standard describes the following four layers.

- **Media access control (MAC):** Defines how the medium is accessed, including frame format, token handling, addressing, algorithm for calculating a cyclic redundancy check value, and error recovery mechanisms. This layer also communicates with higher layer protocols, such as TCP/IP, SNA, and IPX. This layer accepts protocol data units (PDUs) from the upper layer protocols, adds the MAC header, and then passes packets of up to 4500 bytes to the physical layer protocol.

- **Physical layer protocol (PHY):** Defines data encoding/decoding procedures, clocking requirements, framing, and other functions. This layer handles the encoding and decoding of packet data for the cable. It also handles clock synchronization on the FDDI ring.

- **Physical layer medium (PMD):** Defines the characteristics of the transmission medium, including the fiber optic link, power levels, bit error rates, optical components, and connectors. This layer handles analog baseband transmission between nodes on the physical media. PMD standards include TP-PMD for twisted-pair copper wires and fiber-PMD for fiber optic cable. TP-PMD is a new ANSI standard that replaces the proprietary approaches formerly used for running FDDI traffic over copper wires.

- **Station management (SMT):** Defines the FDDI station configuration, ring configuration, and ring control features, including station insertion and removal, initialization, fault isolation and recovery, scheduling, and collection of statistics. SMT is an overlay function that handles the management of the FDDI ring. Functions handled by SMT include neighbor identification, fault detection and reconfiguration, insertion and deinsertion from the ring, and traffic statistics monitoring.

FDDI is a data link layer protocol. The higher layer protocols operate independently of the FDDI protocol. Applications pass packet-level data using higher layer protocols down to the logical link control layer, just as they would do over Ethernet or token ring. But because FDDI uses a different physical layer protocol from Ethernet and token ring, traffic must be bridged or routed on and off a FDDI ring. FDDI also allows for larger packet sizes than Ethernet and token ring. For this reason, connections between FDDI and Ethernet or token ring LANs require the fragmentation and reassembly of frames.

FDDI specifies the use of dual rings. Traffic travels in opposite directions on the rings. Physically, the rings consist of two or more point-to-point connections between adjacent stations. One of the two FDDI rings is called the primary ring; the other is called the secondary ring. The primary ring is used for data transmission, while the secondary ring is generally used as a backup. This implementation, is also called dual-attached ring.

In the dual-ring implementation, stations are connected directly to one another. FDDI's dual counterrotating ring design provides fault tolerance. If the wiring between two stations fails, the ring wraps around the fault (Figure 6.15). Similarly, if a node fails, the ring wraps around the failed node (Figure 6.16). However, one limitation of the dual

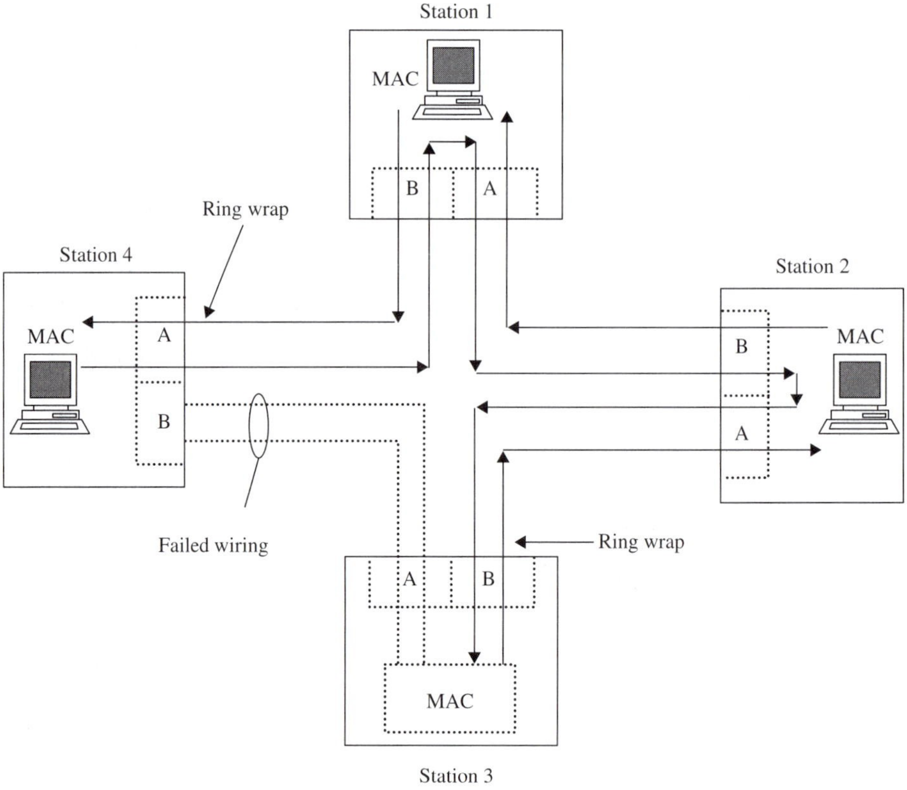

Figure 6.15 Ring wraps around the fault when wiring fails.

counterrotating ring design is that if two nodes fail, the ring is broken in two places, effectively creating two separate rings. Nodes on one ring are then isolated from nodes on the other ring. External optical bypass devices can solve this problem of ring segmentation, but their use is limited because of FDDI optical power requirements (Figure 6.17).

The FDDI implementation called concentrator-based ring fixes the dual-ring problem because concentrators are used to build networks. Concentrators are devices with multiple ports into which FDDI nodes connect. FDDI concentrators function like Ethernet hubs or token ring multiple access units (MAUs). Nodes are single-attached to the concentrator, which isolates failures occurring at those end stations. With a concentrator, nodes can be powered on and off without disrupting ring integrity. Concentrators make FDDI networks more reliable and also provide network management functions using SNMP. For this reason, most FDDI networks are now built with concentrators.

The copper distributed data interface (CDDI) standard, using STP wiring, allows a FDDI network running at 100 Mbps to be delivered to desktop system over segments of STP cable. The maximum length of the copper segment is 100 m, sufficient for devices to be connected to fiber backbone networks via existing copper cabling.

FDDI implementation over copper wire is possible after solving problems of electromagnetic radiation and interference on the copper wire have been solved. When sig-

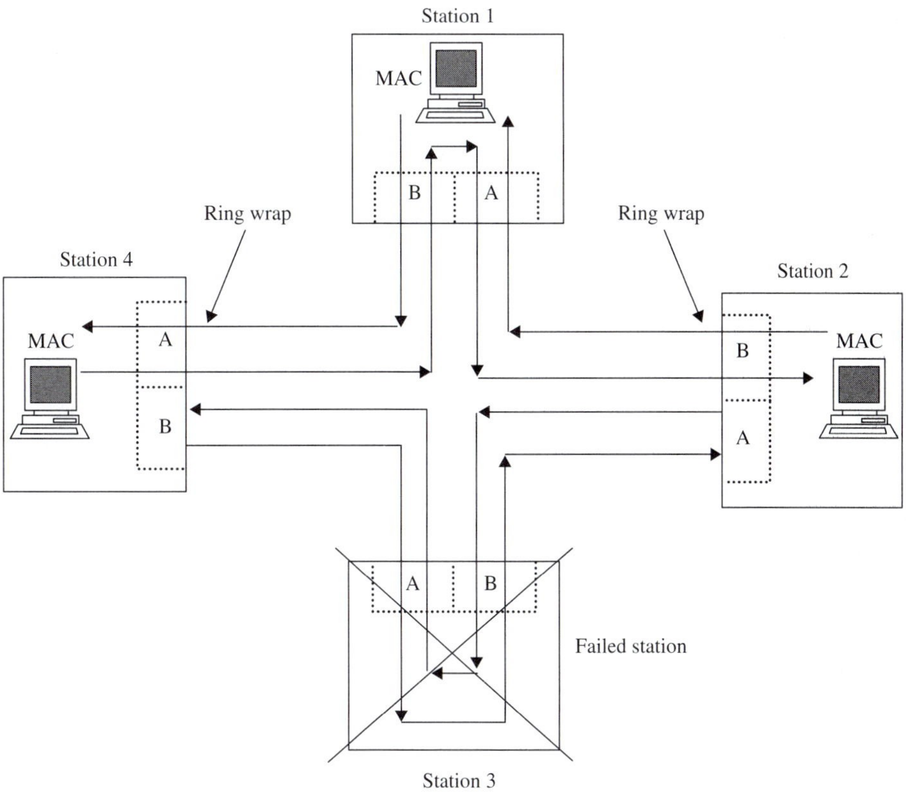

Figure 6.16 Ring wraps around the fault when a station fails.

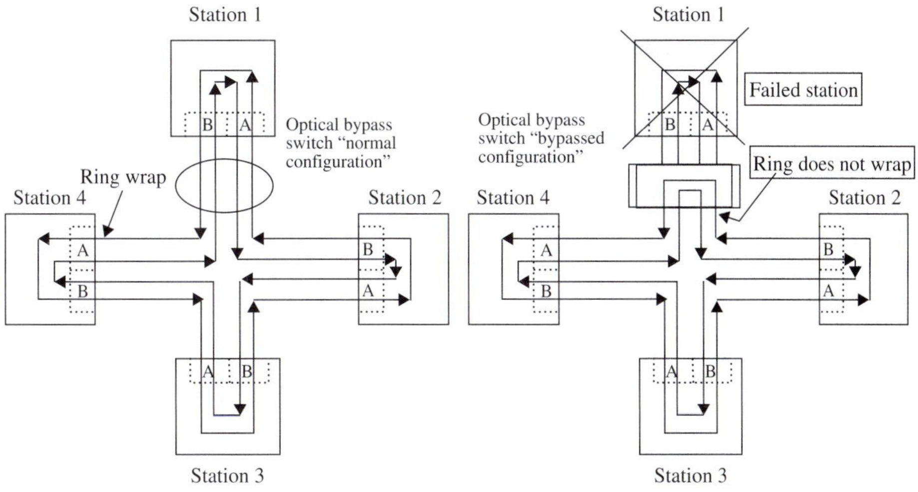

Figure 6.17 Use of an optical bypass switch to avoid multiple rings when more than one node fails.

nals strong enough to be reliably interpreted as data are transmitted over twisted-pair wire, the wire radiates electromagnetic interference (EMI). A FDDI implementation over twisted-pair wire must ensure that the resulting energy radiation does not exceed the FCC specifications.

In FDDI, the following three technologies reduce the radiation of energy:

- **Scrambling:** When no data are being sent, FDDI transmits an idle pattern that consists of a string of binary ones. When this signal is sent over twisted-pair wire, the interference is concentrated at the fundamental frequency spectrum of the idle pattern, resulting in a peak in the frequency spectrum of the radiated interference. By scrambling FDDI data with a pseudo-random sequence prior to transmission, repetitive patterns are eliminated. The elimination of repetitive patterns results in a spectral peak that is distributed more evenly over the spectrum of the transmitted signal.

- **Encoding:** Signal strength is stronger and interference is lower when transmission occurs over twisted-pair wire at lower frequencies. An encoding scheme called MLT3 reduces the frequency of the transmitted signal. MLT3 switches between three output voltage levels so that peak power is shifted to less than 20 MHz.

- **Equalization:** Equalization boosts the higher frequency signals for transmission over UTP. Equalization can be done on the transmitter (predistortion), at the receiver (postcompensation), or both. One advantage of equalization at the receiver is that compensation can be adjusted as a function of cable length.

In June 1990 ANSI established the Twisted Pair–Physical Medium Dependent (TP-PMD) Working Group to develop a specification for implementing FDDI protocols over twisted-pair wire. ANSI approved the TP-PMD standard in February 1994.

6.8 ASYNCHRONOUS TRANSFER MODE (ATM) NETWORKS

Time division multiplexing (TDM), used in today's phone system, matches well with circuit switching applications, in which the information transfer is periodic. The clocks at network nodes must be synchronized in this mode of transfer. Because data traffic is bursty, it is better to transmit the data in asynchronous mode, which lets applications use bandwidth in a way that is as bursty or as constant as necessary. That data transfer mode is called ATM (asynchronous transfer mode). A comparison of TDM and ATM is shown in Figure 6.18.

Two types of standards body are actively involved in ATM: formal standards bodies and industry forums. Industry forums are independent groups of industry experts, vendors, and users who are concerned with interoperability issues. The four industry forums actively specifying ATM and B-ISDN are the ATM Forum, the Frame Relay Forum, the Internet Engineering Task Force (IETF), and the SMDS Interest Group (SIG). The forums provide valuable contributions to the formal standards organizations because their

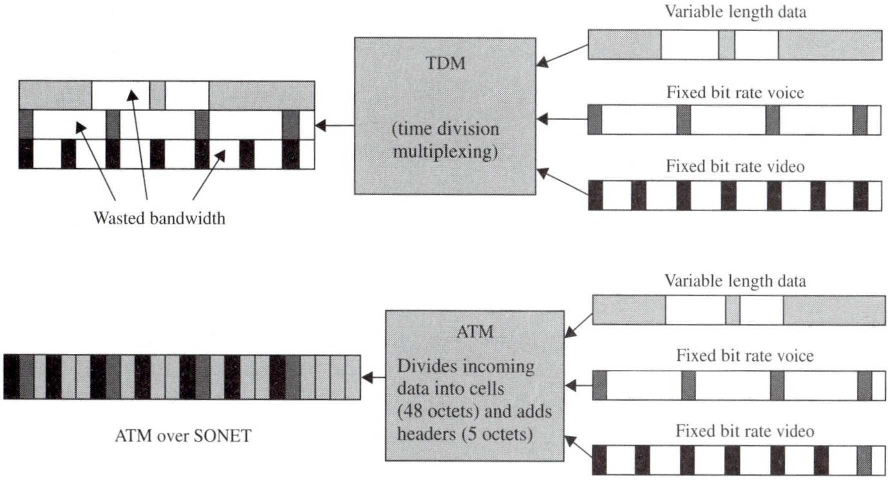

Figure 6.18 Comparison of TDM and ATM. Note that the voice and video signals remain isochronous, but utilization of bandwidth is better in ATM.

knowledge comes from real-world implementations that have been built and tested, rather than just from theoretical knowledge.

The ATM Forum was started in October 1991 by a consortium of four computer and telecommunication vendors. Since its inception, it has seen unprecedented growth, and now it has more than 700 members. The membership is made up of network equipment providers, semiconductor manufacturers, service providers, carriers and, most recently, end users. Strictly speaking, the forum is not a standards body. The ATM Forum is a consortium of companies that writes specifications to accelerate the definition of ATM technology. These specifications are then passed up to ITU-T for approval. The ITU-T standard body fully recognizes the ATM Forum as a credible working group. However, since the ITU-T approval process is relatively slow, the ATM Forum is taking a leading role in developing the standard in parallel.

The tremendous success of the ATM Forum is believed to constitute a strong endorsement of ATM as a technology for the future. Bringing together the best points of TDM and packet switching, ATM provides a versatile, multifunctional platform that can support a variety of services and traffic types. In addition, ATM offers scalability that is difficult to match with other, traditional solutions. These capabilities make ATM popular across a wide spectrum of vendors, carriers, and end users.

6.8.1 ATM Advantages

International standards organizations have recognized that ATM is the best technology to support bursty, broadband applications by transmitting information in short, fixed-length cells. Each cell consists of a header and a payload section. The header section is used for the addressing and control overhead needed to deliver the cell across the network. The payload section is used to carry the information for a particular user. Because the payload contents have no part in switching decisions, ATM is completely protocol independent.

ATM is a connection-oriented transport mechanism. All switches on the path are consulted before a new connection is set up. The switches verify that they have sufficient resources before accepting a connection. The connections in ATM networks are called virtual circuits (VCs). ATM networks use VC identifiers (VCIs) to identify cells of various connections. VCIs are assigned locally by the switch. A connection has a different VCI at each switch. The same VCI can be used for different connections at different switches. Thus, the size of a VCI field does not limit the number of VCs that can be set up. This makes ATM networks scalable in terms of number of connections.

The use of a fixed-length cell in ATM allows switching decisions to be implemented in faster hardware, rather than making routing decisions in relatively slow software. It is also based upon the premise of a short cell length to prevent excessive serialization delay and to prevent one user from blocking others. By combining hardware-based switching with the short, fixed-length cell, ATM offers the potential for extremely low delay and high throughput.

The international standards bodies have chosen a cell length of 53 bytes for broadband services. This represents a compromise between those who wanted longer payloads to reduce addressing overhead and processing load, and those who wanted shorter payloads to improve the quality of low-speed services, avoid congestion, and minimize delay.

In addition to cost and performance benefits, ATM has been recognized as the unifying technology that will be able to support all types of traffic. The use of short, fixed-length cells means that traffic types can be mixed, allowing for a common backbone that can carry voice, circuit data, and frame-based traffic. However, in mixing these traffic types, the need for strict classes of service arises. For example, voice has very different delay and throughput requirements from LAN file transfers or online transaction processing.

6.8.2 ATM Protocol and ATM Layer

ATM is explained by using the layered protocol architecture as shown in Figure 6.19. The physical layer defines the interface with the transmission media, much like the OSI reference model. It is concerned with the physical interface, transmission rates, and how the

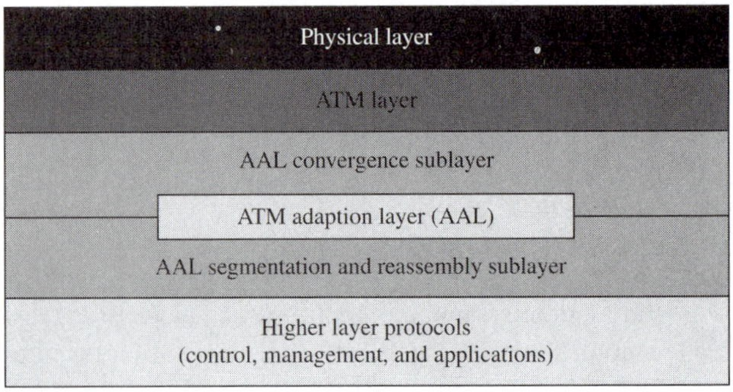

Figure 6.19 The ATM four-layer model. Different types of traffic can be mixed on the same switched network. The AAL convergence sublayer makes sure that every type of traffic—data, voice, and video—receives the right level of service.

ATM cells are converted to the line signal. However, unlike many LAN technologies, such as Ethernet, which specify a certain transmission medium, ATM is independent of the physical transport. ATM cells can be carried on SONET, synchronous digital hierarchy (SDH), T3/E3, T1/E1, or even 9600 bps modems.

The next higher layer (ATM layer) deals with ATM cells. The format of the ATM cell is quite simple. It consists of a 5 byte header and a 48-byte payload. The header contains the ATM cell address and other important information. The payload contains the user data being transported over the network. Cells are transmitted serially and propagate in strict numeric sequence throughout the network. The payload length was chosen as a compromise between a long cell length (64 bytes), which is more efficient for transmitting long frames of data, and a short cell length (32 bytes), which minimizes the end-to-end processing delay and is good for voice, video, and delay-sensitive protocols. Although not specifically designed as such, the cell payload length conveniently accommodates two 24-byte IPX FastPackets.

Two types of ATM cell header are defined: the user-to-network interface (UNI) and the network-to-network interface (NNI). The UNI is a native-mode ATM service interface to the WAN. Specifically, the ATM UNI defines an interface between cell-based customer premises equipment (CPE), such as ATM hubs and routers, and the ATM WAN. The NNI defines an interface between the nodes in the network (the switches), or between networks. The NNI can be used as an interface between a user's private ATM network and a service provider's public ATM network.

The primary function of both the UNI and the NNI is to identify virtual path identifiers (VPIs) and virtual circuit identifiers (VCIs) as routing and switching identifiers for ATM cells. The VPI identifies the path or route to be taken by the ATM cell, while the VCI identifies the circuit or connection number on that path. The VPI and VCI are translated at each ATM switch and are unique for only a single physical link.

The 5-byte ATM header uses the UNI and NNI cell formats. Most of the fields in these two formats are similar except that the UNI cell format includes a 4-bit generic flow control (GFC) field to provide flow control at the UNI level. The NNI level flow control is achieved by using longer VPIs (12 bits vs 8 bits for UNI), thus allowing more virtual paths. Among other fields in the header are VCI (16 bits), payload type (PT, 3 bits), cell loss priority (CLP—1 bit), and header error correction (HEC, 8 bits). The defined PT for both types of cell is user data (with congestion and signaling bits) and management data (encoding different types of management control). The one-bit CLP indicates the priority level of a cell, 1 indicating a higher priority. In case of congestion, cells with 0 CLP may be dropped to meet certain quality of service (QoS) requirements. Finally, the HEP is the CRC field computed over the first 4 bytes of the header.

The purpose of the next layer, the ATM adaptation layer (AAL), is to accommodate data from various sources with differing characteristics. Specifically, its job is to adapt the services provided by the ATM layer to those services that are required by the higher user layers (circuit emulation, video, audio, frame relay, etc.). The AAL receives the data from the various sources or applications and converts it to 48-byte segments that will fit into the payload of an ATM cell. The adaptation layer defines the basic principles of sublayering. It describes the service attributes of each layer in terms of constant or variable bit rate, timing transfer requirement, and whether the service is connection-oriented or connectionless.

AAL consists of two sublayers, the convergence sublayer (CS) and the segmentation and reassembly sublayer (SAR). The convergence sublayer receives the data from various applications and makes variable-length data packets called convergence sublayer protocol data units (CS-PDUs). The segmentation and reassembly sublayer receives the CS-PDUs and segments them into one or more 48-byte packets that map directly into the 48-byte payload of the ATM cell transmitted at the physical layer. The types of AAL service are defined as follows.

6.8.2.1 AAL1

AAL1 is for a constant bit rate, connection-oriented service that requires timing transfer (synchronous), such as voice and video. The CS in AAL1 divides the incoming data from higher layer to 47-byte segments (CS-PDUs) that are passed to the SAR sublayer. The SAR sublayer attaches one byte of header to each CS-PDU and passes them as a 48-byte payload to the ATM layer. The one-byte header consists of a convergence sublayer identifier (CSI, 1 bit), sequence count (SC, 3 bits), CRC (3 bits), and parity (P, 1 bit). The SC field is a modulo-8 sequence number used for cell ordering and for end-to-end error and flow control. The 3-bit CRC is used as a checksum of the preceding 4 bits in the header. Finally the P bit is simply the parity bit for the preceding 7 bits in he header.

6.8.2.2 AAL2

AAL2 is for a variable bit rate, connection-oriented service that requires timing transfer (synchronous), such as compressed voice and video. In this case, the CS divides the incoming data into 45-byte segments, allowing a 1-byte header and a 2-byte trailer. The header consists of 1 bit of CSI for signaling and 3 bits of SC as AAL1. In addition, a 4-bit information type (IT) field identifies the data segment in the cell as the beginning, middle, or end of the message. The 2-byte trailer consists of a 6-bit length indicator (LI) and a 10-bit CRC over the entire data unit. The LI field is designed to be used with the last 45-byte segment. It indicates the start of padding bytes in the final segment.

6.8.2.3 AAL3/4

AAL3/4 is for a variable bit rate, connectionless service that does not require timing transfer (asynchronous), such as SMDS and LANs. The overhead is added both at CS and SAR in this case. The CS accepts a maximum of 64 KB (65,535 bytes) of data from the upper layer; 4 bytes of header and 4 bytes of trailer are included in the data before it is broken into byte segments that is passed to SAR. A padding of up to 43 bytes may be included before the trailer to ensure 44 byte data segments. The header for CS to SAR segments consists of type (T, 1 byte) set to all zeros, begin tag (BT, 1 byte) used as a begin flag, and buffer allocation (BA, 2 bytes), to tell the receiver what size buffer is needed for the incoming data.

The number of padding bytes would be between 0 and 43, depending on the number of data bytes in the final (or final two) segment(s). When there are exactly 40 data bytes in the final segment, no padding is required because the 4-byte trailer completes the segment. When the number of data bytes in the final segment is between 0 and 39, between

44 and 5 padding bytes may be added, respectively, to bring the total to 44 bytes. However, when the number of data bytes is between 41 and 43 bytes, the number of padding bytes must be between 43 and 41, respectively, to bring the total to 84 bytes. These 84 bytes are divided into 44 bytes in the first segment and 40 bytes with the trailer in the next segment to complete last two segments. The trailer field consists of a 1-byte alignment (AL) field, a 1-byte end tag (ET) as ending flag, and a 2-byte length (L) field indicating the length of the data unit.

The SAR header and trailer in ALL3/4 are each 2 bytes long. The header consists of a segment type field (ST, 2 bits) to indicate the segment by itself or begin, middle, or end segment in the message; 1-bit for CSI, 3 bits for SC, and 10 bits for multiplexing identification (MID), a field that identifies cells coming from different sources but multiplexed with the same virtual connection. The trailer consists of a 6-bit LI and a 10-bit CRC over the entire data unit.

6.8.2.4 AAL5

AAL5 is for a variable bit rate, connection-oriented service that does not require timing transfer (asynchronous), such as X.25 and frame relay. AAL5 simplifies the complex overhead mechanism of AAL3/4 by removing the sequencing and error control mechanism that is not required for many applications. The maximum data in this case remains 64 KB like AAL3/4, but the only overhead added in this case is 8 bytes of trailer at the CS layer, which is divided into 48-byte segments at SAR. A padding of up to 47 bytes is used at CS to ensure 48-byte boundaries. The trailer field consists of 1-byte user-to-user identifier left to the discretion of the user, a reserved 1-byte type (T) field, a 2-byte length (L) field, indicating the start of padding, and a 4-byte CRC over the entire data unit.

6.8.3 ATM Switching

ATM switches provide the switching and multiplexing of cells in the system. Switches are designed to provide virtual path and virtual channel switching by means of VPIs and VCIs.

Switches break the incoming data into fixed-length cells with a header and a payload field. Further, switches address the cell by prefixing it with a logical destination called the virtual path (VP) address. They then assign different virtual channels (VCs) within the VP, depending on the type of data.

Switches also multiplex the cells from various sources together onto the outgoing transmission link and map the incoming VPI/VCI to the associated outgoing VPI/VCI address. The demultiplexing task of switches consists of unpacking the cells from various sources, translating them back into their native format, and delivering them to the appropriate device or port.

6.8.4 ATM Internetworking with Frame Relay

Frame relay connection ATM networks are initially deployed not to provide "native-mode" ATM service interfaces, but to support a range of other service interfaces through various adaptation technologies. Foremost among these interfaces will be frame relay. Frame relay is not simply a stepping-stone on the way to ATM; it is likely to provide a justification for deploying ATM switching and to supply the connection to sites that cannot justify a broadband ATM connection.

When cell switching or ATM is used at either narrowband or broadband rates to implement the frame relay service, a frame relay pad (FRP) supports one or more attached networking devices (typically routers or front-end processors) at the edge of the network. As a frame arrives from an attached networking device, it is simultaneously converted to cells. These cells are transmitted across the entire network without being reassembled or passing through any intermediate server modules. At the far end, the frame is reconstructed in the frame relay pad supporting the destination user device.

Frame relay is an excellent access mechanism to ATM networks. As mentioned earlier, the actual transport mechanism within the network for frame relay is separated from the interface specifications for frame relay. ATM is an excellent network infrastructure for transporting frame relay data. A joint specification for the transporting of frame relay over ATM networks has been prepared jointly by the Frame Relay Forum and the ATM Forum.

Some users and some applications, such as those involving broadcast-quality multimedia at speed of 150 Mbps and above, will require native-mode ATM interfaces. Many other applications, though, will be served very well by frame relay interfaces for a long time. After all, frame relay's capability to provide at least 50 Mbps of statistically multiplexed data throughput with global switched connectivity will far exceed many users' needs well into the foreseeable future. And the capabilities needed by most users today are already available with excellent, cost-effective offerings from both equipment and service providers.

6.8.5 IP over ATM

Operating IP on top of ATM involves two main problems. The first problem is that ATM is connection-oriented; that is, a connection must be established before two parties can send data to each other. Once the connection has been set up, all data between the parties is sent along the connection path. On the other hand, IP is connectionless, and so routers forward each packet independently on a hop-by-hop basis.

The second problem is that QoS is major requirement in ATM networks. IPv4 operates on a best-effort basis, in which routers forward each packet without any throughput/delay guarantee. However, IPv6 improves the addressing capabilities of IPv4, and IP integrated services support real-time IP traffic. There are many issues concerning how ATM networks support both technologies.

Requests for comments (RFCs) by the Internet Engineering Task Force (IETF) and the ATM Forum discuss most of the issues. The ATM Forum specifications can be considered to be the de facto standards for private network ATM deployment. The IETF has focused primarily on aspects of IP interworking over ATM. The work of the IETF has been very influential and provides models for the work of the ATM Forum. Some of the key specifications from these two bodies are as follows:

- Overview and Internet Architecture
  RFC 1932, IP over ATM: A Framework Document
  RFC 1620, Internet Architecture Extensions for Shared Media
  RFC 1633, Integrated Services in the Internet Architecture: An Overview

- Models (according to Section 8 of RFC 1932)

  The classical IP model: RFC 1577, Classical IP and ARP over ATM

  The IP multicast model: RFC 2022, IP Multicast over ATM

  The routing over large clouds (ROLC), next-hop-resolution protocol (NHRP) model: RFC 1735, Nonbroadcast multiple access (NBMA) address resolution protocol (NARP), and the NBMA Next-Hop-Resolution Protocol

  MPOA (Multiprotocol over ATM) of ATM Forum: RFC 1483, Multiprotocol Encapsulation

  LAN emulation (LANE) of ATM Forum

The IP over ATM framework can be understood with the help of several models designed to provide some of the ATM features on IP. The peer model relates ATM protocol layers to TCP/IP layers. It considers the ATM layer to be a peer networking layer as IP and proposes the use of the same addressing scheme as IP for ATM-attached end systems. In this model, ATM signaling requests contain IP addresses, and the intermediate switches route the requests using existing Internet routing protocols like OSPF. This model is considered to be meaningless and eventually was rejected because it complicates the design of ATM switches by requiring them to have all the functions of routers in IP and other propriety network layer protocols like IPX and AppleTalk.

The overlay model views ATM as a data link layer protocol with IP running on top. In this model, which is widely accepted by the committees, an ATM network will have its own addressing scheme and routing protocols. The ATM address space is not logically coupled with the IP addressing space, and there will be no arithmetic mapping between the two. Each end system will typically have an ATM address and an unrelated IP address that are tied together via an addressing resolution protocol.

With the overlay model, there are two ways to run IP over ATM. One treats ATM as a LAN and partitions it into several logical subnets consisting of end systems with the same IP prefix. This is known as classical IP over ATM. However, traffic between end systems in different logical subnets needs to go through a router, which may introduce extra delay and may become a bottleneck. The Next-Hop-Resolution Protocol (NHRP) is designed to solve this problem. It allows an end system in one logical subnet (within the ATM network), resolves the ATM address (from the IP address) to another logical subnet, and establishes an end-to-end ATM connection.

The LAN emulation (LANE) approach used in ATM networking consists of simulating (actually "emulating") popular LAN protocols like Ethernet or token ring. IP runs on top of it just as it runs on top of Ethernet or token ring. LANE allows current IP applications to run over an ATM network without modification. However, as in classical IP over ATM, traffic between different emulated LANs still needs to travel through a router. A combination of LANE and NHRP, called Multiprotocol over ATM (MPOA), solves the problem by creating shortcuts that bypass routers between emulated LANs.

6.8.6 ATM Future

ATM is often termed high-priced technology. However, some studies reveal that when priced in terms of megabits per second rather than in terms of cost per desktop, ATM is technology priced lower than Fast Ethernet, switched Ethernet, and IP switching. ATM also boasts the biggest bandwidth (at about 65 Mbps per desktop) and delivers the lowest desktop-to-server round-trip delay (at about 60 μs vs 100–900 μs with competing technologies).

ATM is considered to be a service architecture that will satisfy the demands of many types of user in the future. However, it remains to be seen that whether users will buy the expensive ATM services or premises equipment associated with the technology. IP, which is way ahead in terms of the user network interface for nonvoice services, may be the best UNI but is not the best infrastructure. ATM provides an infrastructure that is likely to support future applications. However, some researchers suggest that ATM is too complex for our needs. Also, ATM design seems to be more supportive of voice traffic, which is not a big percentage of actual traffic on the network.

Demand for public ATM services is presently almost nonexistent, as evidenced by the lack of deployed ATM-capable equipment. If public ATM services were economical, companies would invest in new ATM access equipment because of its clear technical advantages. However, this service would require carriers to deploy ATM switches at both the central office exchange and interexchange levels, and ATM is a more revolutionary technology in terms of switching changes than was ISDN. Such a major investment by the carriers is unlikely without preexisting demand.

6.9 SONET

SONET (**s**ynchronous **o**ptical **net**work) is a set of standards defining the rates and formats for optical networks specified by ANSI T1.105, ANSI T1.106, and ANSI T1.117 in the United States. A similar standard, the synchronous digital hierarchy (SDH), has been established in Europe by ITU-T. SONET and SDH are technically consistent standards. The SONET standard defines the rates and formats, the physical layer, the network element architectural features, and the network operational criteria. SONET also contains recommendations for the standardization of fiber optic transmission system equipment sold by different manufacturers. It is a synchronous network; that is, a single clock is used to handle the timing and control of network equipment. SONET is a multiplexed transport mechanism that does not involve any switching and as such may be utilized for broadband services such as ATM, FDDI, and B-ISDN.

SONET is becoming popular as the physical medium for running ATM. Many telephone carriers and banks are adopting SONET as a base for their networks. Although SONET seems to be the physical medium of the future, some work still needs to be done to improve its usage. One problem with SONET is its lack of interoperability. There are virtually no SONET circuits that originate on equipment from one vendor and terminate on equipment from another. This may cause problems for network management in the future. Another issue confronting SONET network management is found in the differences between types of carrier-based network management. SONET operations, administration, maintenance, and provisioning is based on OSI's Common Management Information Protocol (CMIP) and Common Management Information System Element (CMISE), while the vast majority of private data networks in the United States rely on TL1 (transaction logic 1).

TABLE 6.1
Digital Stream Data Rates

Stream	Bit Rate	Structure
DS0	64 Kbps	1 voice channel
DS1	1.544 Mbps	24 DS0
DS3	44.736 Mbps	28 DS1

6.9.1 SONET Signals and Architecture

The existing North American digital hierarchy was created to carry digitized voice over twisted wire. Each level in the hierarchy is called a digital stream (DS). The lower level digital streams are multiplexed into the higher digital streams as shown in Table 6.1. The lowest level in the hierarchy, DS0, carries a single voice channel.

The European and Japanese digital hierarchies are closely related to this structure. SONET is a re-creation of the digital transmission hierarchy with a whole family of optical carrier (OC) levels running at speeds ranging from 51.84 Mbps in the United States and 155.52 Mbps in Europe to about 9.9 Gbps. The SONET optical levels are shown in Table 6.2.

When DS0 signals are multiplexed into a DS1 stream, extra bits are added to account for variations in the individual streams. This process is called bit-stuffing. When the DS1 stream is multiplexed into a DS3 stream, bit-stuffing is used again. At this level, it is not possible to recover the DS0 without first demultiplexing the DS1 signal.

Asynchronous multiplexing adds much overhead and requires a large number of multiplexers and digital cross connects. SONET performs synchronous multiplexing at all levels. To effectively do this, it uses a concept called pointers for frame synchronization. Low-level SONET streams can be extracted synchronously from a high-level SONET stream. It is interesting to note that the lowest data rate of 51.84 Mbps in OC-1 is greater than the DS3 service and T3 line (44.736 Mbps). However, OC-1 is designed to accom-

TABLE 6.2
SONET Optical Levels

OC level	Bit Rate (Mbps)	Number of Digital Streams		
		DS0	DS1	DS-3
1	51.84	672	28	1
3	155.52	2016	84	3
6	311.04	4032	168	6
9	466.56	6048	252	9
12	622.08	8064	336	12
18	933.12	12096	504	18
24	1244.16	16128	672	24
36	1866.23	24192	1008	36
48	2488.32	32256	1344	48
96	4976.64	64152	2688	96
192	9953.28	129024	5376	192

modate DS3. The difference in capacity is due to the overhead needed for the optical systems. SONET defines a hierarchy of signaling levels called synchronous transport signals (STSs) corresponding to the optical carriers shown in Table 6.2. Thus for example, STS-192 corresponds to OC-192, and so on.

Many existing networks have not deployed the standards proposed for SONET. Communication between various localized networks is costly because of differences in digital signal hierarchies, encoding techniques, and multiplexing strategies. For example, the DS1 signals consist of 24 voice signals and one framing bit per frame. It has a rate of 1.544 Mbps. DS1 uses a bit from an 8-bit byte for signaling. Therefore, it has a rate of 56 Kbps per channel. When the B8ZS bipolar violation encoding scheme is used, every bit can be used for transmission. Therefore, it has a rate of 64 Kbps per channel. The CEPT-1 signal consists of 30 voice signals and two channels for framing and signaling. It has a rate of 2.048 Mbps. CEPT-1 uses the HDB3 coding technique. Multiplexing procedures may also be different between signals—byte interleaving or bit interleaving.

Transporting a signal to a different network requires a complicated multiplexing/demultiplexing, coding/decoding process to convert a signal from one scheme to another. SONET transmission is based on three basic devices: path-terminating equipment, regenerators, and line-terminating equipment (Figure 6.20).

The path-terminating equipment converts electronic signals to optical signals, while the task of line-terminating equipment is to multiplex and demultiplex the signals from different sources into their proper paths. The line-terminating equipment is also called an add/drop multiplexer, and path-terminating equipment combined with an add/drop multiplexer may also be referred to as an STS multiplexer. The regenerator in the SONET system is actually a repeater that receives an optical signal and regenerates it.

To solve the problem of signal transport between different networks, SONET standardizes the rates and formats. The STS is the basic building block of SONET optical interfaces with a rate of 51.84 Mbps, as explained earlier. STS consists of two parts, the STS payload and the STS overhead. The STS payload carries the information portion of

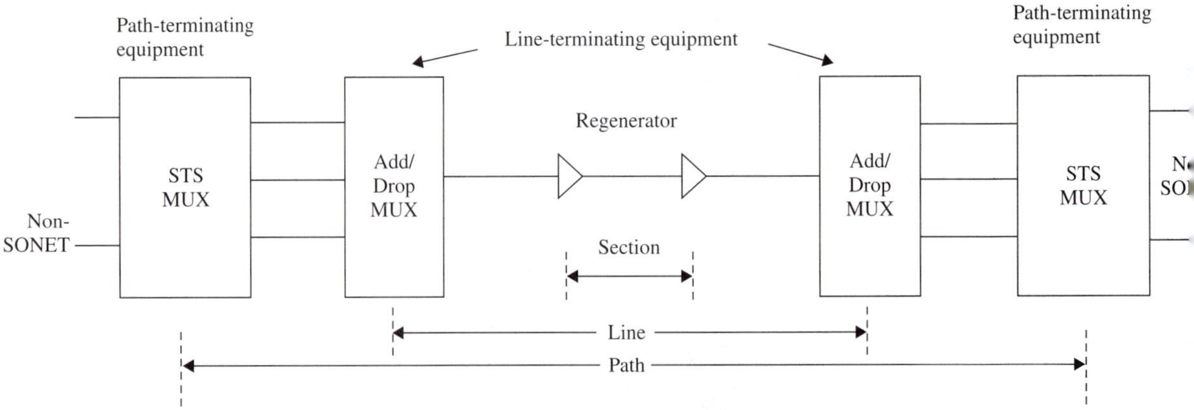

Figure 6.20 A SONET system architecture.

TABLE 6.3
Comparison of the transmission hierarchies of SONET and SDH

Bit Rate (Mbps)	SONET	SDH
51.84	STS-1	
155.52	STS-3	STM-1
622.08	STS-12	STM-4
1244.16	STS-24	
2488.32	STS-48	STM-16
9953.28	STS-192	STM-64

the signal. The STS overhead carries the signaling and protocol information. This allows communication between intelligent nodes on the network, permitting administration, surveillance, provisioning, and control of a network from a central location.

SONET also provides the necessary bandwidth to transport information from one B-ISDN switch or terminal to another. B-ISDN can be used to handle voice, data, and video. An H4 is a digital broadband signal of 150 Mbps used to carry a high-definition TV signal. To provide broadcast quality for such a signal, the signal may be transported by concatenating three STS-1 signals or by using an STS-3 (51.84×3 Mbps) signal. Many of these broadband services rely on ATM that uses short, fixed-length packets to implement a fast packet switching technique. Because of its bandwidth capacity, SONET is a logical carrier for ATM. ATM operates at 155 Mbps; thus an STS-3 may be used to carry the signal.

As mentioned earlier, SONET is compatible with the European transmission hierarchy. SONET is considered to be a subset of SDH, which is a world standard for synchronous transmission. SONET and SDH converge at SDH's base level of 155 Mbps, which is equivalent to three STS-1 signals or a single STS-3 signal. Table 6.3 compares the transmission hierarchies of SONET and SDH. The SDH specification consists of synchronous transport module (STM) that is similar to STS specifications.

6.9.2 SONET Layers and Frames

There are four optical interface layers in SONET: the path, line, section, and photonic layers. The optical interface layers have a hierarchical relationship, with each layer building on the services provided by the lower layers. The photonic layer corresponds to the OSI physical layer. The section, line, and path layers correspond to the data link layer. Each layer communicates to peer equipment in the same layer, processing information and passing it up and down to the next layer. For example, two network nodes exchange DS1 signals: the source node path layer maps 28 DS1 signals and a path overhead to form an STS-1 synchronous payload envelope (SPE), which it hands to the line layer at its site. The line layer multiplexes three STS-1 SPEs and adds the line overhead. This combined signal is then passed to the section layer. The section layer performs framing and scrambling and adds section overhead to form three STS-1 signals. Finally the photonic layer convert the three electrical STS signals to optical signals suitable for OC lines and transmits them to the distant node.

At the distant node, the process is reversed from the photonic layer, where the optical signal is converted to an electrical signal, to the path layer, where the DS1 signals terminate.

We note that transmission rates exceeding 51.84 Mbps are all multiples of STS-1. As far as the actual frame format is concerned, STS-*N* signals are matrices of dimension $9(90 \times N)$. Note that the transport and path overheads are inserted *N* times in an STS-*N* frame. SONET also defines a frame format denoted by STS-NC (concatenated) in which the path overhead occurs only once; STS-3C frames may be used to transmit ATM cells.

STS-1 is the base SONET frame that may be viewed as a matrix of 9 rows × 90 column bytes, with a total of 810 bytes. The first three columns of the matrix are dedicated to transport overhead, which can be further subdivided into section and line overhead. The frame consists of two main areas: transport overhead, occupying the first three columns, and the synchronous payload envelope, occupying the remaining 87 columns. The first column of the SPE is reserved for path overhead. The signal is transmitted byte by byte, beginning with byte 1, scanning left to right from row 1 to row 9. The entire frame is transmitted in 125 μs. This equates to a basic STS-1 transfer rate of 51.84 Mbps, that is:

$$9 \text{ rows} \times 90 \text{ columns} \times 8000 \text{ fps} \times 8 \text{ bits/byte} = 51.84 \text{ Mbps}$$

The synchronous payload envelope runs from column 4 to column 90. Thus the capacity of the STS-1 payload is:

$$9 \text{ rows} \times 87 \text{ columns} \times 8000 \text{ fps} \times 8 \text{ bits/byte} = 50.112 \text{ Mbps}$$

Each STS-1 frame is transmitted starting from the byte in row 1, column 1 to the byte in row 9, column 90. The most significant bit of a byte is transmitted first. The following algorithm shows the transmission sequences.

```
for (row = 1; row <= 9; row++)
          for (column = 1; column <= 90; column++)
               for (bit = 1; bit <= 8; bit++)
               /* bit 1 is the most significant bit */
               /* and bit 8 is the least significant bit */
                    bitTransmit=STSFrame[row][column][bit];
```

6.9.3 SONET Overhead

Before we look at the SONET transport overhead, we examine the overhead of the DS1 signal hierarchy. The percentage overhead *OH* can be defined as follows:

$$OH = C_t - C_i/C_t$$

where C_t is total capacity and C_i is information-bearing capacity. For 24 voice channels, *OH* for DS1 signals is $(1.544 - 1.536)/1.544 = 0.52\%$. As the rate increases, the percentage overhead increases. The additional overhead is used for control bits, alarm and signaling, parity bits, and bit-stuffing. The percentage overheads for DS2, DS3, and DS4 are as follows:

$$\text{For DS2, } OH = \frac{6.312 - (96 \times 64/1000)}{6.312} = 2.7\%$$

$$\text{For DS3, } OH = \frac{44.736 - (672 \times 64/1000)}{44.736} = 3.9\%$$

$$\text{For DS4, } OH = \frac{276.176 - (4032 \times 64/1000)}{276.176} = 6.6\%$$

The percentage of useful capacity decreases as the rate increases. SONET adopts fixed locations for overhead and a fixed percentage for the overhead signal portion independent of system rate. The percentage of SONET overhead is 4.44%, which is found by multiplying 4 columns by 100% and dividing by 90 columns.

As stated, SONET provides overhead information known as transport overhead in the first three columns of the STS-1 frame. The SONET transport overhead is composed of section overhead and line overhead. Section overhead is used for communications between adjacent network elements. Line overhead is for the STS-N signal between the STS-N multiplexers. In addition to the transport overhead, SONET provides path-level overhead that is part of the SPE. Path-level overhead is carried from end to end; it is added to DS1 signals when they are mapped into virtual tributaries (explained next) and for STS-1 payloads that travel end-to-end. Thus, SONET overhead uses a layered approach.

SONET also defines synchronous signals known as virtual tributaries (VTs) to transport lower speed transmission. VTs operate at sub-STS-1 levels. The four defined sizes of VTs are VT-1.5 (1.728 Mbps), VT-2 (2.304 Mbps), VT-3 (3.152 Mbps), and VT-6 (6.912 Mbps). Within an STS-1 frame, each VT occupies a number of columns. Within the STS-1, virtual tributary groups can be mixed together to form an STS-1 payload. To accommodate different mixes of VTs in an efficient manner, the STS-1 SPE is divided into seven groups. A VT group may contain one VT-6, two VT-3s, three VT-2s, or four VT-1.5s. Although VT groups of different types may be mixed into one STS-1 SPE, a VT group may contain only one size of VTs. To synchronize the various low-speed signals to a common rate before multiplexing, bit-stuffing is used.

The first three columns of a SONET frame consist of section and line overheads. The upper three rows of the first three columns are used for section overhead. The lower six rows are used for line overhead. One column in SPE (usually the first) is used for path overhead, which includes the end-to-end tracking information.

The section overhead consists of 9 bytes with following specifications:

A1, A2 framing: 2 bytes provide the SONET framing or alignment; a repeating F6 28 hex pattern signals the receiver for the incoming frames.

C1 identification: One byte number is assigned to each STS-1 signal in an STS-*N* frame according to order of appearance. For example, the C1 byte of the first STS-1 signal in an STS-*N* frame is set to 1, the second STS-1 signal is 2, and so on. The C1 byte is assigned prior to byte interleaving and stays with the STS-1 until de-interleaving.

B1 BIP-8: A bit-interleaved parity (BIP) calculated over all bytes of the preceding SONET frame and used to detect single bit errors.

E1 orderwire: This byte provides a 64 Kbps voice channel for technicians working on problems between regenerators and other SONET equipment.

F1 user circuit: The byte provides a 64 Kbps channel for the user of the SONET equipment. It may be used to download revised firmware to a device.

D1–D3 management bytes: Form a 192 Kbps data communication channel between section equipment that is used for operation, administration, and maintenance (OAM).

The line overhead consists of 18 bytes, apportioned as follows:

H1, H2, and H3 pointer bytes: Used to locate the start of the SPE within the SONET frame.

B2 BIP-8: Provides a BIP check for over all bits except the section overhead.

K1, K2 automatic protection switching (APS): Used to provide signaling between pieces of line-terminating equipment such as multiplexers. APS enables SONET equipment to automatically route around path failures.

D4–D12 line DCC: Forms a 576 Kbps data communication channel (DCC) between line equipment; also used with message protocols for operation, alarm, maintenance, administration, and monitoring.

Z1, Z2 growth: Reserved for future use.

E2 orderwire: Provides a 64 Kbps voice channel between pieces of line-termination equipment used in SONET.

The path overhead is assigned to and transported with the payload from the time it is created by the path-terminating equipment as part of the SPE until the payload has been demultiplexed at the line-terminating equipment. In the case of superrate services, only one set of path overhead is required and is contained in the first STS-1 of the STS-*N*. The 9-byte path overhead consists of the following:

J1 path trace: This byte is to be used to repetitively transmit a 64-byte, fixed-length string that represents the network address of the originator of the SPE payload. The presence of this pattern indicates to the receiver that the source of the signal is still connected to the receiver.

B3 BIP-8: The byte is a path BIP check for all bits of the preceding SPE. It is not tied to the SONET frame contents.

C2 signal label: This byte identifies the SPE within an STS-*N* (FDDI, SMDS, etc.). It is assigned one of 256 possible values.

G1 path status: This byte is used for status information sent back through the network to the sender. That is, it is used to indicate problems and performance of the reverse channel on the SONET link. This allows the state and performance of the two-way path to be monitored at any point along the path or at either end.

F2 user channel: This byte provides a 64 Kbps data channel for the user of the end point SONET equipment. There is no standard usage, but it is a way for the end point customer premises equipment to pass network information at the path level.

H4 multiframe pointer: This byte has several uses. It indicates payloads that cannot fit into a single frame. Currently, it is used only for virtual tributary structure payloads, so it is also called a virtual tributary indicator. This was originally planned for SPEs carrying ATM cells, as an offset to the first byte of the first full ATM cell following the H4 pointer in the SPE. This use is now obsolete.

Z3–Z5 growth: These bytes are reserved for future use.

6.9.4 SONET Fault Tolerance

One of SONET's main advantages over current fiber optic transmission systems is its ring architecture. A SONET network can be organized into several types of ring, each of which provides for automatic network restoration—or self-healing—in the event of a network failure. Self-healing refers to SONET's ability to recognize points of failure in the network within 50 ms and route traffic around them. Self-healing rings are implemented by combining protection switching and uni/bidirectional traffic transports. Protection switching, a backup technique used to prevent loss of traffic due to fiber outages, exists in two flavors: line protection switching and path protection switching.

Line protection switching consists of duplicating the original pair of fibers between point-to-point multiplexers and using the second pair for backup purposes. If the primary pair goes offline or is degraded, the multiplexers at either end detect signal loss and all traffic is routed over the backup pair. Path protection switching assumes that problems with a fiber line are highly unlikely and that any disruption of transmission is a problem with one or more channels on the line. The solution to line problems involves simultaneously transmitting data on both primary and protection fiber pairs. The receiver compares both copies of the data, and the better one is kept. If a channel is disrupted on the primary fiber, the receiving multiplexer can switch it out for its mirror on the protection channel. It is important to note that the "protection fibers" do not have to be actual fibers; a system could have primary fibers running from A to B and use satellite communications as the "protection pair."

There are two primary ring architectures for SONET: bidirectional line-switched rings (BLSRs) and unidirectional path-switched rings (UPSRs). BLSRs are highly survivable and can be implemented with either two or four fibers. In the two-fiber version, the working and protection channels share capacity on both fibers. The four-fiber BLSR supports two types of line switching, providing a high degree of reliability.

UPSRs are a type of survivable architecture in which working traffic is transmitted in only one direction on only one fiber. Duplicating the traffic and sending it on the protection fiber in the opposite direction provides path-level protection. The receiving add/drop multiplexer compares the two copies and uses the better one. In the event of a failure, the loss is detected and the duplicate copy is used automatically.

6.10 DENSE WAVELENGTH DIVISION MULTIPLEXING (DWDM) COMMUNICATION

Wavelength Division Multiplexing (WDM) is a fiber optic modulation scheme where data channels are put on different light wavelengths (or colors) as opposed to different frequencies in FDM. An extension of DWM, called DWDM, is important in enabling service providers to accommodate consumer demand for ever-increasing amounts of bandwidth. Using this technology, it is possible to put data from different sources together on an optical fiber, with each signal carried on its own separate light wavelength. Different data sources such as e-mail, video, multimedia, data, and voice are transmitted over fiber-optic using IP, ATM and SONET/SDH. Using DWDM, separate wavelengths or channels of data can be multiplexed into a light stream transmitted on a single optical fiber. An example WDM implementation is shown in Figure 6.21, with a WDM multiplexer that combines four independent data streams, each on a unique wavelength, and sends them on a fiber; and a demultiplexer at the fiber's receiving end to separate out the data streams.

Since each channel is demultiplexed at the end of the transmission back into the original source, different data formats being transmitted at different data rates can be transmitted together. Specifically, IP data, SONET data, and ATM data can all be traveling at the same time within the optical fiber. This unifying capability allows the service provider the flexibility to respond to customer demands over one network.

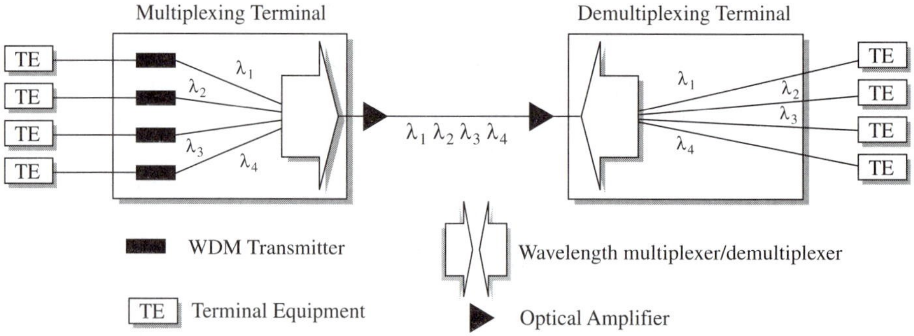

Figure 6.21 A four channel point-to-point WDM transmission system with amplifiers.

The discovery of erbium doped fiber amplifier (EDFA) greatly expedited the applications of DWDM. An EDFA is made of a few meters of optical fiber doped with tiny amounts of rare earth element erbium. The optical signal is injected into this fiber, along with the light from a special "pump" laser that is designed to excite the erbium ions and give a boost to the optical signals. The nice thing about EDFA is that it can amplify all the optical signals regardless of their color, thus in a DWDM system, we do not need a separate amplifier for each channel—greatly simplifying the system and increasing the transmission distance with minimal costs.

Using DWDM systems, the signal is amplified as a group using optical line amplifiers and transported over a single fiber to increase capacity. Each carried signal can be at a different rate (OC–3/12/24, etc.) and in a different format. For example, a DWDM network with a mix of SONET signals operating at OC–48 (2.5 Gbps) and OC–192 (10 Gbps) over a DWDM infrastructure can achieve capacities of over 40 Gbps. A system with DWDM can achieve all this gracefully while maintaining, or even surpassing, the degree of system performance, reliability, and robustness of current transport systems. The technology that allows this high-speed, high-volume transmission is in the optical amplifier. Optical amplifiers operate in a specific band of the frequency spectrum and are optimized for operation with existing fiber, making it possible to boost lightwave signals and thereby extend their reach without converting them back to electrical form.

The basic mechanism of communication in a wavelength-routed network is a lightpath, which is an all-optical communication channel between two nodes in the network, and it may span more than one fiber link. The intermediate nodes in the fiber path route the lightpath in the optical domain using their active switches. The end-nodes of the lightpath access the lightpath with transmitters and receivers that are tuned to the wavelength on which the lightpath operates. For example, in Figure 6.22, lightpaths are established between node pairs on wavelength channels via an active switch that is capable of reusing

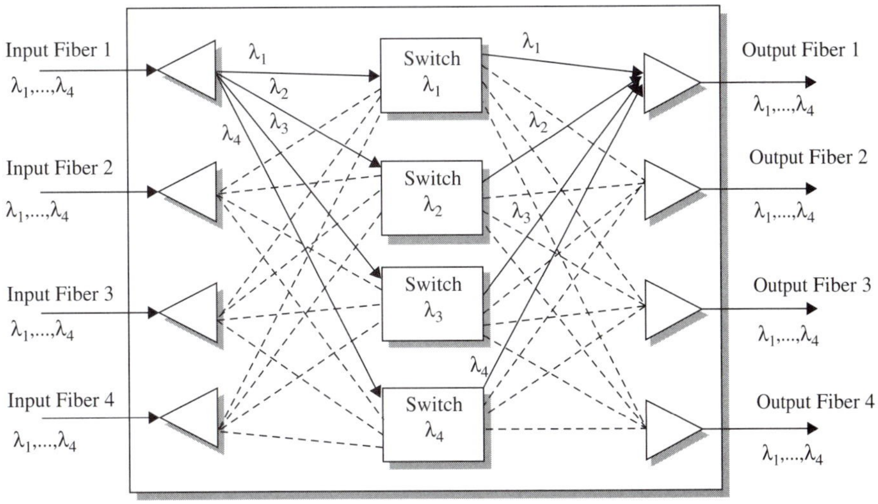

Figure 6.22 A 4×4 active switch (four wavelengths).

wavelengths and allowing n^2 simultaneous connections through itself (n is the number of input/output to/from a switch).

In the absence of any wavelength conversion device, a lightpath is required to be on the same wavelength channel throughout its path in the network; this requirement is referred to as the wavelength continuity property of the lightpath. This requirement may not be necessary if we also have wavelength converters in the network. However, a fundamental requirement in a wavelength-routed optical network is that two or more lightpaths traversing the same fiber link must be on different wavelength channels so that they do not interfere with one another.

Many vendors have developed WDM devices. Lucent's latest DWDM system provides up to 400 Gbps capacity over a single strand of fiber—equivalent to carrying the per-second traffic of the entire worldwide Internet over one fiber. AT&T will be the first to test and deploy the new system. Designed by Bell Labs, Lucent's single-platform system, called the WaveStar OLS 400G, is the first to enable communications providers to grow incrementally from one to eighty wavelengths, or channels. Lucent's new optical networking system can be configured to handle up to eight fibers, each transmitting 400 Gbps, to give communications providers a maximum capacity of 3.2 tbps (or 3.2 trillion bits) of voice, video and data traffic.

Based on WaveStar, Lucent's LambdaRouter uses a series of microscopic mirrors to instantly direct and route optical signals from fiber to fiber in the network, without first converting them to electrical form. This feature allows an operational cost saving of up to 25 percent to service providers and enable them to direct network traffic 16 times faster than electrical switches. Maintaining all-optical traffic through LambdaRouter paves the way for communications network providers to offer customers instant Internet and other high-speed data and video services. For example, a local service provider could use WaveStar system to enable its corporate customer to initiate an HDTV-quality videoconference between two sites at a moment's notice.

6.11 CHAPTER SUMMARY

The latest high-speed and remote access technologies are discussed in this chapter. We start with ISDN and cable modem technologies that are likely to become popular in consumers as the access prices fall. With the limits of communication speed being reached using the conventional modems, it is evident that the network vendors and manufacturers would invest more into high-speed technologies. However, since there are many competing technologies, it is not very clear at this point that exactly which technology is going to take over. ISDN and cable modem seem to have strong prospects, but so do DSL and SMDS. Consumers probably will use the cost/performance trade-off to decide the winner.

Small to mid-sized businesses would not care so much about the operating expenses if a return on their investment could be assured. Other high-speed networking technologies such as Gigabit Ethernet, FDDI, and ATM may be perceived to have more demand in the business world, but with the falling prices one should not be surprised to see consumers taking advantage of these technologies.

With advances in frame relay switching, it is possible to forward data and voice in real time, thus supporting the telephonic conversation. This coupled with voice over IP techniques has caused a significant drop in prices of long distance phone calls. With further availability of bandwidth and more advanced switching techniques, it is not beyond feasibility to expect an affordable commercial videophone system in the near future. The fiber-based SONET system, with its high bandwidth and low error rate, certainly will be helpful.

6.12 PROBLEMS

6.1 What is the type of an ISDN circuit (dial-up, leased, permanent, etc.)?

6.2 Does the basic B channel of an ISDN have a capacity of 128 Kbps?

6.3 Explain the type of connection (connection-oriented or connectionless) and circuit used by each of the following: (a) Frame relay, (b) SMDS, and (c) ATM.

6.4 If an application uses AAL5 and there are 41,000 bytes of data coming in to the CS sublayer, how many data units are passed from the SAR to the ATM layer? How many padding bytes are necessary?

6.5 If an application uses AAL3/4 and there are 22,000 bytes of data coming in to the CS sublayer, how many data units are passed from the SAR to the ATM sublayer? How many padding bytes are necessary?

6.6 When a 1024-byte message is sent with AAL 3/4, what is the efficiency obtained? (What fraction of the transmitted bits are useful data bits?) Repeat the problem for AAL5.

6.7 Which AAL type is designed to support a data stream that has a constant bit rate?

6.8 Which layer in ATM protocols has a 53-byte cell as an end product?

6.9 What kind(s) of transmission media can ATM use?

6.10 What mode of operation is used in ATM? Explain.

6.11 An ATM switch has 1024 input lines and 1024 output lines. The lines operate at the SONET rate of 622 Mbps, which gives a user rate of approximately 594 Mbps. What aggregate bandwidth does the switch need to handle the user load? How many cells per second must it be able to process?

6.12 What is the name of the interface that lies between the user and an ATM switch?

6.13 What is the main purpose of the second ring in the FDDI protocol?

6.13 REFERENCES

BOOKS

Amoss, J. J., D. Minoli, and J.J. Amos, *IP Applications with ATM*. New York: McGraw-Hill, 1998.

Asatani, K., *Introduction to B-ISDN*. New York: Wiley, 1997.

Atkins, J., and M. Norris, *Total Area Networking: ATM, IP, Frame Relay and SMDS Explained*, 2nd ed. Reading, MA: Wiley, 1999.

ATM Forum Technical Committee, *ATM User–Network Interface: UNI Specification*," version 3.1. Englewood Cliffs, NJ: Prentice Hall Software,

Azzam, A. A. *High-Speed Cable Modems: Including IEEE 802.14 Standards*. New York: McGraw Hill, 1997.

Black, U., *ATM: Foundation for Broadband Networks*. Englewood Cliffs, NJ: Prentice Hall, 1995.

Black, U. D., *ATM: Signaling in Broadband Networks*, vol 2. Englewood Cliffs, NJ: Prentice Hall, 1997.

Black, U. D., *ATM Foundation for Broadband Networks*, vol. 1. Englewood Cliffs, NJ: Prentice Hall, 1998.

Black, U. D., *ATM Resource Library*. Englewood Cliffs, NJ: Prentice Hall, 1999.

Black, U. D., and U. Black, *ISDN & SS7: Architectures for Digital Signaling Networks*. Englewood Cliffs, NJ: Prentice Hall, 1997.

Breyer, R. A., and S. Riley, *Switched, Fast, and Gigabit Ethernet*, 3rd ed. New York: Macmillan, 1999.

Chiong, J., *Internetworking ATM: For the Internet and Enterprise Networks*. New York: McGraw-Hill, 1997.

Cicoria, W., J. Farmer, D. Large, and D. Large. *Modern Cable Television Technology: Video, Voice, and Data Communications*. San Francisco: Morgan Kaufmann, 1998.

Cullum, B., and J. Pollock, eds., *ATM Systems Design, The Hitch-hiker's Guide*, 2nd ed. Dallas, Dallas Engineers, 1998.

Cunningham, D., and W. Lane, *Gigabit Ethernet Networking*. New York: Macmillan, 1999.

Flood, J. E., *Telecommunications Switching, Traffic and Networks*. Englewood Cliffs, NJ: Prentice Hall, 1995.

Giroux, N., and S. Ganti, *Quality of Service in ATM Networks: State-of-the-Art Traffic Management*. Englewood Cliffs, NJ: Prentice Hall, 1998.

Goralski, W., *SONET: A Guide to Synchronous Optical Network*. New York: McGraw-Hill, 1997.

Goralski, W., *ADSL*, New York: McGraw-Hill, 1998.

Goralski, W. J., *Frame Relay for High Speed Networks*. New York: Wiley, 1999.

Groom, C. M., and F. M. Groom, *The Future of ATM and Broadband Networking: 2000 to 2010*. Chicago, IL: International Engineering Consortium; Center for Telecommunications Management, 1998.

Guizani, M., and A. Rayes, *Designing ATM Switching Networks*. New York: McGraw-Hill, 1998.

Handel, R., M. N. Huber, and S. Schroder, *ATM Networks: Concepts, Protocols, Applications*. 3rd ed. Reading, MA: Addison-Wesley, 1998.

Hein, M., and D. Griffiths, *Switching Technology the Local Network: From LAN to Switched LAN to Virtual LAN*. Boston, MA: International Thomson Computer Press, 1997.

Humphrey M., ed., *ATM Handbook*. New York: McGraw-Hill, 1999.

Keshav, S., and S. Kesahv, *An Engineering Approach to Computer Networking: ATM Networks, the Internet, and the Telephone Network*. Reading, MA: Addison-Wesley, 1997.

Kesidis, G., *ATM Network Performance*. Amsterdam: Kluwer Academic Publishers, 1996.

Kessler, G. C., and P. V. Southwick, *ISDN: Concepts, Facilities, and Services*. New York: McGraw-Hill, 1998.

Klessig, R., and K. Tesink, *Wide-Area Data Networking with Switched Multi-megabit Data Service*. Englewood Cliffs, NJ: Prentice Hall, 1995.

Kouvatsos, D. D., ed., *ATM Networks: Performance Modeling and Analysis*, vol. 2. London: Chapman & Hall, 1996.

Kumar, B., *Broadband Communications*. New York: McGraw-Hill, 1998.

Kwok, T., *ATM: Private, Public & Residential Broadband Network*. Englewood Cliffs, NJ: Prentice Hall, 1997.

Kwok, T., *ATM: The New Paradigm for Internet, Intranet & Residential Broadband Services and Applications*. Englewood Cliffs, NJ: Prentice-Hall, 1998.

Lewis, C., *Cisco Switched Internetworks; VLANs, ATM & Voice/Data Integration*. New York: McGraw-Hill, 1999.

Marney-Petrix, V. C., and S. Salins, *Mastering Gigabit Ethernet, ATM and Other High-Speed LANs: Self Paced Learning Series*. Fremont, CA: Numidia Press, 1997.

Martin, J., J. Leben, and K. Chapman, *Asynchronous Transfer Mode: ATM Architecture and Implementation*. Englewood Cliffs, NJ: Prentice Hall, 2000.

McDysan, D. E., and D. L. Spohn, *ATM Theory and Application*. New York: McGraw-Hill, 1998.

McDysan, D. E., and D. L. Spohn, *Hands-On ATM*. New York: McGraw-Hill, 1998.

Meter, T. V., *Cisco & Fore ATM Internetworking*. New York: McGraw-Hill, 1999.

Minoli, D., and A. Alles, *LAN, ATM, and LAN Emulation Technologies*. Norwood, MA: Artech House, 1997.

Minoli, D., and J. J. Amoss, *ATM Switching*. New York: McGraw-Hill, 1998.

Minoli, D., et al., *Delivering Voice over Frame Relay and ATM*. New York: Wiley, 1998.

Nye, R. L., *Implementing Frame Relay and ATM in IBM System Environments*. New York: McGraw-Hill, 1997.

Pan, H., *SNMP-Based ATM Network Management*. Norwood, MA: Artech House Telecommunications Library, 1998.

Pattavina, A. *Switching Theory: Architectures and Performance in Broadband ATM Networks*. New York: Wiley, 1998.

Rahman, M. A., *Guide to ATM Systems and Technology*. Norwood, MA: Artech House Telecommunications Library, 1998.

Sackett, G. C. and C. Metz, *ATM and Multiprotocol Networking*. New York: McGraw-Hill, 1997.

Saunders, S., *The Gigabit Ethernet Handbook*. New York: McGraw-Hill, 1998.

Schwaderer, W. D., *Enterprise Networking with Fast Ethernet and ATM*. Milpitas, CA: Adaptec Press, 1997.

Schwartz, M., *Broadband Integrated Networks*. Englewood Cliffs, NJ: Prentice Hall, 1996.

Sexton, M., and A. Reid, *Broadband Networking: ATM, SDH, and SONET*. Norwood, MA: Artech House Telecommunications Library, 1998.

Shah, A., G. Ramakrishnan, and A. Ram, *FDDI: A High Speed Network*. Englewood Cliffs, NJ: Prentice Hall, 1993.

Singh, C., *Gigabit Ethernet Handbook*, New York: McGraw-Hill, 1998.

Smith, P., *Frame Relay: Principles and Applications*. Reading, MA: Addison-Wesley, 1993.

Stallings, W. *ISDN and Broadband ISDN With Frame Relay and ATM*. 3rd ed. Englewood Cliffs, NJ: Prentice Hall, 1995.

WORLD WIDE WEB SITES

The ATM Forum page
http://www.atmforum.com/

Much ATM and other telecommunications information from the University of Manitoba
http://www.ee.umanitoba.ca/~blight/telecom.html

Broadband Bob runs a cable modem newsletter
http://www.catv.org/frame/bbb.html
Cable Modem Info provides much general and very specific information about cable modems
http://www.cablemodeminfo.com
MCNS/DOCSIS specifies the dominant U.S. standard
http://www.cablemodem.com
FAQs on cell relay networks, includes LANs and WANs but mostly ATM
http://cell-relay.indiana.edu/cell-relay/FAQ/ATM-FAQ/FAQ.html
Internet Technology Overview page by Cisco, contains latest information about Internet related high-speed access technologies such as SMDS, SONET, ISDN, frame relay, BGP, IGRP, RIP, OSPF, RSVP, RMON, SNMP
http://www.cisco.com/univercd/cc/td/doc/cisintwk/ito_doc/
Dan Kegel's ISDN page
http://www.alumni.caltech.edu/~dank/isdn/
An ISDN tutorial
http://www.ralphb.net/ISDN/
IP over ATM working group
http://www.com21.com/pages/ietf.html
Fore System's white papers about the some latest networking technology may be found here
http://www.fore.com/products/wp/index.html
High performance networks and distributed system archive contains information about gigabit test-beds and ATM
http://hill.lut.ac.uk/DS-Archive

7

SWITCHING AND VIRTUAL LAN

In previous chapters we have discussed the important determinants of a network's speed and efficiency: the various hardware, software, topologies, protocols, and media used to build the network.

As networks grow in size and complexity, they need to be segmented into components to restrict the amount of traffic and the number of devices on each segment. Repeaters, bridges, and routers are traditional devices for extending, segmenting, and interconnecting networks. When bandwidth became a limiting factor and networks became unmanageable, alternative methods for expanding bandwidth and structuring networks had to be developed. This motivated the development of various types of junction hardware, software, and protocols to make networks more efficient and easier to manage. Switching provides an economic increase in bandwidth while preserving the desktop investment. It also offers greater bandwidth control, access control, and new ways of constructing and structuring LANs. The services that switching and virtual LANs (VLAN) provide cannot be matched within the traditional shared media environment. Starting from hubs, this chapter discusses switching LAN and virtual LAN technologies, which are an essential part of the physical and logical structure of networks.

7.1 HUB TECHNOLOGY

In the pre-LAN mainframe computing era, all enterprise computing was centralized and manageable. The early Ethernet-based LAN systems linked workstations with a single cable. Since all users on the networks shared a common physical data link, a failure of that link could bring everyone on the network down. Token ring LANs, developed as an alternative to the bus architectures, implemented a ring topology and used passive, unpowered wiring concentrators called multistation attachment units (MAU). A MAU can be installed close to workgroups, and easily maintained because it is configured in a star topology and uses flexible twisted-pair cabling. A MAU can detect a broken wire or an out-of-order host in the ring, and automatically isolate the defect without affecting the normal operation of the LAN.

7.1.1 Early Hubs

As the size, complexity, and usage of networks increased, users and managers of LANs needed to find better ways to structure their networks for greater efficiency, performance, manageability, and interoperability. The need to centralize a large number of network connections forced the development of MAUs and multiport repeaters into the first LAN cabling hubs. The cabling hub allowed the combination of several multiport repeaters into a single chassis. This enabled them to share power and signals across a common backplane. Early hubs used proprietary protocols without interoperability.

To overcome the proprietary nature of early LAN cabling hubs, the IEEE standardized a variation of the 802.3 protocol based on Ethernet using unshielded twisted-pair (UTP) cable. The 10BaseT hubs quickly became prevalent because the star-shaped Ethernet LAN supports a structured cable plant and provides strong management capabilities.

7.1.2 Intelligent Hubs and Their Components

Today, intelligent LAN hubs have improved upon 10BaseT hubs by combining multiple protocols, media, networks, and different devices into one system. An intelligent hub also keeps and maintains port statistics and determines port status. Ports on multi-LAN intelligent hubs can be simply enabled, disabled, reconfigured, and automatically segmented when faulty. Intelligent hubs often offer redundancy to provide fault tolerance. Many of the functions traditionally provided by advanced internetworking devices such as bridges and routers are now provided by intelligent hubs, thus overcoming distance limitations and enabling true enterprise-wide communications. Combined with traditional routers, intelligent hubs can provide LANs with access to wide-area networks (WANs). This feature is becoming a must as LANs and WANs continue to migrate toward one common uniform network and become part of the universal Internet.

The hub has become the basic building block for conventional LANs. It provides connections to servers and end stations, as well as such internetworking devices as bridges and routers. Hubs are shared-media devices: all ports, or a group of ports on a hub can be connected to the same LAN segment (see Figure 7.1). Because these ports have access to all traffic as it passes through the LAN, a protocol analyzer or probe attached to a hub port can monitor the traffic on that LAN segment. This greatly facilitates many networking management tasks and makes the new breed of intelligent hubs extremely manageable. Intelligent hubs can be configured and controlled by standard SNMP or vendor-specific and third-party network management systems. In-band management can manage a hub from a station on the network via the network media, while out-of-band management can manage a hub from outside the network media through a special port accessed locally or remotely. Out-of-band management offers the capability of troubleshooting a hub when the network is down. Through the use of intelligent hubs, network redundancy can be implemented, ports can be turned off and on, and traffic can be monitored, all from a single centralized management console.

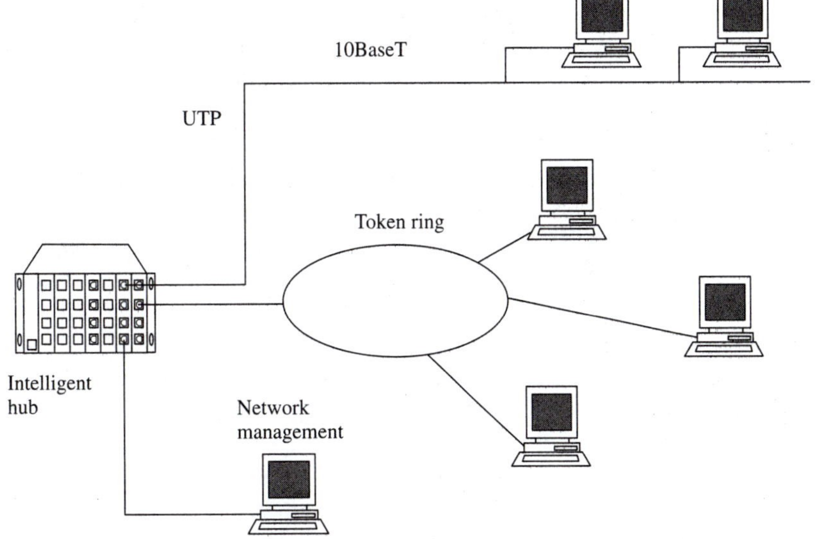

Figure 7.1 Heterogeneous networks can be structured and managed with intelligent hubs.

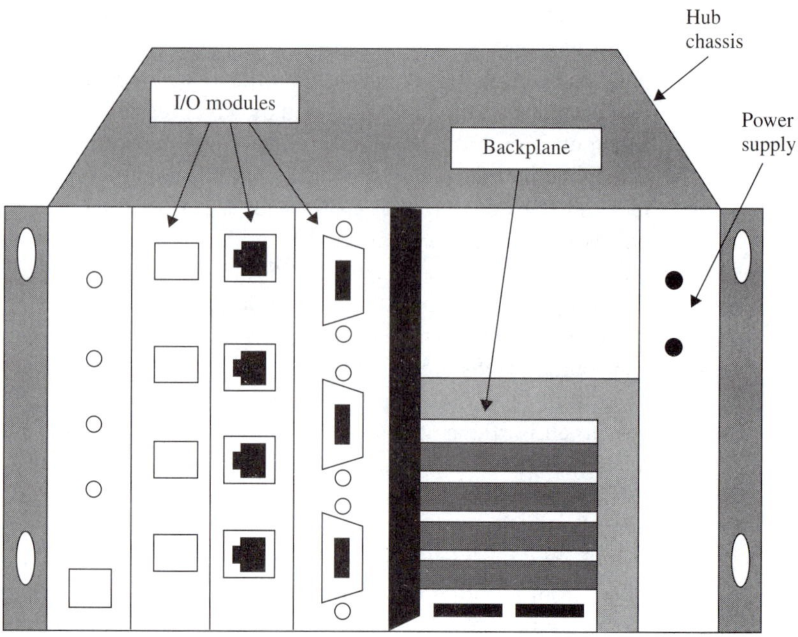

Figure 7.2 Intelligent hub chassis, backplane, and individual I/O modules.

Most hubs have the following basic elements (Figure 7.2).

Hub chassis: This is the main housing unit for a hub. A hub chassis is usually made up of a number of input/output modules. Well-designed hub chassis have interchangeable slots for all the modules and allow hot-swapping—the ability to remove or replace a module without affecting active networks or taking the hub out of service. There should also be enough room for expansion and additional slots for power supplies.

Hub backplane: Located in the rear of the chassis is the hub's true engine—the backplane. All hub modules are plugged into the backplane, which is the path for all LAN signals. Usually the backplane supports multiple LANs of the same protocol type. More advanced backplanes support multiple LANs of different protocol types. This provides a higher degree of configuration flexibility and expandability for future network needs. A passive backplane, which contains no electronic components, is generally less prone to failure than an active backplane.

I/O module: I/O modules occupy the slots on the hub backplane and provide hub functionality for a specific application. I/O modules are the main point of connectivity in a hub. The cabling plant and network topology determines how all I/O modules are combined. Individual computers and hub servers, which provide services such as database management, E-mail gateways, and network routing functions, are connected to the network via I/O modules with specific media and protocol types. Various LEDs on the module indicate port and module status.

Hub management software: The hub's software resides in a separate intelligence module of the hub. The software helps manage the hub network by monitoring ports and network status. Hub management software also interoperates with the broader enterprise management system to control all components in a network and allow communication among diverse devices including hubs, cabling, bridges/routers, servers, and network operating systems. The most widely used hub network management system is based the on Simple Network Management Protocol (SNMP). SNMP compatibility allows a hub to be managed by third-party network management products. The Hub Management Interface (HMI), Novell's de facto industry standard, provides third-party vendors with a platform for developing hardware and hub management tools that can be easily integrated into NetWare networks. Hub management software differs in the features offered and in the level of sophistication with which the hub and related devices are managed and graphically displayed.

7.1.3 Interconnecting LANs and Collapsed Backbone Networks

As LANs proliferate, organizations begin to encounter problems that often stifle network productivity. Many of these problems relate to the unstructured growth of these networks. In addition to different network topologies and protocols, different cabling types span whole buildings and become unmanageable. For example, UTP may have been installed for the telephone system, thick or thin coaxial cable may have been used for Ethernet connections, and fiber optic cable may have been utilized for mission-critical applications. Data communication across the organization becomes difficult. Moving a user within the organization becomes unnecessarily time-consuming, and troubleshooting problems on the networks becomes a big headache for network managers. The solution is a structured wiring design.

First defined by AT&T and IBM, structured wiring schemes connect workstations to wiring closets by means of both a common wiring type and an interface across an entire enterprise. The most widely accepted implementation specifies UTP cable for the horizontal links between hubs and workstations on the same floor. Fast and more reliable fiber optic cable is used for backbone connectivity between the floors of a building or between the buildings of a campus (Figure 7.3). Such backbone networks provide for a variety of networks and support multiple network protocols, such as TCP/IP and IPX. Many carry several types of message traffic, such as voice, data, video, and fax. No matter what medium is used for the backbone, most backbone networks support connections to media of other types. Even wireless LANs can be connected to backbones. Backbone networks generally operate reliably and securely, and usually at speeds high enough to support business operations transparently.

A structured enterprise LAN helps bring order to the network's physical layer. It increases network scalability, and makes it easier to relocate bandwidth and reconfigure the network to adjust for changes in user demand. When structured wiring is combined with hub network management software, network faults can be easily located and remedied.

The five key components of a structured enterprise network are structured LAN cabling, network protocols, intelligent hubs, internetworking devices, and network security and management. A relatively recent development in using backbones to establish inter-

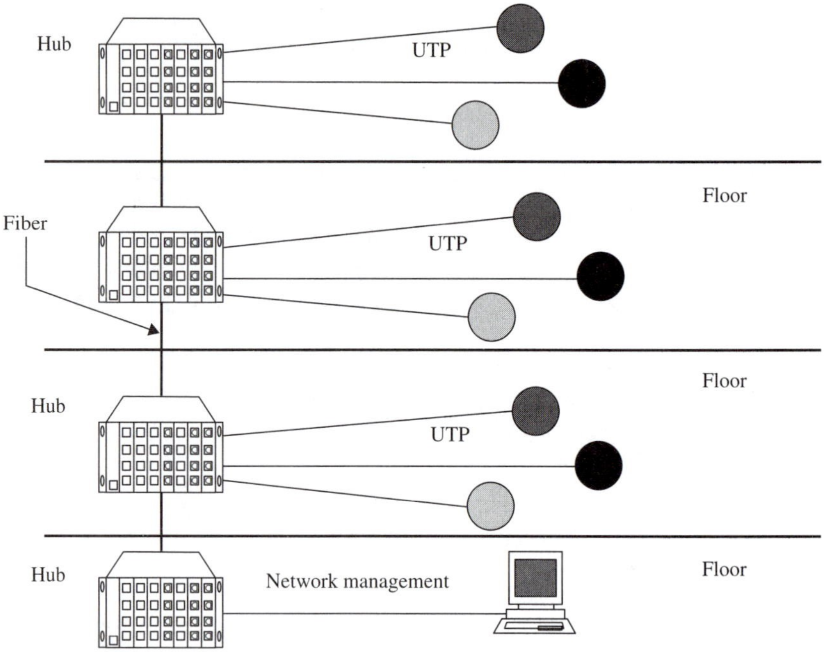

Figure 7.3 Structured wiring system with UTP used in the same floor and optical fiber between floors and buildings.

network connections is known as superhub or collapsed backbone (Figure 7.4). The collapsed backbone structure is most commonly used in a backbone network for a single building. The superhub essentially consists of a connection hub with slots for cabling to all LANs, as well as slots for internetworking devices such as bridges, routers, and gateways.

The collapsed backbone architecture is best for organizations with complex network requirements, where multiple protocols and device types are connected by high-speed routers internetworking various LAN types. The collapsed backbone router possesses the speed and bandwidth to facilitate LAN-to-WAN communication. Routers are centrally located to enhance network control and reliability. In remote locations or on individual floors, traffic is arbitrated and distributed by robust intelligent cabling hubs. In summary, a collapsed backbone network can provide the following features:

- Universal internetworking between different types of LANs and devices
- Network access via the wide area network for remote locations and the outside world
- Centralized network management and troubleshooting for remote LANs to minimize trips to individual wiring closets and remote sites

The collapsed backbone or superhub architecture is the best way to keep centralized networks up and running and to provide efficient internetworking. With built-in fault toler-

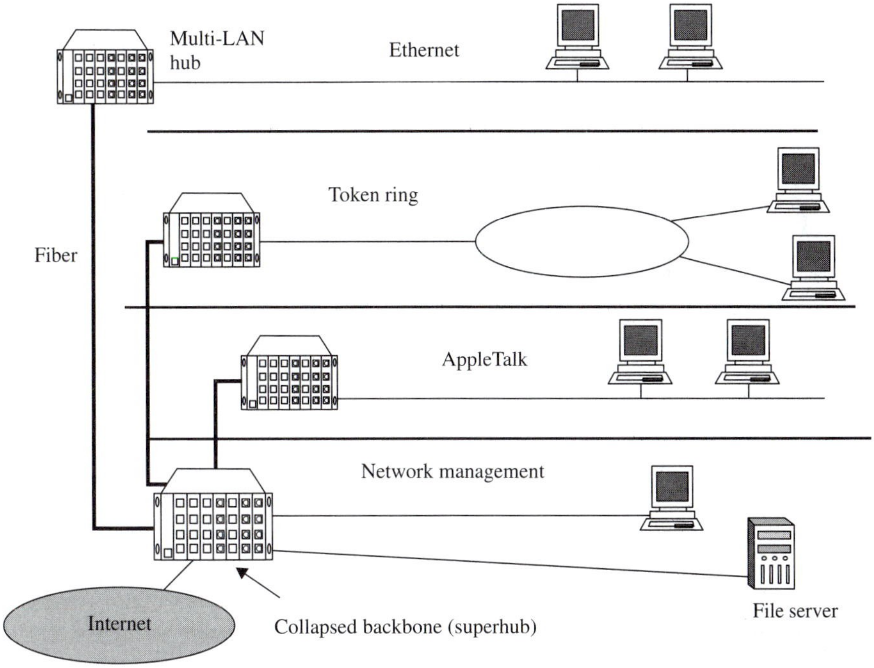

Figure 7.4 Collapsed backbone architecture provides efficient internetworking.

ance capability, collapsed backbone can reduce the risks of creating a single point of network failure. The result is an enterprise network that is both powerful and dependable. The growing popularity of ATM is likely to emerge as the most common protocol used in the enterprise network backbone. In the future, ATM switching superhubs may replace shared backbones.

7.2 SWITCHING TECHNOLOGY FOR LAN AND INTERNETWORKING

Bridges, the first generation of LAN interconnect to pass data up to the data link layer, were fast but dumb. As bridged architectures grew, that lack of intelligence resulted in broadcast storms and instability. Routers, the second generation of LAN interconnect, were smart enough to structure networks. Not long ago, router-based interconnect for point-to-point WANs and shared-media LANs was the industry's mainstream technology. The hubs of today perform many of the functions performed by the routers of the past. Still, the constant demand for higher bandwidth and easier network management is changing hub technology.

What users need today is a scalable client–server network infrastructure that distributes networking capabilities while it centralizes network management. A monolithic, centralized hub architecture that generates redundant traffic and will not scale is no longer anybody's favorite. Emerging high-speed switching technology is replacing the slow and congested router networks of the past. Tomorrow's LAN interconnect will be enhanced by a switched fabric.

A change to switching technology provides an immediate bandwidth boost over shared-media networks. Such a change can be made without having to replace the existing communications infrastructure. Once installed, high-speed switching technology enables users to deploy virtual LANs, which bridge workgroups across wide-area links.

LAN switches are the new LAN-to-LAN and LAN-to-WAN internetworking technology. The basic function of a switch is to simultaneously maintain multiple bridges between network devices by means of high-speed hardware components. Many LAN switches can be considered to be hubs for routing traffic between different types of LAN such as Unix, TCP/IP, Windows/NT, NetWare, and DECnet LANs. These switches are sometimes designed to provide an alternative to bridges and routers. They provide the same functions as collapsed backbones. Organizations often deploy LAN switches because they see them as avenues for increasing bandwidth and as a means for overcoming or bypassing bottlenecks created by the message processing of bridges and routers. More important, switching LANs are stepping-stones toward the more logically structured *virtual LANs (VLANs).*

Switching is not new. However, the mechanical switches found in old PBX cabinets are very different from the high-speed, high-bandwidth electronic switches of today. These switches, either application-specific integrated circuit (ASIC)-based or reduced instruction set computer (RISC) CPU-based, make switched VLAN possible. Traditionally, layer 2 switching is provided by LAN switches, and layer 3 networking is provided by routers. Increasingly, these two networking functions will be integrated into common platforms. There will still be a wide range of platforms providing different performance and capacity for each networking function. But there are fundamental benefits from this integration. Users will be able to reduce the number of networking devices to be supported. Also, the integration of the two technologies enables layer 3 networking to be cost-effectively deployed out toward the network edge and allows its security and quality control capabilities to be more efficiently applied to specific individual users and applications.

Switching offers performance improvements over routing mainly because of the way switches process packets and the way switches are made. Routers must examine multiple packet fields, make substitutions in packet headers, and compute routes on a packet-by-packet basis. In the process, routers introduce latency and congestion. Switches, with cut-through architecture, accept frames, read a simple address header, and move the data along. Unlike routers, which forward at the third layer of the OSI model, the network layer, many switches forward packets at layer 2, the data link layer. This not only reduces the length of the path the data pass through but also enables hardware like ASICs to perform the task. Today, semiconductor manufacturers can put 5 million gates on a single chip. An entire communication switching system can be embedded on a chip with space remaining for a powerful microprocessor, logic, and memory. As the number of processes ASICs can perform increases, the number of components decreases, lowering production time and costs, and improving reliability. Typical latencies of modern switches are one-tenth those of fast routers: 10–50 ms for ATM and fast LAN switches vs 100–500 ms or more for multiprotocol routers. Routers also complicate network management, since each physical segment created by a router must exist as a separate logical subnet.

7.2.1 Switching Architectures

7.2.1.1 RISC CPU-Based Switches

As mentioned earlier, network switches can be based on application-specific integrated circuits (ASICs) or off-the-shelf RISC CPUs with software and shared memory. CPU-based switches rely on software to realize switching. They are, therefore, more flexible and less expensive to develop. Software switches also can be easily upgraded via software downloads. Based on routing technology, software switches often route a variety of protocols. This means they can be used as gateways and can furnish the security and firewall functions usually performed by slower and more expensive routers.

There are some limitations to software switches. The CPU in a software switch usually has responsibilities other than just switching. It must also handle all the processing activities such as interrupt requests and housekeeping tasks. In addition, every feature added to the software switch will require additional CPU cycles. The more features added, the more inconsistent the performance of software switches. That is why an advanced layer 3 switch, which implements virtual LAN routing and performs protocol conversion, requires an extremely powerful RISC CPU.

7.2.1.2 ASIC-Based Switches

In contrast to RISC CPU-based switches, ASIC-based switches have a fixed and shorter latency and lose fewer frames under extreme traffic conditions. ASIC's, features and functions are cast in silicon. Top-performance switches rely almost exclusively on ASIC to process frames. On the other hand, most ASIC-based switches operate only at the data link—MAC (media access control) layer and thus cannot read network layer information. A further negative is the length of time needed to develop an ASIC-based switch. Significant changes or upgrades require the development and distribution of a new chip.

7.2.1.3 Buffering

To minimize frame loss within a busy switch, buffering is essential. This is true for switches of all types, but particularly for those with multiple port speeds. For example, traffic flowing from a 100 Mbps port needs to be held in buffers from time to time to allow a slower 10 Mbps port to catch up. The solution to this mismatch of speeds is buffers: fast memory dedicated to holding frames in queue while a destination port is busy.

There are three major buffer types: input buffers, which hold frames as they arrive from the source port; output buffers, which hold frames at the destination port; and shared-memory buffers, which dynamically allocate buffering to each port.

When the destination port is busy, the first frame in a queue in the input buffers holds up all other frames behind it, causing head-of-line blocking. With a back-pressure algorithm, a flow control mechanism transmits a jam signal to the sending workstation to prevent it from continuing to send frames to ports with full buffers. However, when a network segment with many workstations is attached to single port with back-pressure, all the traffic on the segment can be blocked. Output buffers that are larger than input buffers are often used to avoid such a situation. A more advanced design uses dynamically allocated shared-memory buffers rather than buffers dedicated to each port. A large shared pool of buffers provides the flexibility needed to accommodate changing traffic flows.

7.2.1.4 Cell Switches

In general, a switch keeps its backplane transparent when frames are passed through. In many cases, data frames remain intact across the backplane. Such devices are frame or packet switches. Still, some switches actually decompose each frame into ATM cells before shipping the data across the backplane. The original frame is then reassembled at the receiving port and transmitted on to the destination exactly as it appeared on the input port. Cell-switching devices may very well provide quicker and easier integration into ATM environments than frame switches.

7.2.1.5 Basic Architectures

In terms of use and topology, switching devices are similar to hubs. Structurally, every LAN switch employs one of the following three basic architectures (Figure 7.5):

- Shared-memory switch
- Matrix switch
- Bus switch

Shared-memory switches are built with off-the-shelf RISC CPUs based on router technology optimized for frame switching. Switching is mostly executed by software. When a frame enters the switch, it is synchronized, examined for address information, converted from serial to parallel, and cyclically written to high-speed cache. The software switch searches its address table to find the destination address and establishes the switched connection. Once the switched connection has been established, the frame is read from mem-

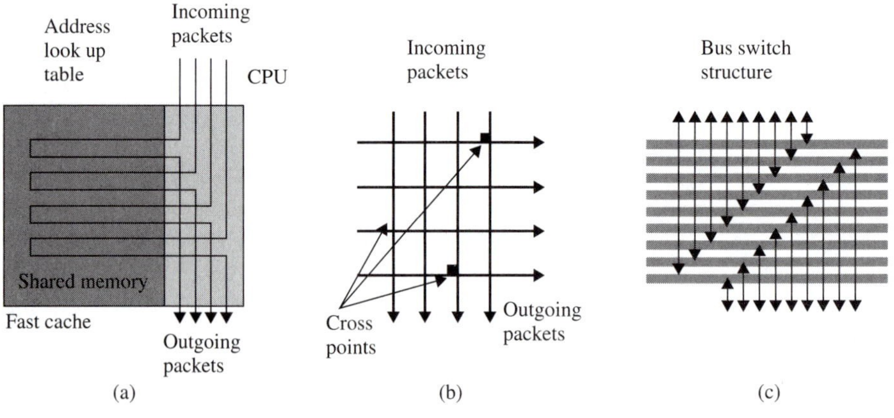

Figure 7.5 (a) Shared-memory switch architecture based on RISC CPU and software design. (b) ASIC-based hardware matrix switch architecture. (c) Bus switch architecture.

ory, reconverted from parallel to serial format, and transmitted via the switched connection (Figure 7.5a). This may sound familiar because it is based on routing technology, or more accurately, the hybrid of bridge and router technology called brouter, which is optimized for frame switching. Software switches do not scale well; performance can drop significantly as more workstations and management features are added. CPU-based switches are inefficient for broadcasts where a single input frame must be placed in multiple output buffers. The CPU must process broadcast frames one by one. Finally, the CPU-based software, running on the CPU and handling shared memory, presents a single point of failure.

Matrix or crossbar switches are based on a straightforward point-to-point hardware structure similar to the old electrical–mechanical telephone exchange switches (Figure 7.5b). As a frame enters the input port, it travels down the matrix until the "intersection" for the right output address is found. It then proceeds through that output port. High-speed electronic gates set up separate connections between MAC addresses for dynamic switching. These connections last only long enough for two ports to transmit a single frame. For its duration, the connection has the full 10 Mbps, 16 Mbps, or even 100 Mbps bandwidth, depending on which protocol—Ethernet, Token Ring, or Fast Ethernet, respectively—is used. The bandwidth need not be shared with the rest of the devices on the network segment. The switch can support several such port-to-port connections simultaneously, and broadcasts are trivial. Matrix switches have a one-to-one mapping from input frame to output frame. As a result, there is no blocking, and matrix switches require a very small overhead for buffering. On the other hand, the one-to-one mapping between input and output complicates the addition of ports or the integration of a fat pipe—a higher-bandwidth input to a matrix switch.

Bus switches (Figure 7.5c) are a more flexible hardware-based architecture than the matrix architecture. They can handle one-to-many and many-to-one transmissions more easily than the matrix switches' inherent point-to-point connection. Expanding a bus switch is relatively easy because the number of input lines does not have to equal the number of output lines. Also, the central bus of the switch provides a more convenient structure for protocol translation than a matrix. The shared-bus architecture is also more manageable because there is one central point over which all the traffic must flow. By using statistical or static time division multiplexing, the switch gives each port its own time slot to send a frame on the bus. Because the time slot is fixed and predictable, this design provides consistent performance under varying load. Generally speaking, shared-bus switches provide a blend of both the software-based switches and the hardware-based matrix switches.

7.2.2 Ethernet Switches

7.2.2.1 Cut-Through and Store-and-Forward

Ethernet switches are the most widely implemented among all the types of switching LAN. Unlike standard-based technologies like ATM, Fibre Channel, and HPPI (high-performance parallel interface), there are no specifications defining how Ethernet switches can control traffic flow. Hence frame loss can be a concern with some specific designs. Ethernet cut-through switches read the destination address before shooting frames onto the network. Transmission starts while the frame is still being received. This type of switch is a good fit for networks that require very low delays, particularly those carrying multimedia applications. The latency of a typical cut-through switch remains constant at about 30 μs regardless of frame size. In contrast, latency of an Ethernet store-and-forward switch, which puts frames into buffers and scrutinizes each incoming frame before the forwarding, increases in proportion to frame size. With a small 32- or 64-byte frame, delays are usually kept within 60 μs or less, which is not a problem. But large frames can have latencies in the milliseconds. However, cut-through switches have a significant drawback in that bad frames can propagate over the network when only the frame header is inspected for address. An Ethernet adaptive cut-through approach lets the switch run in high-performance cut-through mode while it monitors network performance. If a user-specified threshold for bad frames is reached, the switch "adapts" and changes to store-and-forward mode.

7.2.2.2 Classifications of Ethernet Switches

Depending on their functionality and features, Ethernet switches fall into the desktop, workgroup, or backbone classification. Ten to 20 end stations can directly plug into a desktop switch. Workgroup switches sit between isolated desktop switches or intelligent hubs and the rest of the network. They usually support both 10 and 100 Mbps connections and a large number of addresses per port. They typically have two dozen 10 Mbps dedicated ports and one or more high-speed ports for server or backbone links. Each port is connected to 10–20 workstations via a desktop switch or a small hub. To support such configurations, the lookup tables in the switch may contain more than 1000 MAC addresses. Backbone switches are characterized by their modular design and ability to support up to several thousand addresses per switch. The backbone switches can be used to interconnect networks or LAN segments (Figure 7.6).

Some newer workgroup switches also provide ports for token ring and Fast Ethernet connections. The raw power of a switch roughly equals the number of ports multiplied by their speed, for example, 10 Mbps. Hailed as the successor to Ethernet, Fast Ethernet, including 100VG-AnyLAN and Gigabit Ethernet overcomes many of the problems found in higher speed technologies. The 100 Mbps Ethernet is inexpensive, leverages the existing cable plant and management tools, and requires no changes to network operating systems or applications. Some Fast Ethernet autosensing network interface cards can connect network nodes at 10 Mbps and then move to 100 Mbps without having to touch the client. Like other high-speed technologies, Fast Ethernet requires new NICs and switch hubs, and it is available for fewer bus types than the more mature FDDI technology. Some buses simply cannot take full advantage of a channel carrying 100 or more Mbps. The newer Ethernet switches often provide one or two 100 Mbps FDDI or 100BaseT ports to relieve congestion on interswitch or switch-to-server links. Combining a switched 10 Mbps Ethernet with a 100 Mbps Fast Ethernet backbone is a cost-effective way of networking. Today most virtual LANs are implemented with Ethernet switching technology. Powerful servers, LAN switching hubs, dedicated ports, and fast backbones deliver maximum desktop performance at an affordable price.

Switching's strength is in reducing congestion. In terms of performance, a switch has a three- to fourfold improvement over shared-media Ethernet. Therefore, in a client–server

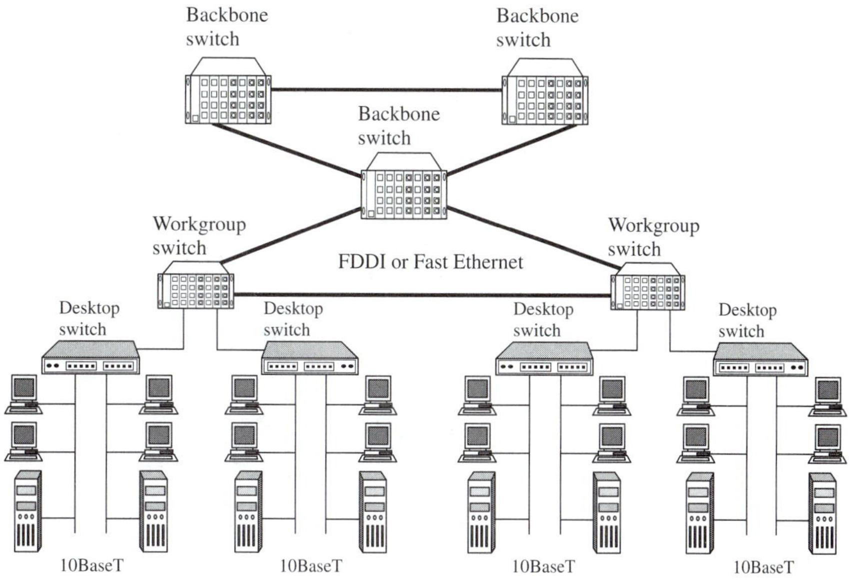

Figure 7.6 Ethernet switch hierarchy.

environment with high average throughput by high-performance workstations, the shared 100 Mbps Fast Ethernet may still perform better than a switched 10 Mbps Ethernet with a 100 Mbps pipe connection to the server or backbone.

7.2.3 Token Ring Switches

It is no wonder that switched Ethernet appeared earlier than its token ring counterpart. When an Ethernet LAN is loaded with end users, the high collision rate causes a deterioration in maximum utilization to 40–70% of available bandwidth. Thus, a 10 Mbps LAN can deliver only 4–7 Mbps. Compare this with a 16 Mbps token ring LAN, which can furnish up to 90% of available bandwidth, or approximately 14 Mbps, even when traffic is very heavy. When Ethernet is fully saturated, most token rings still have bandwidth to spare. However, token ring too is encountering problems. In addition to bandwidth and latency problems, two special problems face virtually every enterprise-scale token ring network: backbone and server–link congestion, and mixed-speed environments.

7.2.3.1 Backbone and Server–Link Congestion

The first problem originates from the source route bridging protocol used by token ring networks. For historical reasons, source route bridging, the standard bridge protocol for token ring networks adopted by the IEEE 802.5 committee, allows a maximum of seven bridges or hops between any two communicating devices. This restriction forced the building of large token ring networks in a strictly hierarchical structure. The extension of the hop limit in source routing from IBM came too late. By the time this extension was released, the seven-hop limit was already built into almost every existing bridge and appli-

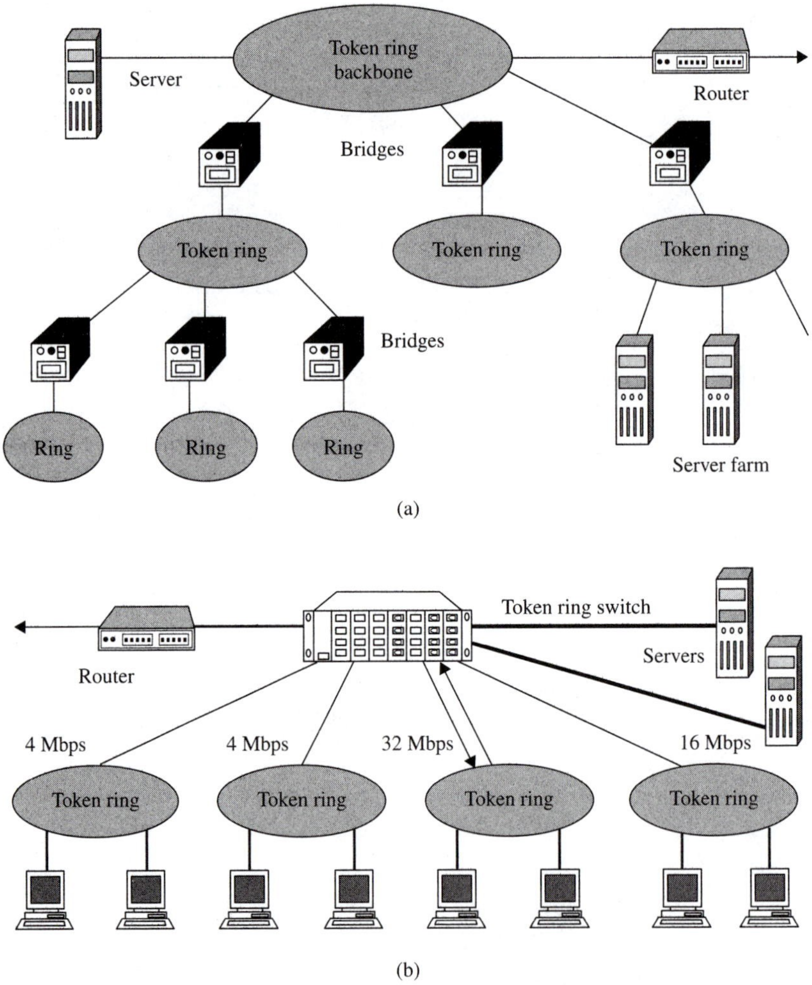

Figure 7.7 (a) A traditional enterprise token ring has a hierarchical structure. (b) A token ring switch replaces the congested backbone and runs at speeds of 4, 16, and 32 Mbps.

cation driver. This design requirement and the resulting hierarchical structure concentrated the traffic in multiple token ring LANs to a few backbone points, which unfortunately run at the same maximal 16 Mbps speed. (Figure 7.7a). Since FDDI was never embraced by IBM, the 100 Mbps protocol with token ring–like topology did not become a popular backbone option for the vast majority of IBM Token Ring users.

By replacing a 16 Mbps physical backbone ring with a token ring switch, the 16 Mbps bottleneck is eliminated and the backbone capacity is effectively raised to the switch's backplane capacity, which is many times greater than 16 Mbps. By adding switches and a dedicated 16 Mbps token ring port to each server in a server farm, where critical servers are clustered in token ring networks, shared bandwidth can be eliminated.

7.2.3.2 Mixed-Speed Environments

The second problem, a mixed-speed environment, is also specific to token ring networks. Unlike traditional Ethernet, which offers a single 10 Mbps speed standard, token ring can run at either 4 or 16 Mbps. But all devices on a ring must run at the same speed. This means that if there is a critical device on the network that runs at 4 Mbps, all 16 Mbps devices must also run at that speed. This is not an unusual occurrence in organization-wide networks. With token ring switches, different ports can run at different speeds to let every device perform at its peak without expensive and often unnecessary upgrades (Figure 7.7b).

7.2.3.3 Enhancing Token Ring Performance

Like their Ethernet counterparts, token ring switches pump up performance by dedicating up to the full bandwidth of the network to individual nodes without having to swap in new adapters or change the current cabling. In this way, token rings not only relieve network congestion but also provide a migration path to ATM.

Most token ring switches rely on a single high-power RISC CPU to handle all forwarding decisions because such switches are cheaper and faster to develop than specialized ASIC-based silicon. While some token ring switches eliminate the need to buffer frames at incoming ports, additional buffering is necessary on the way out because of the token-passing scheme. As token travels around the ring, it stops at each workstation to give it the chance to transmit. This requires the outgoing switch port to hold up the frame until the token has circulated back to the switch. This may cause a significant delay when other nodes are transmitting. The IEEE 802.5 standard—Dedicated Token Ring (DTR), also known as full-duplex token ring (FDX)—eliminates the holding up of frames by eliminating the token altogether. With this technology, each node on the network gets its own dedicated switch connection; the token is disabled. Workstations or servers that employ DTR can communicate at a full 16 Mbps with an attached token ring switch in full duplex. That means that DTR enables devices to transmit and receive data simultaneously at 16 Mbps—effectively doubling the speed to 32 Mbps. This is possible because there are only two transceiver units—the attached station and the token ring switch port—rather than many transceiver units as stations in a traditional token ring LAN. That allows both the end station and the switch port to send data as needed in the transmit immediate mode. Because the switch no longer has to wait for the token to circulate back on the outgoing port, there is no need to buffer outgoing frames. Compared with switched Ethernet, switched token ring does not have the same problem with runt frames, so cut-through switching in token ring does not create any problems. In summary, a token ring switch should have the following features:

- Full-duplex support
- Transparent bridging
- Traffic management
- Ability to upgrade to ATM
- VLAN support

It is difficult to decide whether to go with a switched LAN or to use fast shared-media technologies, such as Fast Ethernet, 100VG-AnyLAN, and FDDI, which offer 10 times the performance of 10BaseT, or even the most advanced switching technology ATM. Each technology has pros and cons. Switching preserves the underlying network infrastructure, access method, and protocol. It also allows the leveraging of existing investments in personnel, training, applications, and management tools. In a client–server environment, where a few nodes (e.g., the file and printing servers) serve as a network's focal point, there is a good chance that the server will become the new bottleneck. Adding a fat pipe—a higher speed connection between the servers and the switch—often becomes inevitable.

7.2.4 FDDI Switches

FDDI was designed as a metropolitan-area network (MAN) technology for network interconnection. It has since found a stronghold in the network backbone market. It is stable, reliable, well understood, and widely deployed. FDDI supports many advanced features. However, since it is more complex and more expensive, it is not economical for desktop applications. When a FDDI backbone of internetworking LANs becomes congested, the best alternative is ATM. ATM is the only scalable high-speed networking technology that spans both LANs and WANs. Nevertheless, for those who currently use a FDDI backbone and cannot wait for ATM, FDDI switches are an effective way to pumps up capacity and preserve investment.

FDDI switches are based on the established FDDI standard. They work with the same FDDI adapters, concentrators, routers, and analyzers. Like the Ethernet LAN switches that preceded them, FDDI switches dramatically improve the performance of shared-media FDDI networks by dividing them into switched, dedicated segments. Generally, FDDI switches are half the cost of FDDI routers and add only microseconds of delay to the network. This is a distinct advantage over the milliseconds of delay for their router counterparts. Although FDDI switches deliver only 100 Mbps of switched capacity compared with 155 Mbps delivered by ATM switches, FDDI switches have higher bandwidth utilization. This is true because FDDI uses a 23-octet header for a frame of up to 4478 octets—a mere 0.5% overhead. In comparison, ATM cells are 53 octets long and use a 5-octet header—almost 10% overhead. Furthermore, the maximum length of a FDDI frame surpasses the maximum frame length of both Ethernet and token ring networks, let alone the short cell length of ATM. The short cell length in ATM requires the segmenting of LAN packets before they travel over the ATM backbone and reassembly on the other end. This makes FDDI more efficient than ATM in transferring long frames. For applications that generate long frames, such as Network File System (NFS), FDDI is a no-compromise solution.

Like other switching technologies, a cut-through FDDI switching port transmits a frame to the destination port as soon as the first 60 octets of the frame has been received, thus reducing the latency to 15 μs. Store-and-forward switching, on the other hand, has a typical 700 μs latency for a 4478-octet FDDI frame. The ASIC-based matrix or crossbar FDDI switches deliver very high throughput with aggregate internal switching capacity of 3–4 Gbps (vs <1 Gbps for a typical shared-memory or time division multiplex FDDI switch).

Like the full-duplex token ring switches, full-duplex FDDI switches allow throughput on a point-to-point FDDI connection between two switches or between a switch and a single end node to be doubled to 200 Mbps. The switch disables the FDDI token and allows a workstation to send frames in two directions at once on dedicated connections.

7.2.5 Switching Network Management

In conventional LANs, hubs provide central points for monitoring networking traffic. In contrast, the ports on a LAN switch, ASIC-based switches in particular, are connected to a switching matrix that provides point-to-point connections between any two ports. Each port can function as a separate LAN segment switching traffic to and from that port only. With this configuration, there is no entry point similar to a shared-media LAN that would allow an analyzer or probe to monitor all traffic in the switch. This imposes special difficulties for LAN management.

The simplest measurement method is to insert a Y-shaped cable linking together the switch, the node under test, and a protocol analyzer. Since most LAN protocols will keep a session alive for the time needed to insert the Y-cable, it can be attached in a midsection without interrupting normal network operation. This inexpensive manual method is commonly used with WAN analyzers. Some low-key traffic reporting approaches use the SNMP agents built into the switches. An SNMP agent works with the management information base (MIB) and reports very basic information such as the number of packets coming into or out of a particular device. But it does not supply more detailed information about broadcasts, multicasts, errors, and the like. One way to improve switching management capabilities is to incorporate analysis ports with embedded remote monitoring (RMON) agents, which can monitor multiple ports simultaneously.

RMON is the network management standard for collecting information about network performance, traffic loads and patterns, faults, and other conditions. RMON probes monitor individual LAN segments, while RMON2 gear tracks events across an end-to-end multivendor network. Dial-up links on RMON probes will allow a third party to link directly into a given probe. This would allow collected data to be fired off directly to the data collector, rather than traversing and perhaps clogging the user's network. With a roving analysis port, a probe or analyzer can be attached to monitor traffic at other ports. The traffic on the port being monitored is then mirrored to the analysis port. The probe provides RMON packet capturing and trend data. Because of the large overhead on memory and performance, an analysis port typically can monitor traffic of only one port at a time. Adding the CPU and memory required to monitor all the ports all the time is cost-prohibitive—calling for a memory add-on of several megabytes just for storing RMON data.

Compared to monitor ports, SNMP's remote monitoring management information base (RMON MIB) and agent approach is more attractive because a full-blown RMON with MIB, and agents can selectively capture important management information at each port. A monitor port must process any and all traffic, which may lead to the bottleneck. Generally speaking, "light monitoring" is preferred for as many segments as feasible, and full monitoring is necessary only on the trouble spots of an enterprise environment. Along with supplying key port statistics such as port status, segment utilization, protocol distribution, number of broadcasts, and collisions for shared-media Ethernet on a per-port basis, RMON agents in switches also supply data on activity or conversations between ports to determine whether bandwidth is being allocated efficiently. For example, per-port statistics might indicate that a server attached to one switch port has sent a million packets to that port, which appears normal, but per-conversation data might reveal that 90% of those packets were sent to a client located across a WAN link. In this case, per-conversation reporting would suggest that the server be relocated to the other side of that WAN link to reduce wide-area traffic.

For conventional LAN internetworking, each LAN segment is assigned its own RMON agent. Switched architectures create special challenges for building RMON agents for the ports. As the number of RMON agents increases, the amount of RMON processing may become overloaded. Distributed management architecture uses RMON subagents to collect data and forward it to a central RMON agent for processing. The central agents then interact with the SNMP management platform. Most current RMON agents are implemented by software and rely on a powerful CPU often shared with the switches. When full RMON agents are installed, management traffic can cause switch performance to deteriorate. The move away from CPU-based architecture toward ASICs will bring faster and less expensive RMON agents to network switching management.

7.3 NON-ATM VIRTUAL LANS

Stations in a LAN discover one another with the help of network protocols using broadcast queries. A network layer device with ports in both LANs, such as a router, facilitates such discovery across two LANs. Since all devices in a LAN receive broadcast messages a very large LAN means excessive number of broadcast messages causing a performance degradation.

Although Ethernet switches by themselves can address some bandwidth performance problems, pure Ethernet switches just create a flat layer 2 architecture that, owing to excessive broadcast propagation and other problems associated with bridge topologies, does not scale beyond a few dozen segments. When VLANs are used to subdivide switched traffic into contained areas, switches bridge, rather than route, traffic destined for different segments within the same virtual LAN. Multiple segments per subnet means fewer routing bottlenecks. Switched virtual networking is a technique that eliminates the logjams associated with a physical LAN topology by creating high-speed switched connections between workstations on different LAN segments. Each defined virtual LAN can include several physical segments per logical subnet, thus bringing flexibility and easing administrative work. A single virtual LAN can connect dozens, even hundreds of LAN users. The ability to include multiple physical LAN segments also gives virtual LANs advantages over multiport routers because each physical segment created by a multiport router must be treated as a separate logical subnet. Traffic going between subnets is subject to a significant added delay because of processing by routers.

VLAN is the broadcast domain of location-independent LANs. Broadcast frames emitted by a workstation are delivered only to the workstations that are members of the same VLAN. This ensures that nonmember workstations are not receiving irrelevant broadcasts. Administrators can populate VLANs with clients and servers that communicate with each other frequently. A well-designed VLAN network system would keep most traffic within the same virtual LAN and would not require routing services for all ports. When necessary, routers interconnect VLANs and filter unnecessary broadcasts between them.

VLANs allow computers to be logically interconnected. Network administrators can define user groups regardless of the physical LAN segment to which they are connected. Users assigned to the same virtual LAN communicate at wire speeds with low latencies and no routing bottlenecks no matter what their physical location in the network. Users on these networks can be close together or separated by a number of switches that create switched virtual circuits linking users together. FDDI, 100-Mbps Fast Ethernet or 100VG-AnyLAN, and 155 Mbps ATM all can serve as the high-speed transport between virtual LAN switches. VLANs provide all the benefits of physical segmentation on logical network segments. They allow efficient separation and management of network traffic and make it easier for administrators to alter the network configuration. The next generation of local backbones will rely heavily on switching and virtual LAN management.

Some VLANs allow switch ports to be assigned to one or more broadcast domains. In essence, this limits certain types of traffic to a specific set of ports. The assignment of ports to VLANs can be dynamic or static based on different protocols or MAC addresses. VLANs also may extend across multiple separate switches. Currently, there are three different approaches for configuring and managing VLANs. Each corresponds directly to one of the lower three layers of the ISO's OSI model. The first approach treats a VLAN as a group of LAN segments; the second views it as a group of Ethernet MAC layer addresses; and the third considers a VLAN to be a group of network layer addresses.

7.3.1 Segment-Based VLAN

At the lowest level, net administrators can create and maintain VLANs based on a group of LAN segments, each of which corresponds to a switch port. The port identifiers are gathered and converted to named groups via a VLAN management application. All workstations on the LAN segments assigned to a VLAN are in the same broadcast domain and can directly communicate with one another. Routers are required for communication between workstations in different VLANs.

The implementation of a segment-based VLAN is relatively simple and inexpensive. No intelligence is needed by the switch to inspect and forward individual workstation frames at the data link and network layers. Once the stations that are to participate in the different VLANs have been determined, the corresponding ports are simply dedicated to those VLANs. A broadcast that comes in a port simply goes out to all other ports in the VLAN. The disadvantage of this type of VLAN is that users cannot freely move around the network because their VLANs are physically limited to a subset of all available LAN segments. A user moving to a LAN segment outside his VLAN has to communicate back with the home server via a router. Furthermore, manual intervention is necessary to set up and maintain each VLAN, which makes it inefficient to support surfing across VLANs for audio, video, and other real-time applications based on the IP multicast Protocol—an increasingly popular addition to the TCP/IP protocol suite.

7.3.2 MAC-Based VLAN

At the data link layer, layer 2 LAN switching is based on a bridged architecture that transmits data using source and destination addresses that are compliant with media access control (MAC) rules. Layer 2 LAN delivers full bandwidth to each port implemented in

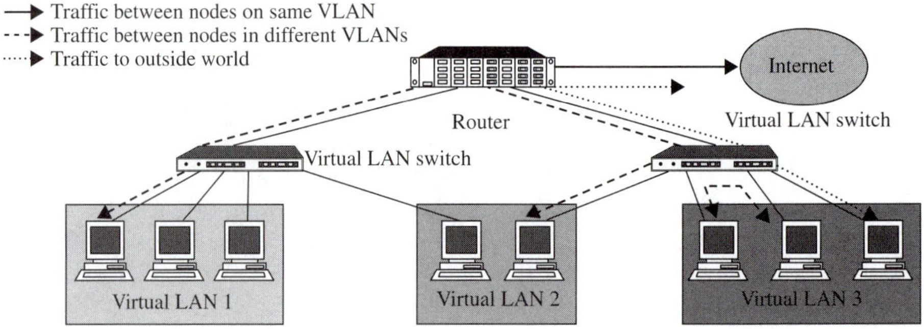

Figure 7.8 Layer 2 switches and router between virtual LANs.

stand-alone multiport boxes and switching hubs. Like all layer 2 schemes, this type of LAN switching by itself cannot establish broadcast and security firewalls. Traffic within a layer 2 VLAN is switched by using MAC addresses, while traffic between virtual LANs is handled by a router that imposes filtering, security, and traffic controls. This limitation makes layer 2 LAN primarily a tool for local workgroups to unite an arbitrary collection of workstations on multiple LAN segments. Routers are used to furnish layer 3 connections and are increasingly deployed between—not within—high-performance workgroups (Figure 7.8).

The MAC-based VLAN allows a workstation to be moved around networks yet remain part of the VLAN. Workstations in different LAN segments can be combined into a single VLAN. Furthermore, workstations in a single LAN segment can belong to different VLANs. When a workstation moves, its VLAN affiliation automatically goes with it because its 48-bit MAC address is still embedded in the computer's LAN interface. This approach to the relocation of workstations is far more flexible than the segment-based method for handling the same situation. But a MAC-address-based VLAN still needs manual configuration because a MAC address is not associated with specific user names, user groups, protocol types, subnetwork addresses, and the like. Besides, users cannot dynamically create MAC-address-based VLANs, making it impractical to implement real-time multicast applications. Most MAC-based VLAN switches are based on the cut-through switching architecture with lower latencies than store-and-forward switches. Such cut-through switches generally lack the intelligence to handle larger virtual LANs. For virtual LANs traversing high-speed backbones, a store-and-forward architecture that dumps frames into buffers and examines each incoming frame is more appropriate.

The standard transparent spanning tree algorithm is often used to handle loops in the layer 2 VLANs. The spanning tree algorithm is defined in IEEE 802.1 to provide distributed routing over multiple LANs connected by bridges. Hardware devices are used to enhance the algorithm for larger VLANs. The maximum number of layer 2 VLANs that can be handled per backbone network ranges from a few hundred to several thousand.

7.3.3 IP-Address-Based VLAN

Based on the network layer, layer 3 VLANs bring basic routing functions to the virtual networking process. Unlike layer 2 VLAN switches, layer 3 switches are protocol-savvy

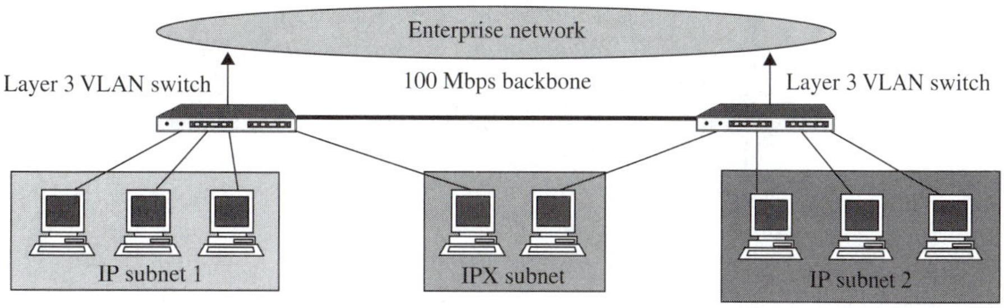

Figure 7.9 An IP-address-based layer 3 virtual LAN includes basic routing functions.

devices that understand the subnet fields of IP and other major network layer protocols. Assigning ports to subnets that correspond to specific protocols configures such VLANs. As with layer 2 VLANs, layer 3 switching ports can be located on the same switch or on multiple switches connected by a backbone. Layer 3 switches use subnet numbers to keep track VLAN traffic across the backbone. Ports associated with one subnet are called a virtual subnet. The terms "virtual LAN" and "virtual subnet" are interchangeable in this context. Traffic within virtual subnets is switched at layer 2. Traffic traveling between virtual subnets is switched at layer 3 without the need for external routers (Figure 7.9).

Every workstation, regardless of protocol, issues broadcasts to request information or announce its presence. The broadcast frame contains the sender's physical address, protocol type, and network (IP) address (for routable protocol). This is everything the network needs to automatically assign the workstation to VLANs. Layer 3 switches read IP addresses and generate Routing Information Protocol (RIP) and Address Resolution Protocol (ARP) messages. They also substitute MAC addresses between subnet hops on the local backbone topology. All this at speeds close to bridging. The IP-address-based switches automatically create VLANs by examining all broadcast and multicast frames. In this way they meet the needs of automatic configuration for VLANs. However, layer 3 VLANs are incapable of handling nonroutable protocols. Layer 3 switches cannot subdivide nonroutable protocols into different virtual LANs. Workstations using the same nonroutable protocol must be placed in the same VLAN, one VLAN for local-area transport (LAT) and one for NetBIOS extended user interface (NetBEUI: see Chapter 11), for example.

Workstations using routable protocols are assigned to VLANs based on type and subnet number: one VLAN for IP subnet 3, another for IP subnet 5, a third for IPX network 8, and so on. Workstations on different segments can be included in the same VLAN, and workstations in the same segment may belong to different VLANs. In other words, IP-address-based VLAN accommodates not only multiple ports per subnet but also multiple subnets per port with much higher performance than MAC-based VLAN. This makes network management much more flexible, allowing for adding, moving, and/or changing of users in a campus network environment. For example, a professor in the Department of Physics moving to an office that is wired into the Department of Mathematics need not change his subnet number or IP address. Because layer 3 VLAN switch ports can accommodate multiple subnets, the physics department's virtual LAN can be simply extended to overlap with the segment that covers the math department.

Filters included with IP-address-based LAN switches can target source and destination addresses, subnet numbers, protocol types, application fields, or any values in a frame. They can block or pass traffic by inspecting user-defined frame fields to fine-tune traffic control inside VLAN firewalls. Protocol-aware VLANs permit greater configuration control of a switch-based network. Network administrators can restrict protocols and network addresses to specific LAN segments and use SNMP-based management applications to acquire and display the global view of the layer 1, 2, and 3 structures of a backbone network.

Automatic VLAN membership based on network-layer attributes also supports IP multicast, which distributes copies of the same traffic to multiple workstations via LAN multicast addresses.

In summary, VLAN is a natural by-product and logical extension of switching technology. In addition to wanting a network performance boost, many users look to VLAN to provide increased network security. Without VLANs, users are given connectivity to the entire LAN to access various servers. A VLAN restricts a user's access to the servers of the VLAN segment. However, VLAN does raise new security issues. For example, if only switch ports, rather than users, can be members of a VLAN, the VLAN will be accessible to anyone using the member's desktop device. Whoever walks up to a member workstation is, by default, able to access the data within that VLAN.

Most existing virtual LANs are based on Ethernet switching. The current technology is still emerging. It has drawbacks, the most important of which is interoperability. Most of today's VLANs are proprietary. They are often tightly coupled to a specific manufacturer's switching fabric and limited to particular networking technologies and protocols. Few VLANs support multiple-protocol LANs with members on, say Ethernet, token ring, FDDI, and ATM. The same holds for logical protocols such as IP, IPX, DECnet, and so on. VLAN takes a big broadcast domain and chops it into smaller domains. While this may enhance security and performance, it sometimes negates the investment made for network management tools. Because most VLANs still must be manually populated via a management console, substantial administrative effort may be required. Also, VLANs may create potential new bottlenecks. The centralized switching works well for small networks but does not scale well. If every user on a large network resides on the same VLAN, the network becomes vulnerable to broadcast storms and security concerns. If several smaller virtual LANs are routed together at a central switch, internetwork congestion quickly becomes a problem. For example, if one workstation belongs to 10 different VLANs, there may well be 10 sets of broadcast traffic flowing across that single switched interface, overloading the switch that physically connects the workstations. The only way to permit efficient scaling is to distribute router intelligence throughout the network. Finally, the current wisdom regarding switched internetworks is *Switch where you can and route where you must*. At present, networks must use routers to link different broadcast domains or VLANs. When packets pass through these routers they slow down, causing a delay of communication between VLANs. Due to the technical complexity, performance, and interoperability issues, the Dynamic Host Configuration Protocol (DHCP) has gained more popularity than VLANs. DHCP fosters mobility by dynamically assigning users IP addresses from a pool of available addresses. It can track mobile users based on their assigned IP address and offers users the benefit of working well in non-switched environments. In addition, the controlled broadcasts provided by flat bridged VLANs are replaced by layer 3 switches.

7.4 ATM VIRTUAL LAN

ATM switches, introduced in Chapter 6, are sometimes called fast packet switches because ATM uses short, fixed-length packets. Touted as the future of networking, ATM brings with it a broad range of promises including LAN/WAN integration, data, voice, and video services, scalable bandwidth on demand, guaranteed levels of performance, and connection-oriented protocols. Hospitals want to use it to deliver X-ray images instantly and directly to the computer displays of individual physicians. Banks and insurance companies look forward to using it to transport the large amount of data they handle. And anyone involved with multimedia applications hopes to take advantage of ATM technology. ATM over optical fiber cable delivers 155 Mbps, and this can scale up considerably. Given the price/performance ratio, ATM is well suited for local-area networking. ATM-based switches will allow high-performance servers to be plugged directly into the switching fabric. ATM also offers a way to tie together different switching devices and protocols.

In terms of desktop deployment, ATM shares the same economic barriers as FDDI. ATM networks require either LAN emulation or that other servers reconcile the connection-oriented nature of ATM with the connectionless services common to most LANs, which means additional overhead, complexity, and cost. Even so, ATM's switching nature makes it the ideal vehicle for constructing high-performance virtual LANs.

7.4.1 ATM LAN Emulation

ATM LAN emulation is a relatively low-level virtual LAN scheme corresponding to the MAC-address-based switched Ethernet virtual LAN. Defining how LAN addresses are mapped to ATM addresses and how to handle control flows, ATM LAN emulation is typically used for both LAN-to-LAN and LAN-to-ATM connectivities. At layer 2, ATM LAN emulation allows bridging between LAN segments only. The ATM Forum and the Internet Engineering Task Force (IETF) are working on MPOA (Multiprotocol over ATM), which will define how to implement routing on ATM networks and make it possible for switched legacy LANs to work with ATM networks.

VLANs based on LAN emulation use an ATM backbone and ATM switched virtual circuits to create high-speed, low-latency broadcast groups for arbitrary collections of end stations on multiple LAN segments. LAN traffic is transported as layer 2 information, while routers are used to link emulated LANs. LAN emulation makes all aspects of ATM, including cell setup and cell segmentation and reassembly (SAR), transparent to the LAN end stations.

LAN emulation defines two major software components: the LAN emulation client (LEC), which acts as a proxy of ATM end stations, and the LAN emulation server (LES), which resolves MAC addresses to ATM addresses. LECs are assigned an ATM address for each attached LAN. They dynamically register the MAC address of locally attached LAN stations with LES. Various devices, including hubs, routers, switches, and ATM-attached end stations, can be implemented with LEC and LES components (Figure 7.10).

When a LEC forwards a frame through the ATM switches to a target LAN station, it sends the LES a MAC-ATM Address Resolution Protocol query message, or a LEC ARP that contains the target station's MAC address. The LES responds to the ARP query

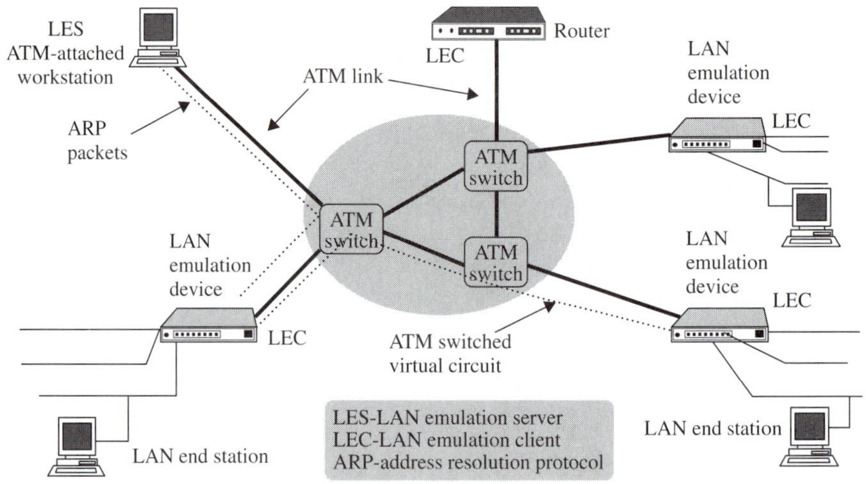

Figure 7.10 In an ATM LAN emulation model, the LAN emulation client (LEC) is the proxy of the ATM station, and the LAN emulation server (LES) resolves MAC addresses to ATM addresses.

with the ATM address of the LEC attached to the target station. The originating LEC then sets up an ATM switched virtual circuit (SVC). When the SVC between LECs is established, MAC frames are converted to ATM cells by standard segmentation and reassembly (SAR) services in each LEC.

A broadcast server is another software component of the ATM LAN emulation structure that forwards LAN broadcast or multicast frames over ATM connections to given VLANs in which LECs can register for each type of broadcast they wish to receive. Like all VLANs, an emulated LAN can be considered to be a zero-hop broadcast containment and security area. Layer 3 subnets can be configured across multiple physical LAN segments. Users can move around within the emulated LAN without changing address configuration. As with LAN switches, ATM virtual LANs using the ATM LAN emulation structure must be manually configured at a management console by grouping physical LAN segments (ports) together. With ATM LAN emulation, routers are typically integrated as a LEC with a single ATM link that contains different virtual connections for each emulated LAN. This is because routers need their own LEC to set up calls across the ATM backbone (Figure 7.10).

ATM LAN emulation is very useful for cable consolidation because a single ATM cable can accommodate many virtual connections for virtual LANs. Attached LANs get full wire speed, which can be dedicated to individual stations or fanned out. ATM links between LEC devices typically run at 155 Mbps with extremely low latencies. With its fast switching virtual connections and meshed architecture, ATM is an excellent tool for overcoming the complexity and limited scalability of switched virtual LANs.

7.4.2 ATM Edge Routers

To overcome the long latencies introduced by traditional routers in ATM LAN emulation virtual LANs, the more advanced ATM edge routers work at layer 3 of the OSI model.

The core of edge router virtual LAN is the ATM router, which is typically constructed by using a high-speed asymmetrical multiple RISC processor architecture with very large shared memory. Each RISC CPU is dedicated to its own tasks: route calculation and management, high-speed forwarding, or ATM connection setup and management. With a look-ahead approach (IETF RFC, 1577 and 1483) that anticipates work currently under way on routing over ATM, some edge router designs avoid the need to define static pipes through the switched structure. This is a deviation from the principle of ATM's dynamic connection and bandwidth allocation. Edge routers weave a fabric of small, modular routers with deterministic latencies of about 50 μs. Edge routers can scale almost indefinitely by exploiting the enormous bandwidth and redundancy of a highly meshed ATM backbone.

As an extension of layer 2 ATM LAN emulation, ATM edge routers map subnet (IP) addresses, as well as MAC addresses, to ATM addresses. This multilayer ARP approach is based on standard ATM multicast services that enable routers to maintain updated routing tables for the open-shortest-path first (OSPF) protocol and for the interior gateway routing and routing information protocols (IGRP and RIP). The multicast feature also allows edge routers to locate LAN subnets that are adjacent to the ATM network.

To locate a LAN IP subnet in an ATM network, the requesting edge router sends out an ARP on the ATM multicast channel dedicated to IP. All edge routers that handle IP receive this ARP query. The remote router that knows how to reach the target subnet responds with its ATM address. After caching the ATM address, the requesting router sets up a switched virtual circuit (SVC) with the remote router. The remote router then forwards the traffic to the target router, which in turn forwards to the local target end station (Figure 7.11).

Virtual subnets (VLANs) are based on layer 3 information and can be used to confine broadcasts and prevent unwanted access. Current router technology allows filters to be installed only at the physical port or segment level. Because ATM edge routers fully integrate routing and switching, virtual subnets can unite collections of physical segments and allow higher level traffic management policies. Sophisticated virtual LANs allow a single subnet to traverse multiple LAN segments on the local backbone. Such subnets are

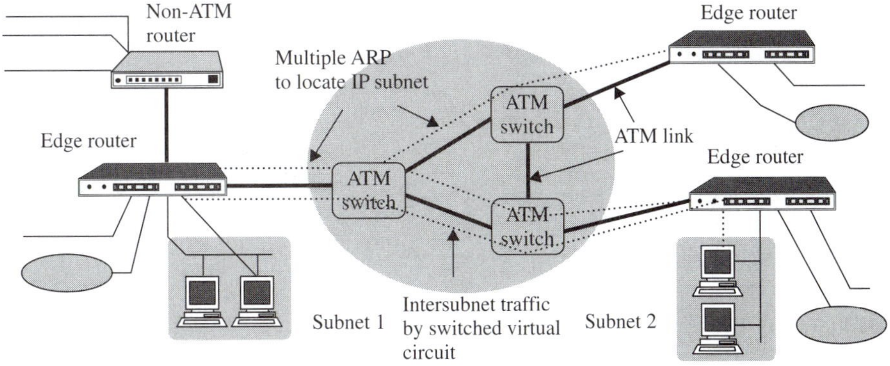

Figure 7.11 An ATM edge router locates virtual LAN subnets by a special ARP. Virtual subnets can be set up across multiple LAN segments anywhere on the backbone.

defined by configuring ports and end stations with standard subnet addresses. ATM edge router VLANs enjoy the same convenience of IP-address- and Ethernet-based VLANs. They also allow devices on different media (FDDI, token ring, Ethernet) to participate in the same virtual subnet. However, like layer 2 VLANs, if users move to segments outside the virtual subnet, the network reconfiguration has to be done manually. Furthermore, since protocols such as OSPF and RIP do not support multisegment subnets, special extensions and treatment are needed.

7.4.3 ATM Virtual Routers

As an alternative to ATM edge router, ATM virtual routers deliver much the same set of features as edge routers (routing, switching, virtual subnets, etc.) but rely on a central route server combined with a number of distributed multilayer switches, which link to individual LANs. The central route server running enhanced IGRP, OSPF, and similar protocols, keeps track of the ATM activities and internetwork topologies, and handles out-of-band, route discovery, and topology updates. The route server even acts as a broadcast server and responds to ARP address queries. Without these chores, the hardware-based intelligent multilayer switches interconnect LANs as fast as a bridge.

Multilayer switches forward on the basis of layer 2 or layer 3 packet fields. As usual, each port on the multilayer switch can be assigned its own subnet address. Multiple ports can also share the same subnet address as with edge routers.

The route server is typically invoked when a session is initiated by a LAN end station. If an IP station tries to log onto a server on a different LAN, the route server intercepts the first packet, locates the target end station, and computes a path between the source and destination. It caches this information and sends a simple forwarding instruction, such as an ATM address, to the initiating multilayer switch. The switch in turn stores the forwarding information in its lookup tables and then uses standard call signaling to set up an ATM connection. If the source and destination end stations happen to be linked to the same multilayer switch, traffic is merely bridged or routed locally; otherwise, the multilayer switch performs MAC address substitution for the layer 3 packet between subnets in the same way as conventional routers do when a packet hops from one subnet to another (Figure 7.12). Once a route has been learned, all subsequent traffic to that destination is forwarded according to information in the multilayer switch's cache tables. Unused cache data are discarded after a specified time period to save resources.

With a virtual router, multilayer switches are the workhorses, while the route server is simply a central route discovery and network management component. Both plug into the backplane of the same chassis. On the surface, an ATM virtual router behaves the same way as a conventional router, performing bridging, protocol processing, routing, filtering, network management, and so on. However, a virtual router shifts infrequent tasks to the central route server while distributing the most frequent tasks to high-speed multilayer switches. This avoids the inefficient software routines of calculating routes on a packet-by-packet basis, and greatly reduces latencies. Furthermore, with conventional routers, two or more hops are required for an end station across the backbone to reach anther end station. The more hops, the more latency and congestion. In contrast, layer 3 packets travel through the virtual router in only one subnet hop, regardless of how many ATM nodes are involved. Virtual routers also help consolidate the management interface.

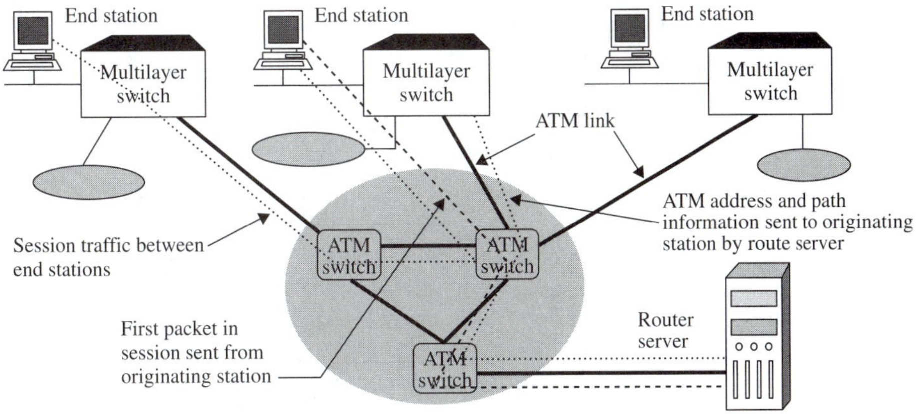

Figure 7.12 An ATM virtual router consists of a central route server and distributed multilayer switches. The route server computes paths and maintains a routing table.

Multilayer switches are low-level hardware components with SNMP agents. The network's active, software-based protocol processing takes place on one or a few centralized route servers, which can be managed with relative ease. Since virtual routers have the same appearance as regular routers, they are easy to interface with existing end stations, routers, bridges, and hubs. Virtual routers also are incrementally scalable: each time an ATM link and a multilayer switch are added, the network gets more forwarding power and more aggregate bandwidth without extra latency.

Finally, virtual routers simplify management but do not eliminate configuration chores. Virtual routers must be manually configured on a port-by-port basis in much the same way as conventional routers. And, as expected, the virtual subnets created by virtual routers must be reconfigured when end users move off the defined group of LAN segments.

7.4.4 ATM Relational Networks

All the approaches to switched virtual networking discussed so far—ATM emulation, ATM edge routers, and ATM virtual routers—are powerful ways to physically and logically structure LANs, but all fall short of *automatic* VLAN configuration. To realize automatic configuration of VLANs, yet retain all the advantages of virtual LANs and virtual subnets, a company introduced a new generation of ATM switched VLANs, called relational networks.

A relational network is built with conventional routers and multilayer relational switches based on a new hardware device that combines a multilayer Ethernet switch and a full-blown five-port ATM switch (Figure 7.13). A critical component of relational switches is the intelligent software that learns the topology of the networks, including subnetwork structure, protocol types, and server locations. The intelligent software, called distributed relation manager, can read layer 3 information to find a packet's source subnet address and the protocol used. The relational architecture inspects traffic and auto-

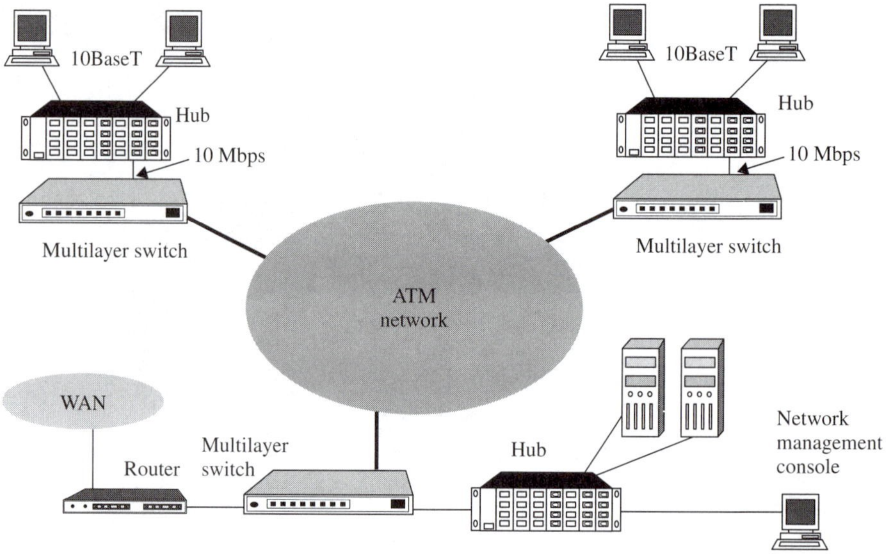

Figure 7.13 ATM relational networks automate configuration for virtual LANs based on protocols and subnet addresses.

matically assigns related groups of users that share a subnet address and protocol to logical subdivisions or relational LANs based on header information. End station moves and changes are automatically sensed and tracked. When a new user is added, it is sensed and automatically included in the right relational LAN. The software also controls and confines broadcast activity to relevant relational workgroups without configuring ports. For example, relational switches will forward NetWare service request broadcasts only to segments with NetWare servers.

Automatically configured virtual subnets can span physical segments without manual reconfiguration. All the end users in the same IP subnet, for instance, are automatically placed on the same relational LAN regardless of their physical location. Also, for each subnet with a single routable protocol, such as IP, NetWare, DECnet, and AppleTalk, and for each unroutable protocol, including local-area transport (LAT) and NetBIOS, a corresponding relational LAN is configured via a relational switch.

With relational networks, firewalls and filtering become smarter. It is possible to instruct relational switches to keep all NetBIOS traffic off a particular relational LAN while admitting IP traffic. This is a higher level filtering for logical user groups than traditional low-level packet filtering based on IP datagram header information and hardware ports, addresses, and protocols.

Although relational networks have built-in multilayer intelligence, they do not manipulate packet headers on a hop-by-hop basis. As mentioned before, conventional routers move traffic among relational LANs in relational networks. They also serve as an on-ramp to the wide-area networks. While the normal routers in relational LANs provide a convenient way of integrating them with installed internetworks, they may also reintroduce routing bottlenecks due to massive traffic between virtual workgroups. The five ATM switch ports can be configured either as user–network interface (UNI) ports for ATM end stations or as network–network interface (NNI) ports for backbones with three different types of backbone topology—daisy chain, ring, and mesh—thereby eliminating a single point of failure. In summary, relational networks use intelligent software and true ATM switching to provide automatic configuration and high throughput with adequate cost-effectiveness.

7.5 IEEE 802.1Q VLAN STANDARD

The IEEE 802.1Q specification establishes a standard method for inserting virtual LAN and MAN membership information into Ethernet frames. The standard also helps provide a higher level of security between segments of internal networks. Sharing VLANs between switches is achieved by inserting a tag field with a VLAN identifier (VID) and/or 802.1p priority information into an Ethernet frame. A VID must be assigned for each VLAN and is numbered between 1 and 4,094. By assigning the same VID to VLANs on many switches, one or more VLAN or broadcast domain can be extended across a large network. The 802.1Q-compliant switch ports can be configured to transmit tagged or untagged frames. If a port has an 802.1Q-compliant device, such as another switch attached, these tagged frames can carry VLAN membership information between switches, thus allowing a VLAN to span multiple switches even across WAN links.

There is a potential problem when 802.1Q-compliant devices are deployed in an older network environment. Network administrators must ensure ports with non-802.1Q-compliant devices attached are configured to transmit untagged frames. Many network interface cards for PCs and printers are not 802.1Q-compliant. If they receive a tagged frame, they may not understand the VLAN tag and will drop the frame. Also, the maximum legal Ethernet frame size for tagged frames was increased in 802.1Q and its companion, 802.3ac from 1,518 to 1,522 bytes. This could cause network interface cards and older switches to drop tagged frames as oversized.

In the case of a network with an ATM WAN, Ethernet switches with ATM uplinks can have a VLAN-to-Emulated-LAN (ELAN) mapping feature that matches 802.1Q VIDs to ATM ELAN names. This offers the benefits of VLAN bandwidth optimization and security being extended between campus buildings or even between remote sites.

VLAN technology is undergoing resurgence in the service provider arena. Web hosting companies are using VLANs to isolate subscriber traffic. MAN service providers also employ VLANs for traffic segregation and security. VLAN technology may also present a solution to the complex wireless LAN (WLAN) roaming problem. The 802.11b wireless LAN standard restricts the transmission range of a single wireless access point within 1000 feet that limits roaming capability. The use of multiple APs enables roaming, but addressing management could become a nightmare because the APs and the client wireless devices associated with the APs would need multiple addresses.

Mobile IP, which uses two IP addresses—a fixed home address and a care-of address that changes at each new point of attachment—is far more complicated and less efficient because of the need for redirecting traffic through the entire network. Creating a VLAN for the APs and the client wireless devices, so they are grouped in one IP subnet, would permit client wireless devices to roam without changing their IP network configuration parameters.

7.6 CHAPTER SUMMARY

Hubs offer cable consolidation and make it possible to physically structure LANs in efficient cost effective designs. Hubs act like an airport hub and disperse data over a network. Routers choose paths for data travel on networks. Today, switches have become networking's linchpin technology.

Ethernet switching and token ring switching enable networks to obtain a boost in bandwidth without upgrading the cable and the network interface for the end stations. FDDI switches, in much the same way, offer dedicated bandwidth for backbone devices. In the future, the difference between cut-through and store-and-forward switching will disappear. The new generation of switches can be programmed to operate in either of the different modes; or the switch can change dynamically between different modes according to network conditions. Congestion control and network management function will also be major features of the new generation of switches. RMON and RMON II functions may be coded into the switches' ASICs.

Switching technology makes virtual LANs and logically structured internetworking possible. VLANs provide LAN segmentation based on how users work together, rather than where they are located. VLAN supports dynamic workgroup configuration and automates changes when users move to new physical locations. This gives managers the ability to create "project groups" and "logical workgroups" with minimal administration and intervention. The scalable VLAN architecture can keep pace with organizational growth as new users are added and new LANs created.

Applications involving image transfer, videoconferencing, and streaming audio/video have put new demands on networks for still higher bandwidth and shorter latencies, motivating continuing migration to newer generations of switching technologies. The migration starts from frame relay moving towards SONET, ATM, and Dense Wave Division Multiplexing (DWDM) for trunk; ATM, Gigabit and 10 Gigabit Ethernet for the backbone; and DSL and cable modem for the last mile. Still, these so called "rich media" applications demand better "traffic engineering."

The implementation of ATM cell switch-based virtual networks largely depends on how quickly vendors resolve interoperability issues across large-scale ATM networks. The Multiprotocol Label Switching (MPLS) is a key development in Internet technologies that provides all IP networks with better traffic engineering, different qualitative Classes of Service (CoS) and quantitative Quality of Service (QoS). MPLS also lays foundation for IP based Virtual Private Networks (VPNs) and addresses the scaling issues faced by the Internet as it continues to grow. (More discussion on MPLS VPN in Chapter 10.)

7.7 PROBLEMS

7.1 Does a hub simulate a bridged LAN with one computer per segment? Explain.

7.2 What are the advantages of using a switch instead of a hub?

7.3 Compare and contrast a shared-medium network and a switched network. What are the relative advantages and disadvantages of each?

7.4 Suppose a bridged network contains three Ethernet (10 Mbps) segments connected by two bridges, and each segment contains one computer. Consider two computers that want to communicate to each other and two computers that want to communicate with a third. What are the maximum and minimum data rates in each case?

7.5 You need to design an Ethernet 10BaseT LAN for a firm with two large rooms in a building. The first room consists of a main hallway with one office; the second room, used for conferences and seminars, is located 150 feet from the first room on the same floor. You may only use 10BaseT technology. Assume that the firm will use 12-port 10BaseT hubs. In the main hallway the single office has 12 PCs. The office in the main hallway is a rectangle 45 feet long and 30 feet wide. The conference/seminar room is square, with 40-foot sides, and has 5 PCs. Draw a diagram of this floor plan showing the devices and cabling you would use to connect the PCs into a LAN. From the following information, calculate the cost of connecting the PCs into a LAN. Explain why your 10BaseT solution would be undesirable.

10BaseT cable costs (UTP): \$0.05/foot

Ethernet 10BaseT hub: \$20/port

NIC for PCs: \$15/card

Installation costs: \$80/PC

You may assume that the average cable run/length for each PC is 12 feet in the office and 16 feet in the conference room.

7.6 Consider two token ring network configurations. In the first configuration there are 24 stations connected in a ring, whereas in the second configuration there are three rings of 8 stations each connected via a three-way bridge. Sketch each configuration. Which configuration is better, and why?

7.7 Assume that three token ring LANs are connected by one bridge and each ring has 100 stations. Also, assume that each link (and its interface element to the station) has a probability of failure p.

(a) Sketch the system, showing the connection of at least one station and its interface on each ring.
(b) What is probability that each ring in part a may fail?
(c) What is the overall system failure probability if the bridge failure probability is q?
(d) If all 300 stations are connected on one ring, for what values of p will the second ring perform better than the first set of three rings sketched in part a if q is considered to be 0.0001?

7.8 Consider two LAN bridges, both connecting a pair of IEEE 802.4 networks. The first bridge is faced with 1000 512-byte frames per second. The second one gets 200 4096-byte frames per second. Explain which bridge will need the faster CPU, and why.

7.9 A department has three Ethernet segments, connected by two transparent bridges. Assuming that the department decides to change the network to token ring, what will happen if the network ends are connected via a transparent bridge to form a closed ring? What are the maximum and minimum data rates?

7.10 What are some of the advantages of using hubs?

7.8 REFERENCES

BOOKS

Black, D. P., *Managing Switched Local Area Networks: A Practical Guide*. Reading, MA: Addison Wesley, 1997.

Breyer, R., and S. Riley, *Switched and Fast Ethernet*. Emeryville, CA: Ziff Davis Press, 1996.

Brown, S., *Implementing Virtual Private Networks (VPNs)*. New York: McGraw-Hill, 1999.

Covill, R. J., *Implementing Extranets: The Internet as a Virtual Private Network*. Oxford: Digital Press, 1998.

Doraswamy, N., and D. Harkins, *IP Sec: The New Security Standard for the Internet, Intranets and Virtual Private Networks*. Englewood Cliffs, NJ: Prentice Hall, 1999.

Fowler, D., *Virtual Private Networks: Making the Right Connection*. San Diego, CA: Academic Press, 1999.

Hein, M., D. Griffiths, O. Berry, and E. Littwitz, eds. *Switching Technology in the Local Network*. Cincinnati: Thomson Executive Press, 1997.

Held, G., *High-Speed Networking with LAN Switches*. New York: Wiley, 1997.

Held, G., *Virtual LANs: Construction, Implementation, and Management*. New York: Wiley, 1997.

Kosiur, D., *Building and Managing Virtual Private Networks*. New York: Wiley, 1998.

Lewis, C., *Cisco Switched Internetworks: VLANS, ATM and Voice/Data Integration*. New York: McGraw-Hill, 1999.

Minoli, D., J. Doyle, and A. Schmidt, *Network Layer Switched Services*. New York: Wiley, 1998.

Murhammer, M. W., T. A. Bourne, T. Gaidosch, and C. Kunzinger, eds., *A Guide to Virtual Private Networks*. Englewood Cliffs, NJ: Prentice Hall, 1999.

Roese, J. J., *Switched LANs: Implementation, Operation, Maintenance*. New York: McGraw-Hill, 1998.

Smith, M., *Virtual LANs: A Guide to Construction, Operation, and Utilization*. New York: McGraw-Hill, 1998.

WORLD WIDE WEB SITES

Information about LAN switching
http://www.netlab.ohio-state.edu/~jain/cis788-97/lan_switching/index.htm

Cisco VLAN Roadmap with whitepaper on VLAN
http://www.cisco.com/warp/public/538/7.html

VLAN information from Ohio State University
http://www.netlab.ohio-state.edu/~jain/cis788-97/virtual_lans/index.htm

VLAN Technology Report from 3Com
http://www.3com.com/nsc/200374.html

Yahoo's VLAN page
http://dir.yahoo.com/Computers_and_Internet/Communications_and_Networking/LANs/Virtual_LAN__VLAN/

8

NETWORK PERFORMANCE

8.1 WHY STUDY NETWORK PERFORMANCE?

A computer network is defined as a "collection of autonomous computers" in which the computers are capable of sending and receiving the messages independently. To satisfy the demands of users, designers, and service providers, the network is expected to meet certain requirements, and the basic requirement is connectivity. Without a connected network, the computers cannot communicate with one another. This basic requirement leads to different network topologies such as bus, star, ring, and tree.

Another network feature is the capability of sharing resources. A network allows the sharing of disk space, printers, and other hardware/software. In addition to being cost-effective, resource sharing allows the incremental growth of the network and provides leverage for the current investment. The network designer can effectively increase the network size without replacing currently utilized resources. Resource sharing is possible because of such advances in multiplexing schemes as time division multiplexing (TDM), frequency division multiplexing (FDM), and wavelength division multiplexing (WDM).

When Arpanet started in early 1970s, many researchers and scientists analyzed the performance of TDM and FDM and proposed efficient protocols to overcome some of the weaknesses. These analyses consisted of computer-based models to determine network throughput and delay under hypothetical workloads. In computer-based models, network-like conditions are produced through computer simulation programs. Many of the current developments in the field of computer networks are attributed to network performance modeling. The benefits can be seen from all three perspectives mentioned earlier: user, provider, and designer. A network user would be interested in having the network functional, with process-to-process connectivity between the hosts. The network should be able to support message streams for video and audio applications. With limited availability of bandwidth, the network designer and provider may not be able to offer complete support for user demands. However, partial support must be provided, with the best possible scenario of resource utilization. This calls for proper network planning and design. A well-designed network provides a reasonably controlled delay for user applications.

One of the most important characteristics of network design is its architecture, which includes the client/server placement in the network. For example, the disk-oriented services in the client will always put more loads on the server and the network than the services requiring occasional server access. In addition, if a WAN connection is provided through the server, heavy traffic is expected to pass through it. Placing multiple servers on the network with dedicated tasks may solve this problem. However, network capacity planning is required to minimize the delay while maximizing the throughput in the network. This in turn demands evenly distributed loads on the links, resulting in lower link utilization. The techniques developed in analyzing network performance help the designer achieve this goal.

There are three types of delay in any network: propagation, transmission, and queuing. Depending on the type of network, one delay may be more prominent than the others. For example, in a fiber-based LAN operating at 10 Mbps, the transmission delay for 1000 bits will be 100 μs, propagation delay between two stations separated by 100 meters will be $100/(2 \times 10^8) = 0.5$ μs, where 2×10^8 is the speed of light in fiber. However, the queuing delay will depend on the protocol used. If an Ethernet-based access mechanism is used, the delay will depend on the active number of stations. Theoretically, the queuing (or buffering) delays in Ethernet may reach infinity under very heavy loads. On the other hand, the delays in a token passing bus or ring may still be finite under heavy loads, but the cost may be higher.

Consider another fiber-based network that contains two nodes separated by 1000 km and sending 1000-bit frames to each other at 10 Mbps. The transmission delay remains the same, but the propagation delay will be 5 ms, which is considerably higher than the preceding case. Also, assuming that the second example illustrates a WAN, the queuing delay will depend on the path taken by each packet and the conditions at intermediate nodes. If the path is congested, then the queuing delay will be higher than the one with less congestion.

Notice that the propagation delay plays an important role for interactive types of connection and may be less important for stream traffic if enough buffering is provided. However, the queuing delay depends on many factors, including the line bandwidth. If the line bandwidth is small, more queuing is expected at the node, since the time to transmit the packet will be higher. Therefore, the queuing and transmission delays may play an important role for both stream and interactive types of traffic. This chapter deals with the subject in detail. Network performance evaluation techniques are discussed to assist readers in determining throughput, delay, and other performance metrics. Analytical as well as simulation techniques are explored and applied to studies of networks of different types.

8.2 ANALYTICAL APPROACHES

The analytical approaches to modeling the network performance are based on queuing theory and probability. The complex queuing phenomena taking place in a network's queues (or buffers) make it very difficult to correctly determine the queuing delay. However, the estimates obtained are very close to the actual measurements and simulation results. We will now look at some of the analytical approaches used to model the network. The section includes several practical examples to illustrate the application of analytical approaches.

8.2.1 Delay Throughput Analysis

As an example of delay throughput calculations, we provide an analysis of the token bus protocol. As illustrated in earlier chapters, in a token bus scheme, a token controls the right to access the medium. Thus the station that holds the token has temporary control of the medium. The token is passed among the active stations attached to the bus. As the token is passed, a logical ring is formed on the bus. We define logical neighbors as the stations involved in token transfer, and thus they are not necessarily physical neighbors. Depending on layout of the network, we can consider two extreme cases: the best case, where the logical neighbors are also the physical neighbors, and the worst case, where the logical neighbors are physically separated by a maximum ring distance.

We assume a fixed logical ring in this model; that is, either the ring is static or, if the ring is dynamic, no addition or deletion of stations is possible. Also, we assume a nonexhaustive service, in which a station after transmitting one awaiting message passes the token (to the next station). If there is no message waiting for transmission, the token is passed without being used. The channel utilization of token bus may be evaluated as a function of the useful data transmission time F, the overhead Y, and the activity factor K. The activity factor is the ratio of the number of active stations to the total number of stations.

For example, in a 10 station network, if there are 3 active stations in a round of token passing, $K = 0.3$; if there are N active stations out of a total of M stations in a round of token passing, C will be

$$C = \frac{NT}{NT + Y}$$

Where Y, the delay overhead in a round, may be evaluated in terms of the end-to-end propagation delay σ. Assuming that the token is passed sequentially on the bus in the same logical order as the physical order of the stations, the total propagation overhead in one complete round will be 2σ. The token passing overhead in one round will be Mt, giving us

$$C_b = \frac{NT}{NT + 2\sigma + Mt} = \frac{1}{1 + 2a/KM + q/K}$$

where a and q are the normalized end-to-end propagation delay and token passing delay, respectively, taken with respect to T, and C_b is the utilization obtained with the best possible order of token passing. However, the actual utilization may be lower owing to a different order of token passing. The lower bound on utilization will be obtained for the worst order of token passing from station $1 \rightarrow M \rightarrow 2 \rightarrow M - 1 \cdots \rightarrow M/2 \rightarrow 1$. If the stations are spaced equally on the bus, then the overhead Y will be:

$$Y = \frac{\sigma}{M - 1}\left[\sum_{i=1}^{M-1} (M - i) + \frac{M}{2}\right] + Mt = \frac{\sigma M^2}{2(M - 1)} + Mt \tag{8.1}$$

The corresponding channel utilization may be found as follows:

$$C_w = \frac{NT}{NT + \dfrac{\sigma M^2}{2(M - 1)} + Mt} = \frac{1}{1 + \dfrac{aM}{2K(M - 1)} + \dfrac{q}{K}} \tag{8.2}$$

We notice the effect of the token passing mechanism in Equations (8.1) and (8.2). The overhead in the best case varies as a function of K and M and has a significant effect when $KM/2$ approaches the normalized end-to-end propagation delay a. In the worst-case token passing order, the overhead due to the token passing order has a significant effect when $2K(M - 1)/M$ approaches a. For $K = 0.5$, for example, the best-case token passing order will cause little overhead, since $M/4$ will be usually much higher than a. For same value of K and large number of stations, however, the token passing overhead will reach a, showing the significance of end-to-end-propagation delay. Figures 8.1 illustrates the dependence of best- and worst-case channel utilizations on a and K. Figure 8.2 shows the dependence of channel utilizations on K. The average token rotation time in terms of the overhead Y and the packet transmission time T may be computed as $NT + Y$, since N frames are transmitted in each round of token passing on an average, with a total overhead time of Y. The token rotation time is a dynamic number, which may change from one round to another depending on the number of active stations N.

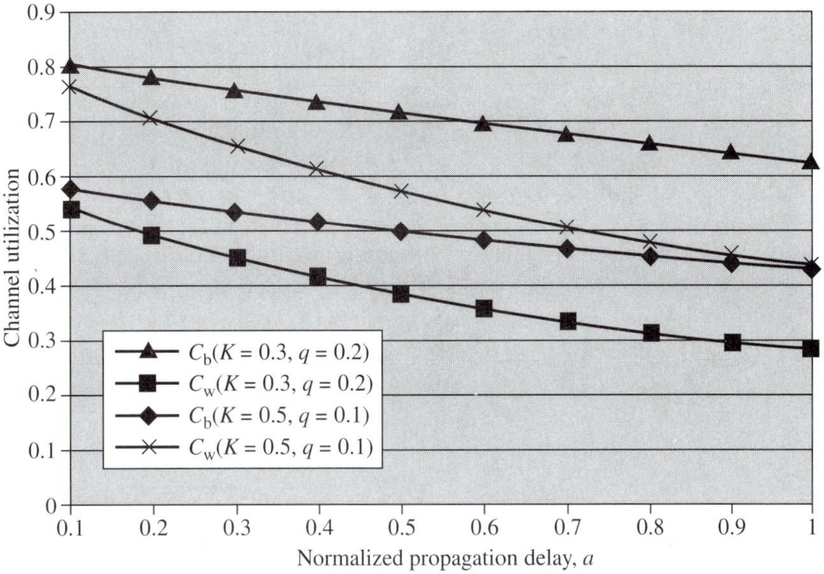

Figure 8.1 Effect of a on channel utilization for $M = 10$.

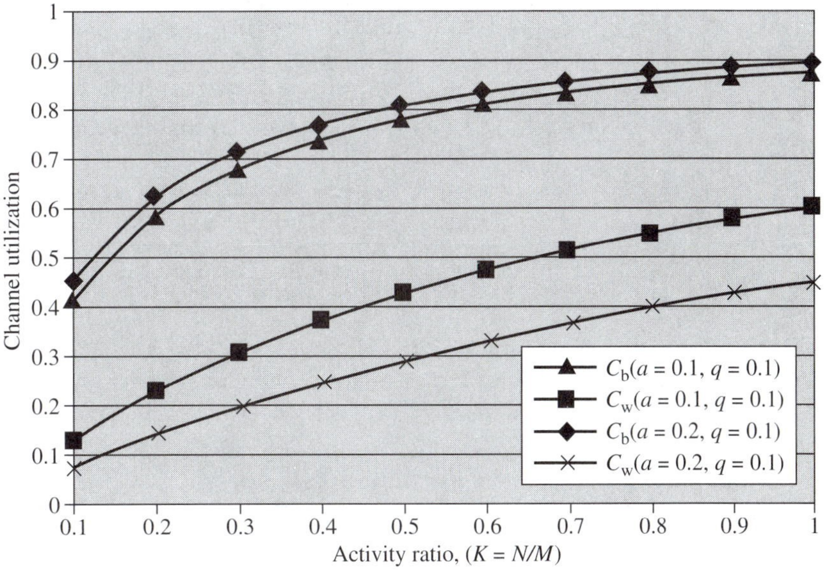

Figure 8.2 Effect of K on channel utilization for $M = 10$.

Example 8.1

Consider a 30-station, 10 Mbps token passing bus system in which there are 15 active stations on average in any round of token passing. If the token is passed explicitly as a separate control packet with 10% size of data frames, what are the channel utilizations and average token rotation times for the best and worst cases of token passing? Consider the data frame to be 1500 bytes and the end-to-end propagation delay to be 10 μs.

Solution: The given values are $M = 30$, $N = 15$, i.e. $K = 0.5$. The normalized token passing delay is 10% of the frame time, that is, $q = 0.1$. The end-to-end propagation delay is 10 μs. Comparing with the packet transmission time T ($1500.8/10 \times 10^8 = 12$ μs), we get $a = 0.833$. From the channel utilization equations for best- and worst-case orders of token passing, we get

$$C_b = \frac{1}{1 + \dfrac{2 \times 0.833}{15} + 2} = 0.763$$

$$C_w = \frac{1}{1 + \dfrac{0.833 \times 30}{29} + 0.2} = 0.485$$

With channel utilizations of 0.763 and 0.485, the effective bit rate is between 7.63 and 4.85 Mbps for any arbitrary order of token passing. The overhead caused by token passing in each round is 36 μs (1.2 μs per token for 30 stations) on an average, whereas the overheads due to propagation are 20 μs (best-case overhead $= 2\sigma$) and 155.17 μs [worst-case overhead $= \sigma \times 30 \times 30/(2 \times 29)$]. The average token rotation time is the useful packet transmission time per round ($NT = 15 \times 12 = 180$ μs), plus the overhead time (best case: $36 + 20 = 56$ μs; worst case: $36 + 155.17 = 191.17$ μs). We also note that the channel utilization may be obtained using the delays as follows:

$$C_b = \frac{180}{(180 + 20 + 36)} = 0.763$$

$$C_w = \frac{180}{180 + 155.17 + 36} = 0.485.$$

The effect of propagation delay is noticeable in this example. As the token passing order changes to worst case, the resulting propagation overhead becomes about eightfold. However, the token passing delay does not cause any significant effect because it stays constant at 36 μs.

8.2.2 Probability Techniques

Network connectivity can be determined by using the principles of combinatorics from probability theory. Combinatorics is the art of accumulating the probabilities of several events to determine the overall system probability. For example, consider a coin toss with

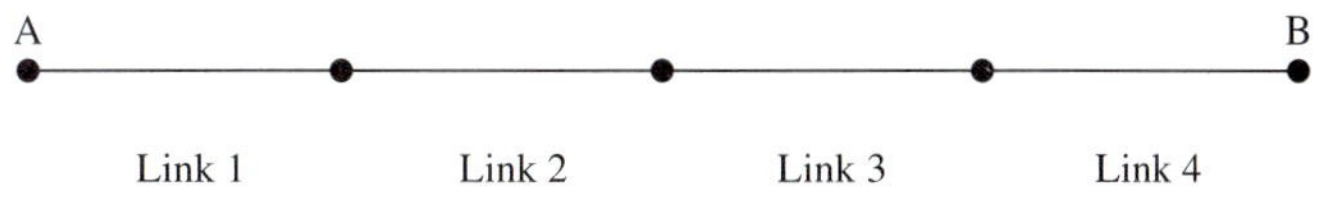

Figure 8.3 A four-link (5-node) network.

50% probability (say $p = 0.5$) of getting a head or a tail. Now assuming the toss method is perfect, we can determine the probability of two consecutive heads (HH), a head followed by a tail (HT), a tail followed by a head (TH), or two consecutive tails (TT). Each of these four events will have 0.25 as the probability of occurrence, that is, $p*p$, $p*q$, $q*p$, and $q*q$, respectively. Here, we assume that p is the probability of getting a head and $q = 1 - p$ is the probability of getting a tail. Now consider an imperfect coin with $p = 0.7$ and $q = 0.3$. The respective probabilities would be 0.49, 0.21, 0.21, and 0.09. Notice that the sum of these probabilities would always be one, which is sometimes a big help in determining the unknown probability in an event space.

Now let us apply this concept to determine the connectivity probability of a simple network. Consider a communication system connecting points A and B via four links, as shown in Figure 8.3. We will determine the probability of no connection between A and B, if each link may fail with probability p. We assume that the link failures are independent of one another. From the binomial probability density function, the probability of no connection is the probability of one or more link failure(s) between A and B. The density function would contain terms p^4, p^3q, p^2q^2, pq^3, and q^4, where $q = 1 - p$. Notice that the sum of these probabilities would be 1. So, the required probability may be simply $1 - q^4$. If $p = 0.1$, we would write

$$\begin{aligned}\text{Probability of no connection} &= 1 - \text{probability of all links operating}\\ &= 1 - (1 - p)^4 = 0.344\end{aligned}$$

Example 8.2

Consider a communication system using four links to connect points A–E as shown in Figure 8.4. What is the probability of no connection between A and C if each link may fail with probability 0.05? Assume that the link failures are independent of one another.

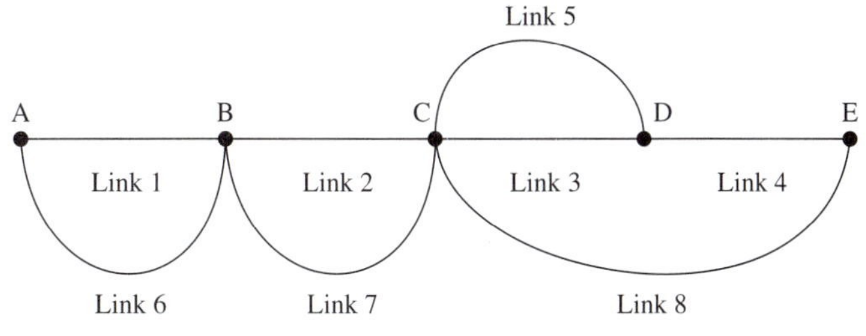

Figure 8.4 Network for Example 8.2.

Solution: From the binomial probability density function, the probability of no connection between A and C is the probability of no connection between A and B (P_{AB}) and the probability of no connection between B and C (P_{BC}). Let p (given as 0.05) be the probability of each link failure. We have

$$P = \text{Probability of no connection between A and C} = 1 - (1 - P_{AB})(1 - P_{BC})$$

where $P_{AB} = P_{BC} = p^2$. Therefore,

$$P = 1 - (1 - p^2)^2 = 0.005$$

Only links 1, 2, 6, and 7 will have an effect on P, since other links do not form the path between A and C.

Let us now consider a model in which we apply combinatorics techniques similar to those just discussed. In combinatorics, we use combination of probability terms from the distribution to determine probabilistic and other useful network-related measures. For example, the probability of no connection in Examples 8.1 and 8.2 is obtained from the binomial distribution, using p and q as the parameters. By using the distribution, we can obtain the probability of m connections if n paths are provided between two points. This kind of system is called an m-out-of-n system. Many physical systems may be classified as m-out-of-n systems. In computer networks, the connectivity of networks may be studied by using this method. As an application of the technique, we analyze the connectivity probability of a light wave network. Fiber optics are becoming media of choice because of the high bandwidth. Also, wavelength division multiplexing (WDM), used for fiber optic media, is claimed to incur minimal losses. We need analytical models to explore the various multichannel light wave network architectures. With the advances in optical computer architectures, it may be feasible in the near future to design novel multichannel light wave network architectures by using many independent high-speed channels simultaneously operating and connecting selected access stations, resulting in a new multimedia telecommunication infrastructure.

Consider a light wave multichannel local-area network (MLAN) in which the WDM technique has been used to share b parallel broadcast channels to which N stations are connected. An 8-node MLAN with 16 shared channels is shown in Figure 8.5. This network is also called a shufflenet network, in which each station is connected to eight incoming lines and eight outgoing lines. To communicate between the stations, one tunes the transmitter and receiver interfaces of the stations to two incoming or two outgoing wavelengths. If b channels are assumed to be connected to each station in MLAN, then the connectivity probability between different nodes may be determined as a k-out-of-b system, where k is the lowest number of channels required between the nodes. Another representation of b channel MLAN for N stations is shown in Figure 8.6.

One application of the probability method can be seen by using the system capacities. Assume that each channel in an MLAN has a bandwidth of W/b bps and the time to transmit a packet on the ith channel is $T_i = bT_0$, where $i = 1, 2, \ldots, b$ and T_0 is the time needed to transmit a packet on a channel with bandwidth W. Also assume that the channel end-to-end propagation delay is A seconds. The normalized propagation delays (with respect to T_i, $i = 0 \cdots b$) are $a_i = A/T_i$.

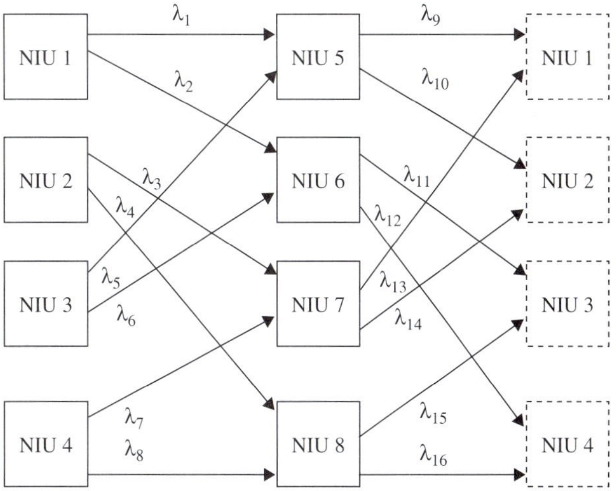

Figure 8.5 A Shufflenet interconnection network modeled as a *k*-out-of-*b* MLAN.

Now consider that the total data traffic offered to the broadcast system, including new and rescheduled packets, is assumed to follow Poisson distribution at a rate of γ packets per second. Measuring time in T_0 units, the offered traffic (traffic during one packet transmission time or normalized traffic) is $G = \gamma T_0$. Assume that channels are randomly chosen, independently of their state, with the traffic offered to the ith channel being Poisson-distributed with rate γP_i, where P_i is the probability of choosing channel i.

The bandwidth loss and fault tolerance of the MLAN may be defined as the expected loss in system's data carrying capacity (in bits per second) due to unreliable or fault-prone interconnections. For normal operation of the system, we assume that W must be at least equal to W_r (the minimum required bandwidth to run the system without any loss of data or excessive delay of traffic). That is, if $W < W_r$, the available bandwidth may not be sufficient to handle the offered traffic. The fault tolerance (n) of the system may be defined

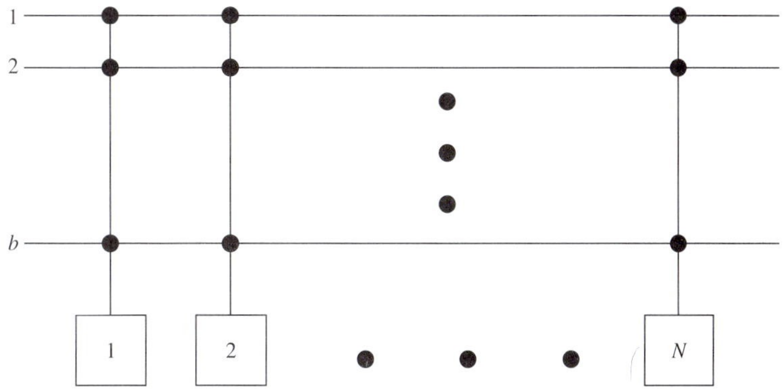

Figure 8.6 A multichannel LAN with b channels and N stations.

as the allowable number of channel failures without any bandwidth loss. In a b-channel system, we assume that the bandwidth W is shared equally between the channels. Thus for a functioning MLAN that can tolerate up to n failures, we have the lost bandwidth due to failure as nW/b. Thus $W - nW/b \geq W_r$ or $n \leq (W_r W)b$. For example, in a 10-channel MLAN ($b = 10$) with total bandwidth of 10 Mbps, we have channel fault tolerance of 3 or less if the required bandwidth is 7 Mbps.

If q is the probability of failure of an channel in an MLAN, the probability that at least one channel is available for communication is $1 - q^b$. We can get the expected bandwidth loss for $W \geq W_r$ by first considering $W = W_r$. In this case, the available bandwidth is sufficient under normal conditions and there is no bandwidth loss if there is no failure. On the other hand, for $W > W_r$, channel failures may or may not cause a bandwidth loss.

Case 1: $W = W_r$. When the required bandwidth is equal to the available bandwidth, any failure would cause loss of bandwidth. The expected bandwidth loss for k failures is given as follows:

$$L = \sum_{i=1}^{k} \binom{b}{i} q^i p^{b-i} \; iW/b$$

Here $p = 1 - q$. For $k = b$, we have

$$L_{k=b} = qW \sum_{i=1}^{b} \frac{(b-1)!}{(i-1)!(b-1-i+1)!} q^{i-1} p^{b-1-i+1} =$$

$$qW \sum_{i=0}^{b-1} \frac{(b-1)!}{i!(b-1-i)!} q^i p^{b-1-i} = qW(q+p)^{b-1} = qW$$

Case 2: $W > W_r$. As mentioned earlier, the system may or may not tolerate faults in this case, since the required bandwidth is less than the available bandwidth and channel failures may or may not cause loss of data packets. If n channels fail, the bandwidth outage would be nW/b as already shown. Thus the available bandwidth would be $(W - nW/b)$. For a system to have no bandwidth loss, we may tolerate up to n channel failures if $(W - Wn/b \geq W_r)$ or $W/W_r \geq b/(b - n)$. The probability of bandwidth loss would be

$$P = 1 - \sum_{i=0}^{n_{\max}} \binom{b}{i} q^i p^{b-i}$$

where $n_{\max}$ is the maximum value of n or the maximum fault tolerance without any bandwidth loss. If we allow more than $n_{\max}$ failures, the bandwidth loss for up to k failures may be calculated as follows:

$$L = \sum_{i=n_{\max}+1}^{k} \binom{b}{i} q^i p^{b-1} \frac{i - n_{\max}}{b - n_{\max}} \frac{iW}{b}$$

Here, iW/b is the amount of bandwidth loss for i channel failure(s). The factor $(i - n_{\max})/(b - n_{\max})$ determines the fraction for the loss from the available $(b - i)$ channels out of $(b - n_{max})$ required channels; that is, $(b - n_{max}) - (b - i) = i - n_{max}$. Here, i takes the value between $(n_{max} + 1)$ and b. As an example, consider the total available bandwidth of 8 Mbps shared equally among 8 channels, in which the required bandwidth is 5 Mbps or the number of required channels is 5. Here $W/W_r = 8/5$ with $n_{max} = 3$ and $k = b = 8$. We get the expected bandwidth loss and the probability of bandwidth loss as follows:

$$L = \left[\binom{8}{4}q^4 p^4(4/5) + \binom{8}{5}q^5 p^3(10/5) + \binom{8}{6}q^6 p^2(18/5) + \binom{8}{7}q^7 p(28/5) + q^8 8\right]\frac{W}{8}$$

$$P = 1 - \left[\binom{8}{3}q^3 p^5 + \binom{8}{2}q^2 p^6 + \binom{8}{1}qp^7 + p^8\right]$$

The first term in the evaluation of L shows the probability of 4 channel failures out of 8 channels. The factor $4/5(W/8)$ indicates that the expected bandwidth loss for 4 channels will be $4W/8$ and the loss factor is 1/5, since one out of 5 required channels has failed. The loss factor changes to 2/5 in the next term, since two required channels have failed. The loss factor increases until it reaches one (5/5) for the last term, where we see that all required channels (5 out of 5) have failed. For $q = 0.1$, we get the expected bandwidth loss as 0.0057 Mbps. However, with the same error probability, if the number of required channels is increased to 7, the expected bandwidth loss is about 73 times higher at 0.417 Mbps, that is:

$$L = \sum_{i=2}^{8}\binom{8}{i} q^i p^{8-i}\left(\frac{i-1}{7}\right)i$$

Figure 8.7 plots the maximum system fault tolerance (n_{max}), using the channel fault tolerance equation with a channel bandwidth ratio (W/W_r) of 10. Also, the foregoing de-

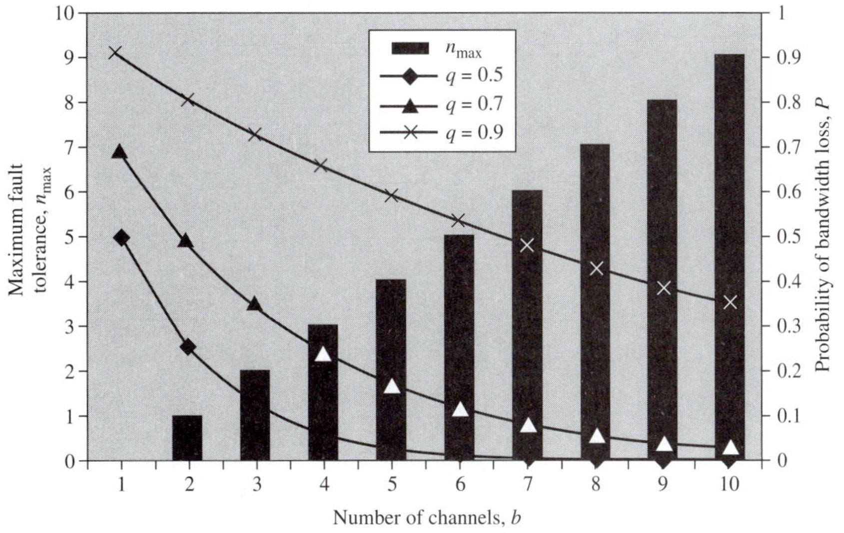

Figure 8.7 P and n_{max} for $W/W_r = 10$.

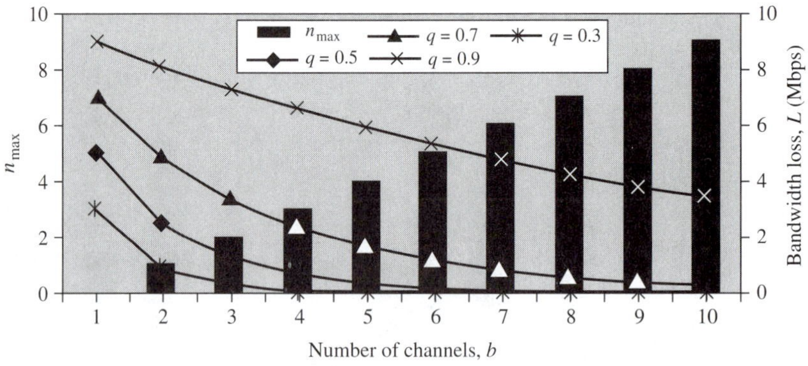

Figure 8.8 Bandwidth loss for $W/W_r = 10$ with a total bandwidth of 10 Mbps.

rivations are used to plot the probability of bandwidth loss P. Note that the probability of loss is initially higher when the number of channels is low, and it falls as the number of channels is increased. Also, the channel fault tolerance n_{max} increases, since there are surplus channels available for transmission. The bandwidth loss plot for different values of q with $W/W_r = 10$ is shown in Figure 8.8. More observations about the system behavior may be made by plotting the number of lost channels as a function of channel failure probability q. Since each channel bandwidth is W/b, the expected number of lost channels is Lb/W. Figure 8.9 shows the corresponding plot for the number of channels, $b = 1$–4.

8.2.3 Queuing Theory Techniques

Queuing theory techniques were used to design the first classical network models. The technique is developed through random arrival and service patterns in the queues. In actual networks, queues are the transmit and receive buffers at the sending and receiving

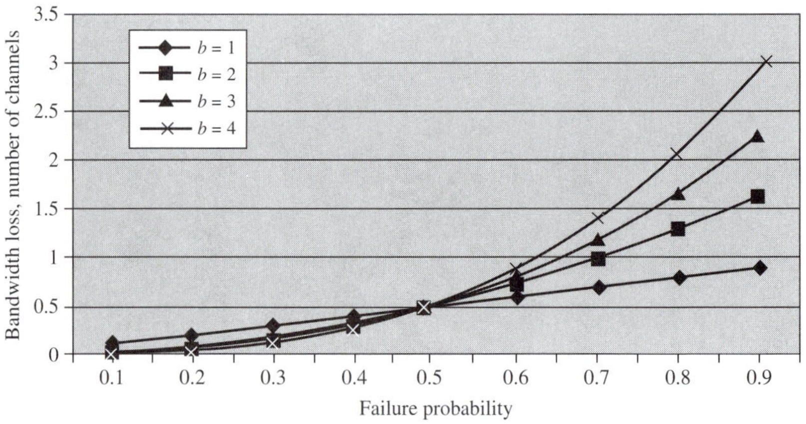

Figure 8.9 Bandwidth loss for $b = 1$–4.

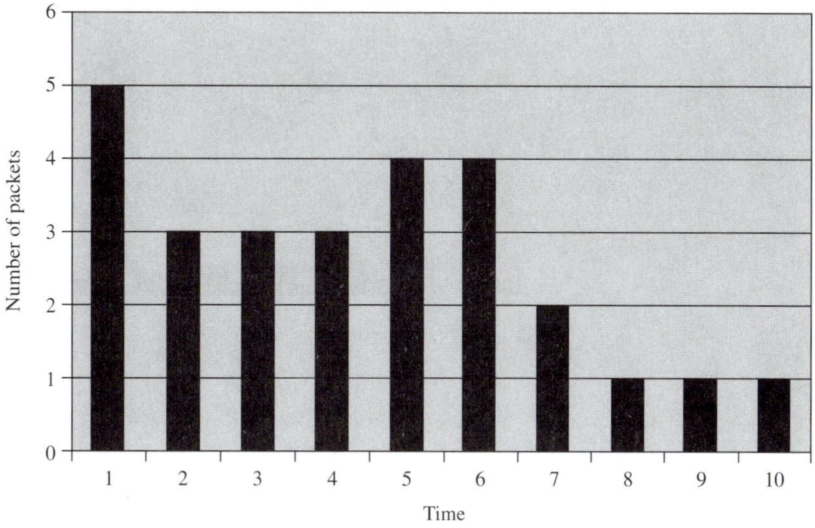

Figure 8.10 Number of packets in a queue.

stations, respectively. Usually the receiving station queue is not a big concern in modeling the behavior of the network, since the queue depends on the processing behavior of the receiving station, not on the network itself. However, the transmitting station queue behavior depends on several network factors such as transmission protocol, transmission capacity, routing/forwarding methods, network congestion, and flow balance. We will examine different queuing models relevant to computer networks in this section. Before we show the different queuing techniques, we will illustrate the possible behavior of a queue within a packet-switched network.

Consider a transmit station connected to a network through a queue (transmit buffer). Ignoring the behavior of the protocol at this time, we may observe the behavior of the queue in terms of the number of queued packets. Figure 8.10 shows the number of packets in the queue as a function of time. We assume that it is possible to have two or more packet arrivals/departures at the same time. Also, we assume that it is possible for arrival and departure to occur simultaneously. We can use Figure 8.10 to find the average time a packet spent in the system during the interval (0,10) if 5 packets were in the system at time 0, 1 packet was in the system at time 10, and 10 packets arrived during the interval 0–10. Also, we can determine the average number of packets in the system.

First we compute the total number of packet-seconds by all packets in the interval (0,10). This will be $5 \times 1 + 3 \times 3 + 4 \times 2 + 2 \times 1 + 1 \times 3 = 27$. The average time spent in the queue during (0,10) by 15 packets is $27/15 = 1.8$ seconds. The average number of packets in the queue is $27/10 = 2.7$.

8.2.3.1 M/M/1 Queues

The arrival and service patterns of a queue may be mathematically modeled by one of several queuing systems. The models differ based on the statistical distributions of arrival

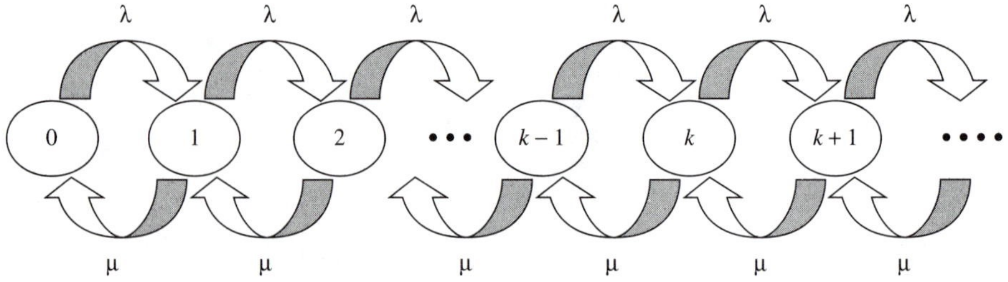

Figure 8.11 State transition diagram for an M/M/1 queuing system.

and service times. A queuing system with exponential distributions of time between arrivals and service with a single service facility is called an M/M/1 queue. It is the simplest of all the queuing systems. The exponential arrival and service distributions are $(1 - e^{-\lambda t})$ and $(1 - e^{-\mu t})$, respectively, where λ and μ are the mean arrival and service rates. Thus an M/M/1 queuing system is a single-server queue with a Poisson arrival process and an exponential distribution for service times.

The M/M/1 queue may be described by analogy to a birth–death process in which either an arrival or a service may take place at a time to change the system population by one. Therefore, if there are k customers in the system at some time, an arrival will cause the population to change to $k + 1$. Similarly, if there is a service event with k customers present in the system, the population will change to $k - 1$. We can use a state transition diagram to model the queue, with the changing numbers of customers in the queue represented by the different states. Figure 8.11 shows the state transition diagram of the M/M/1 queue. State k represents k customers in the queue. The number of customers may vary between 0 and infinity assuming infinite queue capacity. When the queue is empty, there are zero customers in the queue and the corresponding state is 0. As arrivals take place, the number of customers increases and the state changes to higher states occur. The state changes occur to lower states only in the event of service, which is shown by backward transitions.

Assuming an equilibrium state for the process, we may write the flow balance equations for the system. Note that the equilibrium state of the system eliminates the time dependence of state probabilities, since the system is considered to have already passed through the "transient" state and a steady state is assumed in which the state probabilities are no longer time dependent. We get

$$\frac{dP_k}{dt} = -(\lambda + \mu)P_k + \lambda P_{k-1} + \mu P_{k+1} = 0 \qquad k \geq 1$$

$$\frac{dP_0}{dt} = -\lambda P_0 + \mu P_1 = 0 \qquad k = 0$$

Starting recursively from $k = 1$, we get $P_k = P_0\,(\lambda/\mu)^k$. This relation may be used to obtain all state probabilities in terms of state 0 probability P_0. Using the probability property that the sum of all state probabilities must be one, and considering that λ/μ is always less than one, we can get the state P_0 probability as follows:

$$P_0 \frac{1}{1+\sum_{k=1}^{\infty}\left(\frac{\lambda}{\mu}\right)^k} = \frac{1}{1+\frac{\lambda/\mu}{1-\lambda/\mu}} = 1 - \frac{\lambda}{\mu}$$

Putting $\rho = \lambda/\mu$ and using the foregoing relation for P_k, we get

$$P_k = (1-\rho)\rho^k \qquad k = 0, 1, 2, \ldots$$

An important measure of a queuing system is the average number of customers in the system. It is the expected value obtained by using the state transition diagram: that is, the average number of customers is given by summing the products over all states, where the product in the kth state is the expected number of k customers in the system given as kP_k. Thus we get

$$\overline{N} = \sum_{k=0}^{\infty} kP_k = (1-\rho)\sum_{k=0}^{\infty} k\rho^k$$

Here the sum of series may be evaluated as $\rho/(1-\rho)^2$, which gives

$$\overline{N} = \frac{\rho}{1-\rho}$$

Little's result for a queuing system shows the relationship between the average number of customers in the queue $\overline{N}$, the arrival rate λ, and the average time spent by each customer in the queue T, that is, $\overline{N} = \lambda T$, or

$$T = \frac{\overline{N}}{\lambda} = \left(\frac{\rho}{1-\rho}\right)\left(\frac{1}{\lambda}\right) = \frac{1/\mu}{1-\rho} = \frac{1}{\mu-\lambda}$$

The expected number of customers in service for the M/M/1 system is based on the probability of a random variable c, representing the number of customers in the queue, taking the value 1. Note that c may take the value 1 or 0, since there may be at most one customer in service in the M/M/1 queue. We get

$$P[c=1] = \sum_{k=1}^{\infty}(1-\rho)\rho^k = 1 - P_0 = \rho$$

$$E[C] = 0 \cdot p[c=0] + 1 \cdot p[c=1] = \rho$$

Since the amount of service time for a customer is exponentially distributed with parameter μ, the average time spent in service by a customer is $1/\mu$. Accordingly, the average waiting time and average number of customers waiting in the queue (excluding service) are:

$$E[\text{number in queue}] = N_q = \frac{\rho}{1-\rho} - \rho = \frac{\rho^2}{1-\rho}$$

$$E[\text{waiting time}] = \frac{1}{\mu-\lambda} - \frac{1}{\mu} = \frac{\lambda}{\mu(\mu-\lambda)} = \frac{\rho}{\mu-\lambda}$$

Example 8.3

What is the utilization of a server modeled as an M/M/1 queuing system that has average of eight jobs waiting in the queue (**not in service**)?

Solution: Here N_q is 8. According to the M/M/1 equation,

$$N_q = \frac{\rho^2}{1 - \rho}$$

Solving the quadratic equation for ρ, we get

$$\rho^2 + 8\rho - 8 = \Rightarrow \rho = 2\sqrt{6} - 4 = 0.899$$

Assuming M/M/1 queuing behavior, the Little's result may be used in calculating the queuing delay in networks. It is noticed that calculation in real networks using the M/M/1 assumption provides a very good estimate of the queuing time. According to the foregoing equation for E[waiting time], the queuing delay in a queue is the reciprocal of the difference between the service and arrival rates. Since μ must be greater than λ for $\rho < 1$, T will always be positive. An arrival rate that is higher than the service rate indicates an unstable situation in which the queue keeps developing and a steady state is never reached. In real networks it means that the network buffer gets full, since the transmission rate is not enough to handle the traffic.

Consider a transmission mechanism with transmission rate C of 10 Mbps and an exponential distribution of packet sizes with average packet size P of 1000 bits. The average time to transmit each packet P/C will be 100 μs. Therefore, the average service rate for the queue is C/P, or 10,000 packets/second. Furthermore, consider that the packets are generated at random with exponential distribution of time between generations. According to the M/M/1 queue results, if the mean time between packet generations is λ, the average queuing time will be $1/(C/P - \lambda)$. For example, for an arrival rate of 5000 packets/second with mean packet size of 1000 bits over a 10 Mbps line, the mean queuing delay will be $1/(10{,}000 - 5000) = 0.2$ ms. A plot of the mean queuing delay is shown in Figure 8.12 for different values of packet sizes. The increase in packet size results in the queue becoming full faster, as seen in the figure. This may be intuitively obvious, since larger packets occupy the line for longer periods than shorter packets, causing a longer queuing delay for other packets arriving in the queue.

Mean packet delay $\bar{d}$ and mean number of hops per packet $\bar{n}$ are among the performance measures useful in evaluating the routing in a network. Given a traffic matrix for source–destination pairs in a network, it is important to estimate the general forwarding behaviors in terms of $\bar{d}$ and $\bar{n}$. Mean packet delay $\bar{d}$ is the amount of average delay $\bar{n}$ that a packet may face in the network. The mean number of hops is the number of hops on an average a packet has to travel in the network. If T_i is the average queuing delay on a line, then the mean packet delay is given as follows:

$$\bar{d} = \frac{\sum_{i=1}^{n} \lambda_i T_i}{\lambda}$$

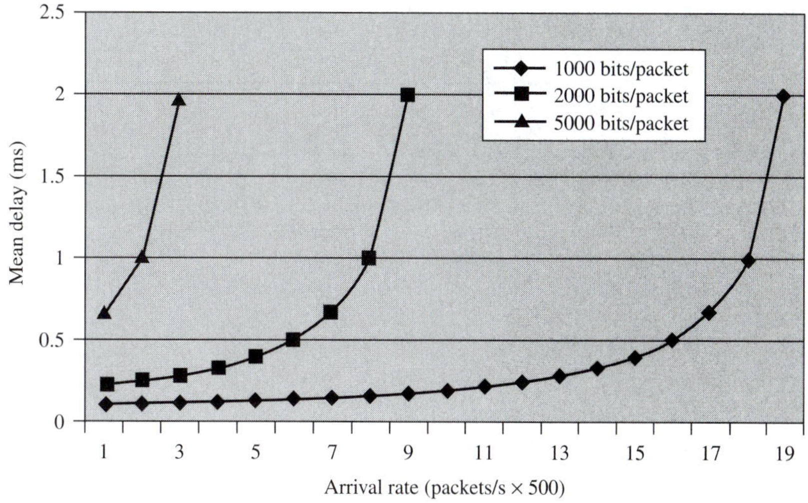

Figure 8.12 Mean queuing delay.

Here n is the number of links in the network, λ_i is the traffic flow on the ith link, and λ is the total traffic flow on all the links in the network. Notice that since each traffic entry in the routing matrix may traverse one or more hops, the flow between a source–destination pair may be repeated on several links. Therefore, the total traffic between all source–destination pairs γ may not be same as λ. Consider a three-node network with nodes A, B, and C connected via links AB and BC. If the traffic from A to C is 10 packets/second, then the traffic on links AB and BC are 10 each. The total traffic on all links in this case will be the sum of flows on AB and BC (i.e., 20 packets/second), whereas the total traffic between all source–destination pairs is 10 packets/second. The mean number of hops traveled by each packet will be the ratio λ/γ, which in this case will be 2, since all traffic is traveling two hops (AC via AB and BC). In general

$$\bar{n} = \frac{\lambda}{\gamma}$$

Example 8.4

Given Table 8.1, a routing matrix (containing the paths and only nonzero nonsymmetric traffic entries, in packets per second), find the mean queuing delay/line, the mean packet delay, and the mean number of hops per packet. Each link is assumed to be full duplex and has a bandwidth of 100 Kbps. Mean packet size is 1000 bits. Ignore the propagation delay.

Solution: There are nine nodes in the network. Notice that some nodes do not have any generated traffic and some nodes do not have any traffic meant for them as a destination. However, all the listed nodes are participating in traffic forwarding. Also, notice that the traffic is nonsymmetrical (i.e., the traffic from A to B is not

TABLE 8.1
Traffic Flow Matrix Including the Routing Information

To/ From	A	B	C	D	E	F	G	H	I
A		10 (AB)	14 (ABC)	20 (ABCD)				7 (ACDGH)	
B					8 (BCDE)		6 (BFG)		6 (BFHI)
C					10 (CDE)				10 (CDEI)
D								6 (DGFH)	
E									
F									
G	9 (GDCA)	8 (GFB)							
H				14 (HFGD)	3 (HIE)				
I	10 (IHA)		15 (IEC)						

same as traffic from B to A). By inspecting the traffic flows in the routing matrix we can figure out the total traffic between source–destination pairs γ and the total traffic on all links λ. To obtain γ, we add all the traffic values on the routing matrix (i.e., the total traffic between all source–destination pairs is 156 packets/second. The line delay is calculated by means of the M/M/1 queue delay result derived earlier. The service rate C/P for each link is obtained as 100 Kbps/1000 bits per packet = 100 packets/second. Using the traffic matrix of Table 8.1 and the M/M/1 delay equation for the queues, we derive Table 8.2.

TABLE 8.2
Traffic Flow and Delay Calculations

Link Number	Link Label	Traffic Flow, λ_i (Packets/s)	$T_i = 1/(\mu - \lambda_i)$ (ms)	Weighted Delay, $\lambda_i T_i$
1	AB	10 + 14 + 20 = 44	17.86	785.84
2	BC	14 + 20 + 8 = 42	17.24	724.08
3	CD	20 + 7 + 8 + 10 + 10 + 9 = 64	27.78	1777.92
4	AC	7 + 9 = 16	11.91	190.56
5	DG	7 + 6 + 9 + 14 = 36	15.63	562.68
6	GH	7	10.75	75.25
7	DE	8 + 10 + 10 = 28	13.89	388.92
8	BF	6 + 6 + 8 = 20	12.5	250
9	FH	6 + 6 + 14 = 26	13.51	351.26
10	HI	6 + 3 + 10 = 19	12.35	234.65
11	EI	10 + 3 + 15 = 28	13.89	388.92
12	GF	6 + 8 + 14 + 6 = 34	15.15	515.1
13	HA	10	11.11	111.1
14	EC	15	11.77	176.55

The total traffic flow, the mean packet delay, and the mean number of hops per packet are obtained as follows:

$$\lambda = \sum_{i=1}^{14} \lambda_i = 389$$

$$\bar{d} = \frac{\sum_{i=1}^{14} \lambda_i T_i}{\lambda} = \frac{6532.83}{389} = 41.88 \text{ ms}$$

$$\bar{n} = \frac{\lambda}{\gamma} = \frac{389}{156} = 2.49 \text{ hops}$$

8.2.3.2 M/M/∞ Queues

In this extension of M/M/1 queues, we consider an infinite number of servers, that is, a system in which there is always a server available for each arriving customer. The state transition diagram will be same as before, except that the service rate will change with the number of customers in the system as shown in Figure 8.13. Again using the flow balance equations and the probability property, to derive the state probabilities, we get

$$P_k = P_0 \prod_{i=0}^{k-1} \frac{\lambda}{(i+1)\mu}$$

$$P_0 = \frac{1}{1 + \sum_{k=1}^{\infty} \prod_{i=0}^{k-1} \frac{\lambda}{(i+1)\mu}} = \frac{1}{1 + \sum_{k=1}^{\infty} \left(\frac{\lambda}{\mu}\right)^k \frac{1}{R!}} = e^{-\lambda/\mu}$$

Thus

$$P_k = \frac{(\lambda/\mu)^k}{k!} e^{-\lambda/\mu} \qquad k = 0, 1, 2, \ldots$$

The final equation is the Poisson distribution. From the result of the Poisson distribution, the average number of customers in the system is found to be λ/μ and the average wait-

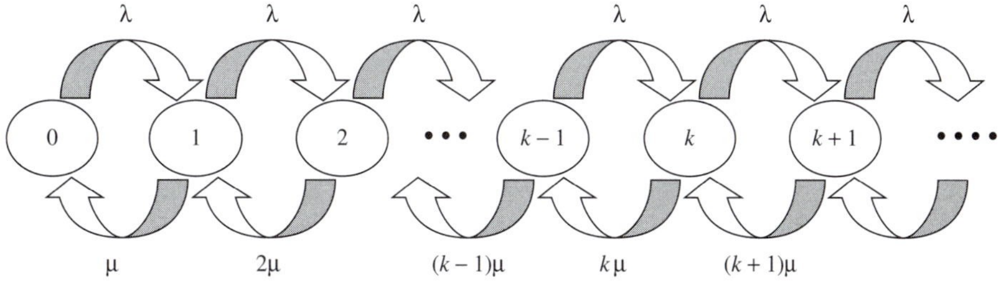

Figure 8.13 State transition diagram for M/M/∞ queuing system.

ing time $1/\mu$. This result appears to be obvious, since each arriving customer has his own server, which means that his time in the system will be only his service time, which is $1/\mu$ on average.

8.2.4 Markov Models

A stochastic process is a family of random variables observed at different times. Such processes are concerned with observations that cannot be predicted precisely beforehand, but the probabilities of the different possible states at some point in time may be specified. The process is defined on a specified probability space, where the state is a value assumed by a random variable, and the state space of the process is the set of all possible values of the random variable. A Markov chain is a special type of stochastic process in which the future probabilistic behavior of the process is uniquely determined by its present state. This type of behavior is considered to be memoryless because the knowledge of the present decouples the past from the future. Behavior of many computer systems including computer networks falls into this category. In this section, we explore the behavior of computer networks through Markov models. A Markovian process with a discrete state space is also called a Markov chain. However, if the time space is continuous, we refer to the process as a Markov process. Needless to say, the continuous state models are not of much interest in the field of computer networks.

8.2.4.1 Discrete Time Markov Model

A discrete time Markov model is defined in terms of a set of transition probabilities between the discrete states of the model. The concept is very similar to the one for the queuing system or birth–death process, except that transitions are possible between any states. In the discrete time model, transitions are associated with transition probabilities between states. The $n \times n$ matrix of transition probabilities between n states is called the transition probability matrix. For an n-state system, the transition probabilities may be arranged as follows:

$$P = \begin{bmatrix} p_{11} & p_{12} & \cdots & p_{1n} \\ p_{21} & p_{22} & \cdots & p_{2n} \\ \cdot & \cdot & \cdots & \cdot \\ \cdot & \cdot & & \cdot \\ \cdot & \cdot & \cdots & \cdot \\ p_{n1} & p_{n2} & \cdots & p_{nn} \end{bmatrix}$$

The element p_{ij} of the transition probability matrix is the probability that the system will go to state j in one step if the present state is i. For example, for a three-state system the transition probability matrix will be

$$P = \begin{bmatrix} p_{11} & p_{12} & p_{13} \\ p_{21} & p_{22} & p_{23} \\ p_{31} & p_{32} & p_{33} \end{bmatrix}$$

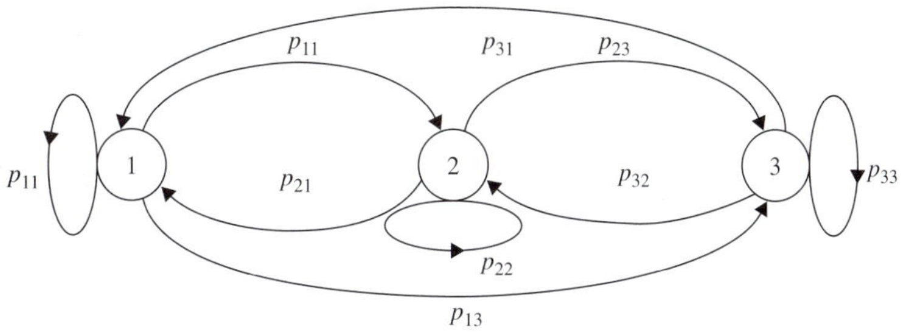

Figure 8.14 State transition diagram for a three-state Markov chain.

The corresponding state diagram is shown in Figure 8.14. If P_0 is the initial probability vector, then the state probabilities after k steps is given by the vector $\Pi_k = P_0P^k$. The matrix P^k is the kth step transition matrix obtained by multiplying the matrix P to itself k times. The initial probability vector contains the initial state probabilities for all states in the chain. At steady state no more changes in state probabilities are expected when P is multiplied by the current probability vector. In other words, the steady state probability vector π remains unchanged if multiplied by P, that is,

$$\pi P = \pi \tag{8.3}$$

where π is the steady state probability vector with steady state probabilities for all states in the chain. For a three-state Markov chain, the vector π is $[\pi_1\ \pi_2\ \pi_3]$, where π_1, π_2, and π_3 are the state 1, 2, and 3 steady state probabilities, respectively. Equation (8.3), with the probability condition $\sum_{i=0}^{n} \pi_i = 1$, may be used to determine all the steady state probabilities.

Example 8.5

A network server is always in one of three states: A, B, or C, where state A is file service request (FSQ), state B is print service request (PSQ), and state C is Internet service request (ISQ). The server moves only within these states to satisfy all the requests it gets, and it can change state only once per millisecond. If it is in city A, it remains in that state for that millisecond interval with probability 0.6 or goes to state B with probability 0.4. If the server is in state B, it remains in that state for the millisecond interval with probability 0.4 or goes to state C with probability 0.6. If the server is in state C, it remains in that state for the millisecond interval with probability 0.5 or goes to state A with probability 0.5. Consider a Markov model for this problem with a steady state condition for the Markov chain.

(a) Draw the Markov chain and show all the probabilities, including the transition probability matrix, and states.
(b) What is the probability that the server is in state C?

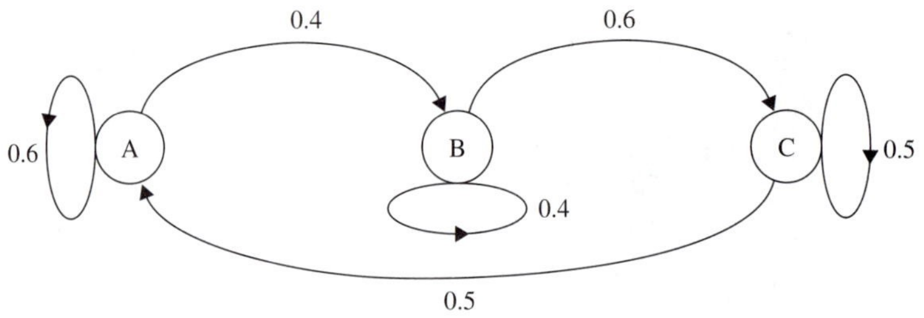

Figure 8.15 State transition diagram for Example 8.5.

Solution: (a) The three states A, B, and C are defined as the server being in the states A, B, and C, respectively.

The transition probability matrix is

$$\mathbf{P} = \begin{bmatrix} 0.6 & 0.4 & 0 \\ 0 & 0.4 & 0.6 \\ 0.5 & 0 & 0.5 \end{bmatrix}$$

Solution: (b) At steady state, we must have $\pi\mathbf{P} = \pi$. Equations from the first two columns are:

$$0.6\pi_1 + 0.5\pi_3 = \pi_1 \Rightarrow \pi_3 = 0.8\pi_1$$

$$0.4\pi_1 + 0.4\pi_2 = \pi_2 \Rightarrow \pi_2 = 0.66\pi_1$$

Using the normalization equation, we get

$$\pi_1[\,1 + 2/3 + 4/5\,] = 1 \qquad \Rightarrow \qquad \pi_1 = 15/37 = 0.406$$

$$\pi_2 = 10/37 = 0.27$$

$$\pi_3 = 12/37 = 0.324$$

The probability that the server is in state C at some time is $\pi_3 = 12/37 = 0.324$.

As a case study, we extend the MLAN model presented in Section 8.2.2. We now formulate a discrete time model for a three-channel system. The Markov chain is shown in Figure 8.16. The transition probability matrix is

$$\mathbf{P} = \begin{bmatrix} p^3 & 3qp^2 & 3q^2p & q^3 \\ 0 & p^2 & 2qp & q^2 \\ 0 & 0 & p & q \\ 0 & 0 & 0 & 1 \end{bmatrix}$$

Here q is the failure probability of a channel as mentioned before, and $p = 1 - q$. It is noticed that state S0 is an absorbing state and at steady state (i.e., a state in which no more network change is expected), the probability of state S0 would be 1. (The steady

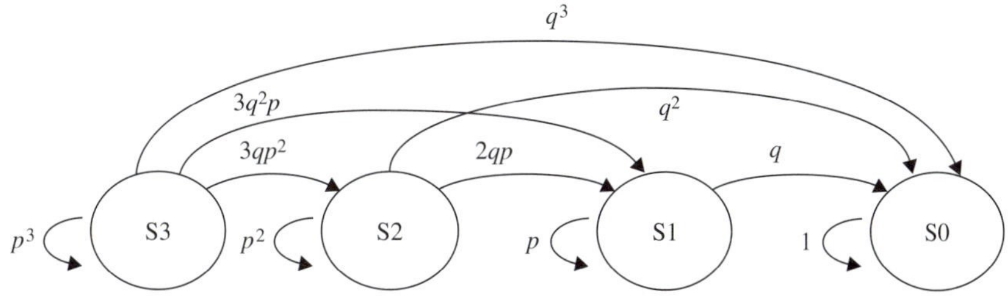

Figure 8.16 State transition diagram of a three-channel MLAN.

state probability of a state becomes unity if there is no outgoing transition from the state.) The transient state probabilities allow us to determine the state probabilities after k steps of transition. In general, if the state probability vector at the kth step is

$$\Pi_k = [P_3(k) \quad P_2(k) \quad P_1(k) \quad P_0(k)]$$

where $P_i(k)$ is the kth step probability of being in state i, with $i = 0, 1, 2, 3$. The $(k + 1)$th state probability is determined in terms of kth state probabilities as $\Pi_{k+1} = \Pi_k P$. For example, the probability of no failure (state S3 probability) after 3 discrete steps would be p^9, if we start in state S3.

Let us now introduce the concept of repair in MLAN. A channel in a WDM network may fail as a result of interface failure or some other problem. Channel repair is like fixing the channel after it has failed, either by replacing it with a standby extra channel or by replacing the faulty interface. With this modification, the state transitions may go back and forth between the states representing channel failures and repairs. Also, this leads to nonzero steady state probabilities of being in states other than failed state. Let r be the probability of a channel repair and $s = 1 - r$. The new state diagram is obtained as shown in Figure 8.17, and the transition probability matrix is given as follows:

$$\mathbf{P} = \begin{bmatrix} p^3 & 3qp^2 & 3q^2p & q^3 \\ p^2r & p^2s + 2qpr & 2qps + q^2r & q^2s \\ pr^2 & 2prs + r^2q & ps^2 + 2qrs & qs^2 \\ r^3 & 3r^2s & 3rs^2 & s^3 \end{bmatrix}$$

Using the transition probability matrix and state vector, we can get the state probabilities after k steps as shown earlier. For example, we get the state 0 probability after 2 steps as $s^5q = s^3 * s^2q$ if the process is assumed to start in state 1.

8.2.4.2 Continuous Time Markov Model

A Markov process is considered to be a continuous time stochastic process because it develops in time in a manner controlled by probabilistic laws. In particular, for studying the network performance, we are interested in discrete state Markov processes and assume that the system can exist in one of n discrete states. Transitions may occur from one state to another as determined by the transition rates between different states. In that sense, this

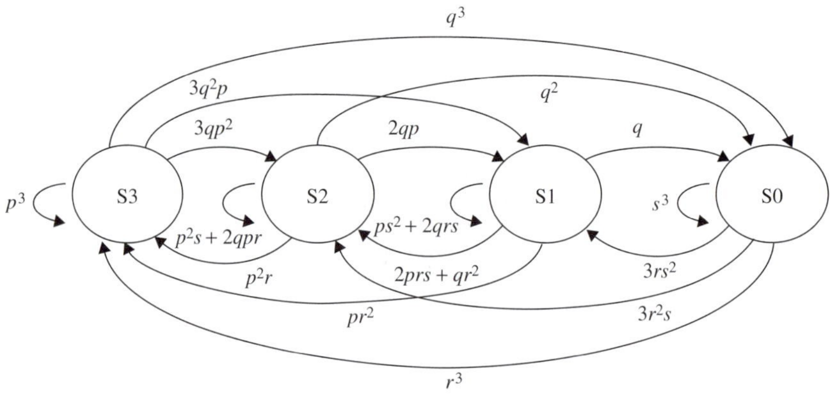

Figure 8.17 State transition diagram of a three-channel MLAN with repair.

process is very similar to the birth–death queuing system we studied earlier. The transition rate matrix consists of transition rates between any state in the process. However, there is no transition to a state itself like the discrete time case. The transition rate matrix includes the rate of departures from a state i to a state j in the process. The derivation is very similar to the derivation of a birth–death process, in which the state flow balance equations are first obtained. Then the equations are solved recursively to obtain the state probabilities. However, the transition rate matrix is used to obtain the state probabilities. Also, the system reliability models are obtained by using this method. In Section 8.2.5, we illustrate the method as applied to the evaluation of system reliability.

8.2.5 Reliability Models

One prominent advantage of a computer network is its fault tolerance. When a computer fails in a networked environment for reasons such as power failure or disk failure, the interrupted task may be undertaken by other computer(s) in the network. A user may not notice a break in service at all. The ability to tolerate faults in the network may be measured in terms of "reliability." A reliable network will give long service, without interruptions, with least overhead and cost. Reliability is defined as the probability that a system will adequately perform its specified purpose for a specified period of time under specified environmental conditions. If λ is the failure rate of a component following an exponential distribution of failures, the reliability of the component is defined as the probability of successful operation of the component at some time t. Following the exponential distribution, the reliability of the component after time t is given as $e^{-\lambda t}$. For example, if network adapters made by a certain manufacturer have the average failure rate of 10 per year (using the exponential distribution), then the reliability of an adapter after 3 months (time = 0.25 year) is $e^{-10(0.25)} = 0.082$.

The time-dependent state probabilities of an n-state Markov process can be found by solving the Markov differential equations for that process. The general form of the state equations is

$$\dot{X} = AX$$

where X is the n-state probability vector, and the dot refers to the time derivative of vector X. The transition rate matrix A has general form

$$\mathbf{A} = \begin{bmatrix} -\sum_{j=2}^{n} \lambda_{1j} & \lambda_{21} & \cdots & \lambda_{n1} \\ \lambda_{12} & -\sum_{\substack{j=1 \\ j\neq 2}}^{n} \lambda_{2j} & \cdots & \lambda_{21} \\ \vdots & \vdots & \vdots & \vdots \\ \lambda_{1n} & \lambda_{2n} & \cdots & -\sum_{j=1}^{n=1} \lambda_{1j} \end{bmatrix}$$

The solution of the differential equation with some initial conditions $X(0)$ at time zero is found as follows:

$$X(t) = e^{At}\, X(0)$$

The evaluation mainly relies on the evaluation of the time-dependent state transition matrix. There are various approaches to solving the matrix, and without going into detail we will show one standard technique, the "resolvent matrix method." In this method a matrix $[sI - A]$ is obtained, then its inverse is taken, and Laplace transformation is used to obtain the time-dependent state probabilities. In this equation, I is the identity matrix and s is the Laplace transform variable.

In a computer network with several alternate data flow paths between computers, the interconnections in the network may be unreliable because they are vulnerable to hardware or software failures. When redundant interconnections are used to build a network, the network will not fail completely if there is a link failure. However, the network is left in a degraded mode of operation. In the degraded mode, the network is expected to operate with fewer interconnections and remains capable of data processing. However, it may be less reliable than before. The concept of reliability may be used to study a network's ability to survive in the presence of faults.

Another related measure, called availability, is the probability that a system may survive in the presence of faults and repairs. The system is considered to have a repair facility that may repair the failed components to bring them back in service. The availability evaluation technique is similar to the reliability evaluation, except that the process must have the ability to repair the faults.

As an application of the described techniques, let us consider again the multichannel communication network already introduced. We now formulate a continuous time Markov model of an MLAN by assuming that channel failures and repairs are exponentially distributed with mean rates λ and μ, respectively. The four-state model for three channels in the system is shown in Figure 8.18. The states S3 to S0 represent 0 to 3 channel failures, respectively. In this model, we can obtain the steady state probabilities by forming the flow balance equation for each state. In our model, system availability is defined as the probability that the system does not enter a complete failure state. In Figure 8.18, it is the probability of not being in state S0.

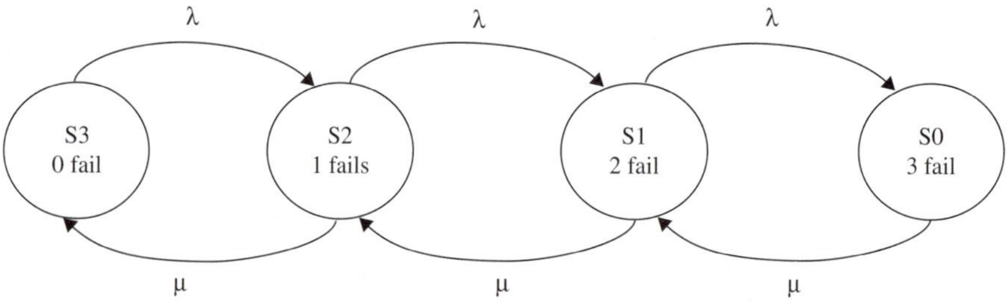

Figure 8.18 State transition diagram of a continuous time Markov model.

It may be noted that the process is actually a birth–death process with finite population (M/M/1/K system). The steady state probabilities are given by

$$P_k = \frac{1 - \lambda/\mu}{1 - (\lambda/\mu)^{K+1}} \left(\frac{\lambda}{\mu}\right)^k \qquad \text{for } 0 \le k \le K$$

The availability of the system is found to be $1 - P_0$. Now simplifying the model to a two-state system corresponding to a single channel system and $\lambda = 0.2$, $\mu = 2$/year, the transition rate matrix A will be

$$\mathbf{A} = \begin{bmatrix} -0.2 & 2 \\ 0.2 & -2 \end{bmatrix}$$

The corresponding $[sI - A]$ and its inverse are found as follows:

$$[sI - \mathbf{A}] = \begin{bmatrix} s + 0.2 & -2 \\ -0.2 & s + 2 \end{bmatrix} \Rightarrow [sI - A]^{-1} = \frac{1}{s(s + 2.2)} \begin{bmatrix} s + 2 & 2 \\ 0.2 & s + 0.2 \end{bmatrix}$$

Taking the first term's expansion and then taking the inverse Laplace transform, we get

$$\frac{s + 2}{s(s + 2.2)} = \frac{1}{11}\left(\frac{10}{s} + \frac{1}{s + 2.2}\right) \Rightarrow \frac{1}{11}(10 + e^{-2.2t})$$

Using the same procedure for other terms, we get

$$e^{At} = \left(\frac{1}{11}\right)\begin{bmatrix} 10 + e^{-2.2t} & 10 - 10e^{-2.2t} \\ 1 - e^{-2.2t} & 1 + 10e^{-2.2t} \end{bmatrix}$$

Assuming state S1 (no channel failed) as the starting state, the initial vector $X(0)$ will be $[1\ 0]^T$, and the time-dependent state probability vector will be $e^{At}X(0)$. The two probabilities in the vector are P_{S1} and P_{S0}, which give us

$$\text{Availability} = P_{S1} = \left(\frac{10}{11} + \frac{1}{11}e^{-2.2t}\right)$$

$$P_{S0} = \left(\frac{1}{11} - \frac{1}{11}e^{-2.2t}\right)$$

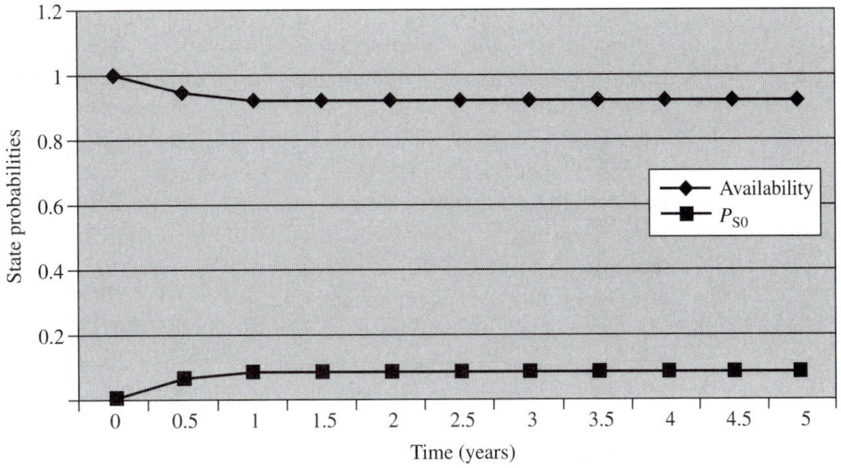

Figure 8.19 State probabilities for a single-channel network.

The two state probabilities are plotted in Figure 8.19. Notice that steady state behavior is reached in about 2.5 years. The steady state probability of states S1 and S0 is 0.91 and 0.09, respectively, in accordance with the steady state M/M/1/K equations shown earlier. The importance of time-dependent probabilities or transient probabilities facilitates the study of system maintenance. However, it is often very difficult to compute these equations for bigger systems. The applications of reliability techniques for studying network performance is just developing, and with applications of reliability tools it is expected that fail-safe networks may be designed, incorporating other improvements as well.

It is noted that as the systems become more complex, it is virtually impossible to obtain a general time-dependent expression for system reliability and availability. However, other measures of interest, such as MTTF (mean time to failure) and MTBF (mean time between failures) may be obtained by using the finite Markov chain concept. These two quantities may be used effectively to schedule system maintenance and to study the general failure behavior of the system.

8.3 SIMULATION

In general, a simulation can be any action that mimics some reality. The computer-based simulation models are computer programs that execute a set of rules to mimic a system. The computer-based simulation models of systems are classified as either discrete or continuous. Note however that the terms "discrete" and "continuous" refer to the model, not the real system. In fact, it may be possible to model the same system with either a discrete or a continuous model. In most simulations, time is the major independent variable. Other variables included in the simulation are functions of time and are the dependent variables. When used to describe a model, *discrete* and *continuous* refer to the dependent variables.

Although simulation has proven to be a valuable tool, it has suffered because developing and validating the computer program usually takes several times as long as is needed to formulate the basic simulation model. When animation techniques are used, model development time is reduced considerably, since the graphical reports that show the hardware layout, software data flow, device utilization, and several other performance measures build confidence in both the analysis and the model. Most of the packages available for network simulations provide animation capability that allows the user to see the network in operation. In this section we will discuss and apply the several simulation techniques used to model the computer networks. First, however, we must consider the role of random number generation in simulation.

To have a computer program simulate a process or a system, we need to be able to generate values that represent the various random values associated with outcomes. For example, if we wish to generate the numbers from 2 to 12 to simulate the roll of a pair of dice in an unpredictable order, we need an algorithm to generate numbers randomly and independently based on a distribution. After a set of random numbers has been generated, it may be tested for distribution and independence. To provide an error bound on the estimate calculated from the collection of output values from a series of simulations, we must know something about the relationships between the individual simulations. If the collection of n output values x_k are independent and identically distributed with a normal (Gaussian) distribution, then an interval $\pm\epsilon$, within which the actual mean would fall with probability p, is given by Student's t distribution normalized by the experimental variance s, that is,

$$\epsilon = t_{n-1,p}\left(\frac{s^2}{n}\right)^{1/2}$$

Here $t_{n-1,p}$ is the Student t distribution with $n - 1$ degrees of freedom and confidence level p.

Example 8.6

Five simulations of a computer network were done to find the average message delay (in milliseconds) between two points A and B. The data were collected as follows:

First simulation run:	3.2, 4.3, 5.1, 4.3, 2.3, 4.2
Second simulation run:	2.2, 3.3, 5.7, 3.3, 5.3, 4.8
Third simulation run:	3.0, 4.7, 5.4, 4.1, 2.1, 4.6
Fourth simulation run:	2.8, 5.3, 4.1, 3.3, 4.3, 6.2
Fifth simulation run:	3.7, 5.3, 3.1, 4.3, 5.3, 5.2

Calculate the experimental mean, the unbiased experimental variance, and the 98% confidence interval using Student's t distribution.

Solution: Let $\overline{X_k}$ be the sample mean for the kth simulation run. We get

$$\overline{X_1} = 3.9 \qquad \overline{X_2} = 4.1 \qquad \overline{X_3} = 3.98 \qquad \overline{X_4} = 4.33$$

$$\overline{X_5} = 4.483$$

The experimental mean $\overline{X}_n$ and the unbiased experimental variance s_n^2 are

$$\overline{X}_n = 4.159 \qquad s_n^2 = \frac{1}{n-1}\sum_{k=1}^{n}(X_k - \overline{X}_n)^2 = 0.05921$$

Looking at the value of the random variable distribution for Student's t with 4 degrees of freedom and probability 0.98, we find $t_{4,0.98} = 3.747$. The 98% confidence interval is calculated by $\epsilon = t_{n-1,p}\,(s^2/n)^{1/2}$; that is, $\epsilon = 3.747\,(0.05921/5)^{1/2} = 0.4078$.

The 98% confidence interval is $4.159 \pm 0.4078 = (3.751, 4.567)$.

8.3.1 Continuous Simulation

In a continuous simulation model, the state of the system is represented by dependent variables, called state variables, which change continuously over time. The model is constructed by defining equations for a set of state variables whose dynamic behavior simulates the real system. The models are frequently written in terms of the derivatives of the state variables in the form of differential equations. In some cases the differential equations may be solved to determine an analytical expression for the state variable. However, in many cases of practical importance, an analytical solution will not be known and simulation may be necessary to obtain the involved integrations. The simulation method divides the time axis into small time periods and the dynamics of the state variables are described by specifying an equation that calculates the values of state variables at period $k + 1$ from the value of the state variables at period k.

It is noted that since digital computers are technically discrete in their operation, the methods of simulating continuous systems involve the solution of differential equations in which the time is divided into small discrete steps. For higher accuracy, the step sizes should be small, resulting in more computations. Often, there has to be a trade-off between accuracy of state variables and computer run time.

8.3.2 Discrete Event Simulation (DEVS)

A discrete system contains objects—people, equipment, raw materials, and so on. These objects are often known as entities. The state of a system is described in terms of the numeric values assigned to the attributes of the entities. An event is defined as a cause or reason that changes the state of a system, and thus a system can change its states only at event times. A discrete event simulation model can be formulated in the following ways:

By defining the changes in state that occur at each event time

By describing the activities in which the entities in the system engage

By describing the process through which the entities in the system flow

The relationship between an event, an activity, and a process is shown in Figure 8.20. Downward arrows show the events. The first event is an arrival event at time t_1 that increases the number of entities in the system by 1. After waiting for some time period, the newly arrived entity is involved in a service activity shown by the start and end of service events at times t_2 and t_3, respectively. This sequence of events is referred to as a process, and it may include one or more events. The access protocols used in networks may

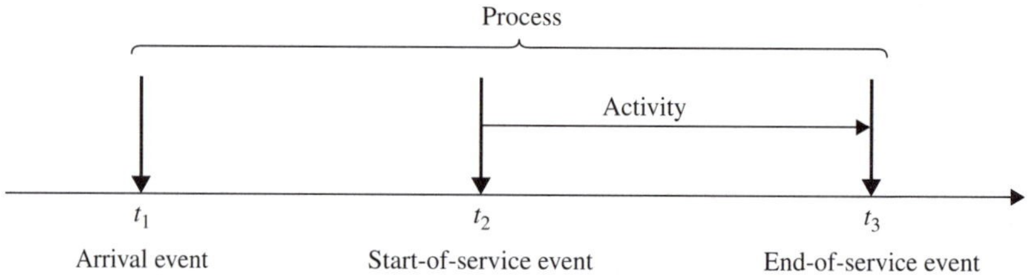

Figure 8.20 Relationship between an event, an activity, and a process.

be described in the form of discrete events. Proper scheduling of these events by calling the event codes at different times provides an accurate protocol operation.

Computer networks belong to a class of physical systems that can be studied effectively by means of discrete event computer simulation models. The networks of interest usually involve complexities that extend beyond the restrictive assumptions required for analytical methods of study. With simulation, network performance can be examined over a wide variety of network topologies, traffic loads, and operating conditions to obtain results where analysis is not feasible. In developing simulation models of communication networks, therefore, it is important to choose a suitable simulation platform. In this section, we present several simulation models of network systems using the DEVS techniques.

8.3.3 Web-Based Simulation

Web-based simulation represents a convergence of computer simulation methodologies and applications on the World Wide Web (WWW). The convenience of running a web-based simulation makes it very attractive especially since the browsers are now capable of running applets. The simulation applets may be run on the client side without requiring the simulation packages to be loaded at the client side, while some simulations execute on the server-side (using CGI scripts for parameter entry and model invocation). Animation applets have been developed for systems such as digital logic, CPU/disk, transportation, oscillators, virtual memory, hysteresis loop, gas turbine, elevator etc.

There are many commercial and non-commercial web-based simulation packages to facilitate the development of simulation models. "Simjava" is a process-oriented discrete event simulation package for Java with animation facilities that can be used very effectively to model the error scenarios in communication protocols, as shown in our example simulation below. Some other Java-based packages are JSIM (a Java-based query driven simulation package), DEVS-JAVA (a Java implementation of discrete event simulation) and DESMO-J (a Java-based process-oriented modeling language derived from DEMOS and Simula).

`simjava` is a toolkit for building working models of complex systems developed by Fred Howell and Ross McNab at University of Edinburgh, UK. It is based around a discrete event simulation kernel and includes facilities for representing simulation objects as animated icons on screen. Simjava simulations may be incorporated as "live diagrams" into web documents. A `simjava` simulation is a collection of entities each running in its own thread. These entities are connected together by ports and can communicate with each other by sending and receiving event objects. A central system class controls all the threads, advances the simulation time, and delivers the events. The progress of the simulation is recorded through trace messages produced by the entities and saved in a file.

A `simjava` simulation contains a number of entities, each of which runs in parallel in its own thread. An entity's behavior is encoded in Java using its `body()` method. Entities have access to a small number of simulation primitives that can be used effectively to schedule events, wait for events, hold the entity and create animation and traces. The following primitives are used most commonly:

- `sim_schedule()` sends event objects to other entities via ports.
- `sim_hold()` holds for some simulation time.
- `sim_wait()` waits for an event object to arrive.
- `sim_select()` selects events from the deferred queue.
- `sim_trace()` writes a timestamped message to the trace file.

8.3.3.1 TCP Congestion Control Simulation

Consider a TCP connection over a link in which a host Station A sends data to another host Station B. Due to transmission interference and high errors in wireless communication assume that every n-th transmission from H1 is lost (or corrupted). For example, if n is 4 and H1 transmits sequences 1, 2, 3, 4, 5, 6, 7, 8, 9 then the transmission number 4 and 8 will be dropped by the data link layer. H1 limits the amount of data it sends using the TCP slow start and congestion avoidance mechanisms. However, it does not implement fast retransmit or fast recovery.

We make the following assumptions in our simulation:

Headers and ACKs are of size 0; the RTT is set by the user that is preset to 1 sec. The retransmission timeout is set to RTT + 0.01 sec; the data in each frame is 1KB; the link BW is 100 KB per second; there are no other losses beside the ones mentioned above (however, every n-th frame is lost irrespective of whether it is the original trm or retrm); H2 sends ACK for every frame it receives but may send cumulative ACK for previously buffered frames; H2 has buffer space to advertise infinite receiver window; the cwnd at H1 does not increase when a duplicate ACK is received.

Figures 8.21–8.25 displays screen snapshots of the simulation run. We display the following in the simulation window:

(a) Input window that includes number of rounds intended and the RTT;

(b) Running values of simulation time, frames sent/acknowledged, cwnd, sthresh and throughput as number of frames successfully transmitted per unit time;

(c) Sliding bar to control the speed of animation, sliding to left increases the simulation speed; and

(d) Simulation trace that displays an animation of frames/ACK movement over the transmission path.

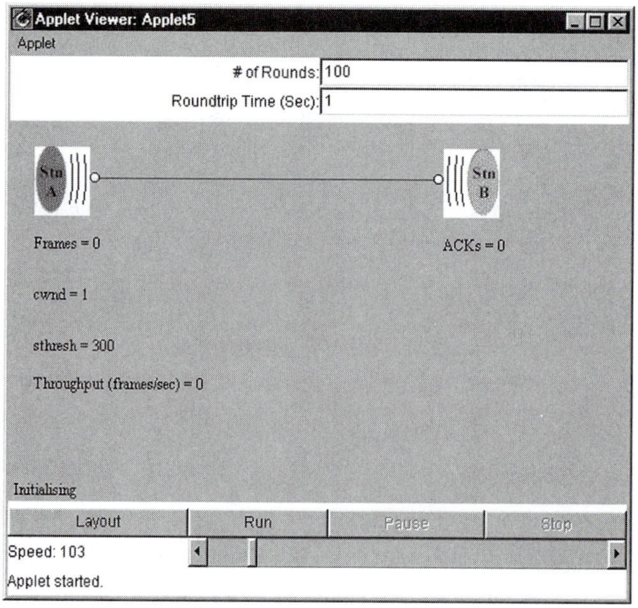

Figure 8.21 Initial screen for TCP Congestion Control Simulation.

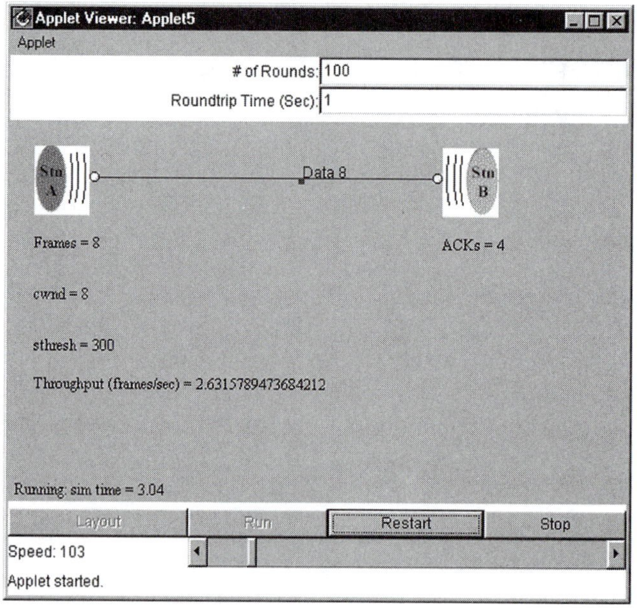

Figure 8.22 Screen snapshot at 3.04 sec displaying data frame 8 being transmitted to Station B.

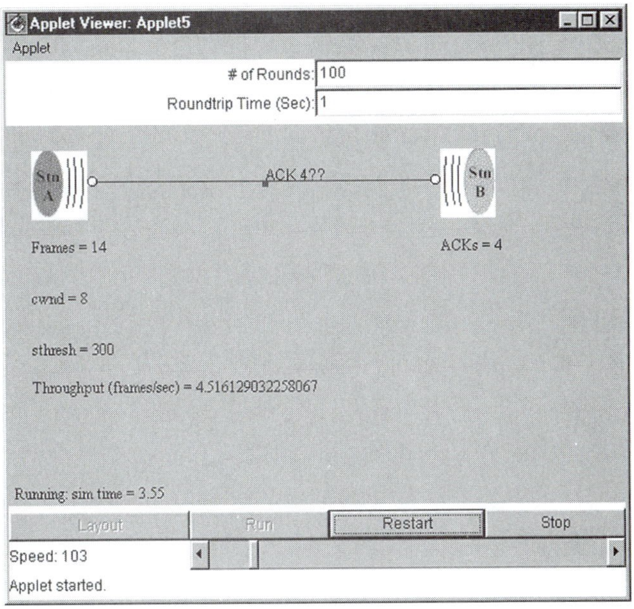

Figure 8.23 Transmission loss of frame 5 is indicated by a duplicate ACK 4 by Station B.

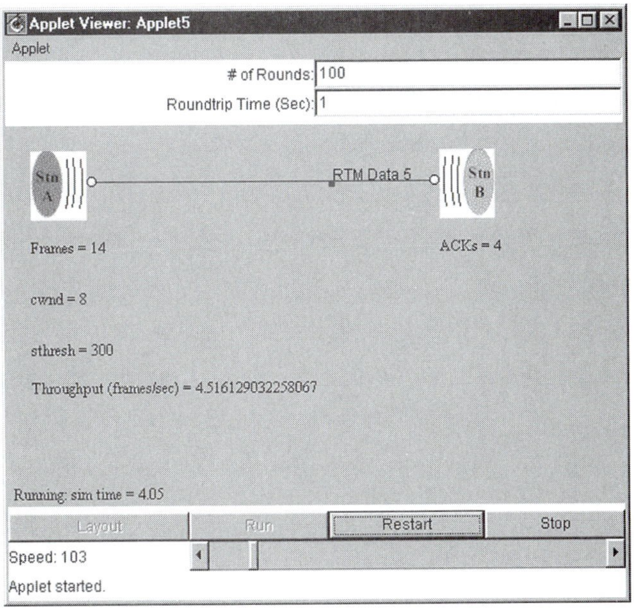

Figure 8.24 Retransmission of frame 4 as the result of previous error in frame 5.

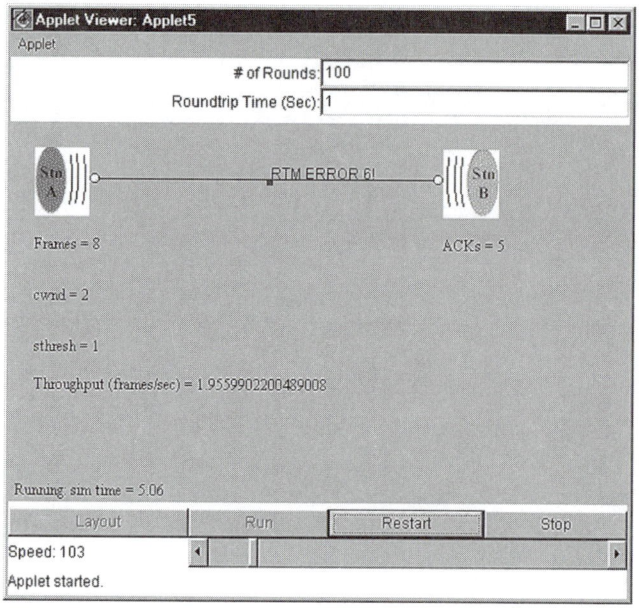

Figure 8.25 Every n-th frame from Station A to B has an error that may also occur in retransmission of a frame (# 6 in this case).

The following algorithm is used to implement slow start and congestion avoidance under error conditions:

```
snd_wnd: Advertised window by receiver (for flow control)
   We assume that snd_wnd is set to MAX window size since no
flow control is implemented
cwnd: Set by sender (for congestion control); sthresh: The
threshold window

 cwnd = 1; HighWin =1; sthresh = MAX;
 Repeat for number of rounds intended

   for (int j=1; j<=cwnd; j++) {
     sim_hold(0.01); // Frame trm delay
     if error { // Trace the packet error and schedule trm
       sim_schedule(io, rtt/2.0, Retrm);
       sim_trace(1, "S io RTM Data " +Retrm); // trace the
trm
       sim_trace(1, "P "+Retrm+ " " +cwnd+ " " +sthresh+ "
"
                    +(HighWin/Sim_system.clock())); // Display
values

       AckTime[Retrm] = Sim_system.clock(); // Record the
transmit time
```

```
        }
        else {
          sim_schedule(io, rtt/2.0, HighWin);
          sim_trace(1, "S io Data " +HighWin);
          sim_trace(1, "P "+HighWin+ " " +cwnd+ " " +sthresh+
" "
                        +(HighWin/Sim_system.clock())); // Display
values
          AckTime[HighWin] = Sim_system.clock(); // Record the
transmit time
        };
        HighWin++;
    }
    sim_wait(ev);                  // Wait for ACK to come back
    if (cwnd < sthresh) cwnd = 2*cwnd; // Slow start
      else cwnd++;
    Get the tag of ACKed frame;
    Compare the frame trm time with timeout value;
    if timeout
      set sthresh to 0.5*win, cwnd to 1 and set retrm flag;
```

8.3.4 Simulation Models

We present simulation models of network systems to illustrate the application of DEVS techniques just explained. The intent is to investigate and develop simulation performance models of communication protocols in a LAN environment. The first communication protocol considered for modeling is the token bus scheme. Next, simulation approaches are shown to model circuit-switched and packet-switched networks. The simulation languages used are SLAM, COMNET III, and OPNET Modeler. However, the method may be used to develop models in other DEVS languages.

8.3.4.1 Simulation of a Token Bus Scheme

The essential components in token passing protocols are a bus, packets, a token, and transmit/receive queues. The bus is the medium used by a station holding the token if there is at least one packet in its transmit queue. The input parameters of the system are bus propagation delay and packet/token transmission times. We restrict our simulation model to a determination of channel utilization and packet queuing delay.

The system initialization process is shown in the flowchart of Figure 8.26. The process sets up the variables such as end-to-end propagation delay and token/packet transmission times. The system initialization event also determines the first station in the token passing sequence by assigning a value to the variable XX(8), and EVENT2 is called to start the token passing mechanism. We assume that a token may not be lost and can be generated only during the initialization phase. Every station in the network functions according to the flowchart of Figure 8.27. A packet is created randomly at each station, and the control is passed to EVENT1. EVENT1 queues the packet in a transmit queue and returns the control to the station where the operation is terminated. However, EVENT1 is not responsible for scheduling the service.

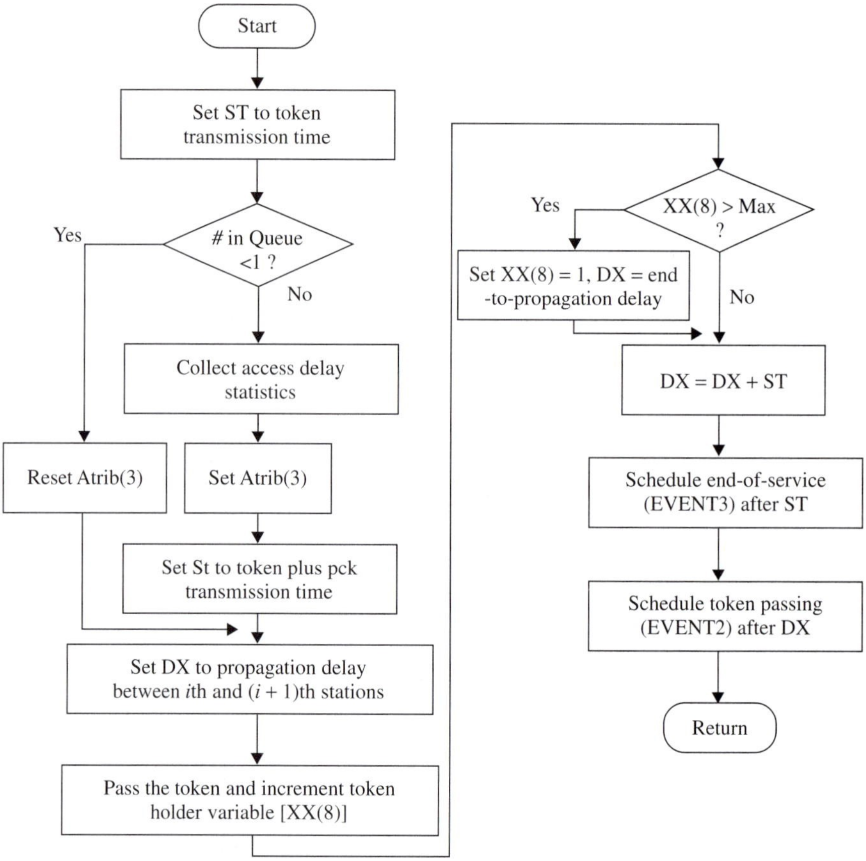

Figure 8.26 System initialization process for token bus.

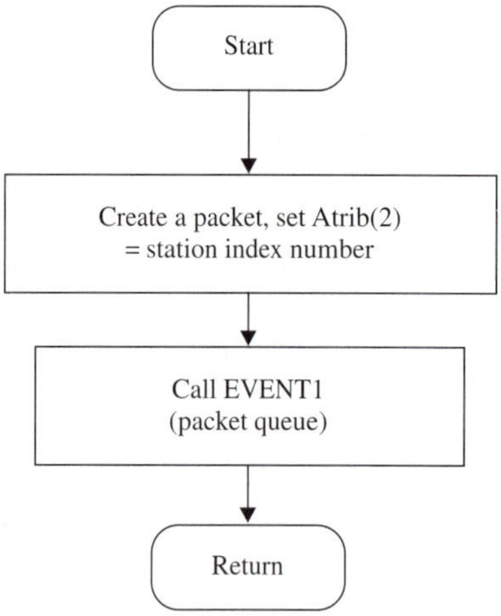

Figure 8.27 Packet creation and EVENT1 calling sequence.

The main simulation task consists of scheduling the packet removals from transmit queues and passing the control over to the next station. Thus, for example, if station 1 has the token, a packet from its transmit queue will be removed and the token will be held for packet plus token transmission time. Then, after an additional delay equivalent to the propagation time, the token is passed to the next station in the logical ring. Now assuming that a station has nothing to transmit, the station holding the token will keep the token for the token transmission time only. The end of transmission is therefore scheduled after the packet plus token delay or just the token transmission delay. A station holding the token schedules its passing to the next station after DX, which contains the propagation delay plus either the packet and token or the token transmission delay.

EVENT2 controls the operation of token passing. Its flowchart is shown in Figure 8.28. A variable ST is initially set for the token transmission time. The transmit queue of station [XX(8)] is then checked for a transmit packet, and if there is a packet, ST is modified to account for the packet transmission time. An attribute value [ATRIB(3)] is set to distinguish a packet with data (and token) from a packet with no data. The propagation delay is set accordingly, and the variable XX(8) is incremented, which determines the next station in the token passing sequence. We assume a 40-station network with token passing sequence in numerical order from 1 to 2 to 3, and so on. The last station in the logical ring (MAX) completes the logical ring by setting XX(8) back to 1. The event op-

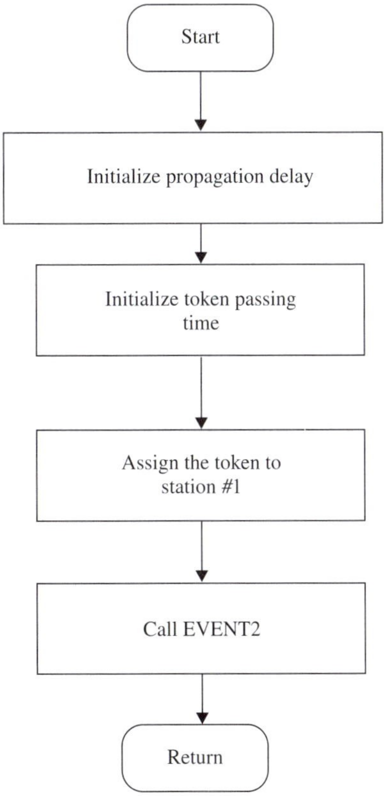

Figure 8.28 EVENT2 for token bus operation.

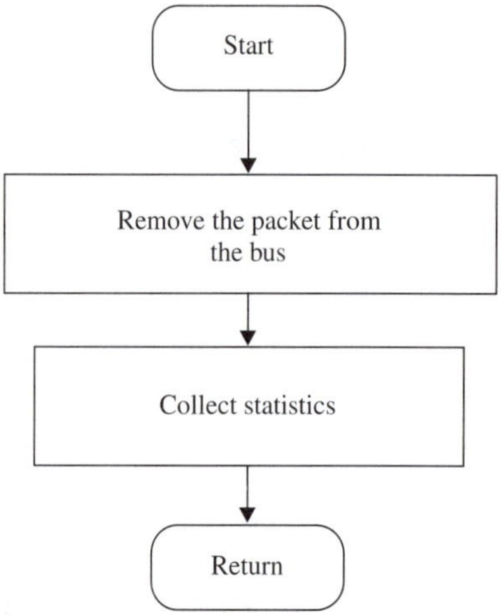

Figure 8.29 EVENT3 for token bus operation.

eration ends after the end-of-service event (EVENT3) has been scheduled after ST and the token passing event (EVENT2) after DX. The end-of-service event removes the packet from the medium. The packet consists of both the data and token portions. The token portion is always occupied, whereas the data portion will be empty if there was nothing in the transmit queue of a station holding the token. The channel utilization and system delay statistics are collected at this time, as shown in Figure 8.29.

The channel utilization simulation results are obtained for the model and a comparison is made with the analytical results presented in Section 8.2.1. The simulation results have a 95% confidence level, with the upper bound on confidence interval as ±0.06. With a confidence level of 99% the confidence interval upper bound becomes ±0.09. Notice that the analytical and simulation results are within about 5%. Table 8.3 shows simulated access delay results for the token bus. A comparison of analytical and simulation results is shown in Figure 8.30.

8.3.4.2 A Circuit-Switched System Model

COMNET III is a network simulation tool designed specifically to simulate computer and communication networks. Its predefined icons allow the development of a network model

TABLE 8.3
Token Bus Best-Case Access Delay: $M = 40$, $a = 0.01$, $q = 0.1$

Arrival Rate (Packets/s)	Simulation (ms)	Arrival Rate (Packets/s)	Simulation (ms)
500	8.1	800	20
600	9.7	830	30.3
700	10.2	850	41.2
750	14.8	880	175

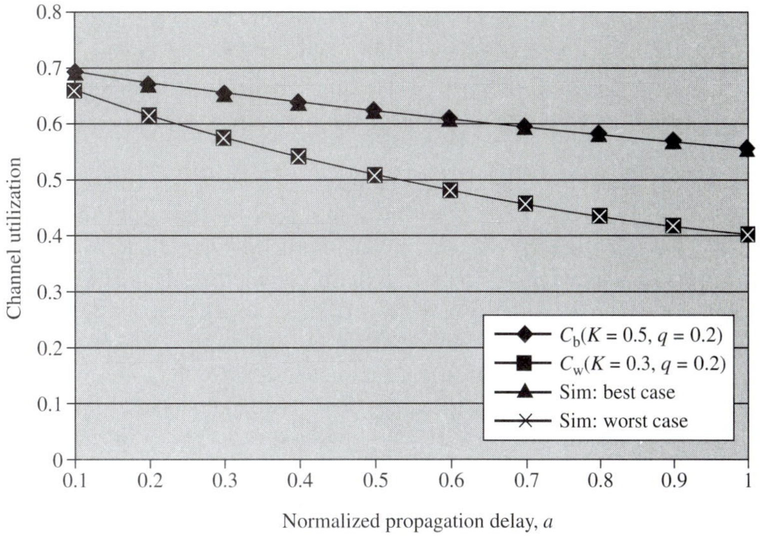

Figure 8.30 Analytical and simulation results of token bus.

by making an interconnection with links of several predefined types. In addition, users may define their own nodes, links, and types of service performed. Nice features of COMNET III are the ability to do animation, the automatic use of default values, and flexibility in choosing the output statistics. The animation feature, which allows the user to see the transmission activity on the screen, is useful for verifying the simulation. However, the depth of simulation is somewhat limited because the application models running on the nodes are not fully controllable by the user: the complex processes running at the nodes could not be designed. Also, the default values of variables may sometimes cause confusion about the input variables provided to the simulation. Future enhancements will probably include more detailed animation (packet/frame level), increased flexibility and control of process design, and a better user interface.

We present a COMNET III model of a circuit-switched system between two nodes, Mt. Pleasant (MP) and San Francisco (SF), connected by point-to-point unidirectional faulty T1 link (MP to SF) and 64 Kbps unidirectional link (SF to MP). We assume that MTBF (mean time between failures) for the faulty link is 10 (exponential distribution) and the MTTR (mean time to repair) is 5 (normal distribution with standard deviation = 1). Also, we assume that both nodes use circuit-switched calls with exponential interarrival times with mean 10 seconds for an average duration of 5 minutes (normal distribution; SD = 2). We use the default parameters for the nodes and perform the simulation for 60,000 seconds using COMNET III. We will determine the following:

Default node parameters

Link blocking probabilities for both links

Link utilizations for both links

Trace of all the calls with the numbers generated at the call sources, number processed at the nodes and links, and number of lost call

Measurement of the actual mean time between arrival of calls at both sources and comparison with the input parameters

For the simulation to be done, first we need the user interface to design the model. The connected model is shown in Figure 8.31. Most of the icon labels and parameters may be changed by double-clicking on them. After the network connections have been set, a routing table must be set up to make sure that all calls from MP to SF take the faulty T1 link and all calls from SF to MP take the 64 Kbps link. The default is to use both links as a bidirectional connection and use the best available path for routing. The next step is to set up the simulation parameters and the required output statistics, which may be done via the pulldown options from "Simulate" and "Report" on the toolbar at the top of the screen (Figure 8.31). We mention five aspects of this procedure.

1. Some of the default node parameters are 10,000 Kbps bandwidth, 100 MB of storage and use of single processor.

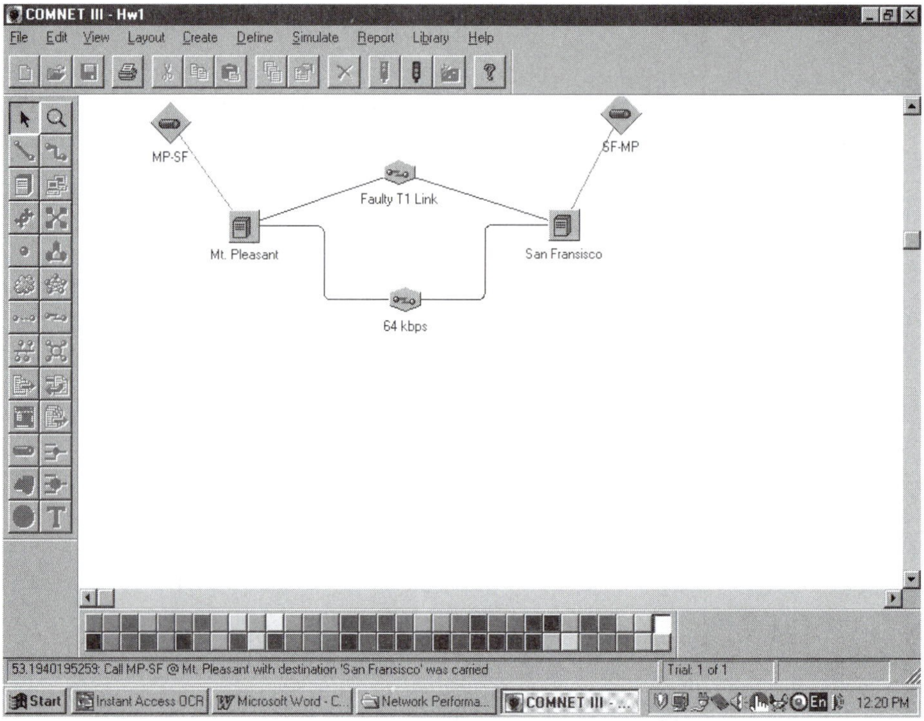

Figure 8.31 Screen snapshot of circuit-switched simulation in COMNET III.

2. The link blocking probability ratio is obtained from the number of calls carried and blocked. On the T1 link, 6036 calls were attempted and there were 2835 calls blocked because of either traffic or link unavailability. So the blocking probability on T1 is 2835/6036 = 0.47. Similarly, the blocking probability on 64K line is 5730/5920 = 0.97.
3. The simulation report gave the T1 and 64 Kbps link utilizations as 52.8 and 96.71%, respectively.
4. MP processed a total of 6226 calls (6036 attempted/sent and 190 received). SF processed a total of 9121 calls (5920 attempted/sent and 3201 received).
5. The total number of calls attempted from MP and SF during 60,000 seconds was 6036 and 5920, respectively, with mean time between arrivals as 60,000/6036 = 9.94 and 60,000/5920 = 10.14 seconds. Because of the random behavior of the simulation, these times are different from the input value of 10.

8.3.4.3 Star Connected Model

OPNET is a software tool built on C++ to simulate large communication networks with detailed protocol modeling and performance analysis. Its features include graphical model specification, a hierarchical object-based modeling approach, and an integrated data analysis tool. There are several tools available in OPNET that allow the users to do the following:

Create/edit networks and nodes.

Create/edit processes running at nodes.

Create/edit packet formats.

Specify the simulation output via the probe editor.

Analyze the simulation results by producing charts.

Define mathematical processes for use in the analysis tool.

The availability of several tools makes OPNET very flexible and provides the ability to model virtually any type of communication system up to the process level. However, learning the tool itself may be somewhat difficult for beginners in the field of modeling and simulation. It is helpful if the user has some experience in programming in one of the object-oriented languages such as C++ or Java. Figure 8.32 shows an OPNET simulation of a star model, a model of a four-node network connected via a hub. The processes running at each node are either transmit or receive. The utilization characteristics obtained with the model are shown in Figure 8.33.

8.3.4.4 Wireless Simulation Models

Ns is a network simulation program available from UC Berkeley (http://www-mash.cs.berkeley.edu/ns) for many platforms including Unix, linux and Windows. It has an animation support that facilitates the study of complex networks including mobile systems. Some other advantages of ns are

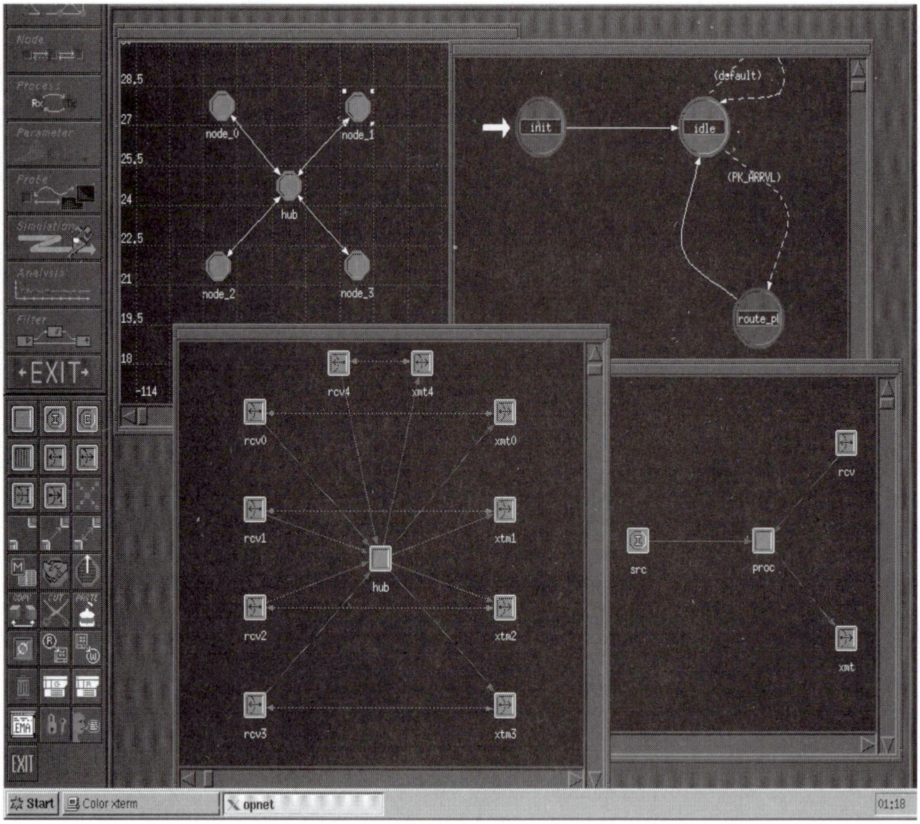

Figure 8.32 Screen snapshot of a star model from OPNET Modeler.

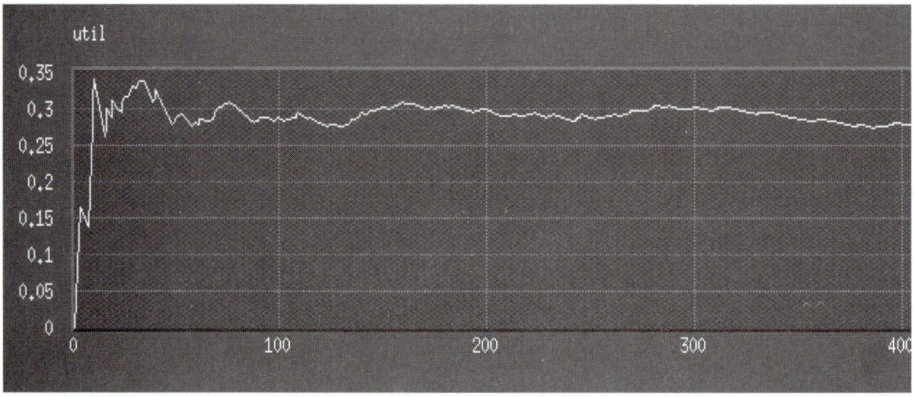

Figure 8.33 The presentation of star model utilization results by means of OPNET Modeler.

- Contributed examples of models are available. These models help understand the performance issues in different type of networks.
- The code is continuously updated (current version is ns-2), and the distribution files contain most of the contributed models.
- It is available at no cost

Ns is based upon tcl and tk, graphic programming languages. (See the Perl/tk FAQ at http://w4.lns.cornell/edu/,pvhp/ptk/ptkTOC.html.) The ns installation process may appear to be cumbersome since its operation depends upon the right choice of tcl/tk versions. However, there are all-in-one packages available too but the user must be familiar with the system to take care of occasional glitches. Ns-tutorials that are part of package documentation help the user get started. Once installed properly, the package is started by typing ns <tcl script>. The script consists of commands to create a network topology and events consisting of traffic sources and destinations. Figure 8.34 shows a simple tcl script. The first three lines set up the ns simulator and tracing for network animator called nam. The finish procedure runs the animator, so a user does not need to take any action to start the animator. The main ns code follows the procedure. There are two ns nodes established in this example. The nodes are connected by a 1 Mbps full-duplex link generating 500 byte packets every 0.005 sec. The simulator is invoked by the command $ns run. The animator window is open automatically after the script is run and the network is displayed. Using the different buttons on the screen, as shown in Figure 8.35, controls the animation. Another network animation snapshot is shown in Figure 8.36.

```
set ns [new Simulator]
set nf [open out.nam w]
$ns namtrace-all $nf

proc finish { } {
       global ns nf
       $ns flush-trace
       close $nf
       exec nam out.nam &
       exit 0
}

set n0 [$ns node]
set n1 [$ns node]
$ns duplex-link $nO $n1 1Mb 10ms DropTail
set cbrO [new Agent/CBR]
$ns attach-agent $nO $cbrO
$cbrO set packetSize_ 500
$cbrO set interval- 0.005
set nullO [new Agent/Null]
$ns attach-agent $n1 $nullO
$ns connect $cbrO $nullO
$ns at 0.5 "$cbrO start"
$ns at 4.5 "$cbrO stop"

$ns at 5.0 "finish"
$ns run
```

Figure 8.34 tcl script to simulate a two-node network.

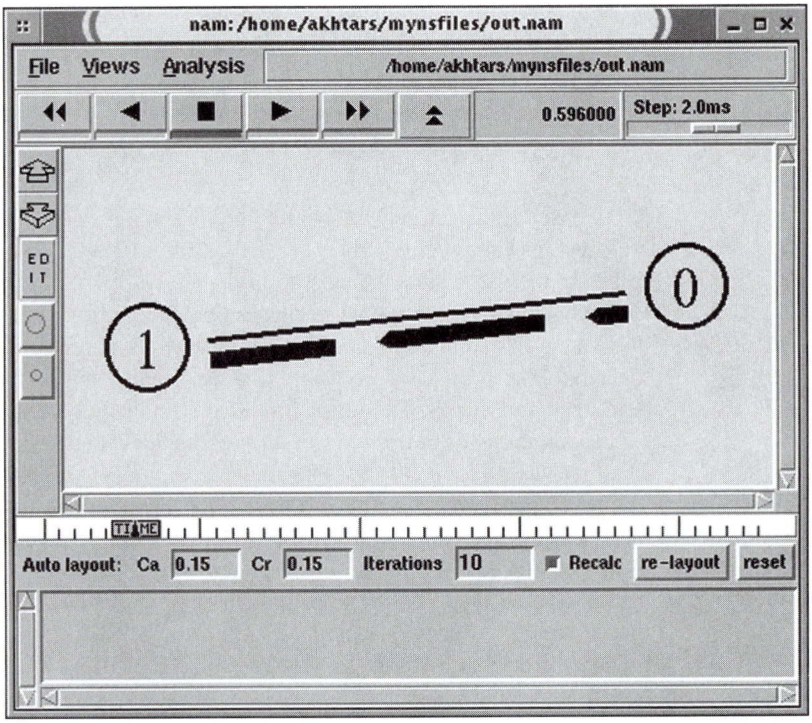

Figure 8.35 Screen snapshot of a two node network simulation.

One of the several extensions to ns is called Monarch developed at CMU. (http://www.monarch.cs.cmu.edu). This extension provides new elements to facilitate physical, data-link and network layers implementations in the simulation. For instance, it is possible to construct details of wireless subnets, LANs, or multi-hop adhoc networks. Additionally, a visualization tool called ad-hockey is provided to create input scenarios and view the trace playback of the simulation. However, it is advisable to get the proper version of Perl/tk for it to run properly since problems are encountered when trying to run with other versions (older or newer).

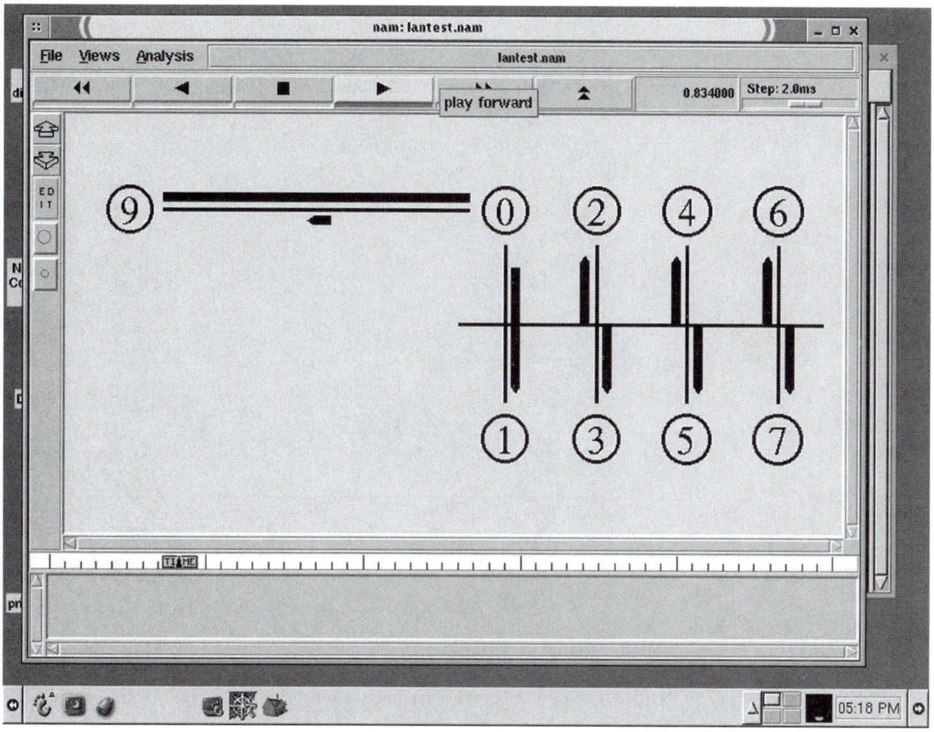

Figure 8.36 A screen snapshot of LAN simulation.

Ad-hockey does not invoke in the simulator automatically. Commands must be typed to run ns on the scenario files created by ad-hockey. The output trace file created by ns is post processed by viz-trace tool to create a .viz file, which is used by ad-hockey as an input. The construction tool allows the user to create mobile nodes and the path to be followed by these nodes. Mobile node characteristics are predefined and may be determined by changing commands in the corresponding tcl script. In addition, several files may also control movement and communication patterns. Message, data and routing protocol packet types are predefined. Figure 8.37 shows the use of ad-hockey to create a mobile system scenario. It shows a mobile scenario of three nodes when each is following a pre-determined path. A connecting line shows the communication between nodes.

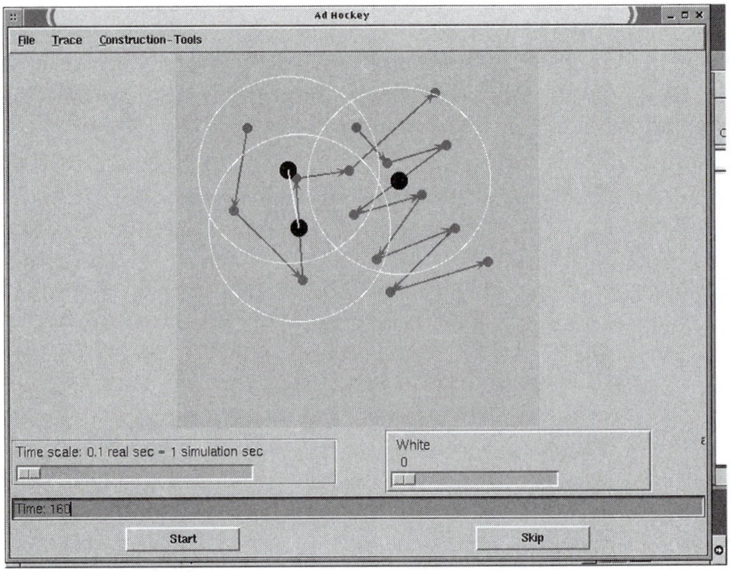

Figure 8.37 Ad-hockey screen snapshot showing a mobile system scenario.

8.4 NETWORK TRAFFIC MONITORING

The theoretical concepts of network performance modeling and simulation are not very useful or important unless they are applied to practice. With the growth of networks in the recent years it is important for network administrators and practitioners to understand the performance issues. Similarly, it is vital for the network modelers to understand the real network issues. This section clarifies some of the current network measurement and modeling practices. The tools mentioned are constantly being updated, and it is not our intent to show how to use these tools. We make an effort to show the features of the tools that may very well be available in many other tools at a different price level. Our intent is to show how the available features may be used to reinforce the theoretical concepts of network modeling and simulation.

Probably the best and most inexpensive approach to traffic monitoring uses Windows-based applications available on the PC. This approach is further enhanced if the PC is connected to the Internet via a domain name server (DNS). Also, access to administrator privileges may be needed to execute the Windows NT–based traffic monitoring tools. However, most of the commands we cover in this section do not require administrator privileges.

8.4.1 Windows-Based Traffic Monitoring

Arp: This command displays and modifies the IP-to-physical address translation tables used by the Address Resolution Protocol (ARP). The IP address is the Inter-

```
C:\WINDOWS>arp -a

Interface: 141.209.141.30
  Internet Address      Physical Address      Type
  141.209.131.2         08-00-20-85-95-77     dynamic

C:\WINDOWS>
```

Figure 8.38 Use of the Arp command for address resolution.

net address of the form abc.cps.cmich.edu, and the physical address is the associated 48 bits (12 hex digits) NIC (network interface card) address uniquely assigned to each adapter. Figure 8.38 shows few example runs of this command.

Netstat: This command provides several protocol statistics and TCP/IP network connections. The protocol options may be selected by the `-s` option to choose from TCP, UDP, or IP. The `-a` option displays all the connections and listening ports. The `-e` option displays the Ethernet statistics. The `-p` option shows connections for the specified protocol. An interval may be selected to redisplay the current statistics after regular intervals (in seconds). Pressing Ctrl-C stops the automatic redisplay. If the interval is not given, Netstat will display the results only once. Figure 8.39 shows the use of this command.

Ping: Ping is the most convenient way to measure the round-trip time to a specified destination given as an IP address. The `-t` option of Ping performs a continuous ping until interrupted. It returns three values corresponding to maximum, minimum, and average times. The `-l` option allows a packet size to be put as an option. The `-j` and `-k` options allow the loose source route and strict source route along a specified host list. The `-i` option allows a time-to-live (TTL) parameter to be included in the command. Figure 8.40 shows an example of running the Ping command. The Windows Ping command is very much similar to the Unix Ping command. However, in Unix, the command may have different options.

Route: The Route command manipulates the network routing tables. When used for a destination node, the `-f` option of the command clears the routing table of all gateway entries. The commands within Route allow addition, deletion, and modification of routes, in addition to printing out the route information. Since this command manipulates the routing table, administrator privileges may be needed to run it.

```
C:\WINDOWS>netstat

Active Connections

Proto  Local Address      Foreign Address          State
TCP    omair:1058         ccserver0.cps.cmich.edu:nbsession ESTABLISHED
TCP    omair:1102         users.cps.cmich.edu:nbsession ESTABLISHED
```

Figure 8.39 Use of the Netstat command.

```
C:\WINDOWS>ping www.yahoo.com

Pinging www6.yahoo.com [204.71.177.71] with 32 bytes of data:

Reply from 204.71.177.71: bytes=32 time=71ms TTL=244
Reply from 204.71.177.71: bytes=32 time=68ms TTL=245
Reply from 204.71.177.71: bytes=32 time=71ms TTL=245
Reply from 204.71.177.71: bytes=32 time=70ms TTL=245

C:\WINDOWS>ping -l 1024 www.yahoo.com

Pinging www6.yahoo.com [204.71.177.71] with 1024 bytes of data:

Reply from 204.71.177.71: bytes=1024 time=83ms TTL=245
Reply from 204.71.177.71: bytes=1024 time=81ms TTL=245
Reply from 204.71.177.71: bytes=1024 time=81ms TTL=245
Reply from 204.71.177.71: bytes=1024 time=81ms TTL=245
```

Figure 8.40 Use of Ping to measure round-trip time.

Tracert: This command allows tracing the route to a destination node. The route along the path to destination is printed. The `-h` option allows a value from the user for the maximum number of hops to search for the target. Also, the route to destination may be selected by using the `-j` option with the host list. Figure 8.41 shows the use of Tracert command. This command is similar to the Unix Traceroute command. Figures 8.42 and 8.43 show two indirect uses of Unix Traceroute via a CGI script.

Winipcfg: This command is very useful for finding out a PC's IP configuration running under Windows. Specifically, if the PC is allocated a new address at each signup, this command gives the IP address of the PC. In addition, it also provides the adapter 48-bit address. Figure 8.44 illustrates the use of Winipcfg.

```
Tracing route to cps.msu.edu [35.9.20.20]
over a maximum of 30 hops:

  1   2 ms     2 ms     1 ms 141.209.128.3
  2   4 ms     2 ms     2 ms fo027-r1.snmp.cmich.edu [141.209.14.1]
  3   5 ms     3 ms     3 ms fo027-merit-r1.snmp.cmich.edu [141.209.4.1]
  4   34 ms    54 ms    25 ms hssi6-0-0.msu.mich.net [198.108.22.10]
  5   22 ms    16 ms    13 ms cc2-gw-fd10.msu.edu [35.9.6.1]
  6   19 ms    19 ms    16 ms eng-gw.msu.edu [35.8.100.14]
  7   16 ms    12 ms    16 ms cps-gw.cps.msu.edu [35.9.21.10]
  8   13 ms    15 ms    15 ms cps.msu.edu [35.9.20.20]

Trace complete.
```

Figure 8.41 Using Tracert to trace the route to a destination node.

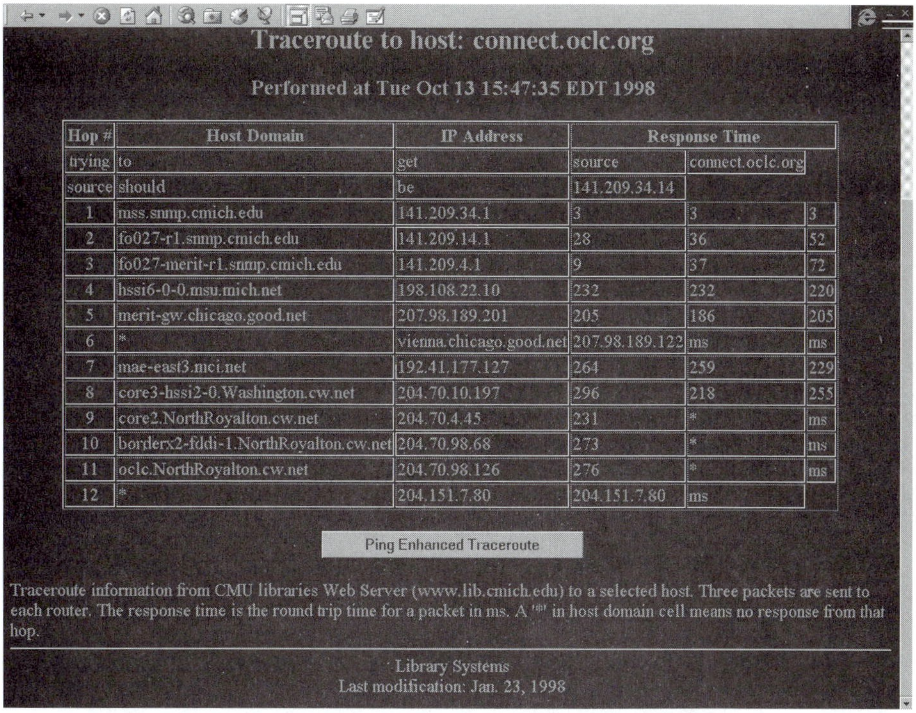

Traceroute to host: connect.oclc.org

Performed at Tue Oct 13 15:47:35 EDT 1998

Hop #	Host Domain	IP Address	Response Time		
trying	to	get	source	connect.oclc.org	
source	should	be	141.209.34.14		
1	mss.snmp.cmich.edu	141.209.34.1	3	3	3
2	fo027-r1.snmp.cmich.edu	141.209.14.1	28	36	52
3	fo027-merit-r1.snmp.cmich.edu	141.209.4.1	9	37	72
4	hssi6-0-0.msu.mich.net	198.108.22.10	232	232	220
5	merit-gw.chicago.good.net	207.98.189.201	205	186	205
6	*	vienna.chicago.good.net	207.98.189.122	ms	ms
7	mae-east3.mci.net	192.41.177.127	264	259	229
8	core3-hssi2-0.Washington.cw.net	204.70.10.197	296	218	255
9	core2.NorthRoyalton.cw.net	204.70.4.45	231	*	ms
10	borderx2-fddi-1.NorthRoyalton.cw.net	204.70.98.68	273	*	ms
11	oclc.NorthRoyalton.cw.net	204.70.98.126	276	*	ms
12	*	204.151.7.80	204.151.7.80	ms	

Ping Enhanced Traceroute

Traceroute information from CMU libraries Web Server (www.lib.cmich.edu) to a selected host. Three packets are sent to each router. The response time is the round trip time for a packet in ms. A '*' in host domain cell means no response from that hop.

Library Systems
Last modification: Jan. 23, 1998

Figure 8.42 Using Unix Traceroute via a CGI script.

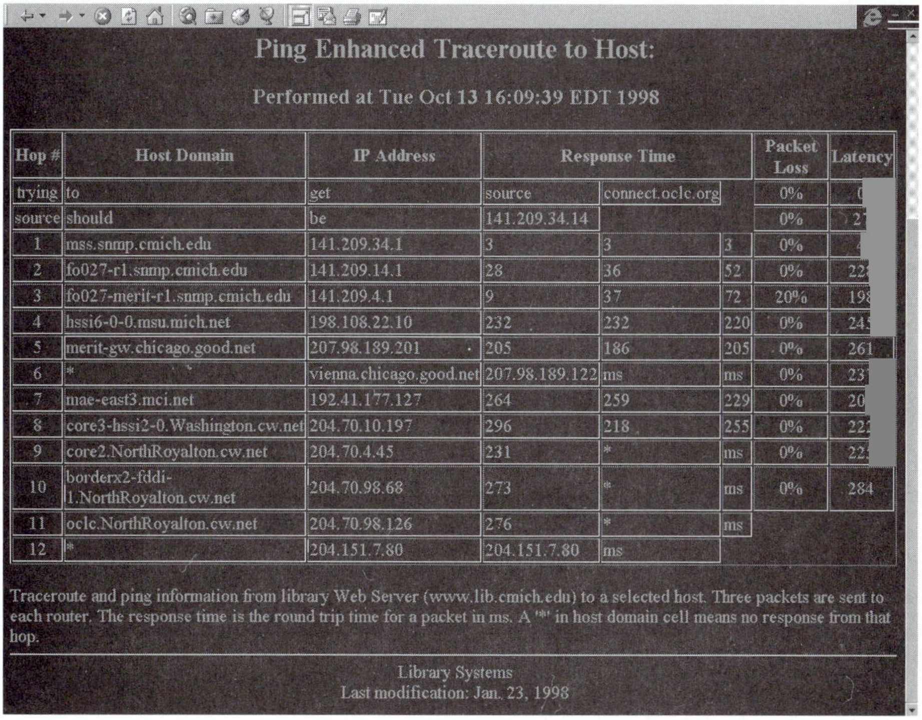

Ping Enhanced Traceroute to Host:

Performed at Tue Oct 13 16:09:39 EDT 1998

Hop #	Host Domain	IP Address	Response Time			Packet Loss	Latency
trying	to	get	source	connect.oclc.org		0%	[illegible]
source	should	be	141.209.34.14			0%	[illegible]
1	mss.snmp.cmich.edu	141.209.34.1	3	3	3	0%	[illegible]
2	fo027-r1.snmp.cmich.edu	141.209.14.1	28	36	52	0%	[illegible]
3	fo027-merit-r1.snmp.cmich.edu	141.209.4.1	9	37	72	20%	[illegible]
4	hssi6-0-0.msu.mich.net	198.108.22.10	232	232	220	0%	[illegible]
5	merit-gw.chicago.good.net	207.98.189.201	205	186	205	0%	261
6	*	vienna.chicago.good.net	207.98.189.122	ms	ms	0%	[illegible]
7	mae-east3.mci.net	192.41.177.127	264	259	229	0%	[illegible]
8	core3-hssi2-0.Washington.cw.net	204.70.10.197	296	218	255	0%	[illegible]
9	core2.NorthRoyalton.cw.net	204.70.4.45	231	*	ms	0%	[illegible]
10	borderx2-fddi-1.NorthRoyalton.cw.net	204.70.98.68	273	*	ms	0%	284
11	oclc.NorthRoyalton.cw.net	204.70.98.126	276	*	ms		
12	*	204.151.7.80	204.151.7.80	ms			

Traceroute and ping information from library Web Server (www.lib.cmich.edu) to a selected host. Three packets are sent to each router. The response time is the round trip time for a packet in ms. A '*' in host domain cell means no response from that hop.

Library Systems
Last modification: Jan. 23, 1998

Figure 8.43 Using Ping-enhanced Unix Traceroute via a CGI script.

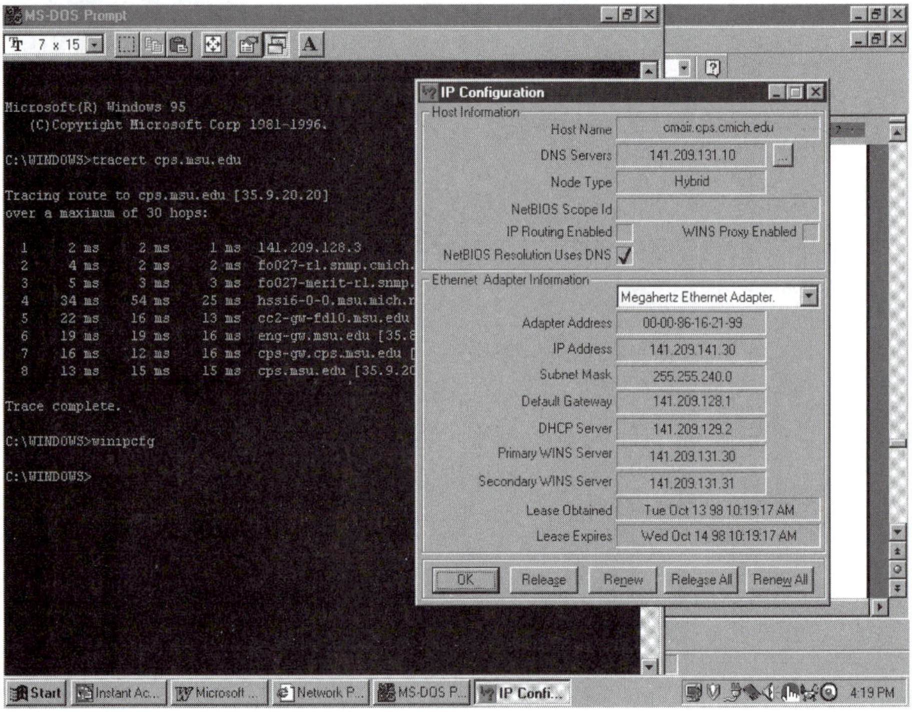

Figure 8.44 Using Winipcfg.

8.4.2 Ethernet Traffic Monitoring

8.4.2.1 DOS-Based EtherVision by Triticom

EtherVision is a low-cost, real-time Ethernet LAN monitor. EtherVision monitors and displays the activity of any Ethernet local-area network. It is intended for use in the monitoring, support, and management of networks. A single workstation running EtherVision can monitor all the traffic present on the Ethernet it is attached to. It is a DOS-based program that takes advantage of a site's existing PCs and network cards and works with all Ethernet media types. Therefore no other hardware is necessary. Some of its features are as follows:

1. EtherVision monitors and displays traffic from all stations present and operating on the Ethernet segment it is attached to.
2. It has several real-time display modes available for monitoring traffic. These include Station Pairs, Skyline, Statistics Summary, and Real-Time Protocol Distribution monitoring screens. There are two screens'-worth of Protocol Distribution information, for a total of 30 different protocols.
3. It can alert the user to conditions about the current state of the network, using "alarms" such as network idle, frame error, broadcast/second, and intruder detection.

4. It can generate several reports at a timed interval specified by the user. Among the types of report that can be generated are station statistics, station bandwidth, frame distribution, network summary, and network utilization.
5. It supports a variety of adapters from vendors including Novell, Western Digital, 3Com, Intel, and Gateway.

Some limitations include the following:

Maximum number of source or destination addresses	1024
Maximum number of station filters	1024
Maximum error count	65,535
Maximum byte count	2,147,483,640
Maximum frame count	Up to byte count
Maximum lines in Network Event Log	200
Maximum number of vendor IDs	175
Network idle time alarm	1–9999 seconds
Error alarm increments	1–9999 errors
Network utilization alarm	1–99%
Alarm frame count interval	1–9999 seconds
Individual station alarms	
Idle time	1–9999 seconds
Error counts	1–9999 errors
Utilization (% of total frames)	1–99%

We tested the demo version of EtherVision (Figures 8.45 and 8.46), which facilitates the exploration of options and operations by allowing you to navigate through all configuration and operational menus. A simulated environment is provided so that you can see how EtherVision operates in real time. The demo version of EtherVision is a simulation and does not actually perform monitoring on your network. It does not require that any Ethernet adapter cards be installed in the station where the demo is running.

The opening screen snapshot of EtherVision is shown in Figure 8.45. You can use the Monitor Traffic option to observe the source and destination traffic. The traffic monitor allows you to measure traffic from different sources or traffic to different destinations in terms of frames or bytes. The station ID may be shown in terms of interface value (Ethernet address) or the ASCII-ID allocated to the stations. The traffic may be displayed in terms of counts or a skyline chart showing the real-time traffic. Counts for short, long, and corrupted packets are provided in addition to the distribution of the frame sizes.

8.4.2.2 Windows-Based JETLAN Traffic Monitor

JETLAN by Chevin Software Engineering (UK) is one of several Windows-based traffic monitoring software packages that allow automatic node recognition, custom statistics,

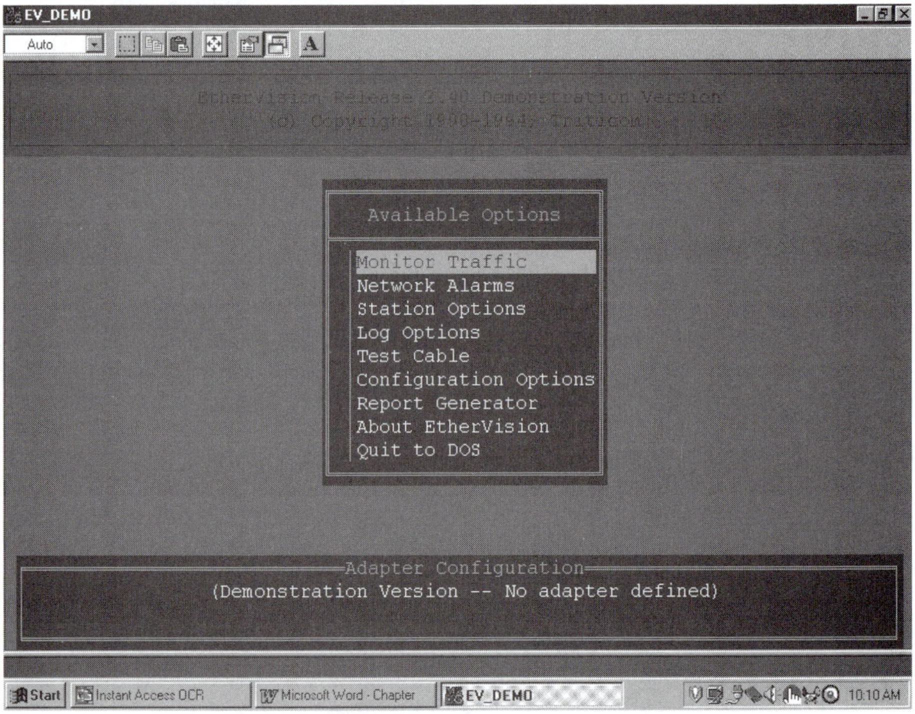

Figure 8.45 Opening menu of demo version of EtherVision.

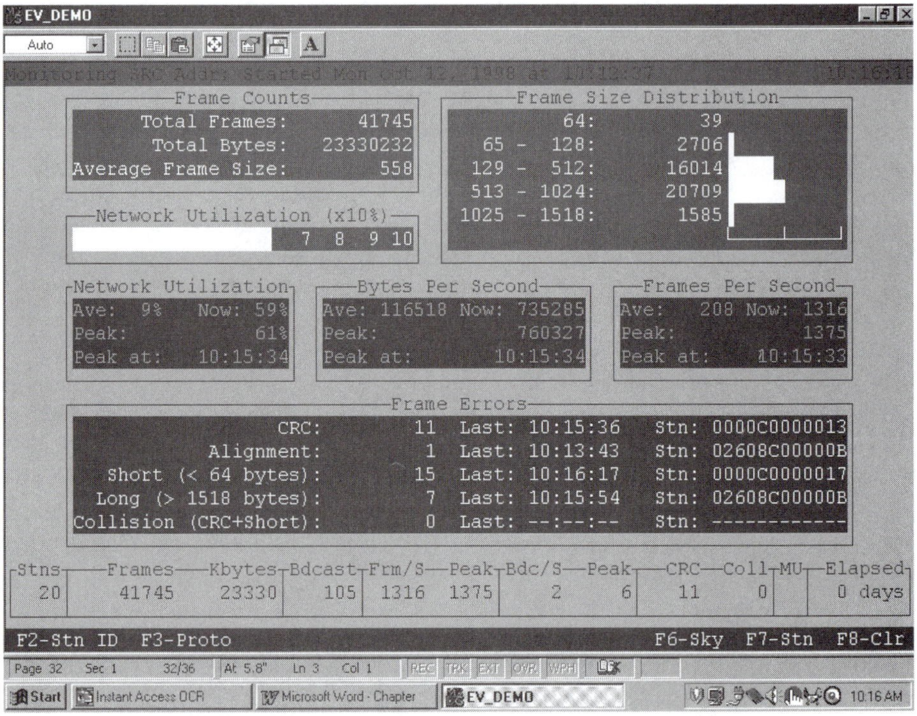

Figure 8.46 Screen snapshot of EtherVision showing the traffic statistics.

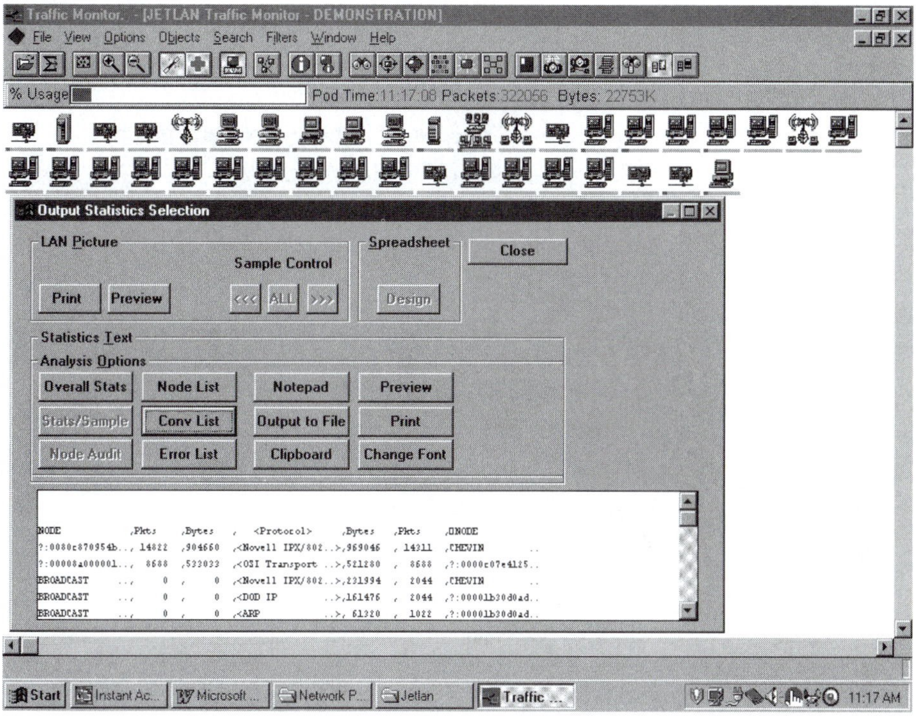

Figure 8.47 Screen snapshot of JETLAN Traffic Monitor.

error statistics, general and output statistics, node information and marking, node movement on the screen, statistics after filtering, and so on. A "stand-alone" personal traffic monitor for single-segment networks of up to 1000 nodes features "real-time" traffic monitoring, statistics collection and interpretation, output to printers and spreadsheet programs, address management, and user-defined types and groups. Figure 8.47 shows a screen snapshot of the options available to help you obtain the output statistics from the traffic monitor. A sample set of output statistics is shown in Figure 8.48.

In general, the screen shows:

- The LAN topology
- Conversations between nodes

```
Ordered by Packets Sent
NODE              ,Pkts Snt,Byts Snt,Pkts Rec,Byts Rec,Errors  ,Num
Conversations
CHEVIN            , 34416  ,  2479K , 33373  ,  2141K ,     0 ,    5
?:aa0004000134    , 31239  ,  1890K , 28808  ,  1728K ,     0 ,   20
?:08002f000a8a    , 19140  ,  2136K , 15312  ,974400  ,     0 ,    2
..
..
BROADCAST         ,      0 ,      0 ,  3476  ,309392  ,     0 ,    3
```

Figure 8.48 Output statistics generated from JETLAN Traffic Monitor.

- Nodes generating errors
- Autodiagnosis of network problems
- Autodetection of node type
- Graphical representation of node types and groups
- Statistics collection and interpretation
- Statistics breakdown by protocol, by errors, by node, by type, by group
- Statistics breakdown by sample period
- Total packets and bytes to and from each node
- Details on each protocol, including average packet size

Output options include the following:

- Output to printers
- Output to Windows Clipboard and to file
- User-configurable spreadsheet options
- Spreadsheet output by sample period
- Spreadsheet output by node

In addition to the basic monitoring facilities, JETLAN Traffic Monitor provides full monitoring of remote segments and intersegment traffic. The Protocol Analyzer feature allows the capture of packets off the LAN and decodes them, showing the details of the protocols and the contents of the packet. The Protocol Analyzer for Windows works in conjunction with the Traffic Monitor software. Although the Traffic Monitor is very powerful, at times a problem involving the actual contents of the packets may occur. This type of problem requires capturing packets and looking in detail at their contents. JETLAN Traffic Monitor works with the following protocol suites:

IBM Protocol Suite
Novell NetWare Protocol Suite
XNS Protocol Suite
TCP/IP Protocol Suite
NFS Protocol Suite
ISO Protocol Suite
DECnet Protocol Suite
Banyan VINES Protocol Suite
AppleTalk Protocol Suite

8.5 CHAPTER SUMMARY

With an increase of network traffic it is becoming important to study network performance at different levels of detail. Network performance can be evaluated using analytical, simulation, and monitoring (measurement) approaches. The analytical models consist mainly of queuing theoretical and stochastic models involving latencies and other time-related factors. Because of the complexity involved, however, these models are based on assumptions and often do not include the complete real network scenarios. However, these are useful in understanding the behavior of the networks under various conditions.

Simulation models are often more relaxed than the analytical models and include the level of details to represent the real scenarios more effectively. Writing network simulation programs can be a tedious job. For this reason, there is a tendency to use specialized network simulation packages. These packages are often suitable for a variety of network solutions such as LANs, WANs, mobile networks, and interconnection of networks. In addition, protocol models of networks may also be developed by using state diagrams similar to finite state machines. This allows a study of networks before implementation without using the hardware resources. Further, these packages may duplicate existing network architectures within the simulation tool to allow whatif studies. With this baselining approach, proposed network development and growth may be studied before expensive equipment is deployed.

An engineering approach to study the network performance usually involves monitoring, measurement, and testing of existing networks. There are a variety of hardware and software devices to help this task. These devices permit users to obtain performance measurements such as latency, throughput, and bandwidth utilization.

We conclude from this chapter that a successful networking system does not necessarily rely on one approach. A combination of these approaches with an effective network management is what makes the system more reliable, with fewest expected failures.

8.6 PROBLEMS

8.1 If packets arrive at a queuing system such that the time between arrivals has an exponential distribution with a rate of 2 per second, what is the probability that the time between successive arrivals is between 400 and 500 ms?

8.2 Given the following routing matrix (containing the paths and only nonzero nonsymmetric traffic entries in packets/second), find the mean delay/line, the mean packet delay, and the mean number of hops per packet. Each link is assumed to be full duplex and has a capacity of 100 Kbps. Mean packet size is 1000 bits. Ignore the propagation delay.

From	**To (A, B, C and D)**		
A	10(AB)	14(ABC)	20(ABCD)
B	8(BCD)	6(BDA)	6(BCD)
C	10(CDB)	10(CDA)	
D	6(DCBA)	8(DB)	10(DBC)

8.3 A LAN has three printers A, B, and C. One of the three printers is always busy, and the print jobs are assumed to be fixed at one minute (i.e., the printers may change state once every minute). If printer A is busy, it remains busy at the next minute with probability 0.2 or printer B becomes busy with probability 0.8. If printer B is busy it remains busy at the next minute with probability 0.5, or printer C becomes busy with probability 0.5. If printer C is busy it remains busy at the next minute with probability 0.6 or printer A becomes busy with probability 0.4. Consider a Markov model for this problem with a steady state condition for the Markov chain.

(a) Draw the Markov chain and show all the probabilities, including the transition probability matrix, and states.
(b) Which printer is most busy? Explain, using the state probabilities obtained?
(c) What is the probability that printer A is busy?

8.4 The queuing model shown in Figure P8.4 consists of two queues and three packets. The first queue has a single server with a service rate of 10, irrespective of number of packets in the queue. The second queue has three servers, each with a service rate of 5. Assume that the model has a product form solution. Using a continuous time Markov chain to compute the state probabilities, find the average number of packets in both queues.

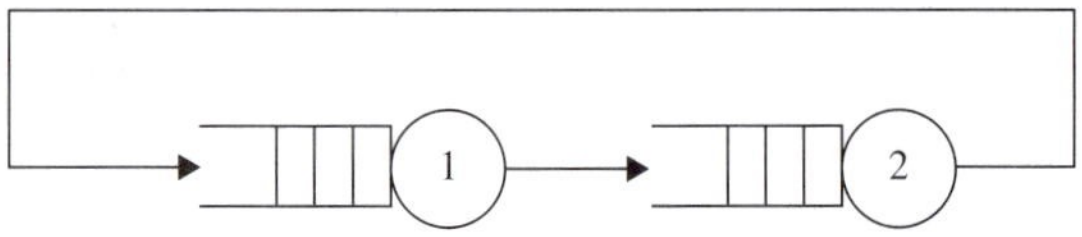

Figure P8.4

8.5 Consider an M/M/∞ queue that has a packet arrival rate of 10 per hour and a service rate of 15 per hour. What is the probability of zero packets in the queue?

8.6 Consider a closed queuing network in which there are two customers doomed forever to cycle between two queues Q1 and Q2. The queue service rates are μ_1 and μ_2, respectively.

(a) Show the system Markov chain. Find the transition rates of the chain.
(b) Find the steady state probability of each state.
(c) Find the expected number of customers in Q1 and Q2 if μ_1 and μ_2 are 10 and 8, respectively.

8.7 The channel associated with a given network is monitored, and the following data are obtained: bit rate, 1 Mbps; average packet length, 1000 bits/packet; steady state input, 500 packets/second; average number of packets stored in a section of the network, 10 packets.
Considering an M/M/1 model for the network:

(a) What is the traffic intensity or utilization factor?
(b) What is the average time delay in the section of the given network?

8.8 Consider an M/G/1 queuing model for a single-channel communication system. Assume Poisson arrivals and a general service process with coefficient of variation 0.3. The arrival and service rates are 10 and 20 packets/second, respectively.

(a) What is the expected packet size in bits if the channel (serving the queued packets) has a capacity of 10 Kbps?
(b) What is the expected number of packets on the channel [i.e., the number of packets in service (or the traffic intensity)]?
(c) What is the total expected number of packets in the system on the basis of one of the Pollaczek–Khinchine formulas?
(d) What is the total delay per packet, (i.e., the queuing delay plus the delay in service), from one of the Pollaczek–Khinchine formulas?

8.9 Consider a 1 km, 1 Mbps token bus system in which the token is passed on a broadcast bus. Assume that the token is passed down the logical ring through N active stations in a way that causes a minimum amount of propagation overhead. Also, assume that there are total of 10 stations, which are distributed evenly on the bus with equal propagation delay between two neighboring stations. If the average transmission rate from each active station is 100 packets/second and the signal propagation speed is 10,000 km/s, what are the channel utilizations (percentage) for packet sizes of 100 and 1000 bits?

8.10 Consider a token ring with a ring latency of 200 μs. What are the ring latencies in number of bits for 4 and 100 Mbps rings if the packet size is 1000 bits? What are the ring utilizations and effective throughputs in each case?

8.11 Suppose you are designing a sliding window protocol for a 1 Mbps point-to-point link to the moon, which has a one-way latency of 1.25 seconds. Assume that each frame carries 1024 bytes of data. How many bits do you need for the sequence number?

8.12 To transmit packets with virtual circuit transport, we first set up a virtual circuit and then we transmit the packets. The network is lightly loaded, and our packets do not face any queuing delay. The virtual circuit setup time is 400 ms. The packets travel over a path that goes through 10 nodes, and the links transmit at 56 Kbps. Each packet has 400 bits of data, a header of 5 bytes to indicate the virtual circuit number and the packet sequence number, and a trailer of 2 bytes that contains bits used for error detection. When we use datagram transport, no virtual circuit is set up, but each packet needs a header of 10 bytes instead of 5 to indicate the full destination address and source address, in addition to the packet sequence number. These packets also have the 2-byte trailer. Assume that datagrams also happen to follow the same path through 10 nodes. How long does it take to transmit N packets when using virtual circuit transport and when using datagram transport? For what values of N is it faster to use virtual circuit transport?

8.13 Consider a multichannel network in which there are several channels connected between each pair of nodes. Within the given network, assume that nodes A and B are connected by a three-channel link, with each channel failure and repair rate as 1 and 2 per year, respectively. Determine the following measures, considering a Poisson process and developing a birth–death model.
(a) The probability of no channel operating.
(b) The probability of all channels operating.
(c) The AB subsystem reliability as the probability of at least one channel operating.
(d) The expected number of channels operating.

8.14 Consider a two-channel multichannel network, with each channel failure and repair rates as 1 and 2 per year, respectively. Assume that the failures are covered with probability c. A channel failure is considered to be not covered when the failure cannot be repaired, causing a permanent failure. Also assume that each channel has its own repair facility. Develop a Markov chain model for the network and determine the following:
(a) The probability of no channels operating at steady state.
(b) The probability of all channels operating at steady state.
(c) The network availability as the probability of at least one channel operating at steady state.
(d) The flow balance equations as differential equations.

8.15 Use ping to determine the round-trip time to each of these four hosts. On the SUN system, the command is "`ping -s hostname`". If you're not familiar with Ping, use "`man ping`" to find out how to use the "`-s`" switch. Report the round-trip time for each of these hosts.
(a) www.cps.cmich.edu
(b) cs.stanford.edu
(c) www.mit.edu
(d) ewww.hawaii.edu
Can you justify the round-trip times to each of these hosts by considering just the speed of light? Compare (approximate) distances to delays.

8.16 Briefly describe the basic operations of a network analyzer. When would you need one?

8.17 Write and test a program to show the output as a sequence of ASCII character transmissions on a transmission line at regular intervals. Consider a line 3 km long and data transmission speeds of 1, 3, 5, 7, and 10 million characters per second.

Sample input/output: (The output assumes that a 1 km line can accommodate 10 characters and the transmission speed is 1 character/μs.)

Enter the data you wish to transmit:

```
My Data
```

Line at 4, 6, 8, 10, and 15 μs:

```
SOURCE              DESTINATION
   ↓        ↓
   D yM
   taD yM
   ataD yM
        ataD
```

8.18 OPNET simulation:
Run the simulation using OPNET (or another similar network simulation package) of an M/M/1 queue using the following parameters:

Interarrival rate: 1.0
Queue service rate: 9600 bps
Packet size: 9000 bits
Simulation seed: 430
Duration: 3000
Update interval: 200

Determine the following:

Plot of mean delay vs time.
Comparison with theoretical result
Effect of changing the packet size and queue service rate

8.19 Consider two network nodes Mt. Pleasant (MP) and San Francisco (SF) connected by point-to-point unidirectional faulty T1 link (MP to SF) and 64 Kbps unidirectional link (SF to MP). Assume that the MTBF for the faulty link is 15 (exponential distribution) and the MTTR is 10 (normal distribution; SD = 1). Also, assume that both nodes use circuit-switched calls with exponential interarrival times with mean 10 seconds for an average duration of 3 minutes (normal distribution; SD = 2). Use the default parameters for the nodes. Perform the simulation for 60,000 seconds using COMNET III (or another similar network simulation package). Determine the following:

(a) Default node parameters.
(b) Link blocking probability ratio for both links.
(c) Link utilization ratio for both links.
(d) Blocking probability ratio for each call source using the calls attempted and carried out.
(e) Call disconnection probability ratio for each call source.
(f) Trace all the calls and figure out how many calls were generated at the call sources, how many were processed at the nodes and links, how many were lost, and so on. Draw a picture if necessary.
(g) Measure the actual mean time between arrival of calls at both sources and compare with the input parameters.
(h) Repeat the simulation with two reruns and explain the differences. What could be the cause of differences (if any)?

8.20 Consider the network simulation of Problem 8.19. Rerun the simulation by setting up the user-defined routing table between MP and SF so that all calls from MP to SF take the faulty T1 link and all calls from SF to MP take the 64 Kbps link. (Do not use two separate network connections.)

Determine the following:

(a) Plot the link blocking probabilities as a function of arrival rates of 1, 3, 5, 7, 9, 11, 13, 15, 17, 19, and 21 seconds at each call source. (There must be two plots, one for each call source.)
(b) Plot the link utilizations as a function of arrival rates as given in part a.
(c) Change the arrival rate at SF to achieve a 100% link utilization of the 64 Kbps link. Also, make sure that the blocking probability for the link is zero.
(d) Change the call processing rate of SF node in part c to see the effect on link utilization. Plot link utilization as a function of node call processing rate.
(e) Comment on the results in parts a–d.

8.21 Simulate a five-node circuit switched network (Figure P8.21) with the following available circuits and paths:

	To				
	A	**B**	**C**	**D**	**E**
A		2 (AB)	3 (AC)	(ACD)	1 (AE)
B	2 (BA)		1 (BC)	1 (BD)	(BDE)
C	3 (CA)	1 (CB)		1 (CD)	(CDE)
D	(DEA)	1 (DB)	1 (DC)		1 (DE)
E	1 (EA)	(EDB)	(EDC)	1 (ED)	

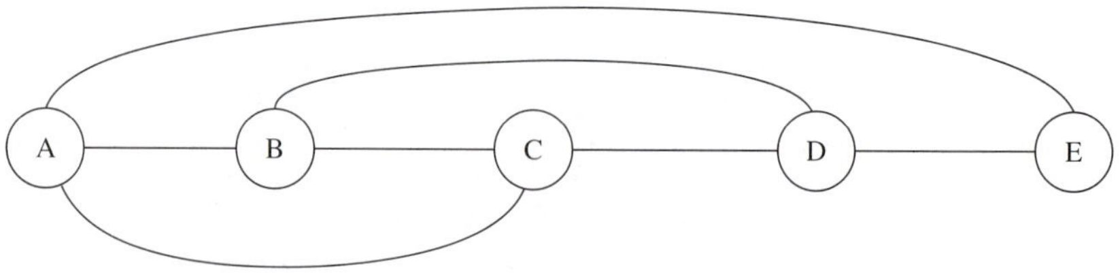

Figure P8.21

Assume that each node is producing the calls at 10-second intervals (average of exponential distribution) for random neighbors. Compare the results with the case of routing table destinations obtained from the accompanying table. (Divide the generated traffic equally among four destination nodes.) Also, compare the results with the optimal path choice (instead of user-defined choice) of paths.

8.22 You are to compare three LAN protocols (Ethernet, token bus, and token ring) being considered for a proposed office network (Simulation Runs, described shortly, explains the statistics to be compared). The proposed network will connect five client computers for faculty and an existing LAN to a new network server. The existing LAN is a laboratory of computers arranged on an Ethernet. It will be linked to the proposed office network by a bridge. System characteristics are as follows:

Network traffic from client computers: Each of the five client computers will generate traffic to other nodes according to the following distribution: 80% to server, and 20% to the existing laboratory LAN. Message interarrivals use the default distribution [Exp(10.0)], and message length is uniformly distributed between 1 and 1000 bytes.

Network traffic from existing LAN: The existing LAN laboratory will generate traffic only to the new server. Use default settings for message interarrival and size. One icon will suffice to represent all stations.

Network traffic from the server: The server will respond to each message it receives. Message size is uniformly distributed between 1 and 2000 bytes.

The existing LAN: The existing LAN laboratory should be modeled as a CSMA/CD network. Set the minimum frame size to 64 bytes and the maximum frame size to 512 bytes. Set the bandwidth to 10 Mbps. Connect it to the new LAN using a bridge with default parameters.

The new LAN: Try three different protocols: 10 Mbps Ethernet, 10 Mbps token bus, and 4 Mbps token ring. Set the minimum and maximum frame sizes to 64 and 512 bytes, respectively.

The simulation runs: Run the simulation for 0.1 second. Repeat this for all three configurations. Compare the following measures for the new LAN: number of frames transmitted, channel utilization, average frame delay, number of collisions (Ethernet only), and average message delay for messages from the server to the other nodes. Summarize any conclusions you reach based on this limited study. Turn in the chart containing the five measures just listed for all three protocols, and your stated conclusions.

8.23 Use the Traceroute utility to trace the routes to following destinations:

(a) http://www.cse.msu.edu

(b) http://www.stanford.edu

(c) http://www.hawaii.edu

(d) http://www.monash.edu.au

Connect to the four web sites listed in a–d. Determine how many hops each of the web sites is from your computer system. Run the programs at three different times in a 24-hour period and tabulate (or plot) the access times. Repeat your observations and plots for the Ping-Enhanced Traceroute as well. What is the difference? Comment about the result.

8.7 REFERENCES

BOOKS

Bolch, G., S. Greiner, H. de Meer, and K. Trivedi, *Queueing Networks and Markov Chains: Modeling and Performance Evaluation with Computer Science Applications*. New York: Wiley, 1998.

Cahn, R. S., *Wide Area Network Design: Concepts and Tools for Optimization*. San Francisco: Morgan Kaufmann, 1998.

Coombs, C. F., *Communications Network Test and Measurement Handbook*. New York: McGraw-Hill, 1997.

Danthine, A., O. Spaniiol, et al., (Editors) *High-Performance Networks for Multimedia Applications*. Dordrecht: Kluwer Academic Publishing, 1998.

Edmead, M. T., and P. Hinsberg, *Windows NT Performance: Monitoring, Benchmarking, and Tuning*. New York: Macmillan, 1998.

Flood, J. E., *Telecommunications Switching, Traffic and Networks*. Englewood Cliffs, NJ: Prentice Hall, 1994.

Gelenbe, E., and G. Pujolle, *Introduction to Queuing Networks*. 2nd ed. New York: Wiley, 1998.

Gross, D., and C. M. Harris, *Fundamentals of Queueing Theory*. 3rd ed. New York: Wiley, 1997.

Hancock B., *Advanced Ethernet/802.3 Network Management and Performance*. 2nd ed. Oxford: Digital Press, 1995.

Haverkort, B. R., *Performance of Computer Communications System: A Model-Based Approach*. New York: Wiley, 1999.

Held, G., *LAN Performance: Issues and Answers*, 2nd ed. New York: Wiley, 1997.

Higginbottom, G., *Performance Evaluation of Communication Networks*. Norwood, MA: Artech House, 1997.

Kenyon, T., *High Performance Data Network Design*. Oxford: Butterworth-Heinemann, 1999.

Kesidis, G., *ATM Network Performance*. Dordrecht: Kluwer Academic Publishing, 1996.

Kiefer, R., *Test Solutions for Digital Networks: Basic Principles and Measurement Techniques for PDH, SDH, ISDN and ATM*. San Francisco: Morgan Kaufmann, 1998.

Kouvatsos, D., *ATM Networks: Performance Modeling and Analysis*. Vol. 2. London: Chapman & Hall, 1996.

Metzner, J. J., *Reliable Data Communications*. San Diego, CA: Academic Press, 1997.

Onvural, R. O., *Asynchronous Transfer Mode Networks: Performance Issues*. 2nd ed. Norwood, MA: Artech House, 1995.

Pattavina, A., *Switching Theory, Architectures and Performance in Broadband ATM Networks*. New York: Wiley, 1998.

Puigjaner, R., *High Performance Networking*. London: Chapman & Hall Publishers, 1998.

Robertazzi, T. G., *Computer Networks and Systems: Queueing Theory and Performance Evaluation*. New York: Springer Verlag, 1994.

Stanley, W. D., *Network Analysis with Applications*, Englewood Cliffs, NJ: Prentice Hall, 1999.

Van As, H., *High Performance Networking*. Dordrecht: Kluwer Academic Publishers, 1998.

Walrand, J., *Advanced Computer Performance Modeling and Simulation*. New York: Gordon & Breach, 1998.

WORLD WIDE WEB SITES

EtherVision, a real-time Ethernet LAN Monitor by Triticom
http://www.triticom.com

JETLAN, network traffic monitoring software by Jaguar Communication
http://www.jaguarcomms.co.uk/jetlanw.html

Manufacturer of the network performance and simulation tool COMNET III
http://www.caciasl.com

Manufacturer of the network performance and simulation tool OPNET
http://www.opnet.com

Network Performance tools link from National Computer Science Alliance (NCSA) at University of Illinois, Urbana-Champagne
http://www.ncsa.uiuc.edu/People/vwelch/net_perf_tools.htm

9

NETWORK MANAGEMENT

When the network stops, everything comes to a halt. For this reason, establishing and maintaining an efficient network management that includes fault management is always the network administrator's top concern. Fault management encompasses failure detection and isolating the source of problem, such as the server, the application, the database, the network interface card, or a misconfigured router. Fault management also includes diagnosing the problem and starting recovery actions to restore service to end users.

However, in the long run fault management is not enough. Just finding problems and fixing them doesn't prevent the same failures—or new ones—from occurring in the future. Network administrators, who must keep the network up and running, can easily find themselves trapped in a troubleshooting mode. Only proactive performance management can prevent such entrapment.

Network performance management is an ongoing process. The first step in any performance management strategy is to monitor the network over time to establish a baseline—a profile of how the network behaves under normal conditions. Once you've established what's normal, it's much easier to spot impending problems before they turn into network catastrophes. Baselining involves capturing snapshots of a network from various perspectives taken at several OSI layers to arrive at a complete picture.

For instance inexpensive, handheld cable testers are most often used to detect cable breaks and other problems at the physical layer. Similarly, traffic monitors and probes are used to monitor the Ethernet and Token Ring protocols. Hubs often include embedded modules for monitoring those protocols.

Software-based internetwork monitors, as well as sophisticated router management applications, can analyze network and transport layer protocols. Protocol analyzers, either stand-alone or networked analysis applications, are capable of decoding data at all seven layers.

Baseline profiles include plots of utilization and throughput with respect to time. In addition, error counts, types with relation to protocols at various layers are measured. It is observed that network administrators new to the baselining process are often frustrated by the lack of published information available on the topic. For instance, there are no widely accepted texts defining precisely what is "normal" utilization for an Ethernet LAN, or whether a router management application should issue an alert if the number of errors per thousand packets exceeds one. Hard and fast guidelines are scarce in part because each network is truly unique, having its own individual requirements. An Ethernet utilization peak of 50% may signal big trouble in most networks—but there are configurations in which that statistic is not all that abnormal.

Due to the trends of larger, more complex networks supporting more applications and more users in an organization, it has become evident that a large network cannot be put together and managed by human effort alone. The complexity of such a system dictates the use of automated network management tools.

The principal driving forces for an investment in network management can be listed in terms of the following needs:

- **Controlling corporate strategic assets:** Networks and distributed computing resources are increasingly vital resources for most organizations. Without effective control, these resources do not provide the payback that corporate management requires.

- **Controlling complexity:** The continued growth in the number of network components, users, interfaces, protocols, and vendors threatens management with loss of control over what is connected to the network and how network resources are used.

- **Improving service:** Users expect the same or improved service as the organization's information and computing resources grow and become distributed.

- **Balancing various needs:** An organization's information and computing resources must provide a spectrum of users with various applications at given levels of support, with specific requirements in the areas of performance, availability, and security. The network manager must assign and control resources to balance these various needs.

- **Reducing downtime:** As an organization's network resources become more important, minimum availability requirement approaches 100%. In addition to redundant design, network management has an indispensable role to achieve an ensuring high availability.
- **Controlling cost:** Resource utilization must be monitored and controlled to enable essential user needs to be satisfied at a reasonable cost.

To control costs, standardized tools are needed that can be used across a broad spectrum of product types—including end systems, bridges, routers, and telecommunication equipment—that can serve in a mixed-vendor environment. As a result, ISO and IETF developed a set of standards that can be used to effectively manage the networks of today and tomorrow. The management protocols are designed to make sure that network protocols and devices not only operate but operate efficiently. They allow managers to locate problems and make adjustments by exchanging a sequence of commands between the stations.

Although there are no standards for defining "normal" with respect to collected data, there are standard definitions for what types of data to collect. The specifications describe management information bases (MIBs) that can be queried by management systems supporting the Simple Network Management Protocol (SNMP).

Even with the availability of standard protocols, the task of performance monitoring is still a time-consuming process. The standards are merely a starting point for collecting performance data—they do not, for example, describe how the data should be interpreted; nor do they provide mechanisms for developing trend reports. Network administrators must invest months, or even years, in learning how to turn collected data into meaningful information that can support network optimization, capacity planning, and other long-term activities.

Today, there are only a few applications on the market designed for turning raw network performance data into statistical analysis reports. These products compress data collected over time, storing the data in specialized databases capable of producing sophisticated trending and analysis reports. The downside to products of these types is the significant learning curve associated with mastering the statistical analysis and reporting features.

A more recent development in this area has been the emergence of specialized service providers who analyze a customer's network and deliver hard-copy performance reports, either once or on an ongoing basis. These "out-tasking" services specialize in performance monitoring and are therefore different from "outsourcing" firms seeking to take over the entire responsibility for running and managing a network. An out-tasking service provider takes care of installing the performance monitoring software, compressing and analyzing the data, and delivering this material in report form to the customer. For this, the customer typically pays a small fee each month for each router or other device that is required to be monitored.

9.1 SNMP

9.1.1 Introduction

As the Internet grows, it becomes more difficult to manage (i.e., to monitor and maintain). It is now evident that a network management protocol needs to be developed for the entire system.

In late 1990s there were three approaches to managing Internet traffic:

- **Simple Gateway Monitoring Protocol (SGMP):** This was getting some use outside of development sites.
- **High-Level Entity Management System (HEMS):** While this had some novel concepts, it was not used in the real world.
- **The OSI protocols:** Common Management Information Protocol (CMIP) over TCP.

The Internet Activities Board (IAB) discussed these three approaches and eventually decided to take the following steps toward standardization.

1. Upgrade SGMP slightly as a short-term solution, to reflect what had been learned since initial development. This upgrade was called the Simple Network Management Protocol (SNMP).
2. Give the OSI-based approach (CMIP over TCP) extensive scrutiny and experimentation, in the hope that one day it would become the long-term solution.
3. Drop HEMS altogether—not because it was in any way inferior, but because it simply was not receiving support outside the development sites.

A common framework was set up to tie SNMP and CMIP together to ease the future transition. In fact, SNMP was originally designed and implemented as an interim specification for communicating with network devices while the OSI specification was being finalized and being implemented during the late 1980s. But things have not worked out that way. By 1993, when OSI finally matured, SNMP had a 3-year head start and had already been implemented in hundreds of products. SNMP is now the de facto standard in Internet management. It manages the routers and hosts called *objects*, which have formal definitions according to a formal language called Abstract Syntax Notation 1, specifically designed for the definition of PDU formats and object definition irrespective of the type of management used.

9.1.2 SNMP Model

The specifications of the Internet Standard Management Framework consist of the following:

- A data definition language
- Definitions of management information (or MIB)

- A protocol definition
- Security and administration

Over time, as the framework has evolved from SNMPv1, through SNMPv2, to SNMPv3, the definitions of each of these architectural components have become richer and more clearly defined, but the fundamental architecture has remained consistent.

Basically, the SNMP model for managing networks is based on three pieces of software: agents, MIBs (management information bases), and management stations (or managers). Figure 9.1 shows the schematic representation of the SNMP model. Agents are pieces of software that run at each network device. They fetch information from a database called the management information base, which is also stored at the device.

Management stations (managers) allow the retrieval and display of information gathered from a device's agent and MIB. A management station can also control (or "set" in SNMP terms) those devices.

SNMP exchanges network information through messages, technically known as protocol data units (or PDUs). From a high-level perspective, the message (PDU) can be looked at as an object that contains variables that have both titles and values.

SNMP employs five types of PDU to monitor a network: two deal with reading terminal data, two deal with setting terminal data, and one, the trap, is used for monitoring network events such as terminal start-ups or shutdowns. SNMP casts all options in a fetch–store paradigm. Conceptually, SNMP contains only two commands that allow a manager to fetch a value from a data item or store a value into a data item. All other operations are defined as side effects of these two operations. Table 9.1 shows the four operations offered by SNMP version 1.

Get retrieves specific management information from the named object, in the community context. If the object does not exist, the noSuchName error is returned.

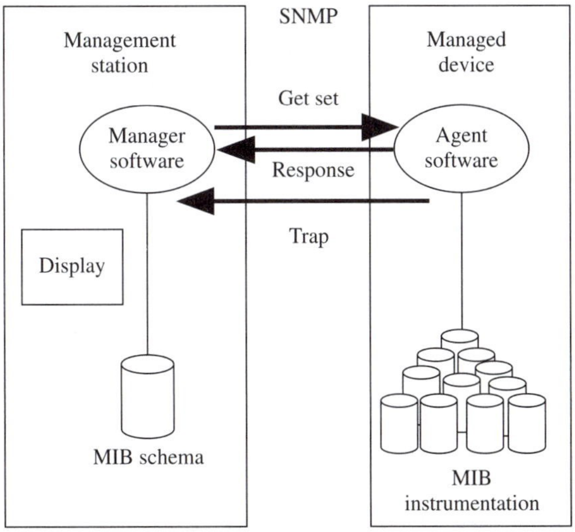

Figure 9.1 SNMP model.

TABLE 9.1
The Four Commands of SNMPv1

Command	Meaning
Get	Fetch a value from a specific variable
Get-next	Fetch a value without knowing its exact name
Set	Store a value in a specific variable
Trap	Send notifications to the management station

Get-next starts to traverse the community profile tree (which contains all objects in the community), by doing a Get on the object lexicographically following the named object. By continuously calling Get-next with the name of the object returned by the preceding iteration, the tree is traversed in its entirety. Once the end of the tree has been reached, the error noSuchName is returned.

Set updates the named objects with variables supplied. As well as the noSuchName error, it is possible to get readOnly if the NMS's access level is too low, or badValue if one of the variables was out of range.

Trap allows a managed object to warn its NMS about things like authentication failures, node initialization, and loss of neighboring node.

9.1.3 Structure of Management Information

An MIB describes information that can be obtained and/or modified via a network management protocol. This information enables systems on a network to be managed. Each entry in a MIB is called a MIB variable. For example, one common MIB variable is *sysDescr*, which describes system hardware and software. For example, if a management station is used to retrieve the *sysDescr* variable from a Macintosh computer, the answer may be something like "Macintosh Quadra 800, System 7.1."

In addition to the MIB standard, which specifies specific network management variables and their meanings, a separate standard specifies a set of rules used to define and identify MIB variables. The rules, known as the structure of management information (SMI), define the model of managed objects and the operations that can be performed on the objects, as well as data types that are permitted for the objects. Objects are unambiguously identified (or named) in SNMP by assigning them an object identifier (OID). Globally unique for all space and time, OIDs are a sequence of nonnegative integers organized hierarchically. For ease of use, a textual name is associated with each sequence element, or component, of an OID. The last component name is used by itself as a shorthand way of naming an object. All textual names of objects defined by IETF working groups are, by convention, made unique by using a different prefix for objects in each new MIB. SNMP uses an encoded form of the numeric value, not the textual name.

The root of the object identifier hierarchy is unnamed but has three direct descendants managed by ISO, CCITT, and jointly by ISO and CCITT. The descendants are assigned both short text strings and integers for purposes of identification. The text strings are used when humans need to understand object names, while computer uses the integers to form compact, encoded representation of the names. Figure 9.2 illustrates perti-

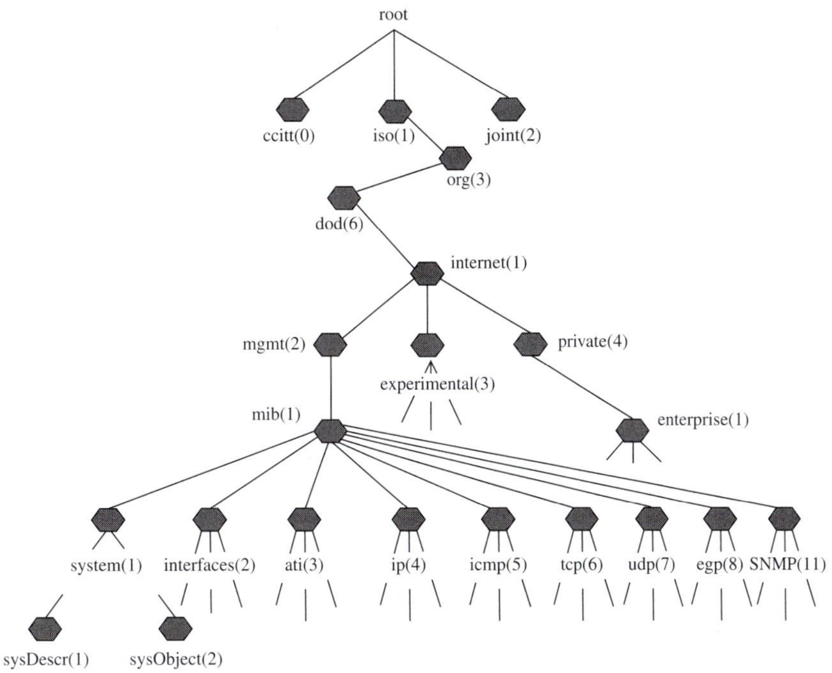

Figure 9.2 MIB OID tree for SNMP.

nent parts of the object identifier hierarchy and shows the positions of the node used by SNMP.

The name of an object in the hierarchy is the sequence of numeric labels on the nodes along a path from the root to the object. The sequence is written with periods separating the individual components. For example, the name 1.3.6.1 denotes the node labeled *internet*. The MIB has been assigned a node under the *internet* management subtree with label *mib* and numeric value 1. Because all MIB variables fall under that node, they all have names beginning with the prefix 1.3.6.1.2.1. Suppose one wanted to identify an instance of the variable *sysDescr*. The object class for *sysDescr* will be:

iso	org	dod	internet	mgmt	mib	system	sysDescr
1	3	6	1	2	1	1	1

The object type would be 1.3.6.1.2.1.1.1 to which an instance subidentifier of 0 is appended. That is, 1.3.6.1.2.1.1.1.0 identifies the one and only instance of *sysDescr*, which should be specified on an SNMP command line to get or set the system information about a remote host.

9.1.4 SNMPv2

Since publication of the original protocol, several proposals have been presented to improve SNMP. In 1992 it was decided to collect these proposals and produce a new standard: SNMPv2. Unfortunately SNMPv2 became far more complex than the original SNMP; whereas the description of the original protocol required, for example, only 35 pages, the description of SNMPv2 occupied about 250 pages. The main achievements of SNMPv2 are improved performance, better security, and the possibility of building a hierarchy of managers.

9.1.4.1 *Performance*

As mentioned earlier, the original SNMP includes a rule, stating that if the response to a Get or Get-next request would exceed the maximum size of a packet, no information would be returned at all. Since managers cannot determine the precise size of response packets in advance, they usually take a conservative guess and request a smaller than optimal value per PDU. Therefore, to obtain all information, managers may be required to issue a large number of consecutive requests.

To improve performance, SNMPv2 introduced the Get-bulk PDU. As opposed to Get and Get-next, the response to the Get-bulk always returns as much information as possible. If the requested information exceeds the maximum size of a UDP packet, the information will be truncated and only the part that fits within the packet will be returned.

9.1.4.2 *Security*

The original SNMP had no security features except for a simple mechanism that involved the exchange of passwords (the term "community string" was used to denote this password). To solve this deficiency, SNMPv2 introduced a full-fledged security mechanism. This mechanism is based on the use of "parties" and "contexts," two concepts that cannot be found in other management approaches. Although the SNMPv2 standards include definitions of both concepts, the definitions are difficult to understand. We present a somewhat simplified view.

Parties have some resemblance to protocol entities. Usually multiple parties are active in a single SNMPv2 subsystem, and these various parties will be configured in different ways. One party may, for instance, be configured such that it is prepared to communicate with every other party in other systems.

Another party may be configured such that it is prepared to interact with only one particular remote party. In such a case, the MD5 authentication mechanism is used to ensure the authentication of the other party. Finally parties may be so configured that they are prepared to interact only with particular remote parties and in addition require that all management information be encrypted according to the data encryption standard (DES) algorithm. A graphical representation of parties is provided in Figure 9.3, which shows three parties being configured in the manager system (Pa1, Pa2, and Pa3) and the agent system (Pb1, Pb2, and Pb3).

To control access to the various parts of a MIB, SNMPv2 has introduced the context concept. Each context refers to a specific part of a MIB. In the example of Figure 9.3, contexts C1 and C2 refer to the two small boxes in the MIB. Contexts may be overlapping and are dynamically configurable, which means that contexts may be created, deleted,

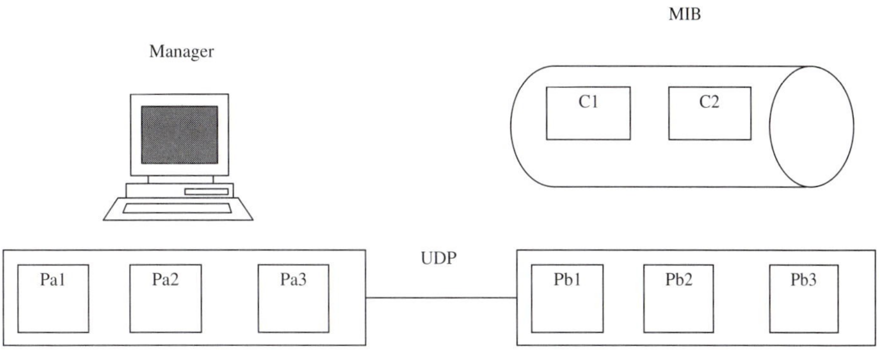

Figure 9.3 Parties and contexts.

or modified during the network's operational phase. Different contexts may be configured for different systems.

To determine the parties that are allowed to perform operations upon parts of the MIB, SNMPv2 has associated with each agent an access control list (ACL). Table 9.2 shows an example of such a list. The first row indicates that party Pa1 (in the manager system) may perform Get operations via party Pb1 (in the agent system) on the part of the MIB that is identified by context C1. The third row shows that Pa3 may do so via Pb3. Pa3 may also perform Set operations on the MIB part identified by context C1.

9.1.4.3 Management Hierarchy

Practical experience with the original SNMP showed that in many cases managers are unable to manage more than a few hundred agent systems. The cause for this restriction is in SNMP's polling nature: the manager must periodically poll every system under his control, which takes time. To solve this problem, SNMPv2 introduced intermediate-level managers (ILMs). Polling is now performed by a number of ILMs under control of the top-level manager (TLM). Figure 9.4 shows an example. Before the intermediate-level managers start polling, the top-level manager tells the ILMs what variables must be polled in which agents. In addition, the TLM tells the ILMs of the events it needs to be informed about. When the ILMs have been configured, they start polling.

TABLE 9.2
Example of Access Control List (ACL)

Remote party	Local party	Context	Operation
Pa1	Pb1	C1	get
Pa2	Pb2	C1	get
Pa3	Pb3	C1	get + set
Pa3	Pb3	C2	get

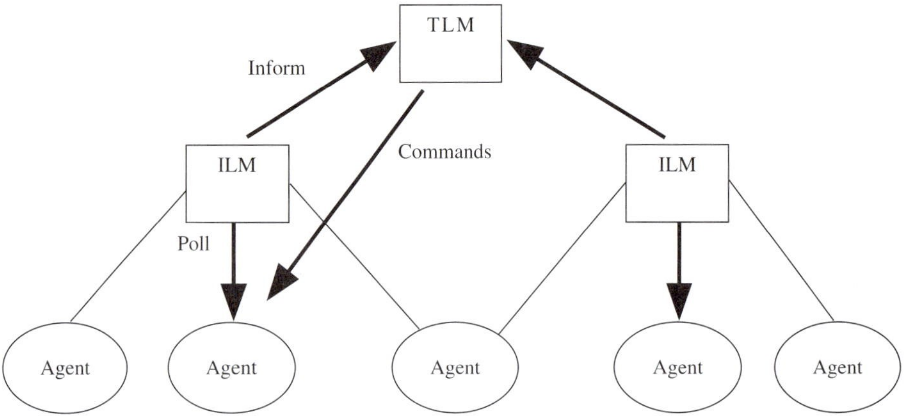

Figure 9.4 Intermediate-level managers (ILMs).

9.1.5 SNMPv3

During the past several years, the SNMPv3 Working Group has been handling a number of activities aimed at incorporating security and other improvements to SNMP. The group has come up with several RFCs (the RFC 2570 series), which includes the following

RFC 2570: Introduction to version 3 of the Internet standard Network Management Framework.

RFC 2571: "An Architecture for Describing SNMP Management Frameworks," which describes the overall architecture with special emphasis on the architecture for security and administration.

RFC 2572: "Message Processing and Dispatching for the SNMP," which describes the possibly multiple message processing models and the dispatcher portion that can be a part of an SNMP protocol engine.

RFC 2573: "SNMP Applications", which describes the five types of application that can be associated with an SNMPv3 engine and their elements of procedure.

RFC 2574: "The User-Based Security Model for Version 3 of the SNMPv3," which describes the threats, mechanisms, protocols, and supporting data used to provide SNMP message-level security.

RFC 2575: "View-Based Access Control Model for the SNMP," which describes how view-based access control can be applied within command responder and notification originator applications.

In general, the working group has considered the following objectives:

- To accommodate the wide range of operational environments with differing management demands
- To facilitate the need to transition from earlier multiple protocols to SNMPv3
- To facilitate setup and maintenance activities

The SNMPv3 Working Group did not start a new work and is using as many concepts, technical elements, and documentation as practical from the earlier activities. Unfortunately, strongly held differences on how to incorporate the improvements into SNMP prevented the SNMPv2 Working Group from coming to closure on a single approach.

SNMPv3 planned specifications consist of Modules and Interface Definitions, Message Processing and Control Module Specification, Security Model Module Specification, Local Processing Module Specification, and Proxy Specification.

9.2 RMON AND RMONV2

As networks expand, the ability to perform remote monitoring becomes more important. Problems can be identified and resolved from a management console, rather than by sending a technician to remote locations, which is expensive and time-consuming. The ability to monitor the performance of remote LAN segments has been made easier with SNMP's **R**emote **Mon**itoring **M**anagement **I**nformation **B**ase (RMON MIB) standard.

RMON, or RMON1, is the remote network monitoring MIB developed by the IETF to support monitoring and protocol analysis of Ethernet and Token Ring LANs. It included more open, comprehensive network fault diagnosis, planning, and performance-tuning features than any monitoring solution on the market at the time. It is an industry standard specification that provides much of the functionality offered by today's proprietary network analyzers and protocol analyzers.

The RMON MIB standards effort started in 1990 with the creation of the RMON Working Group of the IETF. The RMON1 Proposed Standard (RFC 1271) was published in 1991. The first RFC focused specifically on Ethernet. The RMON1 Working Group augmented the initial work with the Token Ring extensions (RFC 1513) in 1993. Owing to high market demand and increasing customer interest, RMON-compliant vendor implementations were rapidly developed and brought to market. With proven, interoperable vendor implementations, the RMON MIB moved to draft standard status in 1994 and was assigned the new RFC number of 1757. Later, RMON MIB version 2 was described in RFC 2021 using SMIv2, and the RMON MIB protocol identifiers were described in RFC 2074.

The RMON standard has enabled explosive growth in the availability and versatility of performance management tools. RMON encourages a greater degree of interoperability among management devices in a multivendor environment. Theoretically, users can now mix and match any RMON-compliant monitor with SNMP management systems that recognize RMON variables.

RMON's greatest benefit to users is that it standardizes data collection done by remote probe devices. Remote monitoring allows network administrators to see traffic on a LAN segment no matter where that segment resides. The segments could be in the same building or in a LAN across the country.

The RMON MIB is organized into nine statistics groups:

1. Segment statistics
2. History

3. Host table
4. Host top *N*
5. Traffic matrix
6. Alarms
7. Filters
8. Packet capture
9. Events

Groups 8 and 7, Packet capture and Filters, are the most difficult groups for a vendor to implement, and they require the most memory—yet they provide some of RMON's richest functionality. Without packet Capture, RMON can help spot potential trouble spots, but it cannot diagnose causes. Because RMON operates primarily at the data link layer, RMON Packet capture is not a substitute for full-featured protocol analyzers—it can, however, act as a valuable supplement.

The key benefits of implementing RMON technology are high network availability for users and high productivity for network administrators. Without leaving the office, a network manager can see the traffic on a LAN segment, whether that segment is physically located around the corner or around the world. Armed with that traffic knowledge, the network manager can identify trends, bottlenecks, and hotspots.

RMON uses monitor or probes attached to network segments to collect statistics. RMON probes provide the advantage of direct attachment to the network segment being monitored. On the other hand, if many segments need to be monitored concurrently, the cost of RMON probes, one for each segment, can become prohibitive. Manufacturers of switching hubs offer RMON support with various levels of coverage. The most complete offer an RMON agent on every port/segment. Others offer RMON on a roving basis, where the network manager chooses which segment to monitor. Finally, RMON could be available on all segments, but the data being gathered on any single segment would be a statistical sample of the segment as opposed to a report of all statistics for each segment.

9.2.1 RMON Model

RMON implementations are generally delivered as a two-part client/server solution. The "client" is the application that runs on the network management station and presents the RMON information to the user. The "servers" are the monitoring devices distributed throughout the remote networks that collect the RMON information and analyze network packets. The monitoring device is commonly called a "probe," and it runs a software program, generally called an RMON "agent." RMON agents can be found in dedicated devices and/or embedded in network infrastructure devices such as hubs and switches. The application and the agent use the SNMP to communicate across the network.

RMON is designed so that the remote probe devices do the data collection and processing. This reduces the SNMP traffic on the network and the processing load on the management station. Instead of continuous polling, information is transmitted only to the management station when required. Many RMON client applications located in various parts of the network can simultaneously communicate with and get information from one

RMON server. The information from a single RMON server can be used for many tasks, from troubleshooting and protocol analysis to performance monitoring and capacity planning. RMON provides valuable statistics on the whole network segment. Contrast this with other SNMP management products, which focus on monitoring and control of a specific network device. While device-specific management tools are important, they do not provide a picture of the health of the whole network segment with all its devices, servers, applications, and users.

With the RMON MIB, network managers can collect information from remote network segments for the purposes of troubleshooting and performance monitoring. The RMON MIB provides the following:

- Current and historical traffic statistics for a network segment, for a specific host on a segment, and between hosts (matrix)
- A versatile alarm and event mechanism for setting thresholds and notifying the network manager of changes in network behavior
- A powerful, flexible filter and packet capture facility, which can be used to deliver a complete, distributed protocol analyzer

Figure 9.5 lists the RMON groups and shows where RMON fits into the ISO and IETF standards.

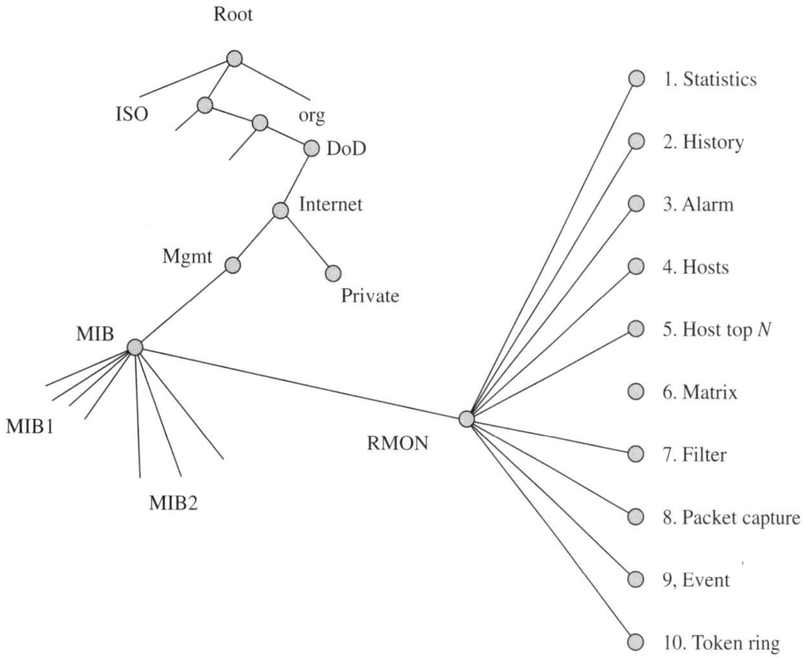

Figure 9.5 RMON MIB tree.

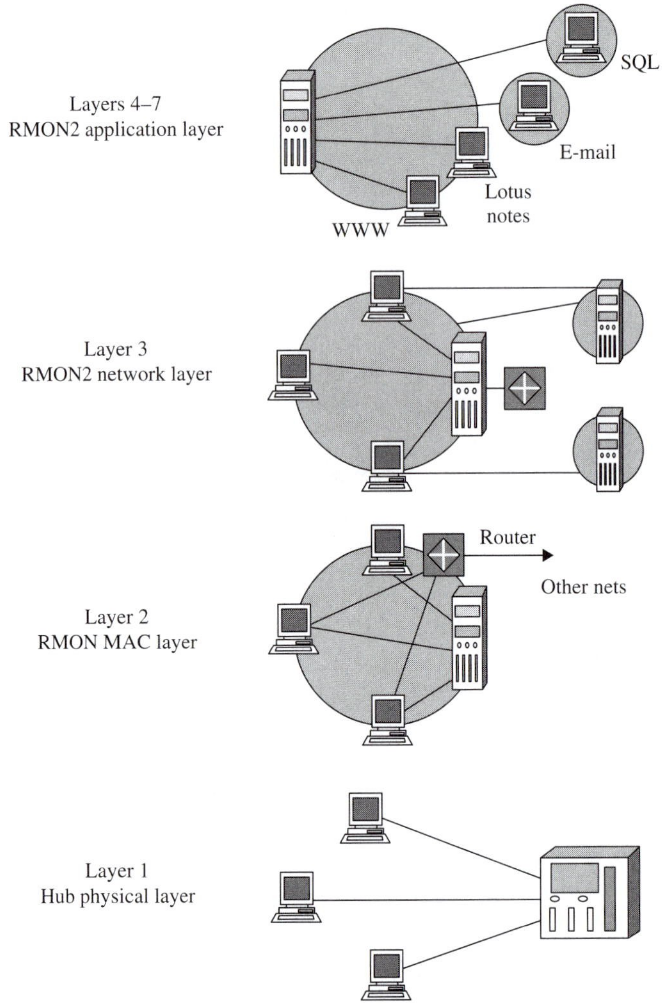

Figure 9.6 The latest infrastructure addition of RMON2.

9.2.2 RMON2

The RMON2 standard is defined by the IETF that covers all seven layers of the OSI protocol stack and overcomes many of RMON's current limitations. However, RMON2 is not a superset or a replacement of RMON1—rather, the two standards complement each other, enabling administrators to use standardized methods for viewing all aspects of network behavior.

A drawback of "RMON"-based probes is that because they view the traffic on the local LAN segment, they are not able to identify network hosts and sources beyond the router connection. For this, a probe/agent must be capable of identifying traffic at the network layer that will provide statistics for all hosts accessing that segment, no matter where they are located or how the network is connected. With RMON2-based agent/probes, all RMON groups map into all the major network-layer protocols such as IP, IPX, DECnet, AppleTalk, Banyan VINES, and OSI, giving a complete end-to-end view of network traffic.

RMON-based monitors collect statistics at layer 2, the data link layer, of the OSI model. The data link layer is primarily concerned with packet formation, addressing, and delivery. While this type of data collection is very useful for problem resolution on a per-segment basis, it makes isolating a problem down to a physical piece of cabling impossible. RMON2 provides the ability to correlate the port or wire information with the end station, delivering tremendous visibility into the network.

With RMON2, any MIB object can be locally tracked and recorded in a historical log. This provides tremendous flexibility for locally logging what is important to the specific user environment for long-term trending. The benefits of RMON2 extend monitoring to the higher levels of the protocol layers. While RMON provides traffic statistics only at the MAC layer of the protocol, RMON2 provides insight into those traffic statistics by specifying the protocol and applications that make up that traffic. Such knowledge is vitally important in managing today's complex client–server networks. Extending the segment view of RMON to the enterprise-wide view of RMON2 provides the network manager with more information and greater knowledge to manage, troubleshoot, and monitor increasingly complex networks at various levels (Figure 9.6). As a result, RMON2 enables a whole new class of applications for client–server analysis, Internet monitoring and design, and network and application layer accounting.

9.3 TMN

New trends in telecommunication networks are producing large software-based, distributed, intelligent networks. Intelligent network elements are being deployed to reduce network operation and management costs. Telecommunication Management Network (TMN), a management network with standard OSI protocols, interfaces, and architectures, was developed by ITU as an infrastructure to support management and deployment of dynamic telecommunications services. It provides a framework for achieving interconnectivity and communication across heterogeneous operating systems and telecommunications networks.

The basic architecture, methodology, and functionality, as well as the generic model standard, have been developed and approved in Committee T1 and its international counterpart, the ITU. The term TMN was introduced by ITU-T. The concept of a TMN is defined by Recommendation M.3010. According to M.3010, "a TMN is conceptually a separate network that interfaces a telecommunications network at several different points." The architecture specifies the methodology of interaction and the semantics of the information models between the telecommunications equipment and operation systems.

The primary concept behind TMN is to provide an organized architecture to achieve the interconnection between operation systems (OSs) of various types and/or telecommunication equipment for the exchange of management information via an agreed architecture with standardized interfaces, including protocols and messages. The TMN model recognizes that several organizations have a large infrastructure of operating systems, networks, and telecommunication equipment already in place, which need to accommodated within its architecture.

TMN provides management functions for telecommunication networks and services and offers communications between itself and telecommunications. Here a telecommunication network is assumed to consist of both digital and analog telecommunications equipment and associated support equipment. Service would consist of a range of capabilities provided to customers.

The TMN model categorizes network management into five functional areas: configuration, performance, fault, accounting, and security management.

The strength of the TMN model is its organization of functional layers along a service provider's business and operational functions. Each management function is focused on a given level without the complexity and detail of other layers. TMN provides an organized architecture that lets various types of OS and telecommunications equipment exchange management information. The weakness of the TMN model is its lack of interfaces for managing IP technology and IP-enabled services. ITU and other standards bodies have only just begun effort to define a network management model for IP technology.

Network element layer: At the lowest layer, the network elements (NEs), are actual hardware devices and systems to be managed (i.e., hubs, routers, switches, probes, and data collection devices). This concept can be extended to situations in which NEs are software elements (e.g., an SNMP agent). Elements at this layer have operations interfaces that allow for control and monitoring of devices by outside control and other levels of management.

Element management layer (EML): Software at this level is technology specific (i.e., WAN switch, core router, access device) and is responsible for provisioning and monitoring a specific set of network elements on a device-by-device or subnetwork basis. EML functions are also responsible for processing information to and from applications at the network management layer (NML).

Network management layer (NML): NML applications are responsible for managing and monitoring the network as a whole. Software at the NML aggregates information from a group of elements and element management systems. It maintains knowledge of the end-to-end network topology (the interconnection of network elements).

Service management layer (SML: This layer provisions, manages, and monitors the network as a resource in support of specific services such as VPN services. SML applications also support SLAs contracted between customers and service-to-service interaction.

Business management layer (BML): The topmost layer of the TMN model, this layer supports the business management functions of the organization. Example applications include customer care, trouble ticketing, and billing.

The relationship between a TMN and the telecommunication network that is managed is shown in Figure 9.7. Exchanges and transmission systems form the interface points

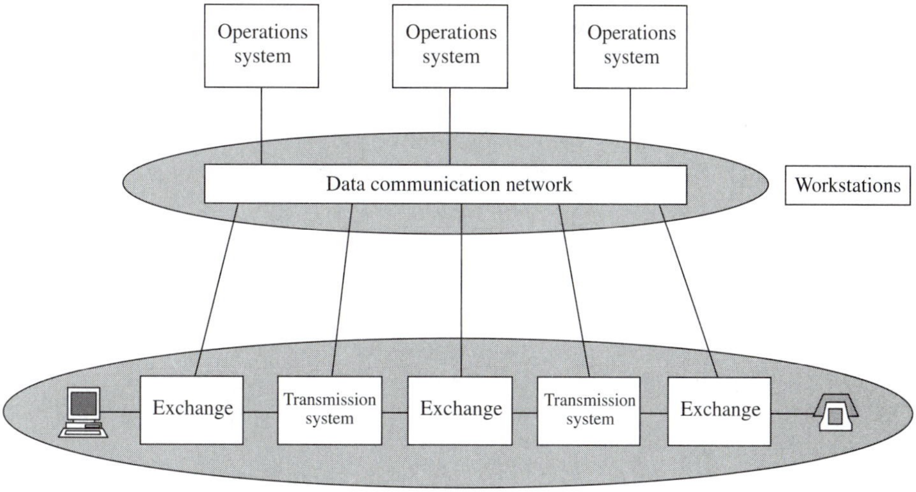

Figure 9.7 General relationship of a TMN to a telecommunication network.

between the TMN and the telecommunication network. For the purpose of management, these exchanges and transmission systems are connected via a data communication network to one or more OSs. The OSs perform most of the management functions; these functions may be carried out by human operators but also automatically. It is possible that multiple OSs will perform a single management function. In this case, the data communication network is used to exchange management information between the OSs. The data communication network is also used to connect workstations, thus allowing operators to interpret management information. Workstations have man–machine interfaces, but the definition of such interfaces falls outside the scope of TMN (workstations are therefore drawn at the border of the TMN).

9.3.1 Physical Architecture

TMN defines a physical architecture to show how function blocks should be mapped upon building blocks (physical equipment) and reference points upon interfaces. The physical architecture defines how function blocks and reference points can be implemented.

TMN's physical architecture defines the following building blocks:

- Network element (NE)
- Mediation device (MD)
- Q adapter (QA)
- Operations system (OS)
- Workstation (WS)
- Data communication network (DCN)

Building blocks always implement the function blocks of the same name (e.g., network elements perform network element functions, mediation devices perform mediation functions, etc.).

A special kind of building block is the data communication network (DCN). Unlike the others, this building block does not implement any function block. In fact, the DCN is used by other building blocks for the exchange of management information; the DCN's task is to act as a transport network.

9.3.2 Interfaces

Interfaces may be regarded as the implementations of TMN reference points that define information exchange between TMN building blocks. Whereas reference points may generally be compared with underlying services, interfaces may be compared with the protocol stacks that implement these services.

Figure 9.8 shows the standard interfaces between TMN components. The suffix F stands for function. For example, OSF is operations system function and so on.

In most cases, reference points and interfaces have a one-to-one mapping. However, no interfaces exist for reference points that:

- Interconnect function blocks that are implemented within a single building block
- Lie outside TMN (*g* and *m* in Figure 9.9); implementation of these reference points is not part of TMN.

The naming of interfaces is also straightforward: an interface gets the same name (this time written in uppercase) as the related reference point. Figure 9.9 shows all possible mappings. Referring back to Figure 9.8 for instance, m reference point can be used to reach the QAF from outside TMN.

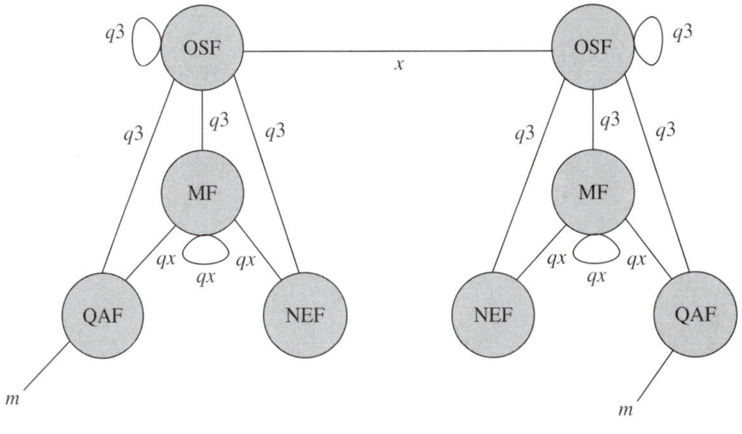

Figure 9.8 Standard interfaces between TMN components.

Reference point	q_x	q_3	x	f	(g	m)
	↓	↓	↓	↓		
Interface	Q_x	Q_3	X	F		

Figure 9.9 Mapping reference points upon interfaces.

9.3.3 Differences Between TMN and OSI

The current TMN architecture, and in particular the part on TMN's information architecture, includes many ideas of OSI systems management. An interesting difference between OSI and TMN management is that OSI has defined a single management architecture, whereas TMN defined multiple architectures at different levels of abstraction.

A second difference between TMN and OSI management is that TMN provides a structure for the multiple levels of management responsibility that exist in real networks; OSI management does not provide such structure. The TMN structure is known as the "responsibility model." The advantage of having such a structure is that it becomes easier to understand and distinguish the various management responsibilities.

A final difference between TMN and OSI management is that unlike OSI, TMN suggests a conceptual separation between the network that is managed (the telecommunication network) and the network that transfers the management information (the DCN). Despite failures in the managed network, TMN management will always be able to access failing components. TMN thus has better fault management capabilities than OSI.

9.4 DIRECTORY SERVICES AND NETWORK MANAGEMENT

In a distributed computer network with several servers and intranet connection, it is important to have a single network logon and a single point of administration and replication. Network users and customers demand features such as a hierarchical view of the directory, extensibility, scalability, distributed security, and multimaster replication. To meet these needs, network operating systems developers such as Novell and Microsoft have come up with their own directory services solutions. While directory services have become critical in today's enterprise networks, they are still largely vendor oriented, placing the goal of a single, unified enterprise directory service well beyond reach for at least the next few years.

Traditionally, directory services have been tools for organizing, managing, and locating objects such as printers, documents, e-mail addresses, databases, users, distributed components, and other resources in a computing system. In their simplest form, directory services are like the white pages of a telephone book. Using specific input (e.g., a person's name) a user can receive specific output (that person's address and telephone number). The purpose of electronic directories is not much different from that of printed directories: that is, to provide names, locations and other information about people and organizations. In a LAN or WAN, this directory information may be used for e-mail addressing, user authentication (e.g., logins and passwords), or network security (e.g., user-access rights). A directory may also contain information on the physical devices on a network (e.g., PCs, servers, printers, routers, and communication servers) and the services available on a specific device (such as operating systems, applications, shared-file systems, print queues). This information may be accessible to computer applications as well as being eye-readable for end users.

Early network directories were most often developed specifically for a particular application. In these proprietary directories, system developers had little or no incentive to work with any other system. But systems users, in an effort to rationalize their ever-increasing workload, sought ways to share access to and maintenance of directory databases with more than one application. This gave rise to the concept of the directory as a collection of open systems that cooperate to hold a logical database of information. In this view, users of the directory, including people and programs, would be able to read or modify the information or parts of it, as long as they had the authorization to do so. This idea grew into the definition of X.500.

In the client–server environment of the application layer of the open systems interconnection (OSI) model, directory functionality (directory administration, authentication, and access control) was initially developed to handle management of e-mail addresses in conjunction with the OSI message handling application (X.400). However, its potential use with many applications was recognized, and therefore it was defined as a separate module or standard: ITU-T Recommendation X.500 (also known as ISO/IEC 9594: Information Technology—Open Systems Interconnection—The Directory). In the X.500 directory architecture, the client queries, and receives responses from, one or more servers in the server's directory service, with the Directory Access Protocol (DAP) controlling the communication between the client and the server.

The current X.500 standard was modified in 1993 from its original form approved in 1988. However, comprehensive and complex coverage generated criticism from many implementers, and as a result, the University of Michigan developed a simpler TCP/IP-based version of DAP, the Lightweight Directory Access Protocol (LDAP), for use on the Internet. LDAP offers much of the same basic functionality as DAP and can be used to query data from proprietary directories as well as from an open X.500 service. Many major suppliers of e-mail and directory services software have expressed interest in LDAP because of its light weight and TCP/IP support, and it is fast becoming a de facto directory protocol for the Internet. Although LDAP started as a simplified component of the X.500 directory, it is evolving into a complete directory service. Developers are using Internet naming services to build layers of security and to add capabilities that exist within X.500 directory components other than the DAP.

While messaging directories move toward LDAP as a standard access protocol, development of proprietary directory-aware applications continues among network operating system (NOS) vendors. NOS and application vendors—namely, Microsoft, Netscape, and Novell—are already coming up with enterprise directory solutions to meet the challenge of unifying and bringing order to diverse server hierarchies. However, an increased reliance on multiple proprietary, vendor-driven directories is expected, with a slow adoption of open directories driven primarily by the need for enterprise directories.

Novell Directory Services (NDS) let companies create and store a single profile for every user that is centrally located in the network. This way, users get access to all their network resources by logging on once, instead of multiple times. It also lets them have, while on the road, the same access and applications configuration that reside on their office desktops. Additionally, Novell's directory technology lets IT departments configure users' desktops automatically from a central location, which eliminates the need, for example, to manually load software on PCs.

Novell launched its NDS version 8 as a scalable directory for companies looking to extend applications beyond their enterprises to the Internet. With NDS version 8, two needs can be handled: the highly distributed enterprise environment that is focused on integration of legacy as well as new systems, and the highly centralized extranet and Internet implementations, which must be scalable.

With NDS 8, Novell has introduced to the market a new type of directory—one that scales to extremely large environments and maintains LDAP performance. They demonstrated an NDS 8 directory tree managing a billion objects—users, servers, printers, applications, and so on—with no degradation of performance. That's several millions more objects than any competing directory or earlier version of NDS. NDS 8 is also claimed to be integrated more easily with Internet standards and applications than any other directory, allowing it to easily interoperate with an organization's existing technology investments and to work with new standards-based applications. In particular, NDS 8 natively supports LDAPv3, which is considered to be a very important directory protocol in the industry. Keeping in view the current Internet trends, it is evident that users want a directory that will support growth not only within their companies but on the Internet as well. NDS 8 is expected as a directory solution for the customers who are interested in the Internet commerce.

Moving forward, Novell plans to offer versions of NDS for Solaris and Windows NT using a common code base. That will eliminate the need for NDS on NetWare even at one location. On the security front, NDS 8 offers improved support for managing public key infrastructures (PKI) and cryptography.

Novell has sold several million user licenses of NDS. But it faces huge competition from Microsoft, whose Active Directory (AD) is expected to become pervasive because it is the basis of Windows 2000, and from Netscape, whose directory server is widely used with its web and e-commerce systems.

The Microsoft Active Directory supports a wide range of well-defined protocols and formats and provides powerful, flexible, and easy-to-use application programming interfaces (APIs). Moreover, the Active Directory provides administrators and users with a one-stop source for resource and management information.

In addition to handling the traditional administrative tasks of the directory services, the Active Directory satisfies a wide variety of naming, query, administrative, registration, and resolution needs. Figure 9.10 summarizes its overall function in the system.

The AD service provides secure interfaces to Internet (http/shttp), mail, and SQL servers. In addition, domain name service (DNS) is integrated into the service. The mail server may operate with the exchange mail or other mail clients.

The architecture of the Active Directory allows it to scale from the smallest of businesses to enterprises supporting international corporations and entire government departments and services.

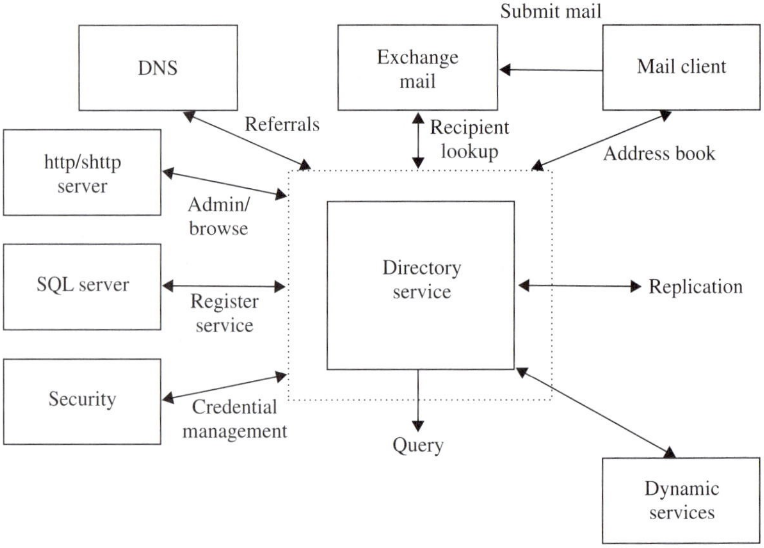

Figure 9.10 The Active Directory is a service provider used to locate all network services and information.

The Active Directory uses a tightly integrated set of APIs and protocols to extend its services across multiple name spaces, and to gather and present directory and resource information that reside on different operating systems and at distant locations. For example, Microsoft today provides a rich set of interoperability components for Novell customers. However, interoperability with NDS and newer versions of NetWare are not established fully. As protocols evolve, it is expected that Microsoft will work with the industry to standardize communications with other environments as well.

The Active Directory includes the following features and benefits:

- Support for open standards to facilitate cross-platform directory services, including support for the domain name service and support for standard protocols, such as LDAP
- Support for standard name formats to ensure ease of migration and ease of use
- A rich set of APIs, which are easy to use for both the scripter and the C/C++ programmer
- Simple, intuitive administration through a simple hierarchical domain structure and the use of drag-and-drop administration
- Directory object extensibility via an extensible schema
- Fast lookup via the global catalog
- Speedy, convenient updates through multimaster replication
- Backward compatibility with earlier versions of the Windows NT operating system
- Interoperability with NetWare environments

Although Microsoft claims that the Active Directory is a next-generation directory service that will solve most of the current directory problems, many other vendors, especially Novell, do not totally agree. AD is actually viewed as a packaged set of enhancements to the NT 4.0 domain structure. It places a directory shell—a directory view—over the NT domain infrastructure, solving some of NT's current limitations. However, because the fundamental domain and trust relationship structures remain, Active Directory cannot provide a truly scalable, highly fault-tolerant, and effective enterprise directory service.

According to Novell, AD, while marketed as a directory service, does not provide many of the cost-saving features found in other directory services, including NDS. Novell claims that until Microsoft adopts a directory model that eliminates legacy domains and trust relationships, NT Server 5.0 will suffer from the same domain limitations found in earlier versions of NT Server.

9.5 WEB-BASED NETWORK MANAGEMENT

Management and monitoring of networks using HTTP tools is called web-based management. This emerging network management technique promises to solve some long-standing net management problems, such as lack of support for multiple operating systems, an integrated universal user interface, and alert correlation.

Web-based management is the application of World Wide Web tools for the management of systems and networks. This includes using HTTP servers and browsers for providing static, dynamic, and interactive content of management information. An HTTP server acting in a management role can provide information in a variety of forms including Hypertext Markup Language (HTML), graphics, executable code, and binary encoded information. Together, this capability allows HTTP to function as a powerful protocol for the management of systems and networks. Moreover, HTTP servers acting in a management role may also be providing nonmanagement content. It is therefore necessary to provide a mechanism for a WWW browser or network management application to have direct access to the management information. However, HTTP is not meant to replace SNMP. HTTP working together with SNMP can provide many benefits, including ease-of-use, zero client-side installation, and security. SNMP is required for the instrumentation of systems and networks.

The web-based management software packages were first created to monitor and manage Internet-related networking tasks. For example, one such product integrates Java and some other technologies to provide real-time network monitoring via web browsers. It can detect dropping packets of a WAN link from a specific router port, forge a link into the router interface, and thus allow an operator to diagnose the dropped packet problems and adjust buffer allocation tables. Without the use of web technologies and standards, this process would require an expensive effort and operator expertise. In Chapter 8 we gave an example that showed how users could get Ping statistics by using their web browsers.

Web browsers are soon to be the standard interface to all net management functions, and the web-based network management technology is rapidly being extended to all the network areas far beyond Internet applications. Also, it is expected that software developers will gain huge savings in developing only one platform-independent Web-based management package, as opposed to porting applications to many different flavors of operating systems and integrating them with proprietary hardware platforms, hence lowering the cost of ownership.

The web-based net management tools also take much less time and effort to deploy, configure, and integrate, as well as to train users, compared with current enterprise-scale management tools. The web-based management products may offer the least expensive and most ubiquitous management, which overcomes some limitations of SNMP in supporting real-time configuration changes to network and system elements. The web-based management tools are improving the state of enterprise management in simplifying and automating remote configuration for all devices of all types, including desktop computers. The web makes management data accessible to everyone in the organization, allowing more informed decisions.

Web-based tools use web technologies such as PERL and the Common Gateway Interface (CGI) programming tools. These tools may be used to obtain displays of traffic loads such as across network segments, the backbone and gateways connecting to the Internet, and other sites. In addition, script files in PERL may be written for web servers and browsers that poll SNMP agents on routers and hubs to prioritize the devices by error rates, for example. The web can also be used to produce reports through probes done for the devices on the manufacturer's web site.

Using the web to perform network management tasks also exposes some limitations and shortcomings of Windows-based operating systems that may keep traditional Unix-based management platforms around for some time. One of these limitations is lack of easy access to stored management data. Another potential problem with web technology lies in the security features of the Java programming language. Java does not allow users or devices to attach to certain port numbers. One of those port numbers may be 161, which is for SNMP queries and commands. Therefore, at least presently, it might not be possible to build something that talks SNMP directly from a Java applet.

9.6 CHAPTER SUMMARY

This chapter deals with several network management approaches, which include SNMP, RMON, TMN, and web-based management. The basic SNMP model consists of a data definition language, definitions of MIB (management information base), a protocol definition, and security/administration features. The newer version of SNMP (SNMPv3) is considered to have more security and authentication features that the earlier versions.

On the other hand, RMON is designed to work with SNMP such that the remote probe devices do the data collection and processing. This results in reduced SNMP traffic on the network and a reduced processing load on the management station. Instead of continuous polling, information is transmitted only to the RMON. The client/server implementations of RMON allow the servers to monitor the remote networks by collecting the RMON information and analyzing network packets.

TMN is recognized to have organized hierarchy with defined task for each level, much like the OSI model. TMN, a management network with standard OSI protocols, interfaces, and architectures, was developed by ITU as an infrastructure to support management and deployment of dynamic telecommunications services.

Finally, the web-based and directory-enabled network management strategy is gaining popularity. Telecommunication device configuration and service provisioning traditionally involve significant manual work. Using increasingly more powerful web-based network management tools, the network and telecommunications industry is making every effort to automate the procedures of equipment configuration and service provisioning. Administration can now be done from any location using any workstation supporting a browser without requiring a dedicated device.

Many Web-based and directory-enabled network management tools are used to administer subscriber self-registration and service activation with appropriate privileges and service policies, through updates to an LDAP directory. An ISP can request, say, to increase their maximum bandwidth from 155 Mbps to 622 Mbps for a specific connection, and receive that increase in a few minutes via a web interface.

There are also tools for DNS and DHCP services and IP address management for various network service providers. The ultimate goal of highly automatic and directory-enabled network management system is to provide an end-to-end equipment configuration and service provisioning without manual operations.

9.7 PROBLEMS

9.1 Explain what fault management is. Specifically name the hardware/software devices to which the fault detection procedures should be applied.

9.2 Name some parameters used for baselining a network.

9.3 Use TMN terminology to name the switching systems, circuits, terminals, and so on that comprise a telecommunications network.

9.4 What do we call a network's managed information and the rules by which that information is presented and managed?

9.5 What TMN functional block can impose a machine-readable, object-oriented structure to an NE's proprietary or legacy information model?

9.6 How many OSI layers are there in a TMN network?

9.7 Is it true that most TMN function blocks can function as both managers and agents? Explain.

9.8 Is it true that that the Q3 interface is the only interface that the QAs, MDs, or NEs may use to communicate directly with the OS? Explain.

9.8 REFERENCES

BOOKS

Bapat, S., *Object-Oriented Networks: Models for Architecture, Operations, and Management*. Englewood Cliffs, NJ: Prentice Hall, 1994.

Berners-Lee, T., R. Fielding, and H. Frystyk, *Hypertext Transfer Protocol HTTP/1.0, RFC 1945, MIT/LCS*. Irvine: University of California, Irvine, May 1996.

Bloommers, J., *Practical Planning for Network Growth*. Englewood Cliffs, NJ: Prentice Hall, 1996.

Feit, S., *SNMP: A Guide to Network Management*. New York: McGraw-Hill, 1994.

Harnedy, S. J., *Total SNMP: Exploring the Simple Network Management Protocol*. Horsham, PA: CBM Books, 1994.

Hazewinkel, H., E. van Hengstum, and A. Pras, *Definitions of Managed Objects for HTTP*. draft-hazewinkel-httpmib-00.txt. Twente, Belgium: University of Twente, April 1996.

Huntington-Lee, J., K. Terplan, J. A. Gibson and J. Gibson, *HP Openview: A Manager's Guide*. New York: McGraw-Hill, 1997.

ITU-T Recommendation M.3010. "Principles of Telecommunications Management Network," 1993.

Leinwand, A., and K. Fang, *Network Management: A Practical Perspective*. 2nd ed. Reading, MA: Addison-Wesley, 1993.

Lewis, L., *Managing Computer Networks: A Case-Based Reasoning Approach*. Norwood, MA: Artech House, 1995.

Miller, M. E., *Managing Internetworks with SNMP: The Definitive Guide to the Simple Network Management Protocol (SNMP) and SNMP version 2*. New York, NY: M&T Books, 1993.

Phaal, P., *LAN Traffic Management*. Englewood Cliffs, NJ: Prentice-Hall, 1994.

Rose, M. T., *The Simple Book: An Introduction to Management of TCP/IP-Based Networks*. 2nd ed. Englewood Cliffs, NJ: Prentice-Hall, 1994.

Rose, M. T., and K. Z. McCloghrie, *How to Manage Your Network Using SNTP: The Network Management Practicum*. Englewood Cliffs, NJ: Prentice-Hall, 1995.

Stallings, W., *SNMP, SNMPv2, and CMIP: The Practical Guide to Network Management Standards*. 2nd ed. Reading, MA: Addison-Wesley, 1996.

Stallings, W., *SNMP, SNMPv2, SNMPv3, and RMON 1 and 2*, 3rd ed. Reading, MA: Addison-Wesley, 1999.

Steedman, D., *Abstract Syntax Notation One (ASN.1): The Tutorial and Reference*. Twickenham, U.K.: Technology Appraisals, Ltd., 1990 (with Errata sheet dated March 14, 1991).

Terplan, K., and J. Huntington-Lee, *Applications for Distributed Systems Management*. New York: Van Nostrand Reinhold, 1994.

Udupa, D. K., *Network Management Systems Essentials*. New York: McGraw-Hill, 1996.

Udupa, D. K., *TMN: Telecommunications Management Network*. New York: McGraw-Hill, 1999.

WORLD WIDE WEB SITES

J. Cellucci, R. Hill, and A. Simon, "You Are Here—New Developments in Directory Services Have Managers Wondering Which Way to Take Their Corporate Networks." *Communications Week* no. 637 (November 11, 1996), Section: CloseUp—Directory Services.
http://www.techweb.com/se/directlink.cgi?CWK19961111S0073

Links to information about X.500-based directory services and other related areas:
http://www.bath.ac.uk/~ccsap/Directory/

X.500 Directories
http://www.nexor.co.uk/users/cjr/x500.html

10

COMMUNICATION AND NETWORK SECURITY

Computer users have always been concerned with data security. When personal computers were indeed personal, such security meant preventing others from using your machine or stealing diskettes. It also meant regularly backing up important files in case data became corrupted. Mainframe security dealt primarily with controlling access to computer rooms and/or tape libraries and performing regular disk backups to tape. With today's distributed systems and global networks, protecting data is far more complicated then passwords, identity cards, and keeping your office door locked. In many cases the door to the data is readily accessible to the world. Anyone can try his or her luck at unlocking it. In addition, many of the people using computers are not aware that their account or their computer provides access to important data. They do not have the same security concerns as those who work with such data. And finally, there are those who see any locked door as a challenge to their own ingenuity. Hackers who initiate viruses have very little control over the ramifications of their work. The destruction spreads almost randomly throughout the world.

This chapter discusses important issues in data security: how to protect data on a network, various methods used to make the data unreadable for outsiders, and how to protect data from viruses. It will become evident that there are no perfect protection schemes and that data security is a never-ending battle as new technologies develop. But there is much that can be done to enhance security and minimize the loss of important data on a network.

After the world's largest network lifted its decade-old ban on business use in the summer of 1991, commercial use of the Internet skyrocketed. Tens of millions of new users joined the Internet. Everywhere the Internet ended, a new network sprouted. The relatively open, interactive nature of the Internet made it easy for new users to find new uses for such a powerful electronic infrastructure. The Internet is now used regularly to transfer data, documents, images, and software. It is also used to advertise, to deliver product patches and support, and to perform transactions with digital money. However, many corporations still refrain from using the Internet because of security concerns. Just how secure can data be that constantly flows through such easily accessed media as phone lines, microwaves, and satellite transmissions?

In today's highly competitive, electronically based business atmosphere, raw information has considerable power and value. Even something as simple as a mailing list can be sold and traded. Computers are the primary tools of industrial espionage and sabotage. They can be used to access or destroy valuable information. There is the familiar tale of hackers breaking into government databases and viewing top-secret design specifications for Stealth bombers. Citibank was the victim of a $10 million fraud perpetrated by Russian hackers. These hacks were done through the use of computers located nowhere near the machines that were improperly accessed. The prevention of such crimes significantly affects the way networks are operated. It affects the way software packages are written and how easy they are to use. And it affects how much bother one must go through to perform networking tasks.

When a computer is connected to a public network, all the data transmitted through the connecting line are exposed to the outside world. An intruder may purchase or build a special network interface card to tap into that line. Unlike other network cards that read only the data intended for that card, this card has a software-programmable address. It can listen in and pick up the address of a frequently accessed machine. It can display on the intruder's screen a duplicate of what is displayed on any of the other machines on the network. Every keystroke is reproduced. Any information displayed is readily available, as if the intruder were looking over the shoulder of any user on the network. The displayed information can be saved for more detailed inspection later. For a skilled intruder, eavesdropping on a network is no more difficult than eavesdropping on a telephone call.

Eavesdropping implies listening only, but in the case of network eavesdropping the intruder can easily modify messages received before passing them on. Forged transmissions can result in the destruction or alteration of vital information. In addition, viruses or software time bombs can be implanted by these techniques.

Recent technological and economic changes have only enhanced the need for network security. Corporate downsizing has reduced the use of the more easily protected mainframe computers. Mission-critical applications like order entry, billing, accounts receivable, and inventory are fairly easy to protect when they are running on mainframes. But when they are shifted to LANs, every unsecured workstation is a potential threat to network security. It is even more difficult to protect data when client–server applications and distributed computing environments, and particularly the Internet, are being used. Moreover, companies have been forced to make it easy for technically unsophisticated users to access network data, further compounding the problem. It is ironic that the renewed interest in network security is an indicator of just how successful the internetworking movement has been. Java, thanks to its popularity and despite its security features, is being used more and more to create potentially damaging hack tools. Recently, a company released a security threat on an Internet web site as a Java applet that can be reproduced by persons without programming knowledge to attack any system connected to the Internet.

Unless those who manage and use networks recognize the importance of network security and are properly trained to maintain it, data on the network will not be protected from intruders and accidental or intentional destruction or modification.

Network security issues fall into the following four categories:

1. Protecting transmitted data—data transmission security
2. Ensuring that users are who they claim to be—authentication
3. Restricting information access to certain people—authorization
4. Preventing destruction and alteration of information—virus prevention

In most cases, the media carrying data cannot be secured. Telephone lines, satellite links, and microwaves are easily accessed by the public. The global Internet is also wide open to sophisticated adversaries. Hence the need for data transmission security, or a means of protecting the data during transmission. The primary security measure consists of ensuring that data are meaningless to anyone other than the right recipient. As we will shortly show, "cryptography" is the most effective method of ensuring data transmission security.

Authentication is the process of verifying the identity of users. To identify a specific user, a system may check the following:

1. Something the user knows—a password. This is the simplest, least expensive, and weakest means of identification.

2. Something the user owns—a token, or a ticket. This is a much stronger means of user identification.
3. Something that is unique about the user—a fingerprint or digital signature. These are mechanisms that rely on verification of unique characteristics peculiar to a single user.

Authorization allows a resource owner to determine who can access that resource and in what manner (read-only, read/write etc.). Authorization also provides a means of determining whether a given user has legitimate access to a resource. This can be managed through the use of firewalls, as discussed in Section 10.3, in addition to conventional directory and file system protection methods. Kerberos, a network authentication system is discussed in Section 10.4.

Popularity of ecommerce and business transactions on the web have made the use of secure socket layer (SSL) protocol a must. Section 10.5 is dedicated to this discussion and also includes an explanation of virtual private networks (VPNs).

10.1 CRYPTOGRAPHY

Cryptography is the process used to make a meaningful message appear meaningless. Complex systems have been used throughout history to protect secret messages from prying eyes. From Roman times to World War II, and from the Cold War to today's Information Superhighway, these systems have been based on various cryptographic algorithms and keys. An algorithm is a set of rules or procedures used to scramble, or encrypt, the plaintext (the original message) to produce *ciphertext* (the scrambled message). An algorithm applies a key to text. Different keys produce different ciphertexts, though the algorithm is the same. Without both the algorithm and the key, the ciphertext remains scrambled and cannot be reversed to plaintext. The advantage of cryptography is that the coded message can be stored in hostile environments and transmitted over hostile communications channels. A good cryptosystem allows the owner of the secrets to control completely who can unravel the protected information.

Encryption uses one secret to protect another. As long as the key remains secret, the message cannot be decoded. The algorithm, however, can be made public. After all, a hardware encryption device can be stolen and reverse-engineered; software encryption code can be disassembled. It is difficult to protect the secrecy of an encryption algorithm for long. Furthermore, developing effective encryption algorithms is difficult. Thus it is impractical to change the algorithm every time suspicion arises that its secrecy is being compromised. In practice, one must always assume that the encryption algorithm is known to the public. Only the key must be kept secret. Even the details of the algorithm used by the official data encryption standard (DES) discussed in Section 10.1.3 were openly published by the U.S. government.

10.1.1 Secret Key Cryptography

Traditionally, a secret key is used to scramble plaintext to ciphertext. The plaintext may take the form of a complete file, or it may consist of interactive online communication.

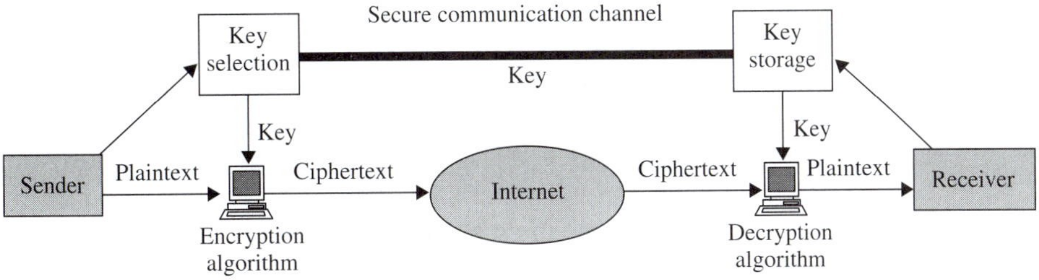

Figure 10.1 Encryption/decryption procedure with a secret key delivered through a secured channel.

The ciphertext (file, message, or image) may be transmitted over nonsecure lines. The secret key, transmitted to the receiver through a secure channel, is then used to unscramble the ciphertext back to plaintext (Figure 10.1).

The key may consist of a short string of characters that can be easily memorized and conveniently changed unless kept in a physically secured computer and transferred through a secured communication channel. In the later case, the key can be a long and randomly selected bit stream. How secure the system is depends on how good the algorithm is, and how well the key is stored and delivered.

All encryption methods are based on mathematics. The easiest *substitution-based* encryption algorithm uses the Boolean exclusive OR (XOR) operation. The plaintext byte and the secret key byte are input to the XOR operation. The XOR makes a bit-for-bit comparison of the two bytes to produce an output byte. The output is the ciphertext. When a sequence of bytes, say, `SECURITY` is selected as the key, this sequence is repeatedly applied to XOR with the plaintext to produce the ciphertext. For example, if the plaintext is `abcdefghi` . . . , then `S` is XORed with `a`, `E` with `b`, . . . , `Y` with `h`, and `S`, again with `i`, and so on. The ASCII output is usually an unreadable sequence of bytes.

Because the XOR operation is completely reversible, if the ciphertext is next XORed with the same key, the output is the original plaintext. This is a fast and easy algorithm. However, if a piece of the message in both the ciphertext and its corresponding plaintext is known, a cryptanalyst can simply XOR the known ciphertext with the known plaintext to obtain the key. The key can then be used to decipher the remainder of the message. If the key is not changed, future messages could also be deciphered.

Another approach uses a randomly arranged set of 26 letters mapped to the normal alphabet. This is called *monoalphabetic substitution*. Each letter in the normal alphabet maps to a corresponding letter in a mapping alphabet. For example, if the following mapping is used:

```
normal alphabet:   ABCDEFGHIJKLMNOPQRSTUVWXYZ
mapping alphabet:  nzaybcxwdveuftgsirjqkplomh
```

`COMPUTER` in plaintext becomes `agfskqbr` in ciphertext. Since there are 26! ways to arrange the 26 letters of the normal alphabet, decryption by brute force (trying every combination) is impractical. A computer using 1 μs for each test would take more than 10^{13} years to try all the possible combinations ($26! = 4 \times 10^{26}$).

However, human languages have patterns that can be recognized very easily. For example, 13.05% of the letters in a reasonably large sample of English language text are the letter *e*, 9.02% are *t*, 8.21% are *o*, and 7.81% are *a*. Spaces show up frequently. In addition, the most frequently used letter combinations (*the*, *a*, *of*, *to*, *on*, *ing*, etc.) can be easily recognized. Educated guessing eliminates most of the possible keys. A brute force approach to the remaining possibilities cracks the code.

To remove the letter patterns, one can use a polyalphabetic cipher. Here multiple cipher alphabets are used in rotation. A short key consisting of numbers is laid repeatedly under the plaintext. The number in the key indicates the alphabet used for the corresponding plaintext letters. For instance, if 1, 2, 3, and 4 represent the following four different mapping alphabets:

```
Mapping alphabet 1: normal:   ABCDEFGHIJKLMNOPQRSTUVWXYZ
                    mapping:  nzaybcxwdveuftgsirjqkplomh
Mapping alphabet 2: normal:   ABCDEFGHIJKLMNOPQRSTUVWXYZ
                    mapping:  hnzaybcxwdveuftgsirjqkplom
Mapping alphabet 3: normal:   ABCDEFGHIJKLMNOPQRSTUVWXYZ
                    mapping:  mhnzaybcxwdveuftgsirjqkplo
Mapping alphabet 4: normal:   ABCDEFGHIJKLMNOPQRSTUVWXYZ
                    mapping:  omhnzaybcxwdveuftgsirjqkpl
```

Here 2143 may be used as the short key, the following mapping list shows how the mapping alphabets are used to change the plaintext `INTERNET` into the ciphertext `wtiaitzr`:

```
Plaintext:            INTERNET
Mapping alphabets:    21432143
Ciphertext:           wtiaitzr
```

Here the first and second *T*s in the plaintext *INTERNET* are mapped into different letters *i* and r, respectively, in the ciphertext because the first *T* uses the fourth mapping alphabet while the second *T* uses the third mapping alphabet. Using a polyalphabetic cipher, a given plaintext letter may be mapped into different letters in the ciphertext depending on its position in the plaintext. Thus the matrix of cipher alphabets comprises part of the key system. The more alphabets used, the more difficult the decryption. With the help of a computer, however, and assuming that an adequate sample set of ciphertext is available, the deciphering job is not as daunting as it might appear. This kind of polyalphabetic cipher system can still be broken by first guessing a length for the key. The ciphertext is then listed in rows of the assumed key's length. If the key length is correct, the ciphertext letters in each column will present the same frequency patterns as normal English usage. They were, after all, encrypted by the same monoalphabetic sequence.

Generally, the longer and more random a key, the harder it is to break. Every additional bit of the key doubles the amount of number crunching needed to crack it. When the key is longer than the plaintext, all possible keys will produce all possible messages. If such a key is not used for further messages to prevent a cryptanalyst from using the redundancy of reoccurring messages for comparison, the key is called a *one-time pad*. One-time pads are the only proven unbreakable cryptosystem. Even with the simplest XOR algorithm, one can convert plaintext safely into a ciphertext bit string by XORing it with a one-time-pad key. The key could be handily chosen from a book starting from page X, line Y, and word Z. To get the plaintext back, the receiver simply performs an XOR with the ciphertext, using the same string from the same book. The difficulty here is in handling the long key. The key may have been transferred over a supposedly secure medium, but there is always the question of whether the medium is secure enough. In addition, small transmission errors can put the process out of sequence, making the created plaintext meaningless.

A transposition-based cipher rearranges the order of the plaintext characters A well-known transposition-based cipher works like a transposition operation on a matrix, but the alphabetic order of the key is used as the column number for the transport matrix. If the key (say `CIPHER`) is 6 characters long, then a matrix is set up with six columns. The plaintext is written across the rows. The columns are then arranged in alphabetic order (`CEHIPR`). The order of the plaintext columns also is changed. The ciphertext is then created by reading down the columns of the matrix. This technique requires a key with no repeating letters. The key `CIPHER` can be used as follows to transform the plaintext *theattackstartsateightam* :

```
Plaintext        Alphabetic       Transport
matrix           matrix           matrix

145326           123456           1 C tari
CIPHER           CEHIPR           2 E ttta
                                  3 H asat
theatt           ttahet           4 I hctg
acksta           atscka           5 P eksh
rtsate           rtatse           6 R taem
ightam           iatghm
```

The ciphertext `taritttaasathctgekshtaem` is created with the output from the transport matrix in the row-directed order.

By using letter frequency statistics, the cryptanalyst realizes that the cipher is a transposition cipher. By then testing different key lengths and possible plaintext content, and by applying brute force trial-and-error techniques, one may find the right order of the columns and decipher the message. From this example and others earlier, it is no wonder that some people think cryptanalysis is part science, part art, and part luck.

Encryption techniques can work along with compression algorithms to further obscure the meaning of the plaintext. Compression drastically reduces the length of the ciphertext and makes its analysis even more difficult. Several publicly available compression products, PKZIP and gnzip, provide adequate compression.

10.1.2 Public Key Cryptography

Over the years, and particularly since the invention of computers, the sophistication of encryption algorithms has increased dramatically, but the basic ideas have remained unchanged. A known encryption algorithm is applied to a secret key. Keeping the key secret is essential. This is not easy when the key has to be transported to the person or machine doing the decryption. However, the public key cryptography (PKC) method presented by Whitfield Diffie and Martin Hellman in their paper entitled "New Directions in Cryptography" (1976) uses a quite different approach that substantially changed the key distribution situation.

A simple mathematical function takes a parameter and yields a result. The square function takes the parameter 5 and produces 25 as the result. A function's inverse takes a result and gives back the corresponding parameter. But in many cases, the inverse function is more difficult than the original function: the square root function, for example. When a function is much more difficult to compute than its inverse, it is called a one-way function. A good example of a one-way function is the telephone book. Given a name, it is easy to find a number, but given a phone number, it is nearly impossible to find the appropriate name. Many mathematical functions are also one-way functions and considerably more difficult than reverse phone books.

Based on this idea, public key cryptosystems use two keys—a *public key* and a *private key*. Data encrypted by one key and a function E can be decrypted only by the corresponding other key and a function D. The system and the keys are designed so that one key (the public key associated with the function E) can be made public, without compromising the other key (the private key associate with the function D). This process works both ways: data encrypted with a public key and function E can be decrypted only with the corresponding private key and function D, and vice versa.

This can be formally expressed as $D(E(P)) = P = E(D(P))$, where P is the plaintext. Since it is practically impossible to deduce the secret key from the public key, the public-key algorithm is called *asymmetrical* cryptography, thus the conventional secret key algorithm is *symmetrical*, since both encryption and decryption use the same key.

With PKC, it is possible for two users A and B to communicate securely without any prior relationship or secret key exchange between them (Figure 10.2). In this method, A and B each have two keys: a public key and a private key. If A wants to send a message to B, B's public key is used to encrypt the message and form a digital envelope $E_B(P)$. Then B uses B's private key, which is known to no one but B, to decrypt the digital envelope and recover the plaintext $P = D_B(E_B(P))$. B's private key is never sent to A or anyone else.

The best-known PKC algorithm is RSA, named after its three inventors, Rivest, Shamir, and Adleman. It is patented with code available free from the RSA web site on the Internet for noncommercial use within the United States. To understand the theory be-

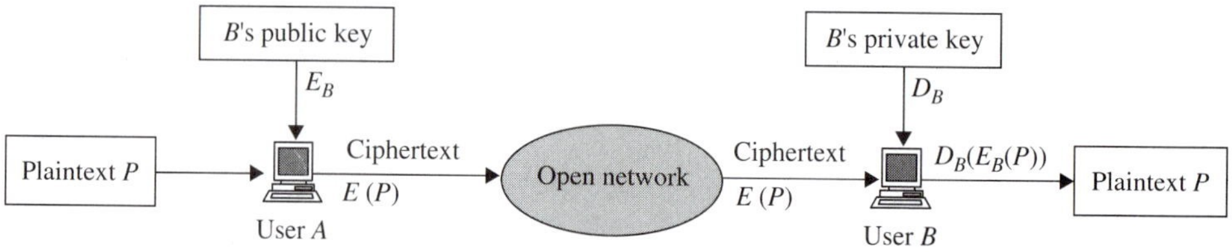

Figure 10.2 Public key cryptography for secure message transmission without distributing a secret key.

hind RSA, consider factoring a one-way function. Given a set of factors, it is very easy to generate a large number, but given the large number, it is far more difficult to find its factors.

RSA gets its security from the difficulty of finding large prime numbers and factoring large numbers—a problem that mathematicians have been unable to solve. With this in mind, the RSA algorithm uses a public key (e), a private key (d), and another number (n), also public. The process of determining these is as follows:

1. Choose two large prime numbers, p and q, where $10^{100} < p, q < 10^{200}$.
2. Compute the product $n = p \times q$ and $z = (p - 1) \times (q - 1)$.
3. Choose a large number $e > p$ and $e > q$, such that e has no factors in common with $(p - 1)(q - 1)$, as the public key.
4. Find the secret key d such that the remainder of $e \times d$ divided by z equals 1. That is, $(e \times d) \% [(p - 1)(q - 1)] = 1$, where % represents modulo (MOD) operation.

Using Euclid's algorithm or the following C program, a very small secret key d will be generated when p, q, and the public key e are given.

```
int secret_key (int p, q, e)
    /* To be simple, p, q, e are all small prime numbers */
  {
   int i;
   long z;

   z = (p-1)*(q-1);
   for (i=1; ; i++)
     if ((z*i+1)%e == 0) break;     /* (e * d) % ((p-1)(q-1))=1 */
   return ((z*i+1)/e);       /* return the secret key */
  }
```

It can be easily verified that the following numbers satisfy the rules except that they are too small:

p	q	e	d
3	5	11	3
7	11	13	37
11	29	37	53
41	47	79	559
47	71	79	1019

The public key (e) can now be used to convert the plaintext to ciphertext. It is converted back to plaintext by applying the private key (d). To encrypt a plaintext P, first convert P into a number regarding P as a bit stream. For example, the ASCII plaintext `ABCD` can be treated as 30313233_{16} or 808530483_{10}. Now compute the corresponding ciphertext $C = P^e \% n$. To decrypt C, simply compute $P = C^d \% n$. Besides e, n must be provided to both communicating parties, thus n should be used along with e as part of the public key.

Since $C = P^e \%n$, we have $C = P^e - kn$, and $C^d = (P^e - kn)^d = P^{ed} + k'n$. In addition, because $(e \times d) \% ((p-1)(q-1)) = 1$, so $ed = k''(p-1)(q-1) + 1$, where k, k', and k'' are all integers. Hence

$$C^d \% n = P^{ed} \% n = P^{k''(p-1)(q-1)+1} \% (pq) = (P \times P^{k''(p-1)(q-1)})\%n = P$$

The plaintext is recovered. The proof of the last step can be found in any book on cryptography or number theory. It also can be proven that the encryption and decryption functions are inverses. That is, the plaintext can just as easily be encrypted with d and decrypted with e.

To show how the RSA algorithm is used, we give a trivial sample with extremely small numbers. To simplify the calculations, we also ignore the rule that e should be larger than p and q. To start, choose $p = 3$, $q = 11$. Then n must then be 33 and z becomes 20. We next select $e = 7$, and can easily find $d = 3$. Using these values, the following procedure shows how the plaintext `COMPUTER` is scrambled to ciphertext, then recovered to plaintext using the MOD function (%).

Plaintext Letter	number	P^7	Ciphertext $C=P^7\%33$	C^3	$C^3\%33$	Plaintext Letter
C	3	2187	9	729	3	C
O	15	170859375	27	19683	15	O
M	13	62748517	7	343	13	M
P	16	268435456	25	15625	16	P
U	21	1801088541	21	9261	21	U
T	20	1280000000	26	17576	20	T
E	05	78125	14	2744	05	E
R	18	612220032	6	216	8	R

This simply creates a monoalphabetic substitution cipher. For a long message M, first divide it into blocks such that each block has a unique numerical representation. A numerical block M_i could be as large as n. That means it may be necessary to divide and group a long plaintext into blocks of k bits, where $2^k < n$. When p and q are large enough, $n = p \times q$ will be larger than 10^{200} or 2^{664}, so blocks of at least 664 bits can be encrypted at once. The encrypted message block is simply $C_i = (M_i)^e \% n$. To decrypt a message, take each encrypted block C_i, and compute $M_i = (C_i)^d \% n$.

The following table, originally posted to Usenet in October 1993, gives some numbers for the expected amount of work required to crack keys of various sizes by factoring. According to research, the time was about 4600 MIPS-years for RSA129 (429-bit key). A MIPS-year is the computation completed by a computer in one year at a speed of 1 million instructions per second.

RSA129 (429-bit key)	4,600 MIPS-years
512-bit key	420,000 MIPS-years
700-bit key	4,200,000,000 MIPS-years
1024-bit key	2.8×10^{15} MIPS-years

This table is based on the multiple-polynomial quadratic sieve (MPQS). Other algorithms under development may have slightly better performance.

For practical purposes, the keys in current systems are limited to 1024 bits. This is far beyond any publicly known computational capability. The bottom line is that the cracking of a 1024-bit key by means of anything like presently known factoring methods will not happen soon. A breakthrough in computer technology or algorithm efficiency that threatens a 1024-bit key will be powerful enough to threaten much larger keys as well. Any successful attack on the algorithm with large key sizes is more likely to come from exploiting other aspects of the system, such as the prime number generation algorithm, rather than by brute force factoring of keys.

PKC has implications far beyond simple data encryption. It presents a means for the secure transmission of important data via public networks. A simple application of PKC is for password protection.

Conventional password protection schemes store passwords in an encrypted form on the host computer. When a user enters his password through an open network, such as telnet, through the Internet, it is not encrypted. Anyone can intercept it by decomposing the datagrams that pass though an intermediate network gateway. In addition, anyone with access to the system memory of the host computer can read the password before it is encrypted and compared with the password file. With PKC, a slightly different logon sequence is used. The host computer keeps a file of every user's public key. When a user logs into the system, the host computer sends the user a challenging random string. The user encrypts the string with his private key and sends the encrypted string back to the host. The host uses the user's public key to decrypt the string sent back and compares it to the original random string. If they are the same, logon continues. The user's private key is never known to the host and is never transported over an unsecured line. If the user's terminal is trusted and secure, no one in the middle, even in the host, can steal the private key.

RSA is not the only PKC method. Both Merkle–Hellman's Knapsacks algorithm and an exponentiation public key algorithm are PKC methods that are patented. These patents, along with the RSA patent, are controlled by a consortium of companies called Public Key Partners.

Public key cryptography has many advantages, but it does not replace conventional secret key cryptography. PKC algorithms are complicated protocols, not ideally suited to encrypt long messages. Some conventional secret key algorithms are 10,000 times faster than PKC when both are implemented in hardware. In addition, PKC has the same key-storage vulnerability as secret key. A common practice is to use PKC to transfer the secret key for another cryptographic algorithm and then use that algorithm to encrypt and decrypt messages. The symmetrical data encryption standard (DES) algorithm is ideal for applications of this kind.

10.1.3 The Data Encryption Standard (DES) and Advanced Encryption Standard (AES)

DES is a widely used data encryption method using a private key that was considered very hard to break. The data encryption standard (DES) has been controversial since the National Bureau of Standards established it in 1977 as the official method of protecting unclassified information. Developed by IBM, it is still in wide use today because it can be easily implemented in silicon. The original DES used a 128-bit (16-byte) key. At the request of the National Security Agency (NSA), the key length was reduced to 7 bytes. What's more, IBM was sworn to secrecy regarding the reasons for choosing the various bit rearrangements, and substitution patterns of the algorithm. The inexplicable modifications of DES by the National Security Agency raise the question of whether the DES was deliberately weakened so that the U.S. government would have an encryption standard that its agents could easily break, while most others could not. This suspicion has given rise to the widespread belief that there is a hidden trick in the standard that allows it to be easily cracked. Though the details of the algorithm are publicly known, no such trick has been reported.

DES is a block cipher. It encrypts data in 64-bit blocks. A 64-bit block of plaintext is supplied to one end of the algorithm and a 64-bit block of ciphertext is obtained at the other end. Both encryption and decryption use the same key and algorithm except that there is a difference in the key schedule. The 56-bit key is usually expressed as a 64-bit number, while every eighth bit is used for parity checking and is ignored.

At its simplest level, the algorithm is nothing more than a combination of the two basic techniques of encryption: confusion and diffusion. The fundamental building block of DES is a combination of a substitution followed by a transposition (permutation) on the text based on the key. This is known as a round. DES has an initial transposition at the beginning, a 32-bit swap operation and an inverse transposition at the end, with 16 rounds between. Each round applies the key and the same combination of techniques to the plaintext block. The algorithm uses only standard arithmetic and logical operations on numbers of at most 64 bits, so it was easily implemented in late 1970s hardware technology. Initial software implementations were clumsy, but current implementations are somewhat better. The repetitive nature of the algorithm makes it ideal for use on a special-purpose chip.

10.1.3.1 Outline of the DES Algorithm

DES operates on a 64-bit block of plaintext. After an initial permutation, the block is broken into a right half and a left half, each 32 bits long. Then there are 16 rounds of identical operations, called function *f*, in which the data are combined with the key. After the sixteenth round, the right and left halves are joined, and a final permutation finishes off the algorithm.

In each round, the key bits are shifted, and then 48 bits are selected from the 56 key bits. The right half of the data is expanded to 48 bits via an expansion permutation, combined with 48 bits of a shifted and permuted key via an XOR, substituted for 32 new bits by using a substitution algorithm, and permuted again. These four operations make up function *f*. The output of function *f* is then combined with the left half via another XOR. If L_i and R_i are the left and right halves at the *i*th iteration, respectively, K_{i+1} is the 48-bit key for round $i + 1$, and f is the function that does all the substituting and permuting and XORing with the key, then at round $i + 1$ the output of left and right halves are, respectively:

$$L_{i+1} = R_i$$

$$R_{i+1} = L_i \oplus f(R_i, K_{i+1})$$

That is, the old right half becomes the new left half; and these operations generate the new right half. These operations are repeated 16 times, making 16 rounds of DES.

After all the substitutions, permutations, XORs, and shifting around, the plaintext is encrypted. With DES it is possible to use the same function to encrypt or decrypt a block. The only difference is that the keys must be used in the reverse order. That is, if the encryption keys for each round are k_1, k_2, . . . k_{16}, the decryption keys are k_{16}, k_{15}, . . . , k_1. The algorithm that generates the key used for each round is a bottom-up right circular shift.

The relationships that come about by using the same key to control a number of encryption steps, especially by means of such a short key, can give rise to a number of hidden interactions. Perhaps there exists a complementary routine that will undo the DES easily by simply rearranging the steps. From NSA's perspective, the advent of unbreakable encrypted communications for computers, telephones, and video would make wiretapping useless. But as we have seen, producing a close-to-unbreakable encryption system is not very difficult. Now that fast computers are widely available, software and hardware implementations of such algorithms are easily accomplished.

So far DES is a quite strong cryptographic protocol. Brute force is the natural choice for determining the key. After all, there are only 2^{56} possible keys. The correct key has to be one of them. Using brute force with networked supercomputers and workstations running for 8 days, a French university student recently cracked the RSA RC-4, a foreign version of the web browser, Netscape Navigator. That software has weaker security (a 40-bit version of DES) than its American version because of export restrictions on encryption algorithms. In fall 1995 the U.S. government outlined a plan called *key escrow* that would eliminate export restrictions on 64-bit encryption software, as long as the government has a way to get the decryption key through a third party. With key escrow, a copy of the key is stored with a certified trusted agent, rather than held by the government.

There have been new attacks against DES. Two attacks called differential cryptanalysis and linear cryptanalysis have been successful against DES and other algorithms. Differential cryptanalysis is a chosen-plaintext attack, which means the attacker not only has the algorithm, some ciphertext, and the unencrypted plaintext but gets to choose what it is. The cryptanalyst compares the differences between pairs of plaintext and corresponding pairs of ciphertext. These differences, along with information about the structure of the underlying algorithm, give the analyst clues about the key. Collecting enough of these differences allows one to find the key more efficiently than would be possible by using a brute force approach.

Although interesting, differential cryptanalysis is still more theoretical than practical. The best chosen-plaintext attack against DES using on the order of 10 terabytes (TB) of chosen-plaintext data has a complexity of 2^{47} compared with the 2^{56} required for the brute force approach.

In an effort to prove its contention that U.S. limitations on exportable encryption are too strict, RSA Data Security Inc. sponsored a joint Internet effort to crack the DES code, offering $10,000 to the first individual or group to decipher its DES-encrypted message. More than 2500 computers worked on the message in a cooperative effort to try every possible key at the rate of 50 trillion keys a day. In less than half a year, the DES encryption code was cracked in June 1997 by a computer powered by a Pentium 90 with 16 MB of RAM, running software developed and made available over the Internet by

Rocke Verser. The relatively modest technical capability of the winning machine underscores the inadequacy of 56-bit encryption.

Linear cryptanalysis is similar to differential cryptanalysis but is a known-plaintext attack—the encryption algorithm, some ciphertext, and the unencrypted plaintext are given, but the attacker has no choice in what they are. This approach looks for linear relationships between selected bits of the plaintext, ciphertext, and the test key. A more impressive record results: the complexity is 2^{43}. However, this still requires too much data to be of practical use.

10.1.3.2 Advanced Encryption Standard

The government's standard encryption algorithm has been the foundation of the entire cryptography world for 20 years, and had been performing unhackable protection for data until 1997, when the first time a nationwide network of computer users broke a DES key in 140 days. With increasing processing speeds, it is practical to break a DES key within a week. It is not inconceivable that before long, given good skill and affordable specialty hardware, a DES key will be broken in one hour. Most cryptographers agree that even though DES is a good and popular algorithm, it is time to find a replacement. Actually, one new version of DES—the Triple DES, developed by IBM—just runs the 56-bit encryption three times. Although it is more secure than the original DES, it is too slow to be acceptable. The U.S. government has proposed a project headed by the National Institute of Standards and Technology (NIST) to find a cryptographic algorithm, called Advanced Encryption Standard (AES), to replace DES as the government new standard.

Three basic technical criteria for AES has been figured out from the document released by NIST to outline recommendations:

- The algorithm must be symmetric, or private, key.
- The algorithm must be a block cipher like DES, rather than stream cipher (the other two types of symmetric key cipher).
- The key length supported by this algorithm could range from 128 to 256 bits. And the algorithm should also support variable blocks of data.
- The algorithms should be implemented in C or Java programming language.

Besides the four technical requirements, AES must be efficient, and the AES algorithm must be made public and royalty free. Selection is a very political process as well as a technical one. NIST intends to have a final algorithm chosen by late 2001 after comment from the cryptographic community. Once the AES algorithm has been developed, any software intended for use by government or for handling with government secure information must support the AES algorithm. The replacement of AES is expected to be the most adopted commercial security algorithm. While DES was designed as a 5- to 10-year standard, AES is expected to be used for 20–30 years.

10.1.4 Serial Encryption

The present trend in information encryption is to create more complex encryption algorithms rather than to increase the key size. Once data have been made "reasonably secure," adding more to the key doesn't significantly improve the security. To enhance the existing encryption algorithms, one can change the key in the middle of the stream to make the algorithm work like a one-time-pad algorithm. The first block decrypted would be a key that would be used to decrypt the next block, which would be the key to decrypt the next block, and so on. The first part of the message is decrypted by using the original key. That first part of the plaintext is then used as the key for decrypting the second part, and so on.

This process is called serial encryption, or stream encryption, or infinite-key encryption. It is often used in two-way communications, but works equally well with files. When used by itself, the regular patterns of the message may cause problems with the encryption of the following blocks. For example, if the plaintext has a string of binary zeros, the following block will not be encrypted at all. So serial encryption should always be used in conjunction with other scramble steps. It does, however, eliminate the problem of repeating the key at regular block intervals.

There are a number of advantages to serial encryption. Serial encryption can take place on a byte-by-byte or block-by-block basis. Even an odd number of bits can be used for the keys, since the plaintext makes a whole stream of bits available. Serial encryption has a variable length key. One restriction is that the first part of the ciphertext must be decrypted first, to provide a new key for the serial encryption. If the new key were to be taken from a part of the plaintext that had yet to be decrypted, the system would lock up. Parallelism, made possible by multiple processor computer systems, will not work if serial methods have been used to encrypt or decrypt messages.

Since serial encryption depends on the security of its first block, one would want to be sure that it couldn't be broken easily. One way to enhance encryption is to insert random numbers into the first block. This reduces regularity. One random pad would decrypt another random pad, which would be the plaintext key to start the serial decryption stream.

Random characters can be added anywhere in the stream as snow. The algorithm, however, must be good at snow removal. Furthermore, "snow" makes the messages longer. Therefore, it should only be used at strategic locations when it will be the most effective.

10.1.5 Link-Level, Network-Level, and Application-Level Encryption

In the OSI network layer model, link-level encryption is the most transparent form of cryptography. The applications, the networking functions, even the device drivers might not know of its existence. Link-level encryption encrypts the entire datagram including the source and destination IP addresses (physical addresses are not encrypted). This guards, to a certain extent, against traffic analysis work that investigates who talks to whom. Link-level encryption is the natural choice for protecting strictly local traffic or for protecting a small number of intensively used but highly vulnerable lines, such as satellite circuits. In the increasingly internetworked world, link-level encryption has some serious drawbacks. Since each site communicates with many other sites, it is not practical to install a dedicated encryption device at both ends of every circuit. Furthermore, data traveling across WANs typically make several hops from router to router, or from switch to switch. With link-level encryption, the data must be decrypted and re-encrypted at each of these points. This exposes the data to security breaches and makes it impossible to apply link-level encryption to the global Internet.

With TCP/IP as the protocol of choice for the Internet, packet encryption at the network layer—IP level—is gaining popularity. In the Network Layer Security Protocol

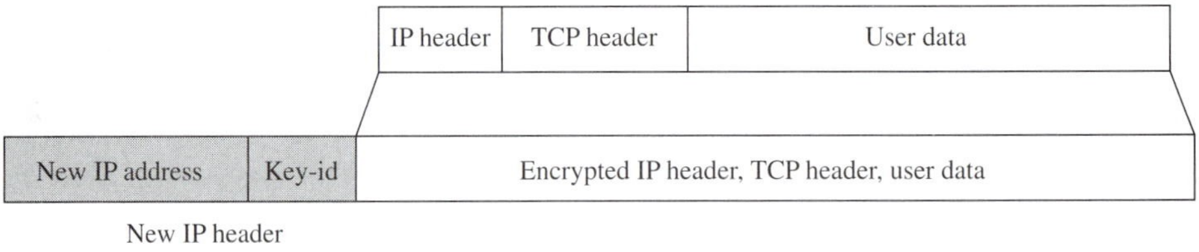

Figure 10.3 Packet encryption at the network layer provides security with both transparency and flexibility.

(NLSP), the packet encryption device at the originating site encrypts the entire packet, including the IP header, and provides it with a new header. This readable new header includes a new IP address and a field called *key-id* (Figure 10.3). The new IP address is the IP address of the security device belonging to the destination site, but it may be different from the original ultimate recipient address. The key-id controls the behavior of the encryption and decryption mechanism. It specifies such information as the encryption algorithm, the encryption block size, the error checking code, and the lifetime of the key.

The encrypted packet can traverse a router-based internetworking environment as is. No modification to the internetworking switching or routing devices is required. Sender and receiver can operate at different speeds without synchronization. At the destination, the encrypted packet is decrypted, the temporary header is removed, and the original IP header becomes readable again, allowing the packet to be delivered to the end user.

With packet encryption, a single encryption device at each site can support any number of virtual circuits across a WAN. Traffic on each virtual circuit can be encrypted with its own distinct key. Because each packet is encrypted individually, encryption keys can be changed as often as needed for more secure communications. The attachment of an unencrypted header onto the original encrypted data and header provides more flexibility in the location of the encryption/decryption devices. For example, the device can be placed in an organization's firewall for incoming and outgoing traffic. Or, it can be placed on an internal segment to allow packets to remain encrypted over certain parts of the internal network. Packet encryption also allows packets to be timestamped. This effectively defends against replay attacks, in which authentic packets are captured and then replayed onto the network later to disrupt service.

When both source and destination addresses of an NLSP communication are attached to the same network, creation of the new IP header can be omitted. This makes packet encryption look like link-level encryption except that it does not encrypt link control fields—checksum, high-level data link control (HDLC), framing, and so on. Like link-level encryption, packet encryption is transparent to most applications.

Performing encryption at the application level is the most intrusive and also the most flexible method. The scope and strength of the protection as well as the user interface can be tailored to meet specific needs. Many well-known encryption packages use application-level encryption: the authentication of Simple Network Management Protocol (SNMP) for example, and some secure electronic mail software packages such as Privacy-Enhanced Electronic Mail (PEM) and Pretty Good Privacy (PGP), to name a few.

Encryption, however, often slows down throughput by as much as 40% because each packet has to be opened and encrypted, and the process is reversed on the receiver's end. This might be a concern for some frame relay applications. Still, if security has a high priority for a particular application, the overall benefits outweigh the delays.

10.1.6 Digital Signatures, Message Digest (MD5), and Digital Certification

Both PKC and secret key cryptography can be used for authentication. MIT's Kerberos system, for example, uses secret key encryption to provide strong authentication. But PKC is a more convenient approach to digital signatures. An effective signature system for electronic documents transmitted over public networks requires the following conditions:

1. The receiver can verify the claimed identity of the sender.
2. Neither the receiver nor anyone else can tamper with the content of the messages.
3. The sender cannot later repudiate the message.
4. The signature must not be reusable. The signature is part of the document, and an unscrupulous person cannot move the signature to a different document.

Sending secure digital signatures through a public network by using the PKC system described earlier involves a rather convoluted use of public and private keys. The procedure is as follows. Assume that A wishes to send B a message with a digital signature. Before the message is sent, A performs two encryptions: first with the private key D_A known only to A, then with the public key E_B available to anyone wishing to communicate with B. The resulting ciphertext $C = E_B(D_A(P))$ is then transmitted through the open network. The receiver then performs two decryptions: first with the private key D_B, the second with the public key E_A. Because $E_X(D_X(P)) = D_X(E_X(P)) = P$ where X represents any user, the plaintext P is the result.

$$P = E_A(D_B\ (C)) = E_A(D_B(E_B(D_A(P)))) = E_A(D_A(P)) = P$$

Since only A knows the private key D_A, nobody other than A can encrypt the message that is decryptable by A's public key. This guarantees that the message came from A. B and others have no ways of tampering with the content of the message before it is decrypted. Neither can the sender A, or anybody else in the middle of the network, repudiate the message because, without all the necessary keys, any tampering with the ciphertext would generate garbage (Figure 10.4).

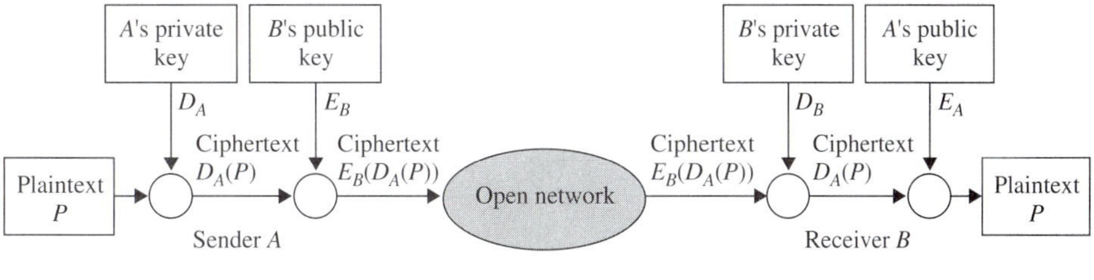

Figure 10.4 Public key cryptography for a secure message transmission with digital signature.

When the message is long, an encryption algorithm, even the hardware implementation of DES, is too time-consuming; and the public key method is several hundred times slower. In such cases, sender A may use a one-way hash function to generate a much shorter code for authentication.

A one-way hash function accepts a variable-length message M as input and produces a fixed-length code $H(M)$—the message digest—as output. Many error-checking code algorithms such as the sequential checksum algorithm, longitudinal parity checking, and the cyclical redundancy code (CRC) algorithm introduced in Chapter 3 fall into this category.

The purpose of a cryptographic one-way hash function is to produce a "fingerprint" of a file, message, or other block of data. A hash function H must satisfy the following conditions:

- H can be efficiently applied to a block of data of variable length and produce a fixed-length output code.
- For any given code c, it is computationally impractical to find the original message M such that $H(M) = c$.
- For any given block M, it is computationally impractical to find another block $N \neq M$, but $H(N) = H(M)$.

The first condition is apparent, and the second condition is relatively easy because the output is much shorter than the original and contains incomplete information about the original message. The third condition is essential. If one can find two messages that generate the same hash output code, the original message may be replaced by a forged counterpart.

When a fourth condition is added

It must be computationally impractical to find any pair of blocks (M, N) *such that* $H(M) = H(N)$

the foregoing hash function becomes a strong hash function. A strong hash function can effectively protect against a sophisticated class of attack known as *birthday attacks*. The name comes from the *birthday paradox*—on average, there must be 183 people in a room for there to be a 50% probability that someone has the same birthday as you. But only 23 people need to be there for two people to share the same birthday with a probability 50%. The goal of a birthday attack is to find any two messages that yield the same output code (have the same birthday). If a brute force attack takes 2^n attempts, a birthday attack will take only $2^{n/2}$.

The publicly available MD5 message digest algorithm developed by Ronald Rivest produces a 128-bit output code with speed, simplicity, and compactness on a 32-bit processor architecture. It first finds a 64-bit representation of the length of the original message. Then 1–512 padding bits and the 64-bit-length code are appended to the original message to yield an integer multiple of 512 bits block (16 32-bit words). A 128-bit buffer that consists of four 32-bit registers with initial values is used for intermediate and final results of the hash function. The blocks are processed by a series of bitwise logical operations to eliminate any regularity in the input data and randomize the output. Source code for the MD5 hash function can be found in RFC 1321.

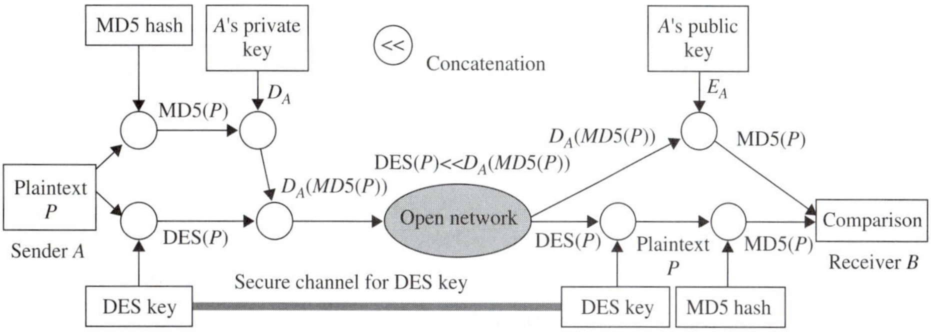

Figure 10.5 One-way hash function MD5 combined with DES and a PKC digital signature sends a long message securely.

The MD5 algorithm has the property that every bit of the hash code is a function of every bit in the input. The complex repetition of the basic logical functions reduces the possibility that two messages with similar regularities will produce the same hash output code. According to Rivest's estimation, about 2^{128} operations are needed to generate a message that produces a given digest. For a birthday attack, about 2^{64} operations are needed to generate two messages with the same MD5 digest.

The MD5 algorithm is used the same way at both ends of a secure transmission. This is basically a fancy checksum, to assure that data have not been modified somewhere en route. The sender A first applies MD5 to the plaintext P, resulting in the output code *MP5*(P). To keep the message secret, A then applies a conventional encryption method such as DES to the plaintext P to get *DES*(P). Next A applies the PKC algorithm and his private key D_A to the message digest code *MD5*(P) and concatenates $D_A(MD5(P))$ to *DES*(P). $D_A(MD5(P)) << DES(P)$ is then transmitted (Figure 10.5).

The receiver B deciphers *DES*(P) and uses the same secret key to get $P = DES(DES(P))$. B then applies MD5 to P to generate MD5(P). This code is then compared to $E_A(D_A(MD5(P))$ received through networks and deciphered by A's public key. If they are identical, B is assured that the message indeed came from A without tampering. This also guarantees that B cannot forge a different message, which replaces the original with an identical MD5 code. The secret DES key can be transmitted through a secure channel, or encrypted with PKC and transmitted through open networks.

In fact, a secret code S shared by the communicating parties in conjunction with the MD5 hash function may replace public key encryption for authentication purposes. In this case, the sender A applies the MD5 hash function to the concatenation of the secret code and the original message, then sends the resulting $MD5(P<<S)$ to B along with the original message $P<<MD5(P<<S)$, encrypted by some conventional encryption algorithm such as *DES*. After separating P from $MD5(P<<S$, the receiver applies MD5 to the plaintext P concatenated with the same secret code S to generate a new $MD5(P<<S)$. This is compared with the $MD5(P<<S)$ received through the communication channel. If they are identical, the message is correct and the sender is authentic (Figure 10.6). Since the secret code S itself is not sent, it is impossible for an adversary to modify an intercepted

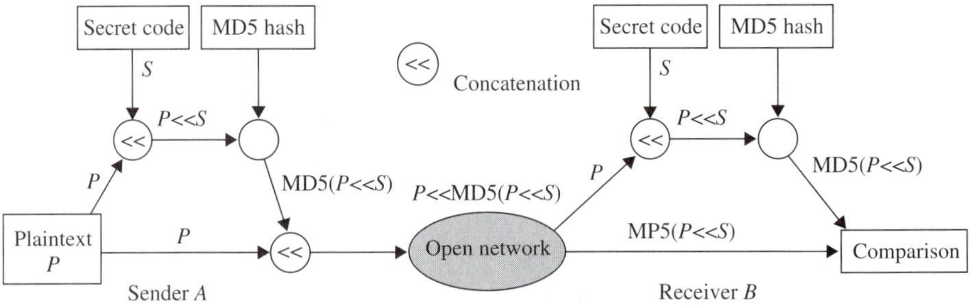

Figure 10.6 A secret code S combined with an MD5 message digest replaces a PKC digital signature for authentication purposes. No encryption is involved here.

message or forge a different one. This secret code technique is adopted for *SNMPv2* described in Chapter 9, Network Management.

10.1.7 Pretty Good Privacy (PGP) for E-Mail

In 1991 Phil Zimmermann released a public domain program package called Pretty Good Privacy (PGP), which is a free software implementation of the RSA algorithm for PCs. It was later ported to Unix, Linux, and VMS under the GNU public license. PGP is well designed, with possibly the most sophisticated key management features available. PGP supports IDEA file encryption, the message digest algorithm (MD5), and RSA digital signatures. IDEA is a symmetrical encryption algorithm similar to DES in overall structure but more secure than DES. It uses a 128-bit key. With PGP, data compression is automatically executed before encryption to reduce file length and eliminate redundancy. PGP keeps as much information secret as possible. The only part unencrypted in the header file is the receiver's ID. The receiver will not learn who sends the message, and when or if it was signed, until the message is decrypted. Complete source code for PGP is available on the Internet via FTP from net-dist.mit.edu. A major update of PGP was released in April 1992 from New Zealand, beyond the reach of the RSA patent.

PGP is a convenient way to secure e-mail exchanges. After obtaining and installing a copy of the PGP code, a simple seed is used to generate a key pair. The software then creates key certificates and prompts for a passphrase that will be used to protect access to the private key. The passphrase should be easily remembered but kept secret. The key certificate is kept offline and physically secured. The public key is then made available to anyone who might wish to communicate with the user. This is where a third-party introducer or digital certificate authority come in.

Once a recipient's public key has been obtained, the next step in sending a secure file is to sign the plaintext form of the file with the sender's private key. Before the file is sent, the sender uses PGP to compress and encrypt the plaintext file, including the sender's signature with the recipient's public key.

To receive the file, the receiver first decrypts the file with his private key. The software will automatically decompress the file. PGP also validates the signature by decrypting the signature certificate with sender's public key. The package provides many options for managing keys and manipulating encryption options.

10.1.8 Write Your Own Encryption Algorithm

People are often discouraged from writing a personal encryption algorithm because of a fear that a small bug in the code will render their decrypted messages meaningless. On the other hand, trusting the security of your transmissions to "experts" can also be a questionable practice.

If you follow the principles outlined here, writing your own encryption system should be easy. For practice, the laboratory manual (part of the Instructor's manual and CD accompanying the book) provides an encryption program written in X86 assembler code. The program incorporates several encryption steps to produce a multiple product cipher and chooses steps that are aimed at thwarting various attack methods. Here are the steps contained in the sample program and some suggestions for designing an encryption system:

1. Initialize the routine so that the start pointers are based on the entire key. For example, use a checksum or CRC of the key to determine where in the key everything starts.
2. Compress the plaintext.
3. Accumulate information from the plaintext for error checking and serial encryption.
4. XOR with the key.
5. Reorder the bits based on another part of the key.
6. Reorder the bytes based on the plaintext.
7. Change the key pointers based on the accumulated information.
8. Change the key order based on the key.
9. Every so often, change the order of the steps in the algorithm.
10. Insert some random snow, especially at the start.
11. Make sure that you can run the steps backward for decryption.
12. Make sure that stepping backward or forward through the algorithm will not produce the key, given a copy of the plaintext and its ciphertext.
13. Include error checking so that any alteration of the encrypted information will raise a red flag.
14. Make sure that changing even a single bit in the key or in the ciphertext will produce garbage.
15. Insert some useful garbage, such as a dummy message, and rescramble the whole thing with a simple, eventually breakable message.

10.2 DIGITAL CERTIFICATE AND PUBLIC KEY INFRASTRUCTURE (PKI)

For a public key system to work effectively, all users must keep their private key secret while publishing their public key on a public network such as Internet. Included with the public key must be some identifying information about the user. This is called digital certification. Why bother with certification?

Let's imagine that A and B negotiate a DES key exchange while an adversary C sits in the middle of the communication channel. C can freely send data to and receive data from both A and B. C also knows the public keys of both A and B. C first pretends to be B and sends A his own public key E_C as B's public key. Then C sends A, a message encrypted by C's secret key D_C requesting a DES key. A would be fooled and would send back the DES key encrypted by C's public key with A's signature $E_C(D_A(DES\_KEY))$. In the middle, C steals the DES key, using his own private key D_C and A's public key E_A.

C can then pretend to be A and may open communication with B by first sending his public key E_C to B by pretending to be A. C may then wrap the DES key with his private key D_C and B's public key E_B and send it to B. The fooled B would then use the DES key to encrypt any data intended for A. Conceivably, a fast-moving C could decrypt all communications from A, then re-encrypt them before sending the messages on to B. Likewise, B's communications could be read by C and rewritten for A. When this kind of man-in-the-middle attack occurs, A and B feel confident that their communications are secure.

Digital certification guarantees that a person (C) cannot provide another person (A) with his own public key (E_C) claiming that it is someone else's (B's). Digital certification involves the intervention of a central trusted authority or disinterested third party that binds a public key to an individual or entity that it has positively identified. This binding mechanism is called a digital certificate, which assures that the owner of a public key is who he claims to be.

A digital certificate is like a passport, driver's license or other piece of valid personal identification. A trusted third party provides a tamper-proof document that binds the information together. A driver's license or passport binds the picture of the bearer with the name and address and other important data. A digital certificate is an electronic credential that binds the public key with specific identifying information, along with the individual's name, address, and unique identifying number. With a passport, information is verified and sealed by a government (a trusted authority) so that it is tamper proof. The government seal attests to the binding of the individual and the passport number. Similarly, a digital certificate is a nonforgeable, tamperproof electronic document that attests to the binding of an individual's identity with his public key. Figure 10.7 shows how to authenticate a client to a server by a certificate through Secure Socket Layer (SSL) mechanism (to be discussed later).

The information contained in the certificate is verified and sealed with the digital signature of a trusted third party, known as a Certificate Authority (CA). A CA issues, verifies, and revokes certificates, just as government issues, validates, and seizes passports. The CA's digital signature attests to the binding of the individual's identity and his public key. The CA, like any individual, has a certificate containing its identifying information and public key so that anyone who needs to authenticate a CA signature can do so.

The PKC systems rely on a central authority to distribute and authorize or notarize individual keys. The X.509 standard includes a model for setting up a hierarchy of CAs. Two parties with certificates issued and signed by a different CA will be able to mutually authenticate each other by relying on the signature of a higher level of CA. In a large organization, the number of certificates required may be too large for a single CA to maintain; different organizational units may have different policy requirements; or it may be

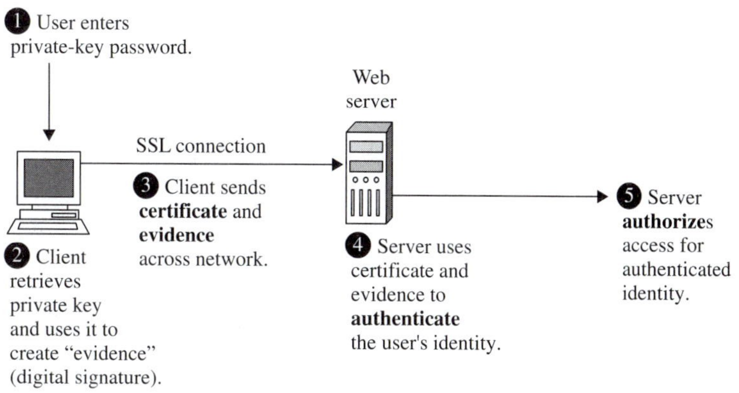

Figure 10.7 Authenticate a client to server using PKC through SSL connection.

important for a CA to be physically located in the same geographic area. For all these reasons, it may be appropriate to delegate the responsibility of issuing certificates to several certificate authorities. In Figure 10.8, the root CA at the top of the hierarchy is authorized to issue a self-signed certificate, that is, the certificate is digitally signed by the same entity. The CAs that are directly subordinate to the root CA have CA certificates signed by the root CA. CAs under the subordinate CAs in the hierarchy have their CA certificates signed by the higher-level subordinate CAs.

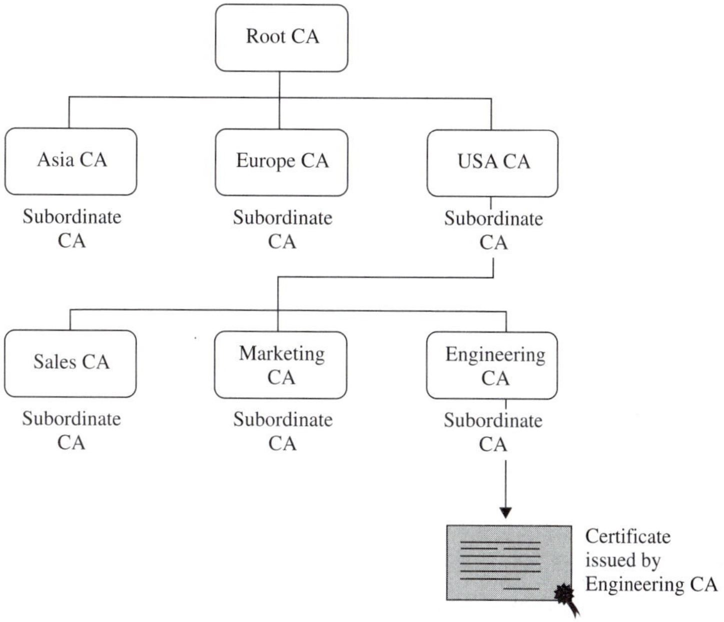

Figure 10.8 A sample hierarchical model of global CA structure.

To be effective, digital certificates must be integrated into the existing information transport and security framework. The most effective vehicle to implement this integration is the X.500 authentication framework, specifically X.509. X.509 is widely recognized as the leading security architecture standard throughout the data communications and networking industry.

The certificate format specified by X.509 consists of eight fields: version, serial number, signature algorithm, issuer, validity, subject, subject public key information, and signature. Any application or device can use the standardized security and authentication services as described in X.509 (or ISO/IEC 9594-8), which is the internationally recognized specification supporting security frameworks as they apply to electronic information transport. The authentication-framework specification within X.509 addresses the handling of public keys via certificates and Certificate Revocation Lists (CRLs).

The standards and services that facilitate the use of public-key cryptography and X.509 version 3 certificates in a networked environment are collectively called Public-Key Infrastructure (PKI). PKI is a central part of the enterprise security architecture. It enables strong distributed authentication, secure communications, and maintain confidentiality, integrity, and non-repudiation services. In any PKI, a CA is a trusted entity that issues, renews, and revokes certificates. To participate in a PKI, a user or end entity, must register in the system. The user submits his/her identification and a newly generated public key to the CA. The CA uses the information provided, and sometimes requiring human intervention, to authenticate the identity. Then the CA issues a certificate, which usually be held on a hard disk, diskette, secure certificate store, or a smart card and associates the user with the public key, and signs the certificate with its own private signing key. End entities and CAs may be in different geographic or organizational areas or in completely different organizations that are linked through an extranet to selected customers, suppliers, and mobile employees via the Internet. CAs may include third parties that provide services through the Internet as well as the root CAs and subordinate CAs for individual organizations. Policies and certificate content may vary from one organization to another. For all these reasons and many others, the deployment and long-term management of any large-scale PKI require careful advance planning and custom configuration.

Certificate management, which covers tasks of certificate issuing, registering, renewing, and revoking, and key distribution and recovering as well, is the most prominent and complex functions of PKI. Netscape, Microsoft, Verisign, Entrust, and Baldimore are currently the major PKI vendors.

10.3 FIREWALLS

As the growth of business on the Internet has exploded, so has interest in security mechanisms. There are several types of risks and attacks that may threaten an organization's normal business, and in particular network operations. These risks and attacks may be unauthorized access and information gathering, illegal disclosure or modification of information, and denial of services. The most frequently engaged methods of attacks include.

- **Spoofing,** which exploits trust relationships by pretending to be someone or something that is actually not by assuming a fake IP address.
- **Malicious data** that is input by various ways, including downloading or email and attachment, to hosts or networks, and then somehow executed on the target machine.
- **Exploiting mis-administrated trust relationships** by some systems coming out of box with some network services turn-on by default, or some software bugs and other system/network vulnerabilities and security holes.
- **Dictionary attack** that takes advantage of the fact that passwords are made up by living people and usually people try to use easy to remember passwords, often words from their native language. One dictionary attack may use a wordlist with as many as 3,000,000 words from different languages and cultures as a guess for password to intrude a target system.
- **Denial of Service (DoS)** is characterized by an attempt to prevent legitimate users of a service by "flooding" the network with massive data or access requests. A well-known example is the "Ping of Death" that may crash systems by sending IP packets that exceed the maximum legal length (65535 octets). A Distributed DoS may be orchestrated by a person who uses a compromised host called handler with a special program running on it for the attack. The handler is capable of controlling multiple agents—also compromised hosts that run a program that generates a stream of packets that is directed toward the intended victim.
- **Social Engineering** is a process of gaining information or opening security holes by talking to people with special skill in conversation, a sound of self-confidence, and a certain amount of bravado or bluster.

Authentication and authorization are the first line of defense. Both end users and management must support their procedures and protocols involving passwords and key systems. Encrypting network traffic is necessary and useful if secure communication with the outside world is desirable. An emerging Internet access control device—a firewall that works as a network intruder-alert system—can effectively prevent security breaches from the outside.

In a car, a firewall is the structure that separates the engine block from the passenger compartment to protect the passengers if the engine catches fire. Though there are many electrical and mechanical connections through the firewall, the passengers remains secure. Translated to network terminology, a firewall protects the machines on one side from unauthorized access or tampering by machines on the other side. All the information flowing to and from a corporate LAN and the Internet is monitored by the firewall. Whatever might cause harm to the corporate LAN is screened out to the extent possible. Firewalls offer another layer of security against intruders from the outside world. Firewalls also provide facilities that monitor and restrict the exchange of information between employees and outside world. Internet firewalls have their roots in control mechanisms and security measures that have long been standard practice in the mainframe community.

Unlike other authentication and authorization mechanisms, which can be added to application programs, Internet access control usually requires changes to basic components

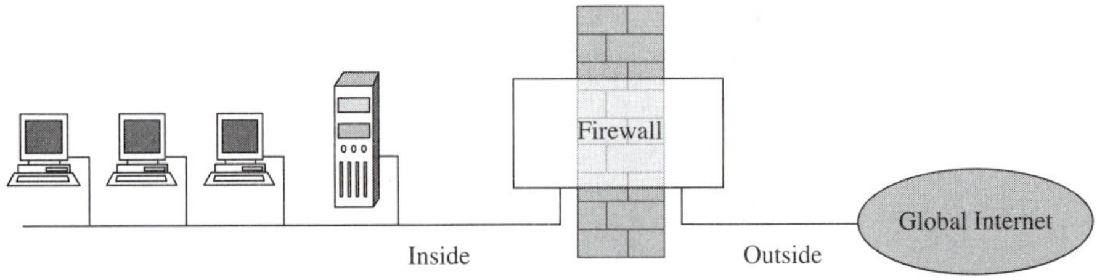

Figure 10.9 A firewall partitions interconnected networks into two regions referred to as inside and outside.

of the Internet infrastructure. A firewall is a block at the entrance to the part of the Internet to be protected. It partitions interconnected networks into two regions, referred to informally as the inside and the outside (Figure 10.9). All the data blocks flowing through the firewall are examined. The most elaborate firewalls block traffic from the outside to the inside, yet allow users on the inside to communicate freely with the outside. Firewalls also provide an important single checkpoint where audit information can be obtained. Firewall technologies can be classified into three groups according to function: packet filtering, application gateway, and circuit gateway.

10.3.1 Packet-Filtering Principle

As we recall from Chapter 5, in the TCP/IP environment, the Transmission Control Protocol (TCP) and User Datagram Protocol (UDP)—the two protocols in the transport layer—provide virtual circuits between source and destination. TCP is the more reliable, while UDP is the more efficient. Both TCP and UDP packets include a source address and a port number as well as the destination address and port number.

TCP applications, such as File Transfer Protocol (FTP), telnet, and Simple Mail Transfer Protocol (SMTP), listen for requested connections on specific well-known ports in the range below 1024 by convention. For instance, telnet uses TCP port 23, SMTP uses TCP 25, and HTTP (or WWW) uses TCP port 80. The UDP is ideal for short query/response sessions such as domain name server (DNS)—UDP port 53, Remote Procedure Call (RPC)—UDP port 111, and Simple Network Management Protocol (SNMP)—UDP port 161, to name a few. Most versions of TCP and UDP for Unix require that only the superuser create a port numbered less than 1024, with the intent that remote systems can then trust the authenticity of information written to those ports. This also makes it easier to restrict outsiders' access to Internet services. When information is returned by a remote server, the client processes generally accept whatever port is assigned by the operating system. As explained later, this arrangement, when combined with the use of a simple packet-filtering firewall, may create the interesting consequence of restricted access for client programs.

The Internet Protocol in the network layer of the OSI model is responsible for delivering data packets, or datagrams, to destinations. IP packets have a header with a source and a destination address in addition to other fields. IP routing chooses a path over which a datagram should be sent. When the source and destination hosts are not in the same network, a router is needed. The router uses source and destination information in the packet

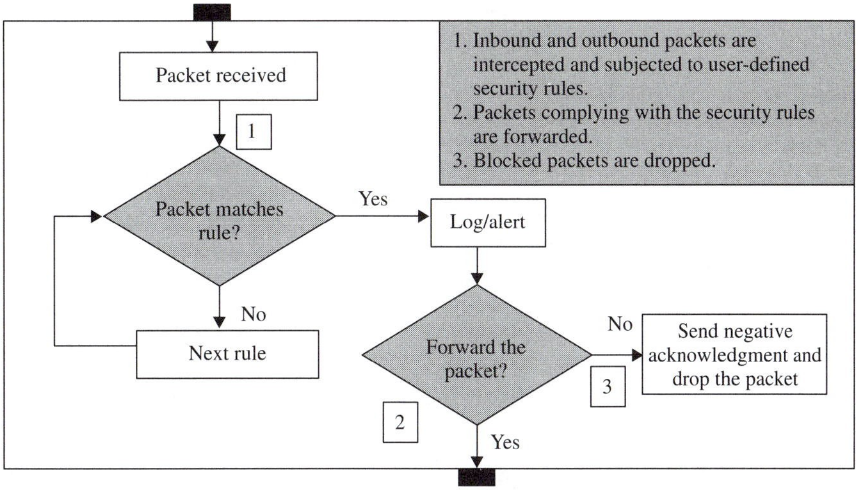

Figure 10.10 Flowchart showing the packet filtering operation.

header to route the packet. Datagrams pass from router to router until they reach a router that can deliver the datagram directly. The routing protocols such as the Routing Information Protocol (RIP) and Open Shortest Path First (OSPF) enable routers to dynamically find, select, and use paths through the Internet.

The usual IP routing algorithm employs a routing table on each machine that stores information about possible destinations and how to reach them. Whenever the IP routing software in a host or a router needs to transmit a datagram, it consults the routing table to decide where to send the datagram.

Since when two or more networks are interconnected, every datagram must go through a router, the header of that datagram must be examined to determine the next destination. Some basic routing rules are applied to the routing table to make the routing decision. It is natural and convenient to set up a checkpoint where routing decisions are made. When the rules are extended for access control purposes, a packet-filtering firewall is created by selectively passing or blocking data packets based on source/destination address or port information in the packet's header (Figure 10.10). There are two important rules for the packet-filtering rule set:

- Rules are tested in order. Testing stops when the first true condition is found.
- There is an implicit deny rule at the end of an access list. If access is not specifically granted by the access list, access is denied.

The packet-filtering mechanism has a stateless nature. This means that the mechanism does not keep a record of interactions or any history of previous packets. Some routers allow a manager to configure separate filter actions for each interface, while others have a single configuration for all interfaces.

The rule set determines how the filter deals with any given packet. A rule can forward the packet, block the packet, pass the packet for processing at another rule, or log the occurrence of the packet. Rules can be applied to packets based on four criteria: packet source or destination network address, host address, source or destination port number, or protocol name or number. Host information can be in the form of names or Internet numbers. Port information is checked for TCP and UDP packets only, because only those protocols use port numbers. These criteria are converted into expressions by a wide array of bitwise, logical, and comparison operators. For example, the following expression:

```
0: ((protocol==udp)&&((srcport==53)||(dstport==53)))
then forward
else 1;
```

checks to see whether the current packet complies with the UDP and that it is either from or to port 53 for DNS. If these conditions are fulfilled, the packet is forwarded; otherwise, the processing continues with the next rule: 1.

Table 10.1 illustrates a packet-filtering rule set specification. In this example, the firewall does the following:

1. Blocks any packets from a particular outside host with IP address 192.244.244.244.
2. Allows inbound e-mail (SMTP, port 25), but only to the mail server.
3. Allows any host in the internal class B network 128.200.0.0 to send e-mail directly to the outside.
4. Allows inquiry about time of day.
5. By default, if none of the preceding rules matches, this one certainly does block anything that is not explicitly allowed.

Because the number of well-known ports is large and growing, a simple act of negligence or omission can leave the firewall vulnerable. In addition, many packets and applications on the Internet do not use well-known ports. To be effective, a firewall that uses packet filtering should block access to all IP sources, all IP destinations, all protocols, and all protocol ports except those computers, networks, and services the organization explicitly decides to make available externally. A packet filter that allows a manager to specify which datagrams to admit instead of which datagrams to block can make such restrictions easy to specify. This stance is of the form *that which is not expressly permitted is prohibited.* Such a paranoid approach greatly reduces the risk that a new or unpublicized security hole will victimize the organization's private networks.

TABLE 10.1
Packet-Filtering Rule Set Specification

In/Out	Action	Source IP	Source Port	Destination IP	Destination Port	Protocol
In	Block	192.244.244.244				
In	Forward			Mail server	25	TCP, SMTP
Out	Forward	128.200.0.0			25	TCP, SMTP
	Forward		37			UDP/TCP Time
	Block					

Obviously, the filtering can be accelerated by ordering the rules such that the most frequently used common types of traffic are checked and processed at the top of the list.

10.3.2 Router-Based Packet-Filtering Firewalls

There are many ways to implement a firewall based on the packet-filtering principle. Almost any router can apply packet-filtering rules to accept or reject packets based on the originating IP address and port. The simplest approach is to use a programmable router that supports packet filtering to create a firewall.

Most commercial router products include a high-speed filtering mechanism sometimes called an access list. In its most basic form, the same router an organization uses to link its networks to the Internet can also be configured to provide simple Internet security through TCP/IP packet filtering. In this way, a system manager can further control packet processing, besides offering normal routing. That is, the manager can specify how the router should dispose of each packet.

When a packet first arrives, the router passes the packet through its packet filter before performing any other processing. If the filter rejects the packet, the router drops it immediately. Most routers do this as the packet leaves the router. This can increase efficiency, since finding and applying the filter rule can be combined with the routing table lookup. On the other hand, some routers can block a packet when it enters the router. This can help prevent the infamous *IP address-spoofing* attack, which sends in packets from the outside with the *incoming* packets bearing the IP address of a trusted *inside* host. The network hacker Kevin Mitnick used this method, among others, to break into numerous credit bureaus, cellular networks, phone companies, and the Supercomputer Center in San Diego. Nevertheless, such measures still will not eliminate all possible address-spoofing attacks because an attacker can simply impersonate an IP address that is trusted but not on an internal network. Finally, a filter on input can also protect the router/firewall itself from attack. Therefore, as a general principle, firewalls should filter out the offending packets as soon as possible and should not trust hosts outside the organization's administrative control.

An alternative to building a packet-filtering firewall is to implement host-based packet-filtering software. This usually requires that the operating system kernel be rebuilt to include additional support functions. The functions provide an interface between the kernel and the loadable filter software modules. After a reboot, the loadable modules are read into the kernel. Digital Equipment Corporation developed *screend*, a kernel modification of Unix that permits a user process to pass on each packet before it is forwarded. The source code for *screend* is distributed free with restrictions. But one needs the source code of other parts of the kernel to build such a packet-filtering firewall. Texas AMU's security tools include software for implementing packet-filtering routers.

The freely distributed Linux operating system supports TCP/IP and also allows a computer running Linux to be configured as a router. Packet-filtering enhancements are made to take advantage of the Linux routing software. To better structure the implementation and enhance ease of use, the packet-filtering package is separated into two parts: one is the modification of the Linux IP software that manages the packet routing. Several functions are added to the original Linux IP software so that when any packet arrives, it is examined before normal routing is performed. The other part of the Linux package is a utility, *ipfw*, that allows users to specify their packet-filtering rule set and provides a user interface.

Through the use of two queues (forwarding and blocking maintained by ipfw), and three types of commands (*Add*, *delete*, and *list*), a user decides whether a packet should be forwarded or blocked. After ipfw has passed a command, it sets the corresponding socket options and variables.

A packet filter is the kernel of a simple but effective firewall. It is also the foundation for more advanced firewalls. Packet filters are efficient and transparent to many protocols and applications. They require no changes in client applications, no specific application management or installation, and no extra software when commercial routing devices are used. With a single, unified packet filtering router, all network traffic is processed and then forwarded or blocked from a single control point.

However, most packet-filtering routers are stateless, understand only low-level protocols, and lack audit mechanisms. The header information most routers manipulate is insufficient for more sophisticated filtering decisions, and the packet-filtering rules are limited. There is also a high performance penalty when many rule instances are used. Lack of context or state information makes packet filtering by routers alone inappropriate for datagram-based protocols such as FTP.

Beside possible security pitfalls such as address spoofing, packet-filtering firewalls create the unfavorable consequence of restricted access for client programs. If an organization's firewall blocks all incoming packets except those destined for ports that correspond to services the organization makes available externally, an arbitrary application inside the organization cannot become a client of a server outside the organization. Preventing outsiders from accessing arbitrary servers in the organization easily leads to a blanket prohibition on packets arriving for an unknown protocol port. As mentioned before, such a firewall has an interesting consequence: it also prevents an arbitrary host inside the firewall from becoming a client that accesses a service outside the firewall. To understand why, recall that although each server operates at a well-known port, a client program does not. When a client runs, the operating system selects a port number not in use. When the client program communicates with a server outside the organization, it generates packets sent to the outside server. Such packets will not be blocked as they leave because they contain the client's protocol port as the source port and the server's well-known protocol port as the destination port. However, when the outside server generates a response, the source and destination ports are reversed. The client's port becomes the destination port and the server's port becomes the source port. The firewall blocks the response because the destination port, which is not well known, cannot be used by outsiders.

Another problem with simple packet-filtering firewalls is associated with the X Windows system. From a user's point of view, X Windows makes good business sense, but unfortunately, it has some major security flaws. The X Windows user simply gives away control of screen, mouse, and keyboard. Remote systems that can gain or spoof access to a workstation's X display can monitor keystrokes and/or download copies of the contents of their windows. With only a simple packet-filtering firewall, it is just too easy for an attacker to interfere with a user's X display. That is why simple firewalls block all X traffic.

On the other hand, in the X Windows environment, the user's terminal, including display, keyboard, and mouse, is a server, while the application program, connected via TCP/IP, is the client. The applications may run on a remote machine on an outside host.

Applications make calls to the server when they need input, output, or the user's attention. When the calls are made from the outside, typical rule sets in a simple packet-filtering firewall block them. This too often means that users are blocked from taking advantage of outside resources.

The more powerful and more advanced application gateway firewalls and circuit gateway firewalls provide more flexibility and permit X traffic with reasonable security.

10.3.3 Application Gateway Firewalls

Many application gateways are just software that runs on a Unix workstation with two network interface cards. Not only do application gateway firewalls, like routers, monitor network traffic, but they are also capable of far more sophisticated analyses of network packets, keeping track of transmissions over longer periods of time and even analyzing the application information transmitted in the packet. Unlike most stand-alone routers, which examine packet headers for addresses and some simple information only, application gateways can look at the package content and try to find out what it wants to do when it reaches its destination. The semantic knowledge inherent in the design of an application gateway can be used for more sophisticated scrutiny.

A special type of application gateway firewall software is referred to as a proxy server or forwarder. This is an application that mediates traffic between a protected network and the Internet. Proxies are often used instead of router-based traffic controls, to prevent traffic from passing directly between networks. Many proxies contain extra logging or support for user authentication. Since proxies must "understand" the application protocol being used, they can also implement protocol-specific security.

Proxy servers allow direct Internet access from behind a firewall. This is done by opening a socket on the server, and allowing communication via that socket to the Internet. The proxy server can be configured to allow specific Internet service requests across the firewall from the computer in a protected network to be directly connected to the Internet. Proxy servers are application specific. To support a new protocol via a proxy, a specific proxy must be developed for that protocol.

Through various proxies developed for each individual service, an application gateway ensures that outside clients do not compromise internal security. These proxies run on the well-known ports connected to the outside network interface card. They collect incoming client requests at the application level and transfer them via the inside card to the actual application servers, including telnet, X Windows, and FTP, running on the internal machines in the private LANs.

The type of filtering used in an application gateway depends on local needs and customs. An FTP proxy might be configured to permit outgoing FTP and block incoming FTP requests. A location with many PC users might wish to scan incoming files for viruses.

Generally speaking, application gateways are not transparent to users, applications, and the gateway host on which they run. Furthermore, they are difficult to configure and manage. Only a few applications are supported, and special tailoring is required. Users must first connect to the gateway or install a specific client application for each application they expect to use. Each application with a gateway is a separate, proprietary piece of software that requires its own set of management tools and permissions. Application gateways are often used in conjunction with other gateway designs, packet filters, and circuit-level relays.

A variation of the application gateway uses a clever stateful inspection method scheme that looks into an incoming envelope and takes notes on just some security-related bits of the data packet. When the reply packet returns from the other direction, the firewall inspects it against the notes on the incoming packet, searching for tampering by a hacker. Because stateful inspection gateways pass along approved packets intact, instead of rewriting them like application gateways, such firewalls barely slow network traffic. Furthermore, because the firewall needs to know little about each new Internet application, this type of firewall can keep pace with emerging applications.

10.3.4 Circuit Gateway Firewalls

A circuit gateway creates a secure virtual circuit between an internal and an external device, and relays TCP connections. The caller connects to a TCP port on the gateway, which in turn connects to a destination on the other side of the gateway. During the connection, the gateway's relay program copies the bytes back and forth. For example, a virtual circuit can be used to connect an outside host with the printer port on an internal computer. The circuit gateway can ensure that only that particular external host can connect to the gateway's printer service without leaving an entry hole, even if the external host is compromised. In some cases, a connection is made automatically. In other cases, a protocol is needed to specify the desired destination and service by host name or numerical IP address. If the connection is successful, data start to flow. The number of bytes copied and the TCP destination are logged for further investigation if this becomes necessary.

Circuit gateways provide a more general implementation, but they support only some TCP/IP applications and no additional protocols. Application and circuit gateways are also well suited for some UDP applications. The client programs must be modified to create the virtual circuit to a proxy process.

Building an application and circuit gateway firewall means writing lots of small programs to open network connections. Several software tools are useful for building an application or circuit gateway firewall. The tools include proxylib, a portable version of the tenth-edition Unix system *connection server* interface and others. SOCKS is a generic proxy system that can be compiled into a client-side application. It has a one-for-one replacement for each standard IP socket call such as `connect()`, `getsockname()`, `bind()`, `accept()`, `listen()`, and `select()`. The replacement routines allow the client transparent use of the SOCKS server while communicating with an external host. This makes it convenient to install. Many major software packages are quickly converted to use SOCKS, including some for PCs and Macintoshes.

The existence of the circuit usually provides sufficient context to allow secure passage through filters. However, with application and circuit gateways, all packets are addressed to a user-level application on a gateway that relays packets between two points. Because each packet must be copied and processed at least twice by all the network layers, both application and circuit gateways slow network performance.

10.3.5 Firewall Architectures

When a host gateway with enhanced security is installed to handle Internet applications and traffic between internal networks and the global Internet, it is called a bastion host.

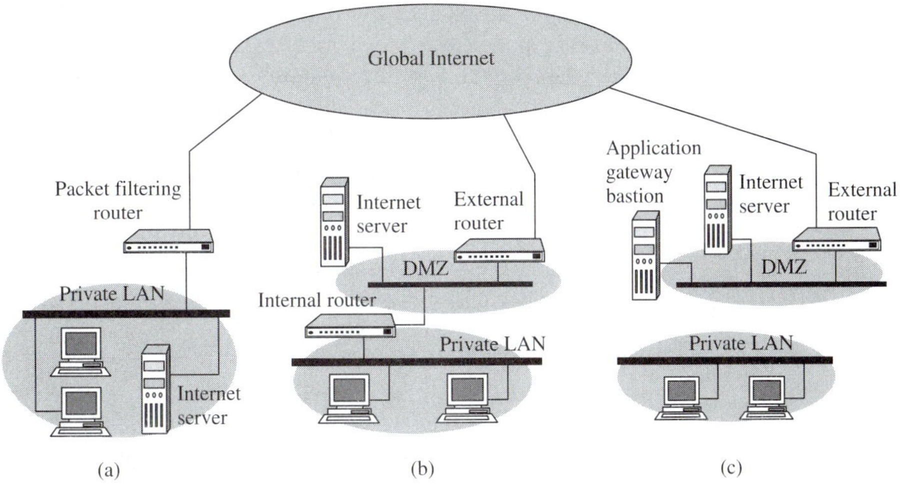

Figure 10.11 Some common firewall architectures: (a) router-based packet-filtering firewall, (b) dual-router gateway with demilitarized zone (DMZ), and (c) router with DMZ and bastian.

An isolated intermediate network is often used as a buffer between the inside and the outside. This offers many more capabilities, including the ability to log all the activity over the gateway. The bastion host computer is a connection between the inside and the outside networks. It has two conceptual barriers. The outer barrier blocks all incoming traffic except packets destined for clients and approved services on the bastion host. The inner barrier blocks all incoming traffic except packets that originate on the bastion host. Each barrier requires a router with a packet filter. While the router-based firewall monitors data packets at the IP level, hosts exert their control at the application level, where traffic can be examined more thoroughly.

An advantage of using such an isolated network is that it simplifies the establishment and enforcement of new Internet addresses, especially for large private networks that may otherwise face the prospect of significant reconfiguration.

Once the security stance has been defined, there are many different ways to configure a firewall. Figure 10.11 shows three of the most common firewall architectures: a router-based packet-filtering firewall, a dual-router gateway with a demilitarized zone (DMZ), and a router with a DMZ and a bastion as the application gateway.

1. The basic router-based packet-filtering firewall is easy to build. It works on a low level and provides basic security services. It is, however, vulnerable to sophisticated attacks. If any one of the allowed hosts on the private LAN has a security flaw such as an old version of *sendmail*, an intruder can break out of that host and access anything on the LAN. This type of firewall does not perform detailed analysis of packets and offers little flexibility;

2. The second configuration uses two routers to establish an intermediate insecure network that acts as a buffer or demilitarized zone. Typically, the DMZ is configured such that it can be accessed by both external and internal networks, but traffic attempting to cross the DMZ is blocked. Some routers include dual LAN ports for the DMZ, where the Internet server is often installed to provide WWW, FTP, gopher, and other Internet-related services for outsiders. Many corporations use the DMZ as a place to store public information about corporate products and services, files to download, bug fixes, and so forth. Several of these systems have become important parts of the Internet service structure. Even if security is breached on one of the servers in the DMZ, the router on the internal LAN adds an extra layer of security.

3. The third configuration uses the more intelligent application gateway combined with a router and a DMZ. The software-based application gateway provides the most flexible and the most secure firewall paradigm. The proxy services software for telnet, FTP, e-mail, and so on may run on the bastion computer for the internal users.

An alternate use of proxy server is called reverse proxy, which can be used outside the firewall to represent a secure content server to outside clients, preventing direct, unmonitored access to the protected server's data from outside firewall (Figure 10.12). It can also be used for replication. So, multiple proxies with large caching space can be attached in front of a heavily used server for load balancing. To set up a reverse proxy, two mappings are needed, a regular mapping and a reverse mapping. The regular mapping redirects requests to the content server. When a client requests a document from the proxy server, the proxy server needs a regular mapping to tell it where to get the actual document. The reverse mapping makes the proxy server trap for redirects from the content server.

Firewalls are not a panacea. They cannot protect against attacks that skirt them. Removable storage devices such as magnetic tapes and recordable optical disks can just as effectively be used to export data. In general, a firewall cannot protect against a data-driven attack. Data-driven attacks encapsulate dangerous codes in an innocuous format, which is then mailed or copied to an internal host where it is decapsulated and executed. This form of attack has occurred in the past against various versions of sendmail. Firewalls do not provide adequate protection against viruses. There are too many ways of encoding binary files for transfer over networks, and too many different architectures and viruses to attempt to search for them all in each packet received.

Firewall policies must be realistic and must reflect the level of security in the entire network. A network is only as secure as its weakest point. A site with top-secret or clas-

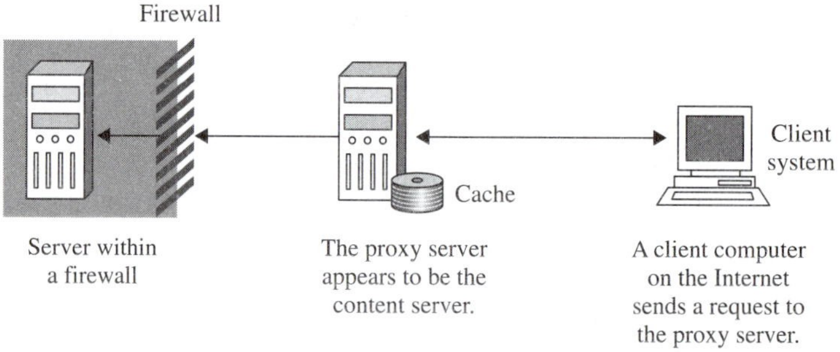

Figure 10.12 A reverse proxy appears to be the real content server.

sified data needs no firewall: it should not be connected to the Internet in the first place. Likewise, systems with highly secret data should be isolated from the rest of the corporate network.

10.4 KERBEROS

Kerberos is a network authentication system for use on physically insecure networks, based in part on Needham and Schroeder's trusted third-party authentication and key distribution protocol.

Named for the three-headed watchdog that guards the gates of Hades in Greek mythology, Kerberos relies on three components to provide network security (Figure 10.13):

- An authentication server (AS)
- A Kerberos database
- A ticket-granting server (TGS)

The three components are all installed on a single, physically secured Kerberos server connected to the network and trusted by all application servers. The Kerberos authentication server verifies any user requesting a network service, such as access to an application on the NFS server. It then issues a ticket-granting ticket (TGT) to the requesting user for that application. The Kerberos database maintains all principals (i.e., users and servers) and their secret keys, as well as network services to which each user has access rights, and an encryption key associated with each service. The TGS issues tickets, encrypted by the application service key, to the user of a particular service.

The client systems for end users, the Kerberos server, and application servers may be located in different sites of an open network.

When a user logs on to a workstation in a network using Kerberos, the client system sends the authentication server a request for credentials. Based on the information in the database, the authentication server determines whether the user has the appropriate access rights to the application being requested. Note that only the user name, not the password,

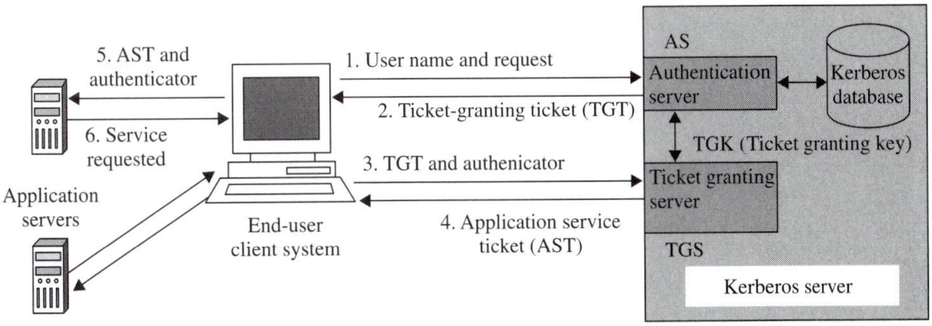

Figure 10.13 Data flow between the Kerberos server, the client system, and the application server.

is sent to the AS to protect the system from possible eavesdropping. After the authentication server has verified the user's identity and authorized a particular application, it creates the credentials—the TGT. The TGT contains the client's name and the IP address of the requesting workstation, and the name of the target application server, along with a timestamp guaranteeing the validity of the ticket for a given amount of time. A session key (SK) (a random string of codes generated by the authentication server) is then attached to the TGT. This is all encrypted with a ticket-granting key (TGK) shared only by the ticket-granting server and the authentication server.

The TGT, along with a copy of the session key, is encrypted again by a standard 56-bit DES key derived from the user's password, and sent back to the requesting user. This is decrypted on the user's system with the user's password and converted to the standard 56-bit DES key, which in turn is used to recover the session key and the TGT, which is still encrypted by the TGK.

Upon receiving the encrypted TGT, the client system issues an authenticator that consists of the user name, the workstation's IP address, and the time that the request originated. The authenticator is then encoded using the session key SK, combined with the encrypted TGT, and sent to the ticket-granting server (TGS). Using the TGK, the TGS decrypts the TGT that also contains the session key. Then the TGS uses the recovered SK to decrypt the authenticator, which yields the time the authenticator was produced and the IP address, user ID and time coded into the ticket. The time check is important because it prevents an adversary from reusing the authenticators and tickets. If all goes well, an application service ticket (AST) is encrypted by the application server key (ASK) (unknown to the user) and sent to the user along with a new session key. Finally, the client requests access to the service by sending the newly acquired AST and an authenticator to the application server. The server uses its private Kerberos key (ASK) to decrypt the ticket. Then, using the session key obtained from the decrypted ticket, it verifies the information in the authenticator, the client name, the IP address, and the current time. If everything still matches, the request to access the service is granted (Figure 10.13). The client must obtain a separate ticket for each service it wishes to use.

In summary, the Kerberos is designed to satisfy the following conditions:

- No clear text passwords or keys are stored on servers and sent over the network.
- Authentication lifetime is minimal and extending it can only compromise the current session.
- Authentication is transparent during normal use.

When a Kerberos-guarded service is requested, the user presents a ticket to the application server to establish his identity, along with the authenticator that proves the ticket was legitimate. Three steps are involved in the authentication process:

1. The client system sends the user name and request to the Kerberos server.
2. The authentication server and ticket-granting server verify the user and the request, then grant the credentials.
3. The user presents the credentials to the application server that provides the specified services.

Two types of credential are used in the process: tickets and authenticators. Tickets are used for the secure transmission of the user's identity and to obtain a temporary session key. Authenticators contain additional information ensuring that the client presenting the ticket and the client to whom the ticket was issued are the same.

A ticket is a sequence of a few hundred bytes and can be embedded in virtually any network protocol. A ticket contains the name of the server, the ID of the user, the IP address of the client system, a timestamp, a lifetime limit, and a random session key. It is encrypted by the DES key of the application server for which the ticket will be used. The ticket may be used by the client to access that particular server until it expires, at which time, another ticket must be obtained. When the application server decrypts the ticket, it should find its own service ID in the ticket. It can verify that the client ID in the ticket matches the client ID in the authenticator, thereby confirming the identity of the client.

An authenticator contains the requesting user's ID, the IP address of the client system, and a timestamp. A client builds authenticators as often as necessary. Each one can be used only once. A client must create a new authenticator for each service requested.

Both ticket and authenticator are encrypted for transmission by a symmetrical encryption method. The keys are different for the password, ticket-granting key, session key, and application server key.

Practically speaking, Kerberos is primarily used in application-level protocols (ISO's OSI model layer 7), such as Telnet or FTP, to provide user-to-host security. It is also used, though less frequently, as the implicit authentication system of data stream (such as SOCK_STREAM) or RPC mechanisms (OSI model, level 6). It could also be used at a lower level for host-to-host security, in protocols like IP, UDP, or TCP (OSI model levels 3 and 4), although such implementations are rare.

The implementation of Kerberos officially uses UDP on port number 88, although many implementations still use the old port number 750.

To use Kerberos, a physically secure Kerberos server must be dedicated to the authentication service. If the Kerberos server is down, all Kerberos-protected application access will be denied. Each Kerberos application and service must be modified or *Kerberized* at the source code level when Kerberos is installed. This can be a significant investment. If passwords are compromised, it is possible for an eavesdropper on the network to use them to decrypt tickets.

Kerberos does not provide authorization or accounting, per se, although applications can use their secret keys to perform these functions securely. Kerberos also does not provide password validation for individual workstations. Password protection can be improved by implementing the PKC challenge/response authentication technique mentioned earlier.

Kerberos is designed for user-to-host rather than host-to-host authentication. It is inefficient for peer-to-peer environments, where hosts have identities of their own and need to access resources such as remotely mounted file systems on their own behalf. It is also important to understand that using Kerberos on multiuser time-sharing machines greatly weakens its protection, since a user's tickets are then only as secure as the "root" account. Furthermore, dumb terminals and most X terminals do not understand the Kerberos protocol, so their cable connections remain insecure.

Kerberos is a network security system, which relies on cryptographic methods for its security. Since Kerberos' encryption system, DES, is not exportable, Kerberos, in its original form, cannot be exported or used outside the United States. As a partial solution to this problem, the Kerberos source code was modified by the addition of `#ifdef NOENCRYPTION` around all calls to DES functions. Compiling this version with the symbol `NOENCRYPTION` defined will use the Kerberos source code to remove all calls to the encryption routines. The result is a system called Bones that looks like Kerberos from an application's point of view but does not require DES libraries. Bones is a system that provides the Kerberos API without using encryption and without providing any form of security. It does allow the use of software that expects Kerberos.

10.5 SECURE SOCKET LAYER (SSL) AND VIRTUAL PRIVATE NETWORK (VPN)

Secure communications and strong authentication are required to ensure that businesses, governments, educations, healthcare, and other organizations may carry out secure transactions. SSL and VPN are two important tools to serve this purpose.

10.5.1 Secure Sockets Layer (SSL) Protocol

SSL, originally developed by Netscape, has been widely accepted on the World Wide Web for authenticated and encrypted communication between clients and servers. The IETF standard, called Transport Layer Security (TLS), is based on SSL.

Originally, Socket was just a special type of file introduced by 4.3 BSD, and it later became one of the prevalent communication APIs for Unix systems. The SSL protocol runs at the Transport Layer that is above TCP/IP and below higher-level protocols such as HTTP or LDAP. It uses TCP/IP on behalf of the higher-level protocols. In the process, an SSL-enabled server can authenticate itself to an SSL-enabled client while allowing the client to authenticate itself to the server, thus establishing an encrypted connection for both machines by means of many encryption methods.

The SSL protocol includes two sub-protocols, the SSL record protocol and the SSL handshake protocol. The SSL record protocol defines the data exchange format. Based on the SSL record protocol, the SSL handshake protocol specifies the initial procedures of exchanging messages between an SSL-enabled server and an SSL-enabled client when they first establish an SSL connection. These procedures include the following.

- Authenticate the server to the client to allow a user to confirm a server's identity. SSL-enabled client software can use PKC to check that a server's certificate and public ID are valid and have been issued by a CA, listed in the client's list of trusted CAs. This authentication makes a vendor trusted by the customers before they send credit card numbers over the Internet.
- Optionally authenticate the client to the server to allow a server to confirm a user's identity using the same techniques as those used for server authentication. This authentication may be required when the server, for example, is a bank sending confidential financial information to its customer.
- Negotiate and agree upon the selected cryptographic algorithms that both client and server support. Many encryption and key exchange algorithms, such as RSA, DES,

Triple DES, MD5, RC2 and RC4, Key Exchange Algorithm (KEY), and RSA Key Exchange, may be used.

- Use public-key encryption techniques to transmit certificates, establish session keys, and exchange symmetric (secret) keys and other shared secrets.
- Establish an encrypted SSL connection that provides a high degree of confidentiality and strong protection from possible tampering.

10.5.2 Virtual Private Network

For many years organizations that need to connect different offices together had to rely on leased lines. These lines are expensive and are typically charged by the mile. A T1 line at the speed of 1.544 Mbps for a coast-to-coast connection can easily cost thousands of dollars a month. Businesses also need to provide their workers with remote access to the resources of corporate intranets as they take to the road, telecommute, or dial in from customer sites. Additionally, business partners may join together in extranets to share corporate information either for a joint project lasting a few weeks or for long-term strategic alliance. All these dynamically changing needs make the lease line approach undesirable.

The ubiquitous Internet provides an inexpensive (and potentially the best) communications vehicle if the communications can be made secure and private. Rather than relying on dedicated leased lines, an Internet-Based Virtual Private Network (VPN) uses the open, distributed infrastructure of the Internet to transmit data between corporate sites.

Organizations using an Internet VPN set up connections to the local connection Points-Of-Presence (POPs) of their Internet service provider, then the ISP ensures that the data is delivered to the appropriate destinations from the closest POPs via the connections the end users use, leaving the connectivity details to the ISP's network and the Internet infrastructure. Internet-based VPNs include measures for encrypting data passed between VPN sites to protect the data against eavesdropping and tampering by unauthorized parties on the Internet.

In addition, VPNs are not limited to corporate sites and branch offices. As an added advantage, a VPN can provide secure connectivity for mobile workers. These workers can connect to their company's VPN by dialing into the POP of a local ISP, which reduces the need for long-distance charges and outlays for installing and maintaining large banks of modems at corporate sites.

Initially TCP/IP and Internet did not require strong security measures or guaranteed performance because of the limited number of users and the type of applications originally designed. However, if Internet VPNs are to serve as reliable substitutes for dedicated leased lines or other WAN links, technologies for guaranteeing security and network performance must be provided.

VPNs need to provide the following three major functions to ensure data security:

- Authentication, that is ensuring that the data originates at the source that it claims
- Access control, that is restricting unauthorized users from gaining admission to the network
- Confidentiality and data integrity, that is preventing unauthorized reading or copying data as it travels across the Internet and ensuring that no tampering occurred

Using a hybrid of application-gateway and packet-filtering methods, a firewall may conceal an internal IP address and share a valid IP address among users on the network to create the VPN over the Internet. The firewall can be combined with integrated encryption based on the DES at the IP level. Digital certificate and PKI, Challenge Handshake Authentication Protocol (CHAP) and Remote Authentication Dial-In User Service (RADIUS), Secure ID card that changes a token number every 60 seconds synchronized with the server, and/or other technologies can be used to authenticate users on a VPN and control access to network resources.

VPN provides both authentication and encryption for each communication channel. The encryption ensures that privacy is maintained for the user data being communicated and for networking information such as addresses and port numbers. Authentication protects against both address spoofing and tampering of data in transit. Moreover, these capabilities do not require any modification of software on the communicating end systems.

The communication channels in VPNs are dynamic, with connections set up based on the organizational needs. In the meantime, a VPN is formed logically, regardless of the physical structure of the underlying network. VPNs usually do not maintain permanent links between the end points. Instead, a VPN is created whenever a connection between two sites is needed, and it is torn down when the connection is no longer needed, making the bandwidth and other network resources available for other uses. Tunneling in VPN allows senders to encapsulate their data in IP packets that hide the underlying routing and switching infrastructure of the Internet from both senders and receivers. At the same time, these encapsulated packets can be protected against snooping by outsiders using encryption techniques.

There are three components of an Internet-based VPN, security gateways (or VPN server), security policy servers, and certificate authorities.

- Security gateways sit between public and private networks, preventing unauthorized intrusions. They also provide tunneling capabilities and encrypt private data before it is transmitted on to the public network. In general, a security gateway for a VPN fits into one of the following categories: routers, firewalls, integrated VPN hardware, and VPN software.
- Security-policy server maintains the access-control lists and other user-related information that the security gateway uses to determine which traffic is authorized. For example, in some systems, access can be controlled via a RADIUS server.
- Certificate authorities are needed to verify keys shared between sites and can also be used to verify individuals using digital certificates. For small groups of users, verification of shared keys might require checking with a third party that maintains the digital certificates associated with shared cryptographic keys. If a corporate VPN grows into an extranet, then an outside certificate authority may also have to be used to verify users from business partners.

For the Internet VPN tunnels, an end point can be either an individual computer or a LAN. So tunnels can be either LAN-to-LAN or Client-to-LAN as shown in Figure 10.14. LAN-to-LAN tunnels have a security gateway at each end of the tunnel that serves as the interface between the tunnel and the protected LAN. In such case, users on either LAN

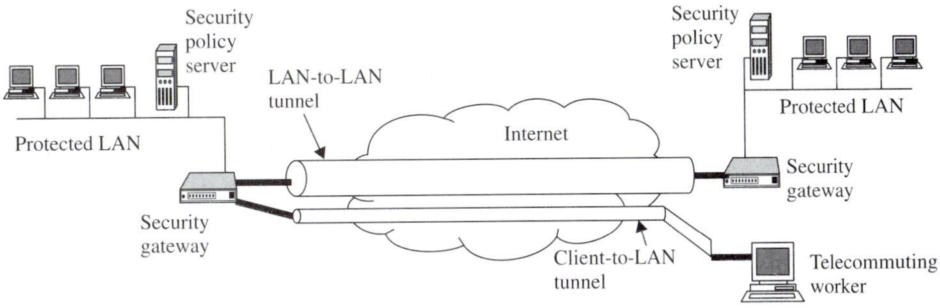

Figure 10.14 VPN with LAN-to-LAN tunnel and Client-to-LAN tunnel.

can use the tunnel transparently to communicate with users at the other end of the VPN. The client-to-LAN tunnel is usually set up for a mobile user who initiates the creation of the tunnel on his/her end in order to connect to the corporate intranet. To do so, special client software is run on client's computer to communicate with the VPN server/security gateway that protects the destination intranet.

Commercial VPN systems often uses DES (Data Encryption Standard) for encryption and the publicly available MD5 message-digest one-way hash algorithm for generating an authentication signature. In order to provide the processing power needed for fast VPN, hardware add-ons with specialized chips for routers and remote-access servers are increasingly used to perform compression and data encryption at speeds T3 (45 Mbps) and above.

Firewalls can be utilized to implement a VPN. When two networks, A and B, are connected and protected by firewalls FA and FB, both implementing VPN, the two firewalls will be configured with a VPN tunnel defined between networks A and B. This tunnel definition includes the type of encryption to be used, the secret key used for authentication and encryption, the hosts on each network allowed to use the tunnel, and the hosts who can be accessed through it. This provides an address based filtering capability. The routing configuration on network A is set up such that routes to the remote network B point to FA. Similarly, network B's routes to network A are set up to point to FB. This can be accomplished by configuring the routers, or by setting up static routes on appropriate hosts. Tunnels can also be configured for dynamically addressed computers.

When a machine on the network A sends packets to a machine on network B, the packets are now routed to FA. The VPN component on the FA system examines all received packets to check if they need to be forwarded over the VPN tunnel. If so, it attaches an authentication checksum to the packet and then encrypts the entire packet including all headers. The encrypted packet is then encapsulated in an outer IP packet and transmitted to FB.

On receiving the packet, FB decrypts the packet and validates the authentication checksum to ensure that FA sent it. If the authentication attempt succeeds, FB strips the outer headers to retrieve the packet that was sent by the originating host. The addressing information in the original packet is then checked to see if the source host is allowed to access the destination specified in the packet. If so, the packet is forwarded to the final destination. Otherwise it is dropped.

In order to create a LAN-to-LAN or client-to-LAN VPN, at least one VPN server is needed. There are many choices for getting a VPN server running on a network as explained below.

- Operating systems, both Unix/Linux and Microsoft Windows Windows NT/2000 servers provide VPN services. These kind of VPN systems are often good low-cost choices for systems that are relatively small and do not have to process a lot of traffic. These solutions can run on existing servers and share resources. They are well suited for client-to-LAN connections.
- VPN services available on many firewall products since firewalls must process all IP traffic. Because of all the processing performed by firewalls, combining tunneling and encryption with firewalls is probably best used only on small networks with low volumes of traffic and without frequent reconfiguration.
- VPN services available on routers because routers, like firewalls, have to examine and process every packet that leaves the LAN. A router can either has add-on software or an additional circuit board with a coprocessor-based encryption engine. The latter is best for situations that require greater throughput. Adding encryption support to existing routers can keep the VPN upgrade costs low.
- Dedicated VPN server hardware specifically designed for the task of tunneling, encryption, and user authentication. These devices usually operate as encrypting bridges that are typically placed between the network's routers and WAN links. Although most of these hardware tunnels are designed for LAN-to-LAN configurations, some products also support client-to-LAN tunneling. Integrating various functions into a single product can be particularly appealing to businesses that do not have the resources to install and manage a number of different network devices.
- Using a specialized outside VPN service provider (outsourcing) such as UUNET

Three VPN protocols have been widely adopted to create VPNs over the Internet to meet different requirements: Point-to-Point Tunneling Protocol (PPTP), Layer-2 Tunneling Protocol (L2TP), and IP Security protocol (IPSec).

10.5.2.1 PPTP and L2TP VPN

PPTP and L2TP are mainly aimed at dial-up VPNs to accommodate mobile users and branch offices for dialing into the protected corporate intranet via their local ISPs (Figure 10.15). PPTP builds on the functionality of Point-to-Point Protocol (PPP), which is the most commonly used protocol for remote access to the Internet.

One of PPTP's main advantage, besides its simplicity, is that PPTP is designed to run at data link layer of the OSI model. Using a modified version of the Generic Routing Encapsulation (GRE) protocol, PPTP encapsulates PPP packets at the link layer. This gives PPTP the flexibility of transmitting data by tunnels over protocols other than IP, such as Internet Packet Exchange (IPX) and Network Basic input/output system Extended User Interface (NetBEUI). Because of its dependence on PPP, PPTP relies on the authentication mechanisms within PPP, namely Password Authentication Protocol (PAP) and Challenge Handshake Authentication Protocol (CHAP).

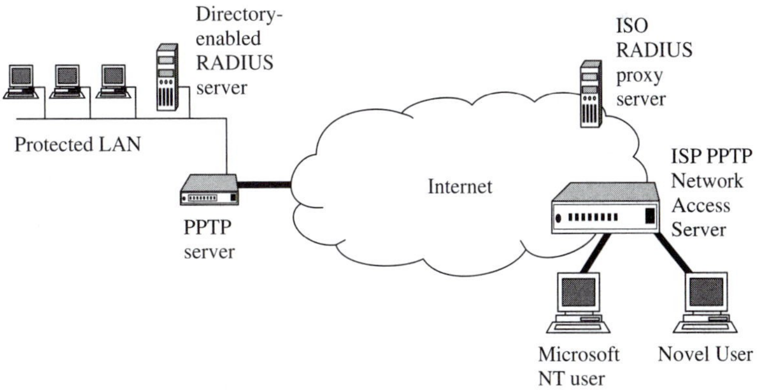

Figure 10.15 A PPTP dial-in VPN structure.

PPTP does have some limitations. For example, it does not provide strong encryption for protecting data nor does it support any token-based methods for authenticating users. However, the IETF-approved L2TP VPN standard is designed to address the shortcomings of PPTP and other older client-to-LAN VPN techniques. L2TP also uses PPP to provide dial-up access that can be tunneled through the Internet to a site. L2TP, however, defines its own tunneling protocol and its transport is defined for a variety of packet media, including X.25, frame-relay and ATM. To strengthen the encryption of the data it handles, L2TP uses IPSec's (to be discussed later) encryption methods. Similar to PPTP, L2TP supports PPP's use of the extensible authentication protocol for other authentication systems, such as RADIUS. L2TP offers the same flexibility as PPTP for handling protocols other than IP, because it is a layer 2 protocol.

10.5.2.2 IPSec VPN

IPSec is designed as a LAN-to-LAN VPN technology for the next generation of IP (IPv6) while keeping it workable for the current IPv4 protocols as well. It operates on the third OSI layer (network layer). IPSec allows the sender (or a security gateway acting on its behalf) to authenticate or encrypt each IP packet or apply both operations to the packet. Depending on different requirements, a packet may be authenticated or encrypted only at transport layer, called as transport mode, or authenticating/encrypting of the entire IP packet may be done for stronger protection, called as tunnel mode.

IPSec may use the majority of standardized cryptographic technologies besides RSA algorithm, such as Diffie-Hellman Key exchanges, DES, and many hash algorithms such as MD5, Hash Message Authentication Code (HMAC), Secure Hash Algorithm (SHA) to provide confidentiality, data integrity, and authentication. The Internet Key Exchange (IKE) is selected as IETF standard for automatically managing and exchanging the cryptographic keys used to encrypt session data.

IPSec realizes its basic functions through defining two headers for IPv6 packets to handle authentication (Authentication Header, or AH) and encryption (Encapsulation Security Payload, or ESP). To be compatible with IPv4, special TCP/IP stacks have to be

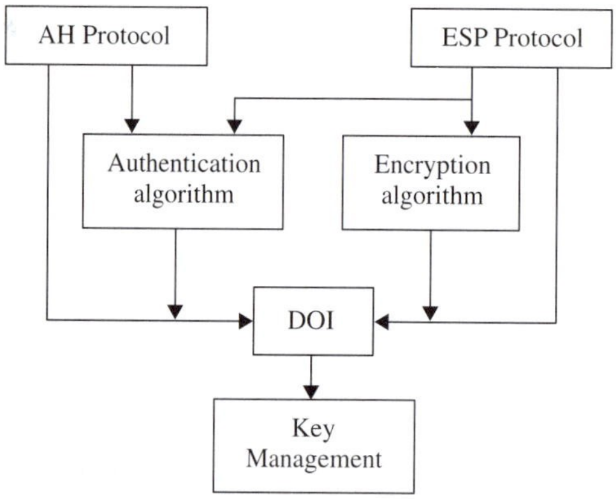

Figure 10.16 IPSec architecture.

used to accommodate these headers when IPSec is implemented. A third component of IPSec is Security Association (SA) that specifies such things as key length, key lifespan, authentication/encryption algorithms, the mode (transport or tunnel mode), and many other parameters when two parties are ready to communicate. The Domain of Interpretation (DOI) is used to simplify and organize the numerous parameters for the specifications with SA (Figure 10-16). The SA specifications are optional since IPSec specifies default authentication and encryption algorisms.

IPSec is often considered the most advanced VPN solution for IP environments, as it includes strong security measures, notably encryption, authentication, and key management, in its standards set. Because IPSec is designed to handle only IP packets, PPTP and L2TP are more suitable for use in non-IP multiprotocol environments, such as those using NetBEUI, IPX, and AppleTalk. IPSec VPN technology has come a long way with increased performance and streamlined management. The IPSec standard makes cross-vendor interoperability possible, but still a concern.

10.5.2.3 Multi-Protocol Label Switching (MPLS) and VPN

All VPN solutions discussed so far (PPTP, L2TP, and IPSec) are not all interoperable and may be tied to one equipment vendor and/or a single service provider. Furthermore, IP is a connectionless protocol so each separate packet moves through the network to its destination along a path determined by a distributed set of routing tables and the current network topology. Such a network is flexible, but the traffic cannot be managed to improve network efficiency, predictability, and stability.

These factors have created strong interest in IP-based VPNs running over the public Internet using standards-based interoperable implementations that work across multiple service providers. The Multi-Protocol Label Switching (MPLS) attaches labels to each data packet and forwards data packets based on these labels. Intermediate MPLS nodes do not need to look at the content of the data in each packet. In particular the destination IP addresses in the packets are not examined, which enables MPLS to offer an efficient encapsulation mechanism for private data traffic traversing through networks thus creating an effective tunnel for standards-based VPNs.

MPLS does not replace IP routing, but works alongside and enhances existing and future routing technologies to provide high-speed data forwarding between Label-Switched Routers (LSRs) together with reservation of bandwidth for traffic flows. MPLS offers enhanced Traffic Engineering with guaranteed QoS providing core technology for converged data and voice networks.

As a packet enters the network, an edge router looks up the destination address of the packet and tags it with a small and fixed-format label that specifies the route and optionally specifies Class of Service (CoS) attributes. At each hop across the network, the packet is routed based on the value of the incoming interface and label, then dispatched to an outwards interface with a new label value. The Label Switched Path (LSP) that a packet takes across the MPLS network is defined by the transition in label values, as the label is swapped at each LSR. As the labeled packet moves across the network, through the processes of label request and label mapping, each router uses the label to choose the destination, and optionally CoS of the packet, rather than looking up the destination address for each packet in a routing table. This decision is a local matter and is based on destination address, the QoS requirements, and the current state of the network. This approach gives MPLS great flexibility that is useful for all kinds applications. As the packet leaves the MPLS network, an edge router uses the destination address to direct the packet to its final destination. Subsequent packets in the data stream are quickly and automatically labeled in this way as anticipated.

The edge routers, also called as Label Edge Routers (LERs), operate at the edge of the access network and MPLS network and play a critical role in the assignment and removal of labels as traffic enters or exits an MPLS network. LERs support multiple ports connected to dissimilar networks such as frame relay, ATM, and Ethernet, and forwards this traffic on to the MPLS network after establishing LSPs. The exiting LER at the egress uses the label signaling protocol at the ingress to distribute the traffic back to the access networks. The set of all packets that are forwarded in the same way is called Forwarding Equivalence Class (FEC). More than one FEC may be mapped to a single LSP. Figure 10.17 shows how two FEC are forwarded through different LSPs.

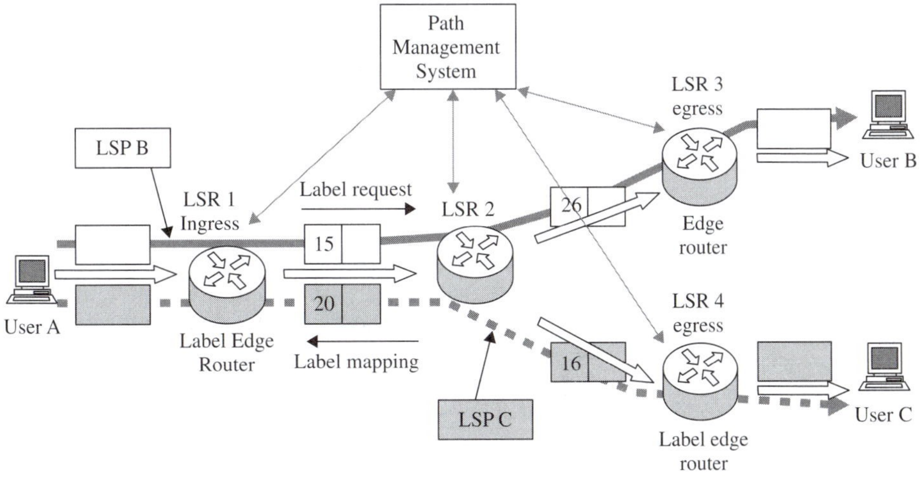

Figure 10.17 LSP B and FLSP C form two tunnels with different sets of labels.

Upon entering the MPLS network at the ingress, edge router LSR 1 adds a label to each packet, determines the FEC it belongs, and the LSP it will follow, then forwards the packet on the appropriate interface for that LSP as shown in Figure 10.17. Note that each packet for User B following LSP B is tagged with label value 15 at LSR1 and will be dispatched out of the interface towards LSR 2, where a new label value 26 is assigned. Packets for User C, following LSP C will be labeled as 20 at LSR 1 and will be re-labeled with value 16 by LSR2, is forwarded to LSR 4.

LSR 2 is an intermediate LSR and it simply receives each labeled packet and uses the pairing [incoming interface, label value] to decide the pairing [outgoing interface, label value], with which to forward the packet. This procedure can use a simple lookup table and, together with the swapping of label value and forwarding of the packet, can be performed in hardware. This allows MPLS networks to be built on existing label switching hardware such as ATM and frame relay. This way of forwarding data packets is potentially much faster than examining the full packet header to decide the next hop. Label edge routers LSR 3 and LSR 4 act as egress LSRs that perform the same lookup, strip the labels from the packets, and deliver them to the final destination by layer 3 routing.

LSPs effectively form tunnels across the entire MPLS network between ingress and egress because intermediate LSRs transited by the LSP do not need to examine the content of the data packets flowing on the LSP. The tunnel thus can have the entire payload, including IP headers, encrypted and safely transmitted without losing direction.

The exact format of a label and how it is added to the packet depends on the layer 2 data link technology used in the MPLS network. For example, a label could correspond to an ATM VPI/VCI, a Frame Relay DLCI, or a DWDM wavelength for optical networking (Chapter 6). For other Layer 2 types, such as Ethernet and PPP, the label is added to the data packet in a special MPLS header between the Layer 2 and Layer 3 headers.

Label distribution, or LSP setup, is the process that populates each LSR with forwarding mapping tables for every LSP. Different Label Distribution protocols, including Label Distribution Protocol (LDP), Constraint-based Routing LDP (CR-LDP), Resource Reservation Protocol (RSVP), Border Gateway Protocol (BGP4), and Open Shortest Path First (OSPF) are allowed to be used under different circumstances. Alternatively, LSPs may be programmed as static or permanent LSPs by programming the label mappings at each LSR on the path using some form of network management protocols such as SNMP with MIBs.

An egress LSR may distribute labels for multiple FECs and set up multiple LSPs. Where these LSPs are parallel they can be routed together by creating a higher-level LSP tunnel between LSRs where the parallelism occurs. This process of placing multilevel labels on a packet is known as label stacking. Labeled packets entering the higher-level LSP tunnel are given an additional label to forward through the network, and retain their previous level labels to distinguish them when they emerge from the higher-level tunnel. Various approaches are used to determine,

- which LSPs to set up to provide connectivity for VPNs,
- how LSRs decide which other LSRs as peers should be included in the LSP tunnel for the VPNs, and

- how different VPNs be mapped into LSP tunnels, i.e. whether a separate tunnel for each VPN, or a single tunnel for all VPNs.

The simplest scheme is to use explicit manual configuration of the VPN peers. When the MPLS network grows in size, alternative schemes can automate the process of discovering VPN peers using a directory service or by overlaying VPN membership information on one or more routing protocols used on the SP network.

Multiple protocols on the VPN can be encapsulated by the tunnel ingress LSR because the data traversing an LSP tunnel is opaque to intermediate routers within the MPLS backbone. Multiplexing of traffic for different VPNs onto shared backbone links can be achieved by using separate LSP tunnels for each data source. Authentication of the LSP tunnel endpoint is provided by the label distribution protocols. And QoS for the VPN data can be assured by reserving network resources for the LSP tunnels.

Four types of VPNs are defined in the MPLS standard:

- Virtual Leased Line (VLL) provides connection-oriented point-to-point links between customer sites. The customer perceives each VLL as a dedicated private connection, although in fact only an IP tunnel exists. The IP tunneling protocol used over a VLL must be capable of carrying different protocols that the customer may use among the sites connected by that VLL.
- Virtual Private LAN Segment (VPLS) provides an emulated LAN among the VPLS sites. As with VLLs, a VPLS VPN requires the use of IP tunnels that are transparent to the protocols carried on the emulated LAN. The LAN may be emulated using a mesh of tunnels between customer sites or by mapping each VPLS to a separate multicast IP address.
- Virtual Private Routed Network (VPRN) emulates a dedicated IP-based routed network among the customer sites. Although a VPRN carries IP traffic, it must be treated as a separate routing domain from the underlying service provider's network, as the VPRN is likely to make use of non-unique customer-assigned IP addresses.
- Virtual Private Dial Network (VPDN) lets customers outsource the provisioning and management of dial-in access to their networks to a third party service provider. Instead of each customer setting up their own access servers and using PPP sessions between a central location and remote users, the service provider provides a shared access server, known as the access concentrator for the users PPP session tunnels for each VPDN. Label Stacking with multi-level labeling is likely used for such VPN architecture.

10.6 NEW TECHNOLOGIES IN NETWORK SECURITY APPLICATIONS

Because of the widespread use of the Internet, numerous companies have begun to utilize this dynamic, versatile, and all-but-ubiquitous medium to reach potential customers.

More and more businesses are seeking ways to connect their users to what is increasingly perceived as a goldmine of information and the avenue to riches. While connecting a corporate LAN to the Internet opens a new world for commerce, it also exposes the business's information assets to hackers, vandals, and criminals. No wonder the demand for innovative security technologies is heating up.

10.6.1 Internet Commerce and Electronic Money

The economics of doing business on the Internet are compelling. Despite the fears and worries about security flaws and frauds, such as the rash of security breaches with the Netscape browser, *Navigator*, the Internet is safer than many people imagine. A business will likely lose much less money on the Internet than through traditional methods of fraud and theft. Merchants, who use the Internet also can lower their costs through reductions in labor force, paper, postage, and inventory, while reaching an international audience at minimal additional expense. With cryptography techniques, digital signatures, and a digital certificate system in place, Internet commerce has grown more quickly than anyone could have anticipated.

Among various types of financial transaction, cash is still the most widely used financial instrument owing to its flexibility. However, in micropayments such as electronic coins, the technology is very complex. Efforts to streamline operations to get the cost per transaction down to the minimum may yet bear fruit.

Electronic cash (e-cash) for use on the Internet or elsewhere is stored on a computer's hard disk or on a smart card and is added or deducted from the user's bank account electronically. To safeguard against forgery, a digital signature and unique identifier, analogous to a watermark and serial number on paper money, assure that the money can be spent only once.

Unlike other forms of electronic payment, using e-cash is anonymous and private. This anonymity appeals to people who do not want to be inundated with direct marketing just because they purchased a product through the Internet. But for governments that fear the process will be used for money laundering, e-cash is problematic.

According to David Chaum's electronic money system, a *digital coin* is a number that can be passed around electronically, like a serial number that has been liberated from its dollar bill. When a bank issues you e-money, it attaches a signature encrypted with its private key. Since everyone can read the signature by using the bank's public key, the users know it is real cash. As soon as a coin is spent, the merchant sends it back to the bank, where the number is retired so the coin cannot be electronically copied and spent more than once. The only hitch is that the bank can track where you spent every coin. So Chaum devised a way for banks to verify a coin's value without knowing the serial number. The concept is explained in rudimentary form by Figure 10.18.

10.6.2 Secure Containers for Electronic Publishing

A host of electronic publishing tool vendors are scrambling to develop technology that will help companies distribute copyrighted material over the Internet without risking a significant loss of income. An emerging technology called *secure container* is being developed to protect the copyright of electronic text, video, or sound files, no matter how

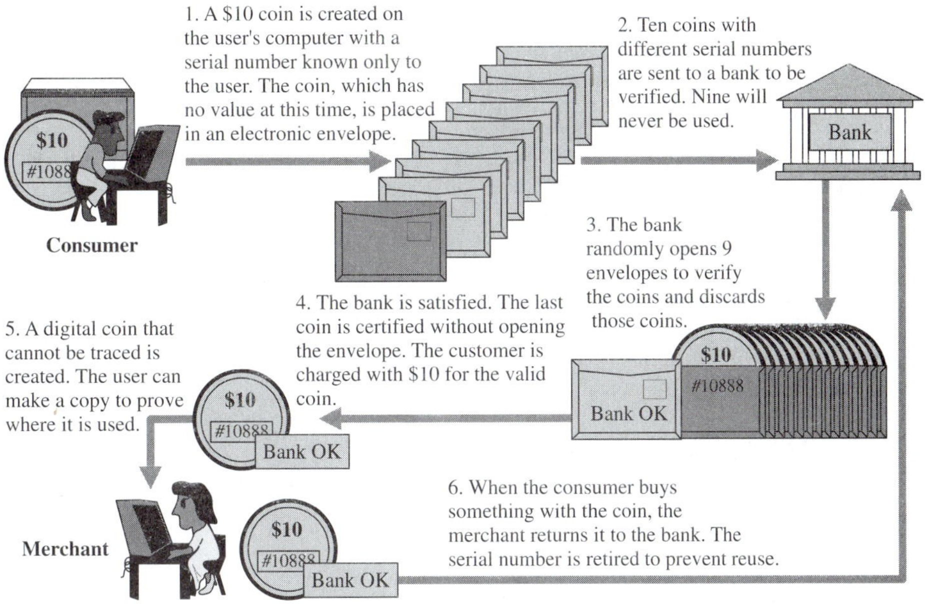

1. A $10 coin is created on the user's computer with a serial number known only to the user. The coin, which has no value at this time, is placed in an electronic envelope.
2. Ten coins with different serial numbers are sent to a bank to be verified. Nine will never be used.
Bank
$10
#1088
Consumer
3. The bank randomly opens 9 envelopes to verify the coins and discards those coins.
4. The bank is satisfied. The last coin is certified without opening the envelope. The customer is charged with $10 for the valid coin.
5. A digital coin that cannot be traced is created. The user can make a copy to prove where it is used.
$10
#10888
Bank OK
Merchant
6. When the consumer buys something with the coin, the merchant returns it to the bank. The serial number is retired to prevent reuse.

Figure 10.18 David Chaum's digital coin concept.

many hands they pass through. The problem with most electronic commerce technology is that once a file has been sold, the originator loses control over it. There is not yet a satisfactory means of protecting electronic content from being used in an unauthorized manner, usually republication and/or redistribution.

One way to address that problem is to add a layer of security to a file that stays with it no matter where it is routed. While it will be several years before the technology becomes mature enough for commercial use, it will provide a more practical electronic commerce model than what exists today. It will also create a publishing environment in which physical barriers disappear while leaving copyrights intact.

Secure container technology creates a cryptographic envelope that not only encrypts files but also adds a layer of rules describing by whom and how the contents can be used. The container and the rule layer remain with the file even when it is passed from one user to another. Each new user must pay for the right to view such a document. Generally, secure container architecture wraps a file in an electronic envelope that will check a user's identity and right to use a document. A billing process will also be launched when the user opens the file. The originator of a document writes rules for its use to protect the copyright and ensure payment. Redistributors can add to the rules and charge extra for the same data (Figure 10.19).

Secure containers are vastly different from current security methods, which allow an owner to charge for a document only once and do little to prevent customers from redistributing information to colleagues or even to their own customers. While the technology is revolutionary, significant efforts will be necessary to establish interoperability standards to make secure container technology widespread and universally accepted.

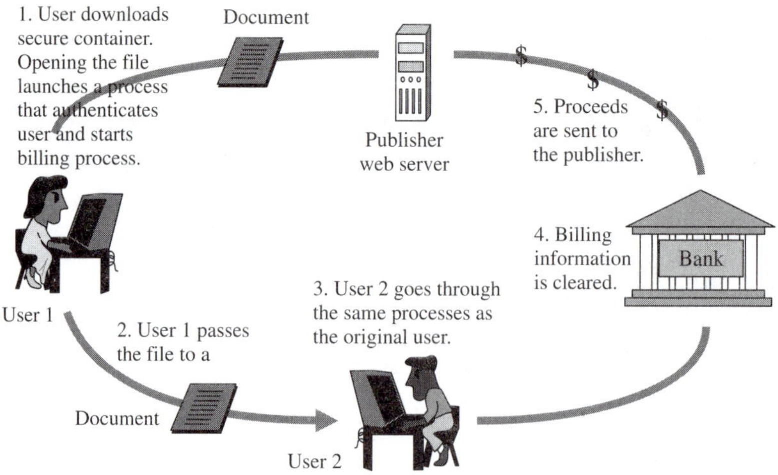

Figure 10.19 A secure container system for electronic publishing and distribution.

10.6.3 Smart Cards and Secure Electronic Transaction (SET)

When a plastic card is enhanced with a magnetic strip that contains limited information, such as the user's name and an account number, the plastic becomes a credit card. The information in a credit card has to be checked and verified with the data stored in a central database, typically in a remote site, through some kind of communications channel.

The computer chip in a smart card can store a thousand times more data than a magnetic strip card. A smart card may someday replace almost everything in your wallet, be it plastic or paper—a driver's license, a health care ID with your medical history and insurance information, an airline ticket with track of record of your frequent-flier miles, credit cards, and even a fistful of dollar bills and loose change. Electronic cash could be conveniently downloaded onto the smart card from your online bank account, then used to buy flowers on the Valentine's Day, and the same card could then be used to feed a parking meter while you buy milk at the corner grocery store. For now, most smart cards in the United States handle only the single task of storing electronic cash, which can be downloaded from the user's bank account. In the years to come, a single smart card will be able to fulfill all the tasks just mentioned. Smart cards represent the biggest leap in consumer and retail transactions since automatic teller machines.

Owing to the power of the microprocessor in the smart card and its larger memory space, as well as its strong security mechanism, a local device can read and authenticate the information in the smart card and authorize its value of use; thus no telecommunication connection is needed to use the card. Thus smart cards became popular in European and many Asian countries first because in Europe phone rates are high, while the low telephone costs in the United States offer l.ttle incentive to use smart cards, with magnetic strip credit cards the remaining plastic of choice.

This situation is about to change, for U.S. businesses see smart cards as a way of offering brand new services, and smart cards are poised for a takeoff in the United States for the following reasons.

- The capabilities of traditional magnetic strip cards are tapped out; in the meantime, the rapid increase in credit card related fraud has created the need for a level of security that smart cards are uniquely able to provide.
- Credit cards are prohibitively expensive, with processing cost reaching 10–20% when the transaction amount is very small (say under $10).
- Smart cards can provide greater confidentiality and privacy than credit cards because they have the ability to allow the user to control who has access to the data carried in the card. A smart card can be used anonymously when it functions as a cash card.
- Smart card technology has matured considerably, and its costs have come down to the point where the technology makes good economic sense in a variety of application areas.
- The Internet provides the best environment and platform for electronic commerce, in particular for electronic cash transactions. As the Internet and electronic commerce ventures gain steam, smart cards provide a crucial link between the World Wide Web and the physical world.

Security is the cornerstone of any transaction system. Both public key and symmetric key encryption, such as the RSA method and DES, are used for smart card security. ISO 7816-7 details standard mechanisms that should be used where security is to be offered.

Secure Electronic Transaction (SET), originally developed by MasterCard and Visa, is a set of authentication and encryption protocols for secure exchange of credit card information over the Internet. A new version of SET, aimed at integrated circuits that are incorporated into smart cards, is being offered as an open industry standard. This version of SET details how bank card transactions on open networks will be secured by means of encryption technology. But smart card users making payments via the Internet will gain security beyond the SET specifications because a variety of tasks, including card authentication, cardholder verification, storage of cryptographic keys, and digital certificate functions, can be performed by the card.

The new specification for smartcards also will work with protocols being developed by the Joint Electronic Payments Initiative (JEPI). JEPI is an alliance of diverse companies that is defining a middleware payment negotiation protocol. Working with the middleware will ensure that smart cards conforming to the SET standard will work with industry-standard computing platforms. So far all the major software publishers have decided to adopt SET in their secure payment applications.

SET security and authentication complement Secure Sockets Layer (SSL), the existing World Wide Web browser-based tool for assuring the security of TCP/IP-based transactions. Both rely on cryptography technologies such as public key encryption and X.509 digital certificates to ensure message privacy, integrity, and authentication. However, when more than two parties are involved in a transaction (e.g., a consumer, a merchant, and one or more banks), SSL cannot guarantee the integrity or confidentiality of messages in the multihop environment.

SET improves on SSL in several ways. It is a multiparty, rather than a point-to-point, protocol. In a SET transaction, the order and card information are encrypted separately, so a cardholder's account information can be channeled to the bank acquiring the transaction without being exposed to the merchant. And because SET can use different levels of encryption for different components, messages can be smaller, allowing faster throughput—up to twice as fast as SSL.

The smart card SET system involves a buyer, a seller, a payment gateway, an acquiring bank, and an issuing bank, as well as the certification authorities. Authenticating the parties—the buyer and the seller—is implemented by digital signatures. Messages are signed by means of an encryption technique based on public and private keys, ensuring both integrity and authentication. Transaction integrity and confidentiality are ensured through message encryption. A digital certificate is required in SET for each party involved in a payment. The cardholder's digital certificate is a virtual card—the equivalent of a physical card. The merchant's certificate has the same effect as a window logo, which represents his or her relationship with the bank.

10.6.4 Random One-Time-Pad and No-Key Encryption

Traditionally, encryption keys have to be used to encrypt and decrypt data. With the data encryption standard, both sender and receiver use the same key for encryption and description. The keys must be carefully created and properly administered. Otherwise the encryption effort is wasted. As technology improves, longer and longer fixed keys will be required to prevent the keys from being cracked by powerful cryptoanalysis methods and devices. Dealing with the various technical and policy issues involved requires a labor-intensive infrastructure. With the widespread uses of networks, problems with key distribution will become much more serious.

The RSA public key cryptography scheme improves the situation greatly but still requires a security infrastructure for distributing, managing, and protecting keys. The Internet has migrated to Pretty Good Privacy (PGP) as its cryptography scheme of choice, in large part because it is free. However, central servers are still needed to manage PGP keys or RSA public keys. In addition, the public key system can be broken, and the performance penalty is heavy.

An old cryptography system—the "one-time-pad" approach, invented in 1917 by Gilbert Vernam, is recently used in the emerging random one-time pad (ROTP) technology, which may release people from the sometime nightmarish task of key management. The ROTP technology uses a random number sequence with a extremely long period (at least 2^{64}) to eliminate the need to assign and manage the traditional keys required to encrypt and decrypt electronic mail messages and Internet ftp sessions.

The technology behind ROTP is a radical departure from existing linear encryption algorithms, which use fixed keys for every session and must be centrally managed in large networks or on the Internet. A product called Power One-Time Pad (POTP) (and possibly others) synchronizes random processes on the two computers as they communicate. The random properties of the two parties become the unbreakable secret they share. In essence, each sequence of bits within each transmission is encrypted with a different key, but the user does not have to know about the keys—the procedure is completely transparent (Figure 10.20).

The communicating parties of POTP first agree to establish a secure channel and share a secret code either manually or automatically. This secret code is valid only for the short initialization procedure. The first encryption and decryption keys are created

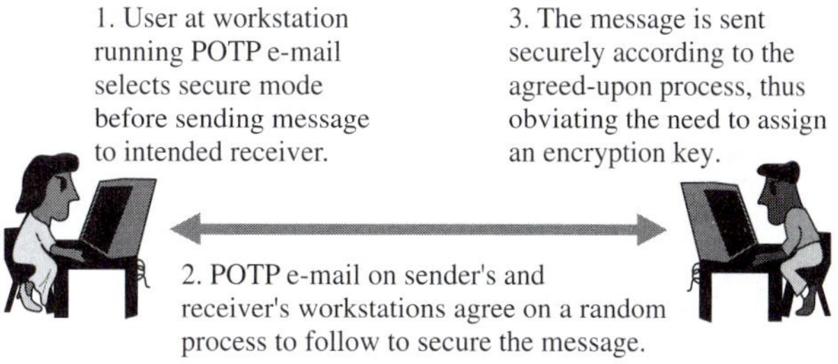

1. User at workstation running POTP e-mail selects secure mode before sending message to intended receiver.
3. The message is sent securely according to the agreed-upon process, thus obviating the need to assign an encryption key.
2. POTP e-mail on sender's and receiver's workstations agree on a random process to follow to secure the message.

Figure 10.20 POTP no-key encryption.

from this shared secret code and are synchronized at both ends. As the communication sessions between the participants proceed, random and synchronized new keys are generated at both ends to encrypt and decrypt messages by simple XOR operations. The key generation is synchronized even though the process is random. No secrets or keys are transferred over communication lines. This entire process caries a very low overhead of less than 1.2% according to the manufacturer.

Serious communications failures such as disconnected, crashed, or noisy lines are immediately identified. This activates an emergency procedure, which creates an immediate, random synchronization point for the participants. The emergency key used for resynchronization is also created by a random process and never used twice. In addition to all the standard identification and access control related information, a hacker would need the POTP *state* for the specific POTP link. The POTP state changes with every session, making the historic states of no value. An earlier POTP state could not be used any more. Any attempt to reuse it would be detected immediately.

To understanding how POTP works, one must look at encryption and communication from a very different perspective. Once the philosophy of continuous randomness becomes clear, conventional computing-intensive and time-consuming encryption methods seem almost silly. User-independent, continuously changing keys provide the strength of PKC without the drawbacks. The one-time pad is theoretically unbreakable, but it used to be impractical because it requires so much computing and telecommunications power. It is becoming usable only as a result of the great advances in computer hardware, memory, and processing technology, in addition to powerful and relatively inexpensive communications capability available today.

10.7 CHAPTER SUMMARY

There are two general approaches to protecting digital information, hardware-based and knowledge-based. Hardware-based methods involve making data physically secure—by locking it in a closet, for instance. Once secured, adversaries must break through the physical barriers to obtain the secrets. However, this approach is very difficult to apply to a global network. The whole network cannot be thrown in a closet each night. Knowledge-based systems depend on using key knowledge to first make the data appear meaningless and then make the meaningless transmission return to a meaningful form. The best information protection systems are combinations of hardware- and knowledge-based systems. Hardware is used to protect the knowledge, which in turn is used to protect the secrets. Among others, the knowledge-based cryptography method and a firewall installed in a physically secure location are the most useful and effective means for network security.

The three basic types of firewall—router-based packet-filtering firewall, application gateway firewall, and circuit gateway firewall—provide different levels and features of network security. Firewalls can enhance security; however, firewalls also significantly change the interactive nature of the Internet culture. A firewall is only one part of the overall security system. A poorly configured firewall can bestow a false sense of security and leave a corporation's information assets exposed. Unless the internal LAN also embodies a solid security strategy, a firewall is merely a cumbersome and porous shell around a crunchy and easily broken paper tiger.

Cryptography is the most universal communication security method. The traditional secret key method is fast and mature, but its key-distribution problem motivated the use of public key cryptography technology. Encryption can be implemented at the link, network, and application levels of the network layer model, with different extents of transparency, convenience, and flexibility.

There are two groups of people in the shadowy world of cryptography. The first group, the cryptographers, scramble or encrypt the plaintext into ciphertext. The second group, cryptanalysts, try to break the ciphertext and reveal its secrets. The battle between cryptographers and cryptanalysts will last forever. In 1977, three leading cryptographers from MIT challenged the world to find the encoded message they had hidden in a 129-digit number. A token award of $100 was offered to anyone who could find the secret. The cryptographers predicted that the attackers would need 40,000,000,000 million years to crack the code and discover the secret.

In 1995 the code was deciphered by an international team orchestrated by Arjen Lenstra, a mathematician at BellCore and assisted by colleagues at MIT, Oxford, and Iowa State University, as well as 600 volunteers, all linked by e-mail on the Internet. Using either a mainframe or a PC, each participant was responsible for sifting through a portion of the enormous haystack of possible solutions. More than 100 quadrillion instructions were processed over 8 months. The results were sent to a mainframe. Based on the preliminary calculations, a supercomputer took only 45 hours to determine the two prime numbers that when multiplied, gave the RSA 129. With that critical number in place, only seconds were taken to extract the three professors' secret message: *the magic words are squeamish ossifrage*. Though the message seems trivial and part gibberish, to those who depend upon the secure transmission of data, it is unmistakably clear "No code is uncrackable."

10.8 PROBLEMS

10.1 What are the three major aspects of network security?

10.2 What are some of the methods that could be used to guarantee the integrity of data and to prevent accidental damage?

10.3 Briefly describe the mechanism used to authenticate the sender of an e-mail message.

10.4 Using the RSA public key cryptosystem,
(a) List five values for d if $p = 7$ and $q = 11$.
(b) Find e if $p = 13$, $q = 31$ and $d = 7$.
(c) Find e and encrypt the message "abcdefghij" if $p = 5$ and $q = 11$.

10.5 Using the mono-alphabetic substitution cipher with +5 shift, encrypt the given text and rewrite in blocks of five letters (ignore the blanks):

In a substitution cipher each letter is replaced

10.6 Using NOMONEY as the key, obtain the ciphertext as blocks of five letters using the transposition cipher for the given text:

pleasetransferonemilliondollarstomyswissbankaccountsixtwotwo

10.7 Break the following columnar transposition cipher, assuming an arrangement of seven rows and four columns. The words have been arranged as groups of five letters, and blanks are ignored in retrieving the plaintext.

BKSEE EDAID GHAKE HOOTR ORTCT TGE
In a substitution cipher each letter is replaced

10.9 REFERENCES

BOOKS

Amoroso, E., and R. Sharp, *PC Week Intranet and Internet Firewall Strategies: Identify Your Security Requirements and Develop a Plan to Protect Your Information*. Emeryville, CA: Ziff Davis Press, 1996.

Anonymous, *Maximum Linux Security; A Hacker's Guide to Protecting Your Linux Server and Network*. Indianapolis: SAMS Publishing, 1999.

Biham, E., and A. Shamir, *Differential Cryptanalysis of the Data Encryption Standard*. New York: Springer-Verlag, 1993.

Birman, K. P., *Building Secure and Reliable Network Applications*. Englewood Cliffs, NJ: Prentice Hall, 1996.

Chapman, D. B., and E. D. Zwicky, *Building Internet Firewalls*. Sebastapol, CA: O'Reilly Associates, Inc., 1995.

Cheswick, W. R., and S. M. Bellovin, *Firewalls and Internet Security: Repelling the Wily Hacker*, 2nd ed. Reading, MA: Addison-Wesley, 2000.

Doraswamy, N., and D. Harkins, *IPSec: The New Security Standard for the Internet, Intranets, and Virtual Private Networks*. Englewood Cliffs, NJ: Prentice Hall, 1999.

Electronic Frontier Foundation, *Cracking DES: Secrets of Encryption Research, Wiretap Politics and Chip Design*. Sebastapol, CA: O'Reilly & Associates, 1998.

Ford, W., *Computer Communications Security: Principles, Standard Protocols and Techniques*. Englewood Cliffs, NJ: Prentice-Hall, 1994.

Garfinkel, S., *PGP: Pretty Good Privacy*. Sebastapol, CA: O'Reilly & Associates, 1994.

Goncalves, M., *Protecting Your Web Site with Firewalls*. Englewood Cliffs, NJ: Prentice Hall, 1997.

Goncalves, M., *Firewalls Complete*. New York: McGraw-Hill, 1998.

Harrison, R., *ASP/MTS/ADSI Web Security*. Englewood Cliffs, NJ: Prentice Hall, 1999.

Icove, D., K. Seger, and W. Von Storch, *Computer Crime: Crimefighter's Handbook*. Sebastapol, CA: O'Reilly & Associates, 1995.

Kaeo, M., *Designing Network Security*. San Jose, CA: Cisco Systems, 1999.

Kaufman, C., R. Perlman, and M. Speciner, *Network Security: PRIVATE Communication in a PUBLIC World*. Englewood Cliffs, NJ: Prentice Hall, 1995.

MacGregor, R. S., A. Aresi, and S. Andreas, *www.security: How to Build a Secure World Wide Web Connection*. Englewood Cliffs, NJ: Prentice Hall, 1996.

Pistoia, M., *Java 2 Network Security*. 2nd ed. Englewood Cliffs, NJ: Prentice Hall, 1999.

Stallings, W., *Protect Your Privacy: The PGP User's Guide*. Englewood Cliffs, NJ: Prentice Hall, 1995.

Stallings, W., *Cryptography and Network Security: Principles and Practice*. 2nd ed. Englewood Cliffs, NJ: Prentice Hall, 1998.

Tung, B., *Kerberos: A Network Authentication System*. Reading, MA: Addison Wesley, 1999.

ARTICLES, PAPERS, AND PUBLIC DOCUMENTS

Bellovin, S. M., and M. Merritt, "Limitations of the Kerberos Authentication System," *USENIX* January 1991.

[source: research.att.com:dist/internet_security/kerblimit.usenix.ps]

Biham, E., and A. Shamir, "Differential Cryptanalysis of DES-like Cryptosystems," *Journal of Cryptology* 4, no. 1 (1991), 3–72.

Biham, E., and A. Shamir, "Differential Fault Analysis of Secret Key Cryptosystems," *CRYPTO* (1997), 513–525.

Chaum, D., "Achieving Electronic Privacy," *Scientific American* August 1992, 76–81.

Chaum, D., "Money Wants to Be Anonymous," *Worth Magazine* October 1995, pp 95–104.

Diffie, W., and M. E. Hellman. "New Directions in Cryptography," *IEEE Transactions on Information Theory* IT-22 (1976) 644–654.

Gong, L., "A Security Risk of Depending on Synchronized Clocks," *Operating Systems Review*, 26 no 1 (January 1992), 49–53.

Koblas, D., and M. R. Koblas, "SOCKS," in *Proceedings of the UNIX Security III Symposium*, Baltimore, MD, September 14–17, 1992, pp. 77–83.

Matsui, M., "A New Method for Known Plain Text Attack of FEAL Cipher," *Proceedings of Advances in Cryptology—Eurocrypt '92*: Hungary: Springer-Verlag, 1992.

Matsui, M. "Advances in Cryptology," in *Proceedings of Advances in Cryptology—Eurocrypt '93: Workshop on the Theory and Application of Cryptographic Techniques Lofthus*. Berlin: Springer-Verlag, 1993.

Matsui, M., "Linear Cryptanalysis Method for DES," in *Proceedings of Advances in Cryptology—Eurocrypt '93: Workshop on the Theory and Application of Cryptographic Techniques Lofthus*. Berlin: Springer-Verlag, 1993.

Molva, R., G. Tsudik, E. V. Herreweghen, and S. Zatti, "KryptoKnight Authentication and Key Distribution System." Proceedings of European Symposium on Research in Computer Security 1992.

[source: jerico.usc.edu:pub/gene/kryptoknight.ps.Z]

Nechvatal, J., "Public-Key Cryptography," NIST Special Publication 800-2. Gaithersburg, MD: National Institute of Standards and Technology, April 1991.

Needham, R. M., and M. D. Schroeder, "Using Encryption for Authentication in Large Networks of Computers," *Communications of the ACM*, 21 no. 12 (December 1978), 993–999.

Presotto, D. L., and D. M. Ritchie, "Interprocess Communication in the Unix System," in *Proceedings of the 10th Usenix Conference*, 8th ed. 1985, pp. 309–316.

Rivest, R. L., "RFC 1321: The MD5 Message Digest Algorithm," 1992.

Schneier, B., Key-Exchange Algorithms," in *Applied Cryptography*, 2nd ed. New York: Wiley, 1996, Ch. 22.

Steiner, J. G., C. Neuman, and J. I. Schiller. "Kerberos: An Authentication Service for Open Network Systems," *USENIX* March 1988.
[source: athena-dist.mit.edu:pub/kerberos/doc/usenix.ps]

Voydock, V. L., and S. T. Kent, "Security Mechanisms in High-Level Network Protocols," *ACM Computing Surveys* 15, no. 2 (June 1983).

WORLD WIDE WEB SITES

More information on the Cramer–Shoup cryptosystem
http://www.zurich.ibm.com/Technology/Security/publications/1998/CS.pdf
DEC CRL X proxy
FTP site: crl.dec.com
Karlbridge: A PC-based packet filtering router kit
(FTP site: nisca.acs.ohio-state.edu, pub/kbridge)
In the United States and Canada, Kerberos is available via anonymous FTP from
ftp.athena-dist.mit.edu
Microsoft security adviser
http://www.microsoft.com/security/
NT security issues
http://www.ntshop.net/
Phil Zimmermann's public domain program package called Pretty Good Privacy (PGP) available from
http://www.rsa.com
Related document
RFC 2440: Open PGP Message Format
Security and encryption information from Carleton College
http://www.carleton.edu/curricular/CS/security.html
Security WWW site list and information organized by Jessica Kelley, Center for Information Technology, National Institutes of Health, Bethesda, MD
(http://www.cit.nih.gov/)
http://www.alw.nih.gov/Security/security-www.html
Socks download via FTP
ftp.nec.com: pub/security/socks.cstc
Socks resources and information
http://www.socks.nec.com
Texas AMU's security tools including software for implementing packet filtering routers
ftp: net.tamu.edu, pub/security/TAMU
Yahoo's links on security and encryption
http://dir.yahoo.com/Computers_and_Internet/Security_and_Encryption/

INFORMATION ON FIREWALLS

ftp.greatcircle.com - Directory: pub/firewalls
(Firewalls mailing list archives)
ftp.tis.com - Directory: pub/firewalls
(Internet firewall toolkit and papers)
net.tamu.edu - Directory: pub/security/TAMU
(Texas AMU security tools)
research.att.com - Directory: dist/internet_security
(Papers on firewalls and break-ins).

11

NETWORK PROGRAMMING

Network programs allow transfer of information between millions of networked computers located all over the world. There are several languages used for writing network programs. Simple web-based programs for information transfer may be written in HTML (hypertext markup language). However, network programs involving TCP/IP components are usually written in Java, C/C++, Perl, and Visual Basic. Additional network programming approaches involve serial/parallel port programming and NetBIOS. Several proprietary network programming approaches, such as NetWare and AppleTalk also are used. In this chapter, we examine some approaches to network programming and present practical examples. However, we caution readers that only a few network programming approaches are introduced here: this chapter is by no means a comprehensive coverage on the topic. A list of references is included at the end for further reading.

11.1 SOFTWARE ARCHITECTURES THAT SUPPORT NETWORK PROGRAMMING

Before we start discussing the different programming approaches, we examine the basic software system architectures that support network programming.

11.1.1 Mainframe Architecture

With mainframe software architectures, all intelligence resides in the central host computer. Users interact with the host through a terminal that captures keystrokes and sends that information to the host. Mainframe software architectures are not tied to a hardware platform. Users can use PCs or Unix workstations to interact with the system. A limitation of mainframe software architectures is that they do not easily support graphical user interfaces (GUIs) or access to multiple databases from geographically dispersed sites. However, recently, mainframes have found a new use as servers in distributed client–server architectures.

11.1.2 File-Sharing Architecture

The original PC networks were based on file sharing architectures, where the server downloads files from the shared location to the desktop environment. The requested user job is then run (including logic and data) in the desktop environment. File-sharing architectures work if shared usage is low, update contention is low, and the volume of data to be transferred is low. In the 1990s, the PC-based LANs changed because file-sharing capacity was strained as the number of online users grew and GUIs became popular (actually, this also made the mainframe and terminal displays appear out of date). PCs are now normally used in client–server architectures.

11.1.3 Peer-to-Peer Architecture

In a peer-to-peer architecture each workstation has equivalent capabilities and responsibilities. This differs from client–server architectures, in which some computers are dedicated to serving the others. Peer-to-peer networks are generally simpler and less expensive, but they usually do not offer the same performance under heavy loads. Peer-to-peer networks aren't nearly as expensive to create as client–server-based networks. All the software you need usually is part of a standard operating system such as Windows 95/98/2000. The affordability often makes peer-to-peer networks ideal for small businesses or home users. However, the main disadvantage of using this type of network is that by placing network control in the hands of end users, security is sacrificed. Therefore, this method strategy is obviously inappropriate for high-security environments.

11.1.4 Client/Server Architecture

In some systems, each computer or process on the network is either a client or a server. Servers are powerful computers or processes dedicated to managing disk drives (file servers), printers (print servers), or network traffic (network servers). Clients are PCs or workstations on which users run applications. Clients rely on servers for resources, such as files, devices, and even processing power. Client/server is a computational architecture that involves client processes requesting service from server processes. The term was first used in the 1980s in reference to PCs on a network. The model started gaining acceptance in the late 1980s. The client/server software architecture is a versatile, message-based, and modular infrastructure that is intended to improve usability, flexibility, interoperability, and scalability in comparison to centralized, mainframe, time-sharing computing.

The client/server architecture emerged as a result of the limitations of file-sharing architectures. This approach introduced a database server to replace the file server; instead, a database management system (DBMS) is invoked to answer user queries directly. The client/server architecture reduced network traffic by providing a query response rather than total file transfer. It improves multiuser updating through a GUI front end to a shared database. A client is defined as a requester of services and a server is defined as the provider of services. A single machine can be both a client and a server depending on the software configuration.

Client/server computing is actually the logical extension of modular programming. Modular programming has as its fundamental assumption that separation of a large piece of software into its constituent parts ("modules") creates the possibility for easier development and better maintainability. Client/server computing takes this concept a step further by recognizing that those modules need not all be executed within the same memory space. With this architecture, the calling module becomes the "client" (that which requests a service), and the called module becomes the "server" (that which provides the service). The logical extension of this is to have clients and servers running on the appropriate hardware and software platforms for their functions.

In client/server architectures, Remote Procedure Calls (RPCs) or standard query language (SQL) statements are typically used to communicate between the client and server. The client is a process that sends a message to a server process, requesting that the server to perform a task. Client programs usually manage the user interface portion of the application, validate data entered by the user, and send requests to server programs. The client process also manages the local resources that the user interacts with, such as the monitor, keyboard, workstation CPU, and peripherals. One of the key elements of a client workstation is the GUI, which is normally a part of operating system.

A server process accepts and fulfills the client request by performing the task requested. Server programs execute database retrieval and updates, manage data integrity, and send responses to client requests. The server-based process does not have to run on another machine on the network. It could be running on the same machine as the client, or it could be running on the host operating system or network file server. A server provides both file system services and application services. We now discuss the several client/server architectures used in the industry.

11.1.5 Two-Tier Architecture

In two-tier client/server architecture, a client talks directly to a server, with no intervening server. It is a good solution for distributed computing when workgroups are defined as a dozen to a hundred users interacting simultaneously on a LAN. It does have a number of limitations. When the number of users exceeds 100, performance begins to deteriorate. This limitation is a result of the server maintaining a connection via some messages with each client, even when no work is being done. A second limitation of the two-tier architecture is that the implementation of processing management services by means of proprietary database procedures restricts flexibility and choice of DBMS for applications. Finally, current implementations of the two-tier architecture provide limited flexibility in moving (repartitioning) program functionality from one server to another without manually regenerating procedural code.

A common error in client/server development is to prototype an application in a small, two-tier environment, and then scale up by simply adding more users to the server. This approach usually results in an ineffective system, as the server becomes overwhelmed. To properly scale to hundreds or thousands of users, it is usually necessary to move to a three-tier architecture.

11.1.6 Three-Tier Architecture

Our final type of client/server architecture is a special one consisting of three well-defined and separate processes, each running on a different platform:

1. The user interface, which runs on the user's computer (the client).
2. The functional modules that actually process data. This middle tier runs on a server and is often called the application server.
3. A DBMS that stores the data required by the middle tier. This tier runs on a second server called the database server.

The three-tier design has many advantages over traditional two-tier or single-tier designs, the chief ones being as follows:

- The added modularity makes it easier to modify or replace one tier without affecting the other tiers.
- Separating the application functions from the database functions makes it easier to implement load balancing.

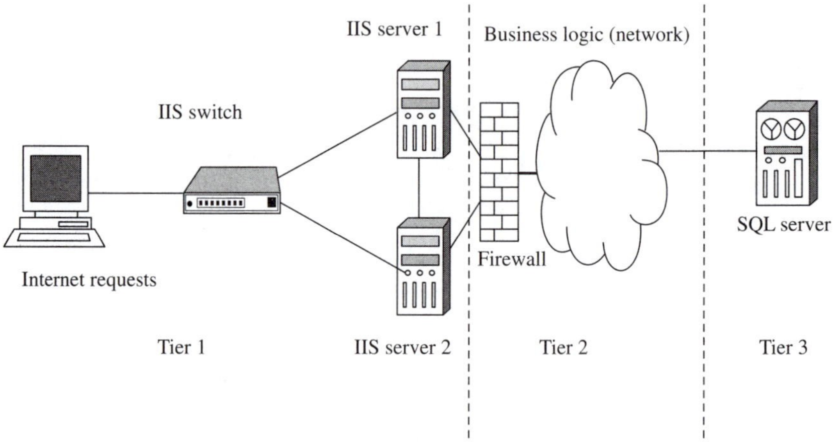

Figure 11.1 A three-tier Internet information server (IIS).

Notice that it is the same three-tier architecture as discussed in chapter 4. However, we now discuss it with reference to a client–server system.

The three-tier architecture emerged to overcome the limitations of the two-tier architecture. It is a client/server architecture in which the middle tier is introduced for the application logic. The middle tier acts like an "agent" between the client and the server. The middle tier provides the following services:

1. Translation services (as in adapting a legacy application on a mainframe to a client/server environment)
2. Metering services (as in acting as a transaction monitor to limit the number of simultaneous requests to a given server)
3. Intelligent agent services (as in mapping a request to a number of different servers, collating the results, and returning a single response to the client)

The existing architecture used by businesses that emphasize e-commerce is a three-tier system (Figure 11.1). Usually the three-tier system consists of an application program that is divided into three major parts, each of which is distributed to a different place or places in a network. The three parts are

- The workstation or presentation interface
- The business logic
- The database and programming related to managing it

In a typical three-tier application, the application user's workstation contains the programming that provides the GUI and application-specific entry forms or interactive windows. Business logic is located on a LAN server or other shared computer. The business logic acts as the server for client requests from workstations. In turn, it determines the data needed (and the data's location) and acts as a client in relation to a third tier of programming, which might be located on another computer that may be a mainframe system. The third tier includes the database and a program to manage read and write access to it. While the organization of an application can be more complicated than this, the three-tier view is a convenient way to think about the parts in a large-scale program.

Some advantages of this architecture are ease in processing of business transactions, development of new applications, and distributed transaction processing. A single architecture does not address all problems. Architecture must be evaluated for fitness to task. The three-tier model provides it. In addition, it provides a reliable platform to test, model, and implement software involving dynamic components such as active server pages. For instance, if two Internet information servers are deployed, one could easily be dedicated for developers while the other might still meet the demands of the customers. However, for higher expected loads it may be desirable to have the switching function implemented on three IISs to result in higher fault tolerance.

The middle tier may be implemented in a number of ways, including transaction processing monitors, message servers, or application servers. It can perform queuing, application execution, and database staging. For example, if the middle tier provides queuing, the client can deliver its request to the middle layer and disengage because the middle tier will access the data and return the answer to the client. In addition, the middle layer adds scheduling and prioritization for work in progress. The most basic type of three-tier architecture has a middle layer consisting of transaction processing (TP) monitor technology: that is, the transaction is accepted by the monitor, which queues it and then takes responsibility for managing it to completion, thus freeing up the client.

Messaging is another way to implement three-tier architectures. Messages are prioritized and processed asynchronously. Messages consist of headers that contain priority information, and the address and identification number. The message server implementation of three-tier architecture connects the server to the relational DBMS and other data sources. The three-tier application server architecture allocates the main body of an application to run on a shared host rather than in the user system interface client environment.

Using technologies that support distributed objects to develop client/server systems holds great promise because these technologies support interoperability across languages and platforms, as well as enhancing maintainability and adaptability of the system. Currently industry is working on developing standards to improve interoperability and determine what the common object request broker (ORB) will be. There are currently two prominent distributed object technologies:

- Common object request broker architecture (CORBA)
- COM/DCOM [component object model (COM), Distributed COM (DCOM), and related capabilities]

The distributed/collaborative enterprise architecture emerged as a software architecture based on ORB technology, but goes further than the CORBA by using shared, reusable business models (not just objects) on an enterprise-wide scale. The benefit of this architectural approach is that standardized business object models and distributed object computing are combined to give an organization flexibility to improve effectiveness organizationally, operationally, and technologically.

11.2 SERIAL PORT AND PARALLEL PORT PROGRAMMING

Serial and parallel port programming provides a low-cost networking environment between two or more computers. A network of computers may be designed, programmed, and operated by using the low-level programming on the ports. Many programmers prefer to deal at the port levels to get byte-level control of data in an experimental setup. In addition, a network laboratory with the state-of-the-art hardware and software may not be affordable in all situations. In such cases, low-level programming may be done to implement a LAN setup.

11.2.1 Serial Port Programming

Since parallel communications requires too many wires, serial communications may be used for transferring data over long distances. Serial data received from a modem or other devices are converted to parallel to permit transfer to the PC bus. The serial data may be transmitted synchronously or asynchronously in simplex, half-duplex, or full-duplex modes. In a synchronous transmission data are sent in blocks, and the transmitter and the receiver are synchronized by one or more special characters called sync characters.

The serial port of the PC is an asynchronous device in which a bit identifies the start of transmission and 1 or 2 bits identify the end without need for any synchronization. The data bits are sent to the receiver after the start bit. The least significant bit is transmitted first. A data character usually consists of 7 or 8 bits. Depending on the configuration of the transmission, a parity bit is sent after each data bit. It is used to check errors in the data characters. Finally 1 or 2 stop bits are sent.

The serial port of the PC is compatible with the EIA standard RS-232C. This standard was designed in the 1960s to communicate a data terminal equipment or DTE (the PC in this case) and a data communication equipment or DCE (usually a modem). The serial port standard specifies 25 signal pins that is compatible with RS-232 standards. The most used connectors are the DB-25 male, but many of the 25 pins are not needed. For that reason, in many modern PCs a DB-9 male connector is used.

The integrated circuits that convert the serial data lines to parallel and vice versa are called UART (universal asynchronous receiver–transmitter). The typical PC UART is the Intel 8251A, an IC that can be programmed like a synchronous or an asynchronous device. Eight data bits (D0–D7) connect the 8251A to the data bus of the PC. The chip select (/CS) input enables the IC when is asserted by the control bus of the PC system. This IC has two internal addresses, a control address and a data address. The control address is selected when the C-/D input is high. The data address is selected when the C-/D input is low. The Reset signal resets the IC. When the /RD is low, the computer reads a control or a data byte. The /WR enables the PC to write a byte. Both signals are connected to the system signals with the same names. The UART includes four internal registers:

THR: Temporary output register

TSR: Output register

RDR: Input register

RSR: Temporary input register

Every character to be transmitted is stored in the THR register. The UART adds the start and stop bits, then copies all bits (data, start, and stop bits) to the TSR. To finish the process, the bits are sent to the line by the TD signal. Every character received from the line RD is stored in the RSR register. The start and stop bits are eliminated and the UART writes this character to the RDR. To finish the process, the character is read for the PC. The serial port may be addressed in two ways: by the 14H BIOS interrupt and by the 21H DOS interrupt. The 14H BIOS interrupt uses four functions to program the serial port. Each function is selected assigning a value to the AH register of the microprocessor. The four functions are as follows:

Function 00H: Initializes the serial port and sets the speed, data and stop bits and the parity parameters.

Function 01H: Sends a character to the specified serial port.

Function 02H: Reads a character from the specified serial port.

Function 003: Returns the state of the specified serial port.

There are three functions in the 21H DOS interrupt related to the operation of the serial port:

Function 03H: Reads a character from the COM1 serial port.

Function 04H: Writes a character to the COM1 serial port.

Function 40H: Sends a number of bytes from a buffer to the specified device (a common outfunction for all files and devices that use a handle access)

The following Borland C++ program (source: http://www.ctv.es/pckits/tutorial.html) shows the communication between two PCs via serial port:

```
//Program to communicate two PC s via serial port
//00H bios function (AL register)

//bits 7 6 5 Baud rate
// 0 0 0 110
// 0 0 1 150
// 0 1 0 300
// 0 1 1 600
// 1 0 0 1200
// 1 0 1 2400
// 1 1 0 4800
// 1 1 1 9600
```

```
//bits 4 3 Parity bits
// 0 0 no parity
// 0 1 odd parity
// 1 1 even

//bits 2 stop bits
// 0 1 stop bit
// 1 2 stop bit
//bits 1 0 Number of bits per data
// 1 0 7 data bits
// 1 1 8 data bits

//Register Dx 0->com1, 1->com2, 2->com3, 3->com4

//Configuration: 9600 bd,no parity,2 stop bits and 8 data
bits
//AL register value is 1 1 1 0 0 1 1 1 -> 0XE7

#include ⟨stdio.h⟩
#include ⟨process.h⟩
#include ⟨conio.h⟩
#include ⟨dos.h⟩
#include ⟨bios.h⟩

#define TRUE 1
#define PARAM 0XA7
#define COM1 0
#define COM2 1

void init_port(void);
char state_port(void);
void send_byte(unsigned char);
unsigned char read_byte(void);
void keyb(void);

int tecla = 1;

void main ( void )
{
unsigned char read_com;
unsigned char read_kb;
clrscr();
init_port();

while(read_kb !p= 'c')
{
read_kb=etch();
send_byte(read_kb);
```

```
read_com=read_byte();
if(read_com!=0){printf("%c",read_com);}
}
}

void init_port()
{
union REGS regs;
regs.h.ah = 0X00;
regs.x.dx = COM2;
regs.h.al = PARAM;
int86( 0X14, &regs, &regs);
}

//return the state of the port

char state_port()
{
union REGS regs;
regs.h.ah = 0X03;
regs.x.dx = COM2;
int86( 0X14, &regs, &regs);
if(regs.h.ah & 0X80) printf("\t EXCEED TIME\n");
if(regs.h.ah & 0X40) printf("\t TSR EMPTY\n");
if(regs.h.ah & 0X20) printf("\t THR EMPTY\n");
if(regs.h.ah & 0X10) printf("\t INTERRUPTION\n");
if(regs.h.ah & 0X08) printf("\t THREAT ERROR\n");
if(regs.h.ah & 0X04) printf("\t PARITY ERROR\n");
if(regs.h.ah & 0X02) printf("\t OVERLOAD ERROR\n");
return (regs.h.ah);
}

//Keyboard handle

void keyb()
{
union u_type{int a;char b[3];}keystroke;char inkey=0;
if(bioskey(1)==0) return;
keystroke.a=bioskey(0);
inkey=keystroke.b[1];
switch (inkey)
{
case 1: keyb=0;
return; /*ESC*/
default: keyb=15;
return;
}
}
```

```
//Send a character to the serial port

void send_byte(unsigned char byte)
{
union REGS regs;
regs.h.ah = 0X01;
regs.x.dx = COM2;
regs.h.al = byte;
int86( 0X14, &regs, &regs);
if( regs.h.ah & 0X80)
{
printf("\t SENDING ERROR ");
exit(1);
}
}

//read a character from serial port

unsigned char read_byte()
{
int x,a;
union REGS regs;
if((estate_port() & 0X01))
{
regs.h.ah = 0X02;
regs.x.dx = COM2;
int86(0X14,&regs,&regs);
if(regs.h.ah & 0X80)
{
printf("\t RECEIVING ERROR");
exit(1);
}
return(regs.h.al);}
else
{
return(0);
}
}
```

Now we turn to a CSMA/CD implementation using Motorola 68000 based microcomputers (PT68K2 systems). However, this approach could be used for any serial port in a PC system. In this case, the network speed is restricted by the limitations of serial ports to 19.2 Kbps. The transmission medium is chosen to be the twisted-pair line owing to the simplicity of installation and maintenance. Since the speed of the network is significantly lower than the limitations of the twisted pair, very little signal disruption is expected. The transmit and receive lines are connected together to provide for a bidirectional link to the network medium. By logically connecting the transmit and receive lines, access to the medium becomes similar to that of an Ethernet interface; in fact, the interface becomes a pure bus medium interface, receiving what it sends.

The presented implementation in C requires the use of some functions written in assembly language. The assembly language functions are needed for displaying the keyboard and network status as well as for transmitting/receiving the data to/from the network. However, these functions are transparent to the user. Users are required only to use the function calls in their programs written in C, resulting in a simpler programming task for users with no assembly language background. The networking concepts covered are data framing and error checking, packet assembly/disassembly and addressing, and protocol implementation.

The required equipment is an interface box to connect PC serial ports on Motorola 68000 based systems, a PC serial port, interfaces and wires. This implementation is an extension of the CSMA protocol. It works similar to the CSMA protocol as long as there is no collision. As soon as a collision is noticed, the transmitter aborts the transmission and retransmits after a random time interval. The protocol suggests that a node should sense the medium before transmitting. Only if the medium is free is a packet transmitted. Packet transmission is completed only if there is no collision. When this technique is employed, collisions are not noticed by the user. In other words, even if a collision takes place at the physical medium, the user will never notice it because the packet retransmission is done at the physical layer.

The next implementation examines the situation of collision detection for CSMA. The physical layer is responsible for sensing the medium before any transmission. Medium sensing is done by observing the medium in receive direction for a finite interval of time A_t. If nothing is received during that period, it is assumed that the medium is free and the transmission is started. The medium speed is 19.2 Kbps. With the medium speed of 19.2 Kbps, the time to transmit 1 byte would be 8/19,200 = 0.4167 ms. Allowing 1 byte time for the ACIA delay, a total waiting time of 2 byte (A_t = 0.8333 ms) should be enough to detect any transmission on the medium. If the medium is busy (i.e., if a character is found on the medium), the transmission is deferred and the medium sensing is repeated after 2 bytes time.

The algorithm design is simple. Waiting can be achieved by introducing a meaningless loop with a high count value. The instruction performed in the loop may take a finite amount of time. If the medium is busy, the NOP (no operation) loop can be entered and executed for 0.8333 ms. At the end, the medium is checked again. If the medium is still busy, the operation is repeated until it is found to be free. A flag called NFY can be used to indicate the status of the medium. If NFY is 0, the medium is busy; otherwise it is free. The flag must be reset to 0 as soon as the PRE byte of a packet is received at a station.

A piece of circuitry is necessary for this experiment to abort the transmission as a collision is detected. When two nodes output simultaneously, an interrupt for the collision detector is generated that informs the involved stations of the situation, and the transfer is aborted. If a receiving node detects a collision, it aborts the reception of the current packet and clears its input buffer. The retransmission of packets is not shown. However, additional circuitry may be needed for requeuing the packet. Since retransmissions are not implemented, if a collision occurs, the message is lost.

The physical layer interface of the program is responsible for sensing the medium before any transmission as in the CSMA scheme. Medium sensing is done by observing the medium in receive direction for a finite amount of time A_t. If nothing is received during that period, it is assumed that the medium is free, and the transmission is started. The medium speed is 19.2 Kbps. With this speed, the time to transmit 1 byte would be 8/19,200 = 0.4167 ms. Allowing 1 byte time for the ACIA (serial port) delay, a total waiting time of 2 bytes (A_t = 0.8333 ms) should be enough as a minimum waiting time before transmission. In other words, each station senses the medium for 2 bytes time and if the medium is free it transmits, thus implementing the carrier sensing.

CSMA/CD is implemented by checking the status of medium before transmission. If the medium is found free, transmission is not done right away. Transmission is started if the medium is found free after waiting for 2 bytes time. When the medium is found busy during carrier sensing, then a station must defer for a random amount of time. That random time must be greater than 2 bytes time, which is the minimum duration for carrier sensing. In our implementation, the maximum random value could be as much as 20–30 bytes time.

A flag (NFY) is used in the program to indicate the condition of the physical layer. If NFY is 0, the medium is busy; otherwise it is free. The flag is reset to 0 as soon as the PRE byte of a packet is received at a receiving station. The CSMA/CD pseudocode with packet assembly is given as follows:

```
/* Interface to application side */
repeat forever
if char at the keyboard and TBF == 0 then
        AI = 1; /* Application flag */
        get the char;
        array[0] = PRE; /* Set preamble PRE */
        get the next char;
        if char = machine ID then                /* check address */
        while char != CR
                place char in the array;
                inc array index;
                get the char;
        /* end while */
        place CR in array;
        inc array index;
        place LF in array; /* Line Feed */
        inc array index;
        place EOT in array; /*End flag */
        AI = 0; TF = 1; TBF = 1;
/* end of if statements */
if DF == 1 then /* Data Flag from buffer */
        DF = 0; AD = 1;
        get the char from buffer;
        display the char string until CR;
        ignore CR;
        ignore LF;
        ignore EOT;
```

```
        AD = 0; RBF = 0;
/* end of if statement */

/* Physical layer interface */
if char at the network and RBF == 0 then
        NBY = 0;   /* assume that the network medium is free
*/
        wait 2 byte times; /* loop */
        if NBY == 0 then      /* if medium is still free */
           BYS = 1;    /* enter transmit state */
           SP = 0;     /* clear packet transmission request */
        else
          wait for a random amount of time (greater than 2
bytes time);
        /* end of if statement */
        NR = 1;
        get the char;
        if char == PRE then continue;
        while char != EOT
              place in array;
              inc array index;
              get the char;
        /* end while */
        NR = 0; DF = 1; RBF = 1;
/* end of if statement */
if TF == 1 then
        TF = 0; NT = 1;
        while not EOT
          send char from transmit buffer to network;
          compare contents of buffer with the characters on
the network;
          if correct seq of chars is on the net, continue
the transmission
              else abort transmission; /* as this implies
collision */
        /* end while */
        NT = 0; TBF = 0;
/* end of if statement */
/* end of repeat loop */
/* end of main */
```

C and C++ are considered to be very effective programming languages for doing low-level interfacing in PCs. However, there is a recent trend toward the use of Visual Basic to access serial ports on PCs. Link designs using serial ports and microcontrollers may be programmed with Visual Basic. Projects can be developed to make a PC-to-PC link, PC-to-embedded-controller link and RS-485 network. The advantages of custom controls and Windows API make this alternative very appealing for the development of communication applications in Visual Basic.

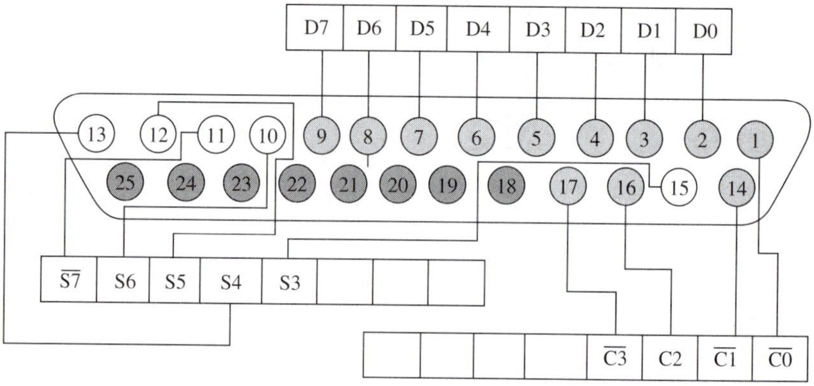

Figure 11.2 Parallel port.

11.2.2 Parallel Port Programming

The parallel port is the most commonly used port for interfacing homemade projects. This port allows the input of up to 9 bits or the output of 12 bits at any one given time, thus requiring minimal external circuitry to implement many simpler tasks. The port is composed of four control lines, five status lines, and eight data lines (Figure 11.2). It is found commonly on the back of any PC as a 25 pin female connector. There may also be a 25 pin male connector, which is a serial RS-232 port and thus totally incompatible. Newer parallel ports are standardized under IEEE standard 1284, first released in 1994.

The parallel port is commonly used to connect a printer to a computer. It is used for the CPU to send data to a printer. This interface drives some input and output signals. The purpose of these signals is to let the computer know the state of the printer and control it. Eight data bits carry all the information sent with each clock pulse. The hardware of this port consists of 8 output data bits, 5 input control bits, and 5 output control bits. The control signals are listed as follows:

- **Outputs**

  Pin 1, STROBE/: Tells the printer when the 8 data bits are ready to be read. Turns to a low logic level when the data are ready.

  Pin 16, INIT/: Reset the printer.

  Pin 17, SLCT IN/: Selects the printer when it turns to a low logic level.

  Pin 14, AUTO FD/: Tells the printer to print an empty line followed by a carriage return.

  Pins 2–9, D0–D7: Data bits.

- **Inputs**

  Pin 10, ACK/: Tells the CPU that the data were correctly received.

  Pin 11, BUSY: The printer sets this line when its buffer is full. The computer will stop sending more data.

Pin 17, SLCT: Tells the computer that a printer is present.

Pin 15, ERROR/: An error has occurred. The CPU stops sending more data.

Pin 12, PE: The printer is out of paper.

MS-DOS supports three parallel ports, called LPT1, LPT2, and LPT3. So we can find three addresses dedicated to these ports in the memory map of the PC. Let's study the addresses dedicated to LPT1 first. Each parallel port uses three addresses of the I/O map. For LPT1 these addresses are 378H, 379H, and 37AH.

378H port: In this address the CPU writes the data to be sent to the printer. It is an output port. The 8 data bits (D0–D7) are latched to appear in the output connector.

379H port: This is an input port. These signals are used by the CPU to know the state of the printer.

37AH port: In this port the computer writes the signals that control the printer. Therefore, it is an output port.

The two short programs that follows are examples of how the parallel port can be programmed. They are all compiled with the Borland C++ 3.1 compiler. The first program shows how to send a byte to the parallel port output addresses. The `outportb()` function sends a byte to a specified I/O port. The first function parameter is the address of the port to write a byte. The second parameter is the value of the byte to send. Both parameters can be defined as variables. In this case, the first parameter must be an unsigned integer, and the second an unsigned character.

The second example shows how to read a byte from the parallel port input address. The main function is only used to show the value of the byte in the screen. The `inportb()` function reads a byte from the specified I/O address of the computer. The parameter must be an unsigned integer (program source: http://www.ctv.es/pckits/tutorial.html).

Program 1

```
#include <stdio.h>
#include <dos.h>
#include <conio.h>

/*******************************************/
/*This program set the parallel port outputs*/
/*******************************************/

void main (void)

{
clrscr();
outportb(0X378,0Xff);
```

```
outportb(0X37a,0Xff);
getch();}
```

Program 2

```
#include <stdio.h>
#include <dos.h>
#include <conio.h>

/****************************************/
/*This function read parallel port inputs*/
/****************************************/

int Read_Input()

{int Byte;

Byte=inportb(0X379);
return Byte;}

void main (void)

{int PP_Input;

clrscr();
PP_Input = Read_Input;
printf("%d",var);
getch();}
```

11.3 NETBIOS PROGRAMMING

Before talking about NetBIOS, it is important to understand what BIOS (pronounced "bye-ose") is. It is the acronym for Basic Input–Output System. BIOS is the built-in software that determines what a computer can do without accessing programs from a disk. On PCs, the BIOS contains all the code required to control the input/output devices. To ensure that the BIOS will always be available and will not be damaged by disk failures, it is typically stored in a ROM chip that comes with the computer. Therefore, sometimes it is called a ROM BIOS. It also makes a computer boot by itself. Since RAM is faster than ROM, though, many computer manufacturers design systems so that the BIOS is copied from ROM to RAM each time the computer is booted. This is known as shadowing. Many modern PCs have a flash BIOS, which means that the BIOS has been recorded on a flash memory chip, which can be updated if necessary.

NetBIOS, for Network Basic Input–Output System, consists of a set of routines for building software applications that augments the DOS BIOS by adding special functions for LANs. Almost all LANs for PCs are based on the NetBIOS. Some LAN manufacturers have extended it to add network capabilities. NetBIOS is also defined as a software interface between computer programs and a local-area network adapter (LANA), or "LAN card." The main purpose of the NetBIOS is to isolate the application program from the actual type of hardware used in the LAN. It also spares the application programmer the details of network error recovery and low-level message addressing or routing.

There are many advantages of using NetBIOS for communication between applications as opposed for instance, to using Winsock. Network adjustment for PCs is simpler (it can be accomplished with only one protocol: NetBEUI (pronounced "net-booey), and designating an **e**xtended **u**ser **i**nterface for Net**B**IOS). NetBEUI is the fastest of all transport protocols). Network administration task becomes simpler because the IP addresses are not required; rather, NetBIOS allows the use of convenient symbolic names. In addition, NetBIOS allows complete platform compatibility between Win32 (95/98/NT), DOS, Win 16, and OS/2.

Peer-to-peer networks are easier to build using NetBIOS than client/server-based networks. The steps usually involve selecting and connecting appropriate cables between the network cards (with or without a hub) and then configuring the network operating system. In fact, the cost of establishing a 10BaseT network between two PCs is very low because a hub is not needed. All you need is a network card for each machine and a special 10BaseT cable called a "crossover" cable, which has a different wire pattern on each end. Plug one end of this cable into each machine's network card. As an alternative, a coaxial cable with BNC connectors on the ends with a "T" and a "terminator" at both ends may also be used.

NetBIOS is based on a message format called server message block (SMB), which is used by DOS and Windows to share files, directories, and devices. Actually many of the network products, including LAN Manager, Windows for Workgroups, Windows NT, and LAN Server, use SMB to enable file sharing among different operating system platforms.

The enhanced version of NetBIOS called NetBEUI is used by many network operating systems, such as LAN Manager, LAN Server, Windows for Workgroups, Windows 95, and Windows NT. NetBEUI was originally designed by IBM for their LAN Manager server and later extended by Microsoft and Novell.

In a NetBIOS implementation of a LAN, the computers on the system are known by names (unique and group names). A unique name is guaranteed by the NetBIOS to be unique across the LAN. A group name added at one computer may also be added, as a group name, at other computers. The names are alphanumeric names, 16 characters in length (no asterisk, *), and should be ASCIIZ form—that is, the last byte is 00h. Up to 16 local names may be added to the NetBIOS for each computer on the network by the use of the Add-name and Add-group-name commands. These names are stored in a local name table and are lost when the computer is turned off, or when the NetBIOS Reset command is issued. A local name can be removed from the local name table by use of the NetBIOS Delete-name command.

Computers on the system can also be known by names designated by the programmer. The commands available in NetBIOS include commands to add and delete names. Computers on a NetBIOS implementation of a LAN can communicate either by establishing a session or by using datagram or broadcast methods. Sessions allow a larger message to be sent and handle error detection and recovery, but the networked computers can communicate on a one-to-one basis only. Datagram and broadcast methods allow one computer to communicate with several other computers at the same time, but are limited in message size (512 bytes or lesser).

The two types of communication within NetBIOS are also called connectionless (datagram type) and connection oriented (session type). In connectionless service, a station requests that the data be sent in the form of a datagram. All other stations are continually checking the network for datagrams, to identify any data meant for them. If such data are found, the message is received. However, there is no form of handshaking, or acknowledgment, so there is no guarantee that all stations on the network will receive the message as intended. Datagram-level data transfers are addressed to a lower level in the LAN adapter card. The initialization requirements for sending or receiving a datagram message are simpler; however, more of the error control and data formatting problems are left to the user program.

The connection-oriented connection is established between two names. The connection is known as a session, as mentioned earlier, and is not necessarily between two stations. The session can be between any two names, including two names on the same station. In this connection, either the message is delivered successfully or an error is returned to the application, so the connection is aware of the transmission, or receive failure. Session-level commands allow user programs to communicate without requiring that they handle the details of network error detection or breaking larger messages into smaller messages to match any physical network limitations. Before using session-level data transfer, the applications must negotiate a "logical connection." This is accomplished by issuing a NetBIOS Call command on one computer and a NetBIOS Listen command on another computer. A session-level data transfer sequence can be grouped into four steps. Each step on one computer in the network must have a complementary step performed on another computer in the network. These steps are as follows:

1. Use the Add-name command to add a unique name to the NetBIOS local name table.
2. Use the Call or Listen command to initiate a session. If a Call is used, then another computer should have performed a Listen.
3. Use the Send and Receive commands to transfer messages.
4. Terminate the session with the Hangup command.

All commands are communicated to the NetBIOS in a format called network control block (NCB). Such blocks are allocated in memory by the user program. An NCB contains several fields. Some fields are used to pass input values to NetBIOS, while others are used by NetBIOS to return results from the command execution. The NCB fields include a command code field and a command result field. The user program is also responsible for setting the necessary input fields of the NCB and initializing the fields not used to zeros. Several fields in the NCB are reserved for output from NetBIOS upon completion of a command.

The NetBIOS command is initiated by setting up the NCB and calling interrupt 5Ch, with ES:BX pointing the the NCB. Control will be returned to the program only when the command has been completed (or has timed out). There are two different methods by which a NetBIOS command can be used—synchronous and asynchronous. Bit 7 of the command byte is set to 0 to indicate a synchronous command. When control is returned, the return_code field of the NCB indicates the initiation or completion status. The value of this field is also returned in register AL. An example of the use of this type of command is adding a name for communication. The program cannot continue until this has been successful or has failed.

The asynchronous command is initiated in the same way as synchronous commands; however, bit 7 of the command byte is set to 1, to indicate that an asynchronous command is required. Control is immediately returned to the program, with the return_code field, and AL, containing the initiation status. It is necessary to check to see when the command has been completed. There are two ways of doing this:

- Poll the command_complete field of the NCB. This contains FFh, if the command is still being executed, and is set to another value, the completion code, once the command has been completed (or an error has occurred).
- There is a 4-byte field (post_address), which is usually set to 0000:0000. This can be used to point to a Post routine. A Post routine is a user routine that is called once the command has been completed. The Post routine is called by the networking software with ES:BX pointing to the NCB that has been completed, and AL containing the contents of the command_complete field. No other registers are defined. The Post routine should be written in the same way as an interrupt service routine—that is, interrupts are disabled, and should not be re-enabled, and the routine should be as quick as possible and terminated with an IRET instruction.

Asynchronous command can be used, for example, in a talk program. An NCB is set up to receive a message (datagram) on a specific name. The POST points to the routine to process the received message and display it on the screen. The software must not wait until a message has been received, since the user individually cannot enter any messages. Asynchronous commands can also be used for multitasking. An asynchronous command can be initiated within the Post routine of a command: for example, once a message has been received in a talk, the NCB must be reinitialized to receive the next message. However, synchronous commands cannot, and should not, be called in a Post routine, or the machine will also definitely crash.

11.4 TCP/IP AND SOCKET PROGRAMMING

The most general mechanism for communication between the processes offered by Berkeley Unix is the socket. Two processes communicate by creating sockets and sending messages between them. A socket appears to the user to be like a file descriptor on which users can read, write, and perform input/output control. In the connection-oriented mode, the file is like a sequence of characters that can be read with as many read operations as needed. In the connectionless mode, a whole message is supposed to be captured in a single read operation. If it is not, what is left over of the message is lost. Sockets can also be used in a single computer system for interprocess communication (the Unix domain) and for communication across computer systems (the Internet domain).

Types of socket vary based on the way socket's address space is defined and the type of communication desired. Typically a triplet that includes domain, type, and protocol defines a socket address. The most common domains are `AF_UNIX` (for Unix path names), `AF_INET` (for Internet addresses), and `AF_OSI` (as specified by international standards for OSI). The various address formats are defined as constants in the file ⟨sys/socket.h⟩, which is essentially included for any socket programming.

Protocol specifies the protocol used. It is usually 0 to indicate the use of the default protocol for the chosen domain and type. The communication channel created with sockets can be connection oriented with a TCP (Transmission Control Protocol) connection (type `SOCK_STREAM` for domain `AF_UNIX`) or with a datagram—that is, UDP (User Datagram Protocol) with IP-level connection (type `SOCK_RAW` for domain `AF_UNIX`). Connection-oriented communication is reliable; that is, the system takes care of errors. Datagram oriented communication is unreliable: messages may be lost or delivered in an order different from the one in which they were sent.

For communicating between sockets, a three-component interlocutor consisting of IP address, port, and protocol is defined. This represents the IP addresses (like w.x.y.z) of IPv4 addresses.

Ports are 16-bit unsigned integers. The first 1024 port numbers are reserved (e.g., port 80 for http). These ports are called well-known ports. On Unix systems, standard uses (ftp, telnet, finger, etc.) of these ports are listed in the files `/etc/services` and `/etc/inetd.conf`. From port 1014 up, things are not too well established. Ports from 49152 to 65535 are private and anybody can use them. These are also called ephemeral ports. The interval 1024 to 49151 is usually left for standard uses. These are called registered ports. At one time the registered ports stopped at 5000; above that, ports were ephemeral. The port 0 is used as a wild card, to request the kernel to find a port for the user.

11.4.1 Interprocess Communication and Pipes

Another way of communication between the programs (running on same or different computers) is via interprocess communication. A simple form of this communication is the pipe on Unix systems. A pipe connects the standard output of one process to the standard input of another process. One process writes into the pipe, the other process reads from the pipe. If the pipe is full or empty, one of the involved processes will have to wait. In Unix systems, a pipe is formed by concatenating a "|" between the commands.

The first process in a pipe should be a source of information. A process that is used in the middle is called as a filter because it modifies the information before passing it to the next process in the line. The standard output of the last process in a pipeline may be redirected to a file, or else it simply appears on the screen. For example, consider the concatenation of Unix commands "who" and "wc" as follows:

```
$ who | wc
```

In this example, the output of `who` is piped to the word count program wc, which counts the lines, words, and characters appearing on its standard input. The resulting display would be the number of users logged onto Unix. As another example, consider the command "`deroff`" used with "`sort`", "`pr`", and "`more`".

```
$ deroff -w report2 | sort | uniq | pr -5 | more
```

In this command sequence, the first command produces stream of words from the file "`report2`", one per line, followed by a sorting (`sort`) of the words with only single occurrences (`uniq`), in five-column format (`pr -5`) displayed one page at a time (more).

The concept of Unix pipes is extended further to allow the system programmers to "fork" subprocesses (or "child" processes) in the programs. Again, there may be certain input passed to a child process from its parent and the child process may itself generate further children with inputs passed to them. The `fork()` system call is used to create a new process in Unix. It creates a new context based on the context of the calling process. This call is however unusual, since it returns both in the process calling it and in the newly created process. Here is an example of how it may be called.

If fork is successful, it returns a number greater than 0 and represents the PID of the newly created child process. In the child process, fork returns 0. If it fails, then its return value will be less than 0. A more efficient version of `fork()`, the fork call `vfork()`, does not duplicate the entire parent context and may be suitable for some situations.

In the following example of fork usage, the parent process prints "`Hello`" to `stdout`, and the new child process prints "`World`". However, the order of printing is not guaranteed. "`Hello`" may or may not be printed before "`World`". Usually a synchronization method is used to ensure an order.

```
#include <stdio.h>

  char string1[] = "Hello";
  char string2[] = "World\n";

  int main(void)
  {

          int PID;

          if ((PID = fork()) == 0)
             printf("%s", string2);
                    /* In the child process */
             else   /* In the parent process */
                printf("%s", string1);
             exit(0); /* Executed by both processes */
}
```

11.4.2 Creation of Sockets

A socket is created with the `socket` system call by specifying the desired address family, socket type and protocol as follows:

```
sd = socket(domain, type, protocol)
```

A typical call to socket may consist of the following:

```
if ((sd = socket(AF_INET, SOCK_DGRAM, 0) < 0)
     { perror("socket"); exit(1);}
```

Here all the variables are of type int. The returned socket descriptor is similar to the file decriptor and is a small positive integer that can be used as a parameter to reference the socket in subsequent system calls. A -1 is returned in case of errors.

When a socket is created, it does not initially have an address associated with it. However, since it is impossible for a remote host or process to find a socket unless it has an address, an address is provided with bind system call, which has three parameters: sock, address, and addrlen. The protocol port number at which a socket will wait for messages is specified by bind. A call to bind is optional on the client side and required on the server side. A typical bind call is:

```
status = bind(int sd, struct sockaddr *address, int addrlen)
```

Here status returns a 0 for success and a -1 otherwise. The descriptor returned by socket() is sd; the pointer to protocol address structure of this socket is address; and the length in bytes of the structure referenced by address is addrlen. It is important to have the address in the correct format, for otherwise the call fails. The call also fails when the address is already in use or the socket already has an address bound to it.

The general form of sockaddr structure in ⟨sys/socket.h⟩ is:

```
struct sockaddr {
  short sa_family; /* address family */
  char sa_data[14];/* upto 14 bytes of direct address */
}
```

The socket address in the Unix domain must follow a Unix path name, which may be up to 108 characters long. When one is using sockets in the Unix domain, it is advisable to only use path names for directories directly mounted on the local disk. The Unix domain allows interprocess communication only for processes working on the same machine. The following structure, sockaddr_un, used to define the Unix address format, can be found in ⟨sys/un.h⟩:

```
struct sockaddr_un {
  short sun_family;              /* AF_UNIX */
  char  sun_path[108-4];         /* path name */
};
```

In the Internet domain, address consists of two parts; a host address (network number and a host number) and a port number as mentioned earlier. The host address allows processes on different machines to communicate, while the port number allows multiple addresses on the same host. The sockaddr_in in the Internet domain is defined in ⟨netinet/in.h⟩. Notice that the address in sockaddr just shown is more general but is compatible with the following address, since both start with same 16-byte field.

```
struct sockaddr_in {
  u_short sin_family;/*protocol id; usually AF_INET */
  u_short sin_port; /* port number */
```

```
  struct in_addr sin_addr; /* IP address */
          /*IP addresses of the current host*/
          /*It is considered a wildcard IP address*/
  char sin_zero[8]; }; /* Unused, always zero */

struct in_addr {
  u_long s_addr;
};
```

11.4.3 Using Internet Library and DNS

Note that the `sockaddr_in` structure (16 bytes length) just presented facilitates reference to the elements of a socket address. It includes `sin_family` (2 bytes), `sin_port` (2 bytes), an unsigned long (4 bytes) IP address, and 8 bytes of `sin_zero` (with all zeros) to pad the structure to bring it to the length of `struct sockaddr` (16 bytes). For Internet addressing, notice that `sin_family` corresponds to `sa_family` in `sockaddr` and should be set to "AF_INET." Also, the `sin_port` and `sin_addr` must be in network byte order. So if a variable `ineta` is declared to be of type `sockaddr_in`, then `ineta.sin_addr.s_addr` references the 4 byte (`u_long`) IP address stored in network byte order.

A network byte order is actually a "big-endian" order in which the memory data are placed in increasing order of memory locations (bits/bytes are numbered most significant to least significant in increasing order). However, a complication is faced when some machines such as IBM-compatible PCs) store the bytes in reverse, or "little-endian," order. For instance, to be able to communicate between an IBM-compatible PC and a Sun workstation, data sent over the network must be concerted to network byte, or "big-endian," order. Routines for converting data between a host's internal representation and network byte order must use the following `include` statements:

```
#include ⟨sys/types.h⟩
#include ⟨netinet/in.h⟩
```

The conversion statements are:

`htons()`	"Host to Network Short (2 bytes)"
`htonl()`	"Host to Network Long (4 bytes)"
`ntohs()`	"Network to Host Short (2 bytes)"
`ntohl()`	"Network to Host Long (4 bytes)"

These functions are macros and result in the insertion of conversion source code into the calling program. On little-endian machines, the code will change the values around to network byte order. On big-endian machines, no code is inserted because none is needed; the functions are defined as null.

An Internet address is usually written and specified in the dotted-decimal notation as already shown. When used in network programs, it should be stored in a structure of type in_addr. To convert between the two representations two functions are available:

```
unsigned long inet_addr(char *ptr)
char *inet_ntoa(struct in_addr inaddr)
```

A character string in dotted-decimal notation is converted to a 32-bit Internet address by means of inet_addr(). The return value is not consistent, and it may be system dependent. The correct return should be a pointer to a structure of type in_addr, but many systems, following an older convention, return only the internal representation of the dotted-decimal notation. The system's main pages usually clarify the situation for the host system on which the function is used.

The other function, inet_ntoa(), expects a structure of type in_addr as a parameter (note that the structure itself is passed, not a pointer) and returns a pointer to a character string containing the dotted-decimal representation of the Internet address. As an example, consider

```
ineta.sin_addr.s_addr = inet_addr("141.261.5.10");
```

Notice that inet_addr() returns the address in network byte order already and there is no need to call htonl(). However, it is not a very efficient way of coding. Imagine if an error is returned from inet_addr with -1: it cannot be detected. Now consider the following code:

```
char *a, *b;
    .
    .
a = inet_ntoa(ineta.sin_addr); /* this is 141.261.5.10*/
b = inet_ntoa(ineta.sin_addr); /* this is 141.101.1.18 */
    printf("address 1: %s\n",a);
    printf("address 2: %s\n",b);
```

will print:

```
141.261.5.10
141.101.1.18
```

Now consider the situation of an unknown IP address of a host. DNS allows a conversion to IP address as we explained earlier. So if "telnet cps215.cps.cmich.edu" is entered at the command prompt, telnet can find out that a connection is needed to 141.209.131.215. Internally in the program, the function called gethostbyname() returns a pointer to the filled structure hostent, or a Null pointer in case of an error. It requires the following statement:

```
#include <netdb.h>
```

The function is defined as follows:

```
struct hostent *gethostbyname(const char *name);
```

A pointer to a structure `hostent` is returned that has the following format:

```
struct hostent {
   char h_name; /* Host name */
   char **h_aliases;/* A NULL-terminated array of
                       alternate names for the host */
   int h_addrtype; /* Address type, such as AF_INET */
   int h_length; /* Address length in bytes */
   char **h_addr_list; /* A zero-terminated array of
                       network addresses for the host
                       given in Network Byte Order */
};
#define h_addr h_addr_list[0];
                       /*First address in h_addr_list */
```

Here is an example program to illustrate the use of the function:

```
#include <stdio.h>
#include <stdlib.h>
#include <errno.h>
#include <netdb.h>
#include <sys/types.h>
#include <netinet/in.h>

int main(int argc, char *argv[])
{
     struct hostent *name;

     if (argc != 2) { /* error check the command line */
          fprintf(stderr,"usage: getip address\n");
          exit(1);
     }

     if ((h=gethostbyname(argv[1])) == NULL) {
                        /* get the host info */
        herror("error: gethostbyname");
        exit(1);
     }

     printf("Host name : %s\n", name->h_name);
     printf("IP Address : %s\n",
          inet_ntoa(*((struct in_addr *)name->h_addr)));

     return 0;
}
```

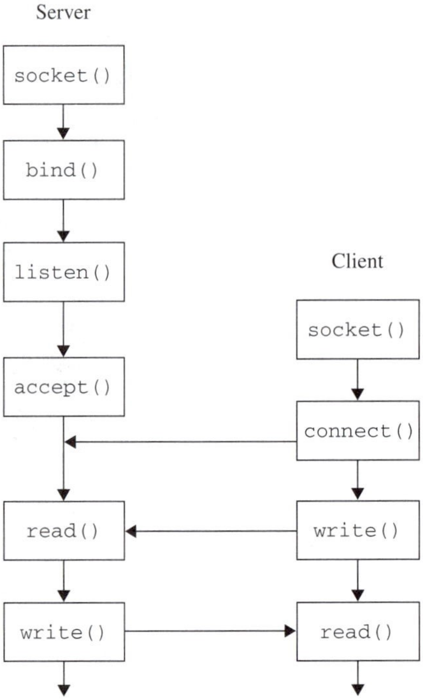

Figure 11.3 Client–server Interaction in TCP.

11.4.4 Socket Connection between Client and Server

Now that we have investigated some Internet-related functions and structures, we introduce a simple client–server program. However, before we look at the programs used for data transfers on the Internet, we need to understand the required operations at the client and server sides.

In connection-oriented (TCP) service, socket is created, a bind is done to a local port, a service is set up with indication of maximum number of concurrent services, connection requests are accepted from connection-oriented clients, messages are received and replied to, and then the connection is terminated. The overall operation on the client side consists of creating a socket, binding it to a local port (if a bind is not used, the kernel will select a free local port), establishing the address of the server, writing and reading from it (or performing `sendto` and `recvfrom`), and then terminating. If the client is not interested in a response, it does not need to use bind. The overall operation of a TCP (reliable byte stream connection) server is shown in Figure 11.3. The following steps are followed:

- **Server**

  Create end point (`socket()`)

  Bind address (`bind()`)

  Specify queue (`listen()`)

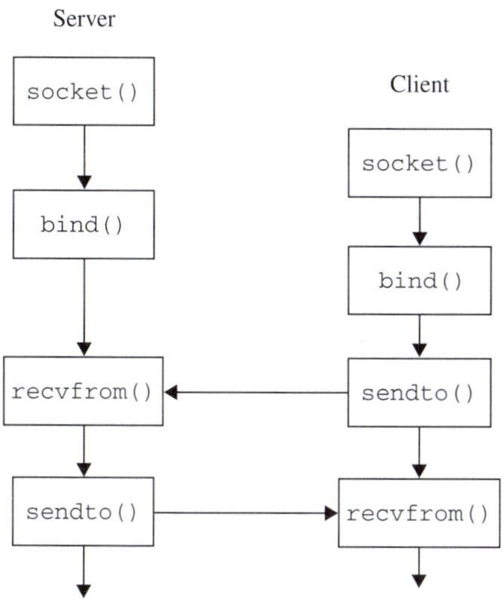

Figure 11.4 Client–server Interaction in UDP.

Wait for conection (`accept()`)
Transfer data (`read() write()`)

- **Client**
  Create end point (`socket()`)
  Connect to server (`connect()`)
  Transfer data (`write() read()`)

Servers in UDP create a socket, bind it to a local port, accept and reply to messages from the client, and terminate. If the server does not need reply to the client, it can just use `read` instead of `recvfrom`. The client follows a similar procedure except that it requests data and subsequently accepts data from the client. The overall operation is shown in Figure 11.4, and the steps involved in connection are as follows:

- **Server**
  Create endpoint (`socket()`)
  Bind address (`bind()`)
  Transfer data (`recvfrom() sendto()`)

- **Client**
  Create endpoint (`socket()`)
  Bind address (`bind()`)
  Connect to server (`connect()`)
  Transfer data (`sendto() recvfrom()`)

The connect system call used by the active end (client) in TCP connection has three parameters: `sd` (socket descriptor), `name` (address of remote socket), and `namelen` (name length). The parameter name is of type `struct sockaddr`, and its interpretation depends on the communication domain, which is `sockaddr_un` for `AF_UNIX` and `sockaddr_in` for `AF_INET`.

The responding server at the other end of the connection performs a listen and an accept followed by the bind. Since `listen()` initializes a queue for waiting connection requests, it has two parameters, `sd` and `qlen` (queue length). The queue length specifies the maximum number of queue connections allowed. The accept call takes the first pending connection request off the queue and returns a socket descriptor for a new connected socket, whose address is given by the parameter "name" and called as follows:

```
new_sd = accept (old_sd, name, namelen)
```

Here `namelen` should be initialized to the size of the address structure being passed. However, upon return it is set to the actual length of the address returned. The `old_sd` remains unaffected and may be used to accept more connections. If there are no more pending connection requests accept blocks.

The data transfer in sockets may be done by system call pairs (`read, write`), (`recv, send`), (`recvfrom, sendto`), (`recvmsg, sendmsg`), and (`readv, writev`). The `send` and `recv` calls are typically used with TCP sockets but may be used with UDP if the sender has already done a connect or if the receiver does not care who the sender is. The `read` and `write` calls may be used with any connected sockets, and (`recvfrom, sendto`) are used with datagram sockets. The `sendto` call allows a specific datagram destination to be specified, while the `recvfrom` call returns the name of the remote socket sending the message. The `writev` and `readv` calls make it possible to scatter and gather data to/from separate buffers. Finally, the calls `sendmsg` and `recvmsg` allow scatter/gather capability as well as the ability to exchange access rights. The calls `read`, `write`, `readv`, and `writev` take either a socket descriptor or a file descriptor as their first argument; the rest of the calls require a socket descriptor.

Now let us look at a TCP client program for the `AF_UNIX` connection domain:

```
#include <sys/types.h>
#include <sys/socket.h>
#include <sys/un.h>

main ()
{
  int sd, ns;
  char buf[256];
  struct sockaddr_un server;
  int fromlen;

  sd = socket(AF_UNIX, SOCK_STREAM, 0);
  strcpy(server.sun_path, "/tmp/foo");
  server.sun_family = AF_UNIX;

  if(connect(sd, (struct sockaddr *) &server,
      strlen(server.sun_path) + sizeof(server.sun_family)))
```

```
    exit();
  write(sd, "hi there!", 9);
}
```

In this client program, first an "sd" is obtained from the socket call. A string "tmp/foo" is then copied to the sun_path in the structure sockaddr_un and the sun_family is set to AF_UNIX. The connect call uses the pointer to structure sockaddr, which is internally interpreted to point to the structure sockaddr_un because of tje AF_UNIX domain. That is why the name length is obtained by adding the lengths of sun_path and sun_family in the sockaddr_un structure (Notice that bind is not used here.) Finally, a write is done via the sd. Now let us look at the corresponding server program:

```
#include <sys/types.h>
#include <sys/socket.h>
#include <sys/un.h>

main ()
{
  int sd, ns;
  char buf[256];
  struct sockaddr_un addr;
  struct sockaddr_un from;
  int fromlen;

  sd = socket(AF_UNIX, SOCK_STREAM, 0);
  strcpy(addr.sun_path, "/tmp/foo");
  addr.sun_family = AF_UNIX;
  bind(sd, (struct sockaddr *) &addr, strlen(addr.sun_path)
                  + sizeof (addr.sun_family));

  listen (sd,5);
  ns = accept (sd, (struct sockaddr *) &from, &fromlen);
  read (ns, buf, sizeof buf);
  printf("server read %s\n", buf);
  close(ns);
}
```

At the server side, the variable addr is used to represent the sockaddr_un, which is used in bind to establish a meaningful socket connection. Then a listen followed by an accept is used to get the new socket descriptor ns of the connected socket. The address of the socket at the client end is returned via from. Next, a read is done in buf via the new socket descriptor ns. Finally the string is printed, followed by the connection close.

The next set of client–server code shows the current time transfer from a server to the client using UDP.

```
/* cludp1.c - Client using datagram service on sockets
 It has two command line parameters: a host name, and a
 port number.
```

```
 It sends the current time to the server, waits for
 and prints the reply, and repeats after waiting
 for 30 seconds.
*/

#include <sys/types.h>
#include <sys/socket.h>
#include <netinet/in.h>
#include <netdb.h>
#include <stdio.h>
#include <string.h>

#define MAXBUF 2 * 1024

void main(int argc, char *argv[])
{
  int sock;                    /* file descriptor for socket */
  long tm;                     /* time */
  struct sockaddr_in name;  /* Address of the server */
  struct hostent *hp;
  char data[MAXBUF];

  if (argc != 3){
     printf("Usage %s servername portnumber\n", argv[0]);
     exit(0);
  }

  /* create a socket */
  sock = socket(AF_INET, SOCK_DGRAM, 0);
  if(sock < 0) {
    perror("Opening datagram socket");
    exit(1);
  }

  /* Host and port come from input arguments */
  /* Fill in the server's UDP address */
  hp = gethostbyname(argv[1]);
  if(hp == 0) {
    fprintf(stderr, " %s: unknown host",argv[1]);
    printf("Error h_errno %d\n", h_errno);
    exit(2);
  }
  bcopy((char *)hp->h_addr, (char *)&name.sin_addr,
      hp->h_length);
  name.sin_family = AF_INET;
  name.sin_port = htons(atoi(argv[2]));
  while (1) {
    /* send the message */
    time(&tm);
```

```
    printf("At Client: Current time is %s\n", ctime(&tm));
    strcpy(data, ctime(&tm));
    if(sendto(sock, data, strlen(data) * sizeof(char), 0,
           (struct sockaddr *)&name, sizeof(name)) < 0)
    perror("Sending datagram message");

    /* Wait for the reply */
    if(read(sock, data,(1024 * 2)) < 0) {
      perror("Error receiving data from server");
      exit(1);
    }
    printf("Data from server %s\n", data);
    sleep(30);
  }
  close(sock);
  exit(0);
}

/* svudp1.c - Server using datagram service with sockets.
  In a loop it prints out the message it receives from a
  client and sends back the current time.
  It also prints out information about the client.
*/
#include <sys/types.h>
#include <sys/socket.h>
#include <netinet/in.h>
#include <netdb.h>
#include <stdio.h>
#include <string.h>

#define MAXBUF 2 * 1024

void main(int argc, char *argv[])
{
  int sock; /* file descriptor into UDP */
  struct sockaddr_in name;
                /* the address of this service */
  int length; /* length of address */
  char buf[MAXBUF]; /* buffer to hold datagram */
  struct sockaddr_in client_addr;
                /* the address of the client */
  long tm; /*time*/

  if (argc != 1){
     printf("Usage: %s\n", argv[0]);
     exit(0);
  }

  /* create a socket */
```

```
    sock = socket(AF_INET, SOCK_DGRAM, 0);
    if(sock < 0) {
      perror("Opening datagram socket");
      exit(1);
    }
    bzero((char *)&name, sizeof(name));
    name.sin_family = AF_INET;
    name.sin_addr.s_addr = INADDR_ANY;
    name.sin_port = 0;

    if(bind(sock,(struct sockaddr *)&name,sizeof(name)) < 0) {
      perror("Error binding datagram socket \n");
      exit(1);
    }

    /* print port number */
    length = sizeof(name);
  if(getsockname(sock,(struct sockaddr *)&name, &length) < 0) {
      perror("Error binding datagram socket \n");
      exit(1);
    }
    printf("Socket port #%d\n", ntohs(name.sin_port));

    length = sizeof(struct sockaddr);
    while(1) {
      if(recvfrom(sock, buf, MAXBUF, 0, &client_addr,
                                               &length) < 0)
      perror("Couldn't read datagram");
    printf("Client Socket Family: %x\n",
                          ntohs(client_addr.sin_family));
    printf("Client Port = %u\n",
                         ntohs(client_addr.sin_port));
    printf("Client IP = %s\n",
                        inet_ntoa(client_addr.sin_addr));
    printf("From Client %s\n", buf);
    time(&tm);
    strcpy(buf, ctime(&tm));
    printf("Going to send to client %s\n", buf);
    if(sendto(sock, buf, MAXBUF, 0, &client_addr,
                                             length) < 0)
      perror("Couldn't send datagram");
    }
  }
```

11.4.5 Blocking, Synchronization, and Timing

By default, all the reading and writing calls are blocking. That is, read and write calls do not return until at least one byte of data is available for transfer and there is enough buffer space to accept the data. For example, if `recvfrom()` is used and there are no data to receive, the call is said to be blocked (i.e., it sleeps) until some data arrive. Some applications need to service (or `accept()`) network connections simultaneously, performing operations on connections as they become enabled. The three techniques used to support such applications are nonblocking sockets, asynchronous notifications, and the `select()` system call; `select()` is used most commonly, nonblocking sockets is less preferred, and asynchronous notifications are rarely used.

When the socket descriptor is created first with `socket()`, the kernel sets it to blocking. If a blocking socket is not desired, then a call to `fcntl()` must be made:

```
#include ⟨fcntl.h⟩
.
.
sd = socket(AF_INET, SOCK_STREAM, 0);
fcntl(sd, F_SETFL, O_NONBLOCK);
.
.
```

The second and third parameters are commands and corresponding arguments found in ⟨`fcntl.h`⟩. Another similar system call, `ioctl()`, uses the constant from ⟨`sys/ioctl.h`⟩ and can be used if device control is needed. In the foregoing example, a socket is set to nonblocking, thus allowing the socket to be "polled" for information. However, this type of nonblocking socket is not used very commonly. It is just like putting a program in the busy-wait state looking for data on the socket consuming CPU time. A more elegant solution for checking to see whether data are waiting to be read is to use the call `select()`.

The `select()` call makes synchronous multiplexing of sockets and file descriptors possible. That is, it can be used to determine when there are data to read or when it is possible to send more data. For instance, by using this call, with some additional calls, a server may keep listening for incoming connections as well as keep reading from the connection it already has. The `select()` call gives the power to monitor several sockets simultaneously. For instance, it may indicate which socket is ready for reading, which one is ready for writing and which one has raised exceptions. It uses several masks in the call corresponding to particular sockets. So, for example, if `writemask` is zero, sockets will not be checked for writing. Here is an example system call:

```
#include ⟨sys/time.h⟩
#include ⟨sys/types.h⟩
int select(int numfds, fd_set *readfds, fd_set *writefds,
 fd_set *exceptfds, struct timeval *timeout);
```

In this case `select()` monitors the file descriptors `readfds`, `writefds`, and `exceptfds`. If it is desired to read from standard input and a socket descriptor, `sd`, the file descriptors to 0 will have to be added, and `sd` set to `readfds`. The parameter `numfds` should be set to the values of the highest file descriptor plus one. In this example, it should be set to `sd+1`, since it is assuredly higher than standard input (0).

When `select()` returns, `readfds` are modified to indicate the file descriptors selected, and ready for reading. The file descriptors may be manipulated (mask bits set, cleared, and tested) by using the four macros defined in ⟨sys/types.h⟩. Using each set of the type fd_set, the following macros may operate:

- FD_ZERO(fd_set *set) Clears a file descriptor set (by clearing all bits in the mask)
- FD_SET(int fd, fd_set *set) Adds fd to the set (by setting bit fd in the mask)
- FD_CLR(int fd, fd_set *set) Removes fd from the set (by clearing bit fd in the mask)
- FD_ISSET(int fd, fd_set *set) Tests to see if fd is in the set (by testing bit fd in the mask)

Finally, let us see the structure timeval in the select() call. It has the following fields:

```
struct timeval {
    int tv_sec;          /* seconds */
    int tv_usec;         /* microseconds */
};
```

This structure is used to indicate the time value to wait in number of seconds (tv_sec) or microseconds (tv_usec). For instance, if the time value is set to 75 seconds and select() has not found any ready file descriptors, it may return with a message to allow continuation of processing. If the fields in the structure timeval are set to 0, select() will time-out immediately, effectively polling all the file descriptors in the sets. If the parameter timeout is set to Null, it will never time-out, waiting as long as necessary until the first file descriptor is ready.

11.4.6 Sample Client/Server Programs

We present two program examples in this chapter. Additional programming projects are included in the laboratory manual. The first program shows the use of pipe() calls for communication between client and server programs. The main program first creates two pipes between a parent process and a child process that it forks. The first set of pipes is used by the client() to communicate to the server(), and the second set is used by the server() to communicate to the client(). The requested service from the server() is of one of the two valid types, ltime (long time) and stime (string time).

```
/* Main program for implementing time service using
   pipe() system call. Recall that pipes only work for
   related proceses. It first creates two pipes for
   exchanging information between the client and server
   processes. It spawns a child using fork() system call
   that acts as the server process. The main process
   itself acts as the client process. The extra file
```

```
   descriptors that are not needed any more are closed by
   each process to conserve them.
 */

#include <stdio.h>

main()
{
    int childpid;
    int pipe1[2], pipe2[2];

    if (pipe(pipe1) < 0 || pipe(pipe2) < 0)
     serr("main: can't create pipe");

    if ((childpid = fork()) < 0)
     serr("main: can't fork");

    else if (childpid > 0) {        /* in parent */

          if (close(pipe1[0]) < 0 || close(pipe2[1]) < 0)
               serr("main: pipe fd close error");

          client(pipe2[0], pipe1[1]);

          while(wait((int *) 0) != childpid)
               ;

          if (close(pipe1[1]) < 0 || close(pipe2[0]) < 0)
               serr("main: pipe fd close error-1");
          exit(0);
      }
      else {         /* in child */

          if (close(pipe1[1]) < 0 || close(pipe2[0]) < 0)
               serr("main: pipe fd close error-2");

          server(pipe1[0], pipe2[1]);

          if (close(pipe1[0]) < 0 || close(pipe2[1]) < 0)
               serr("main: pipe fd close error-3");
          exit(0);
          }
}

/* client(): This is the code for client process. It reads
   a service name from stdin using library function
   fgets() and sends it to the server by writing it
```

```
    to 'writefd' using write() system call. it then
    waits for the reply from server by reading from 'readfd'
    using read() system call. It also writes the information
    received from the server to stdout.
*/

#include <stdio.h>
#define MAXBUF 256

client(readfd, writefd)
int readfd, writefd;
{
   char buf[MAXBUF];
   int n;

   if (fputs("service? ", stdout) == NULL)
     serr("client: stdin write error");

   if (fgets(buf, MAXBUF, stdin) == NULL)
     serr("client: service read error.");

   n = strlen(buf);

   if ('\n' == buf[n-1]) n--;

   if (write(writefd, buf, n) != n)
     serr("client: service write error");

   if (( n = read(readfd, buf, MAXBUF)) > 0)
     if(write(1, buf, n) != n)
          serr("client: data write error");

   if (n < 0)  serr("client: data read error");
}

/* server(): This code is executed by the server process.
   It reads the service request from 'readfd'.
   It checks the nature of the service request,
   as only two services are provided 'ltime'
   and 'stime', for all others the server reports
   an error. It uses read() and write() system
   calls to read from readfd and write to writefd.
   It also uses sprintf() library function to place
   error information into a string to be sent to the
   client.
*/

#include <stdio.h>
```

```
#define MAXBUF 256

server(readfd, writefd)
int readfd, writefd;
{
   char buf[MAXBUF];
   int n;

   if ((n = read(readfd, buf, MAXBUF)) <= 0)
     serr("server: service read error");

   buf[n] = '\0';

   if (0 == strcmp(buf, "ltime"))
     longtime(buf);
   else if (0 == strcmp(buf, "stime"))
       stringtime(buf);
   else
       sprintf(buf, "%s", "server: unknown service\n");
   n = strlen(buf);

   if (write(writefd, buf, n) != n)
        serr("server: data write error");
}

/* longtime(): It invokes the library function time() that
   returns the time elapsed since January 1, 1970
   in seconds, writes it to the string 'buff' using
   sprintf() library function.
*/

longtime(buff)
char buff[];
{

   long lrslt, time();

   lrslt = time((long *) 0);
   sprintf(buff, "%ld\n", lrslt);
}

/* stringtime(): It invokes the library function time()
   that returns the time elapsed since January 1, 1970
   in seconds, converts it to the time 'now' as a
   string using library function ctime() and
   sprintf().
*/
```

```
stringtime(buff)
char *buff;
{

   long lrslt, time();
   char *ptr, *ctime();

   lrslt = time((long *) 0);
   ptr = ctime(&lrslt);
   sprintf(buff, "%s", ptr);
}

/* serr(): This prints the error message 'reason' and
   aborts the process with error exit 1.
*/

#include <stdio.h>

serr(reason)

char *reason;
{
   fprintf(stderr, "%s\n", reason);
   exit(1);
}
```

The following script file shows how the program works.

```
Script started on Mon Aug 02 08:03:53 2000
cps215:pipe $ clntserv
service? ltime
933595446
cps215:pipe $ clntserv
service? stime
Mon Aug 2 08:04:14 2000
cps215:pipe $ clntserv
service? Anytime
server: unknown service
cps215:pipe $ clntserv
service? a c v time
server: unknown service
cps215:pipe $ exit

script done on Mon Aug 02 08:05:03 2000
```

Our next example program is a client–server program that shows the usage of socket calls to communicate between client and server. The server uses a port number to accept a message from the client and prints its own message before printing the client's message.

```
/* Client program */

#include ⟨sys/types.h⟩
#include ⟨sys/socket.h⟩
#include ⟨netinet/in.h⟩
#include ⟨netdb.h⟩
#include ⟨stdio.h⟩

char *message[] =
{
  "This is the client message, line 1\n",
  "line 2\n",
  "⟨end⟩\n",
  0
};

main (argc, argv)
int argc;
char *argv[];
{
  char c;
  FILE *rfp, *wfp;
  char hostname[64];
  register int i,s;
  struct hostent *hp;
  struct sockaddr_in sin;
  char buffer[128];

  if (argc !=2)
  {
    fprintf(stderr, "usage: %s ⟨port number⟩\n",
    argv[0]);
    exit(1);
  }

  gethostname (hostname, sizeof(hostname));

  if ((hp = gethostbyname(hostname)) == NULL)
  {
      fprintf(stderr, "%s: unknown host.\n", hostname);
      exit (1);
  }

  if ((s = socket(AF_INET, SOCK_STREAM, 0)) < 0)
  {
    perror("client: socket()");
    exit(1);
  }
```

```
    sin.sin_family = AF_INET;
    sin.sin_port = htons(atoi(argv[1]));
    bcopy(hp->h_addr, &sin.sin_addr, hp->h_length);

    if (connect(s, &sin, sizeof(sin)) <0)
    {
      perror("client: connect()");
      exit(1);
    }

    rfp = fdopen(s, "r");
    wfp = fdopen(s, "w");
    do
    {
      fgets(buffer, sizeof(buffer), rfp);
      printf("%s", buffer);
    }
    while (buffer[0] != '.');

    for (i=0; message[i]; i++)
      fprintf(wfp, "%s", message[i]);
    fflush(wfp);

    close(s);
    exit(0);
}

/* Server program */

#include <sys/types.h>
#include <sys/socket.h>
#include <netinet/in.h>
#include <netdb.h>
#include <stdio.h>

char *message[] =

{
    "This is the first line of test message\n",
    "And the second line\n",
    "And the third line\n",
    ".\n",

    0
};

main(argc, argv)
int argc;
```

```
char *argv[];
{
  char c;
  FILE *rfp, *wfp;
  int fromlen;
  char hostname[64];
  struct hostent *hp;
  register int i, s, ns;
  struct sockaddr_in sin, fsin;
  char buffer[128];

  if (argc !=2)
  {
    fprintf(stderr, "usage: %s ⟨port number⟩\n",
    argv[0]);
    exit(1);
  }

  if ((s=socket(AF_INET, SOCK_STREAM, 0)) <0)
  {
    perror ("server: socket()");
    exit(1);
  }
  sin.sin_family = AF_INET;
  sin.sin_port = htons(atoi(argv[1]));
  sin.sin_addr.s_addr = INADDR_ANY;

  if (bind(s, &sin, sizeof(sin)) < 0 )
  {
    perror("server: bind()");
    exit(1);
  }

  if (listen(s, 5),0)
  {
    perror("server: listen()");
    exit(1);
  }

  if ((ns = accept(s, &fsin, &fromlen)) <0)
  {
    perror("server: accept()");
    exit(1);
  }

  rfp = fdopen (ns, "r");
  wfp = fdopen (ns, "w");

  for (i=0; message[i]; i++)
```

```
fprintf(wfp, "%s", message[i]);
fflush(wfp);

while (fgets(buffer, sizeof(buffer), rfp) != NULL)
printf("%s", buffer);

close(s);
exit (0);}
```

The following script file shows the working of the client–server program:

```
Script started on Mon Aug 02 07:49:17 2000
cps215:hw4 $ server2 7676 &

[1] 2695
cps215:hw4 $ client2 7676
This is the first line of test message
And the second line
And the third line
.
This is the client message, line 1
line 2
⟨end⟩
[1]+  Done                  server2 7676
cps215:hw4 $ exit

script done on Mon Aug 02 07:50:05 2000
```

11.5 WINSOCK PROGRAMMING

Winsock (or Win Sock) is short for Windows socket, and the terms are normally used interchangeably. Winsock is an open specification that allows for the independent development of network applications. Most of the current Winsock applications use version 2 of Winsock, which is a 32-bit version of original Winsock designed for 16-bit machines.

The authors of Winsock version 1.1 originally limited the scope of the API specification primarily to TCP/IP. However WinSock 2 formalizes the API for a number of other protocol suites such as ATM, IPX/SPX, and DECnet, and allows them to coexist simultaneously. These APIs, sometimes referred to as WSAs (Winsock APIs), consist of a collection of function calls, data structures, and conventions. The WSA provides standard access to the network services of an underlying protocol stack to Windows applications.

WinSock 2 application can transparently select a protocol based on its service needs. The application can use the mechanisms WinSock 2 provides to adapt to differences in network names and addresses. Winsock 2 provides protocol-independent and protocol-specific APIs with a possibility of vendor-specific API additions.

The original design goal of Winsock was to maintain exactly same procedure calls as Berkeley sockets (4.3 BSD), thus allowing easy portability of source code. Additional goals were to provide an application binary interface (ABI) for the TCP/IP suite, and finally provide a networking interface that is well suited to the Windows environment.

Windows DLL (Dynamic Link Library) provide a means to support ABI that allows the executable software written to use Winsock to work on other compatible machines. A Windows application could include BSD-like socket calls that could be dynamically linked at run time with any properly implemented Winsock DLL.

In WinSock 1.1, a single WINSOCK.DLL (or WSOCK32.DLL) provides the WinSock API, and this DLL communicates with the underlying protocol stack via a proprietary programming interface. This works fairly well since WinSock v1.1 supports only TCP/IP and most computers running Windows have only a single network interface. However, the WinSock 1.1 architecture limits a system to only one WinSock DLL active in the system path at a time. As a result, it is not easy to have more than one WinSock implementation on a machine at one time. For example, it may not be very easy to connect a protocol stack from one vendor over an Ethernet connection to another vendor's stack over a serial line.

WinSock 2 adopts the windows open systems architecture (WOSA) model, which separates the API from the protocol service provider. In this model the WinSock DLL provides the standard API, and each vendor installs its own service provider layer underneath. The API layer communicates to a service provider via a standardized service provider interface (SPI), and it is capable of multiplexing between multiple service providers simultaneously. Figure 11.5 illustrates the WinSock 2 architecture. The design shows two parts, the API for application developers and the SPI for protocol stack and namespace service providers. The intermediate DLL layers are independent of both the application developers and service providers. These DLLs are provided and maintained by Microsoft and Intel. Also, it is noticeable that the layered service providers (SPs) would appear in this illustration one or more boxes above a transport service provider (TSP).

11.5.1 Using Winsock

The only required `include` file for using Winsock is `Winsock.h`, although the standard BSD `include` files are provided for compatibility. Most of the standard calls used to establish a socket are same as BSD. However, there are some minor exceptions, noted later.

There are some new calls added for Winsock. For example, you must call `WSAStartup()` as the first call to `Winsock.dll` and `WSACleanup()` when you are finished. Also, `WSAGetLastError()` must be used to obtain the error code value and `WSASetLastError()` to set the error code value.

The Windows version of the `select()` call is `WSAAsyncSelect()`, which specifies events for which the notification is desired in the running application. `Winsock.dll` sends messages for these Windows through windows. The calls to control the blocking behavior are as follows:

`WSAIsBlocking()`: Returns true if `Winsock.dll` is blocking on a socket call.

`WSACancelBlocking()`: Cancels an outstanding blocking call.

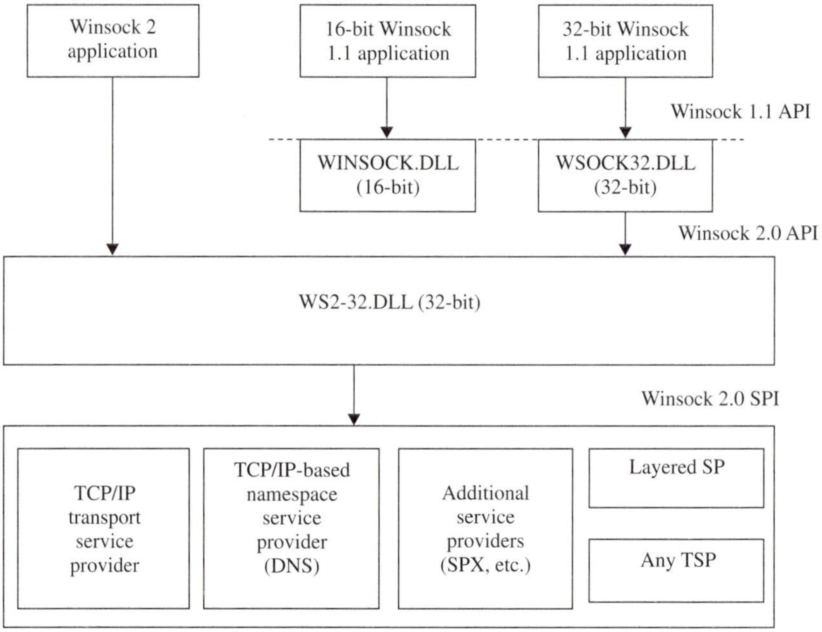

Figure 11.5 Winsock 2 architecture.

`WSASetBlockingHook()`: Specifies a user-defined function to be substituted for the default `Winsock.dll` pseudoblocking function.

`WSAUnhookBlockingHook()`: Restores the default `Winsock.dll` pseudoblocking function.

Additional calls are used for sending asynchronous messages when certain database information is available. For example, `WSAAsyncGetHostByAddr()` requests that an asynchronous message be sent when the host name, address, and alias information associated with the given host address is available.

Winsock does not support many BSD calls. For instance, the functions to open a database file that are available in BSD [`sethostent()`, `getnetent()`, `endservent()`, etc.] are not available in Winsock. These functions were not included because of pending changes to the interpretations of the network and host portions of an IP address. Similarly, the Unix I/O calls that work on sockets as well as files in Unix [`read()`, `write()`, `readv()`, `writev`, `close()`, `fcntl()`, etc.] are not available in Winsock. Also, the scatter/gather send and receive functions `readmsg()` and `writemsg()`, and such signal processing functions as `signal()`, `sigmask()`, `sigblock()`, and `sigpause()` are not available in Winsock. Since signals are specific for Unix systems, an equivalent `WSAAsyncSelect()` may be used to generate socket event notification messages.

There are two renamed functions in Winsock. They are `close()` renamed as `closesocket()` and `ioctl()` renamed as `ioctlsocket()`. This is done mainly to avoid confusion, since these functions may be defined for file operations in Unix systems.

11.5.2 Example Program

This example program, written by Warren Young, uses Winsock calls to get the local IP address. It is a small program shown to illustrate the use of some Winsock functions and can be found at http://www.cyberport.com/~tangent. There are some more Winsock programs at this web site.

This example shows how to get the local machine's IP address(es). Although the majority of machines have only a single IP address, it is becoming common for one machine to have two or more IP addresses. For example, on many machines, there may be one IP address for the Ethernet network card connected to a LAN and one for the modem when it is connected to an Internet service provider.

```
// getlocalip.cpp
// Borland C++ 5.0: bcc32.cpp getlocalip.cpp
// Visual C++ 5.0: cl getlocalip.cpp wsock32.lib

#include <iostream.h>
#include <winsock.h>

int doit(int, char**)
{
    char ac[80];
    if (gethostname(ac, sizeof(ac)) == SOCKET_ERROR) {
        cerr << "Error " << WSAGetLastError() <<
                " when getting local host name." << endl;
        return 1;
    }
    cout << "Host name is " << ac << "." << endl;

    struct hostent* phe = gethostbyname(ac);
    if (phe == 0) {
        cerr << "Yow! Bad host lookup." << endl;
        return 1;
    }
    for (int i = 0; phe->h_addr_list[i] != 0; ++i) {
        struct in_addr addr;
        memcpy(&addr, phe->h_addr_list[i],
                              sizeof(struct in_addr));
        cout << "Address " << i << ": " <<
inet_ntoa(addr)
<< endl;
    }
    return 0;
}

int main(int argc, char* argv[])
{
    WSAData wsaData;
```

```
        if (WSAStartup(MAKEWORD(1, 1), &wsaData) != 0) {
            return 255;
        }

        int retval = doit(argc, argv);

        WSACleanup();
        return retval;
    }
```

11.6 RPC PROGRAMMING

Remote Procedure Call (RPC) allows network applications to use specialized kinds of procedure call designed to hide the details of underlying networking mechanisms. It is a transport-independent mechanism that is able to take advantage of whatever kinds of networking mechanism may be available. It actually implements a logical client–server system supporting network applications that do not require the users to be aware of the underlying network.

The RPC mechanism has two major components. First, a protocol is needed to manage the messages between the client and the server, irrespective of the properties of the underlying network. Second, programming language/compiler support is necessary to support the argument passing between the client and server.

RPC basically operates by connecting functions (or procedures) at the client and server sides. With RPC, the client makes a procedure call that sends data packets to the server, as necessary. When these packets arrive, the server calls a dispatch routine, performs whatever service is requested, and sends back the reply, whereupon the procedure call returns to the client. The overall operation of the RPC mechanism is shown in Figure 11.6.

With the remote procedure call paradigm, a client makes a procedure call to send a data packet to the server. When the packet arrives, the server calls a dispatch routine, performs whatever service is requested, and sends back the reply, and the procedure call returns to the client. Until a reply is returned, the client is blocked and waits for the return reply. In this context, a server is a machine at which some network services are implemented. A service is a collection of one or more remote programs. A remote program implements one or more remote procedures; the procedures, their parameters, and results are documented in the specific program's protocol specification.

It is noted that several network clients may initiate remote procedure calls to several services. Also, to ensure forward compatibility with changing protocols, a server may support more than one version of a remote program. The remote procedure call model is similar to the local procedure call model in which the control winds through two processes: the caller's process and a server's process. That is, the caller process sends a message to the server process and waits (blocks) for a reply message. The call message includes the procedure's parameters, and the reply message includes the procedure's results. When the reply message returns, the caller extracts the results of the procedure and resumes execution.

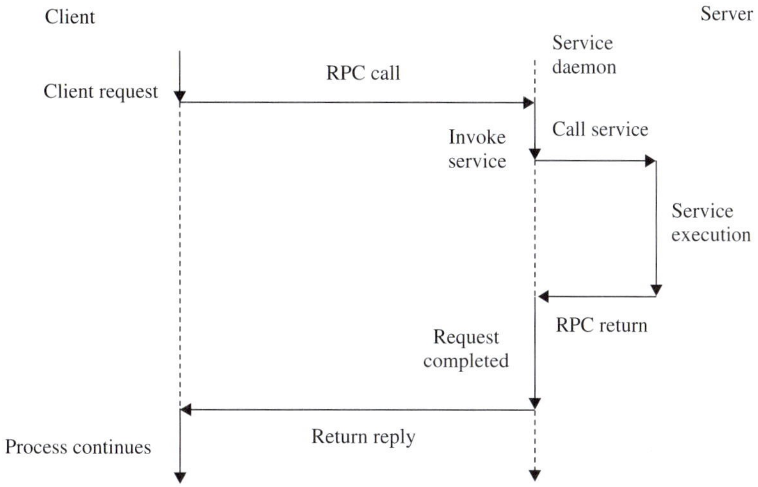

Figure 11.6 RPC operation between two machines.

On the server side, a process remains dormant and waits for the arrival of a call message. When a reply arrives, the server process extracts the procedure's parameters, computes the results, sends a reply message, and then waits for the arrival of the next call message. In this Remote Procedure Call model, only one of the two processes is active at any given time. However, the RPC protocol makes no restrictions on concurrency, and other scenarios are possible.

Programming with RPC produces programs that are designed to run within a client/server network model. Such programs use RPC mechanisms to avoid the details of interfacing to the network and provide network services to their callers without requiring that the caller be aware of the existence and function of the underlying network. For example, a program can simply call `rusers()`, a C routine that returns the number of users on a remote machine. The caller is not explicitly aware of internal use of RPC. The program that uses `rusers()` is as follows:

```
#include <stdio.h>
 /*
 * A program that calls rusers()
 */
main(argc, argv)
        int argc;
        char **argv;
{
        int num;

        if (argc != 2) {
          fprintf(stderr, "usage: %s hostname\n", argv[0]);
          exit(1);
        }
```

```
        if ((num = rusers(argv[1])) < 0) {
          fprintf(stderr, "error: rusers\n");
          exit(1);
        }
        printf("%d users on %s\n", num, argv[1]);
        exit(0);
}
```

For this program to work correctly, the `rusersd` daemon must be running on the remote host. The RPC library routines such as `rusers()` are in the RPC services library `librpcsvc.a`. Thus, the foregoing program should be compiled with

```
$ cc program.c -rpcsvc -lnsl
```

Another RPC service library routine, `rwall()`, which is used to send messages to specified remote machines, may be used in the same manner.

RPC provides a multilevel application programming interface for development of network applications by means of RPCs. At the simplified interface (the highest level), the package provides great transparency but offers only limited control over the underlying communications mechanisms. Program development at the simplified interface can be rapid, and is directly supported by the `rpcgen` compiler.

Interfaces to lower levels of the RPC package provide increasing control over Remote Procedure Call communications. Programs that exercise this control pay for the power in terms of greater complexity of code. Effective programming at the lower levels requires knowledge of computer network fundamentals. In order of increasing control and complexity, these levels are called the top level, intermediate level, expert level, and bottom level. Interested readers may further investigate the details from the provided references.

Now we look into an RPC programming interface that allows an application to use architecture-independent mechanisms to make procedure calls to remote machines. This is called the e**x**ternal **d**ata **r**epresentation (XDR) data encoding. By using XDR language, the byte order differences are resolved, hence allowing the portability of applications. XDR has a set of library routines that allow a C programmer to describe arbitrary data structures in a machine-independent fashion. RPC assumes the existence of XDR and uses it for transferring the data between diverse machines.

The XDR language is similar to the C language, but it is not a programming language and can be used only to describe data. Any program running on any machine can use XDR to create portable data by translating its local representation into the XDR representation; similarly, any program running on any machine can read portable data by translating the XDR standard representations into its local equivalents. This process of converting from a particular machine representation to XDR format is called serializing, and the reverse process is called deserializing.

The details of programming applications to use Remote Procedure Calls can be overwhelming. The writing of the XDR routines necessary to convert procedure arguments and results into their network format and vice versa may be very complex. However, `"rpcgen"` (part of Sun's RPC) helps programmers write RPC applications in many situations. It does most of the dirty work, allowing programmers to debug the main features of their application, instead of requiring them to spend most of their time debugging their network interface code. A compiler that accepts a remote program interface definition written in the RPC language, "rpcgen" and produces a C language output that includes stub versions of the client routines, a server skeleton, XDR filter routines for both parameters and results, and a header file that contains common definitions.

The client stubs interface with the RPC library and effectively hide the network from their callers. The server stub similarly hides the network from the server procedures that are to be invoked by remote clients. The output files of `rpcgen` can be compiled and linked in the usual way. The developer writes server procedures in any language that observes Sun calling conventions and links them with the server skeleton produced by `rpcgen` to get an executable server program.

To use a remote program, a programmer writes a main program that makes local procedure calls to the client stubs produced by `rpcgen`. Linking this program with the stubs created by `rpcgen` creates an executable program.

The input to the `rpcgen` command is the RPC language. The syntax structure that is most commonly used for the `rpcgen` command is where it takes an input file and generates four output files. For example, if the InputFile parameter is named `prog.x`, then the `rpcgen` command generates the following:

`prog.h`:	Header file
`prog_xdr.c`:	XDR routines
`prog_svc.c`:	Server side stubs
`prog_clnt.c`:	Client side stubs

In general, the following steps may be followed in building an RPC with `rpcgen` on Solaris systems:

Step 1: Build and test a conventional application.

Step 2: Decide which functions will be moved to the remote machine.

Step 3: Write an `rpcgen` spec (a `.x` file such as `example.x`) for the remote program.

Step 4: Run `rpcgen` (`$rpcgen example.x`). This will generate some files, including `example_svc.c`, `example_clnt.c`, and `example.h`.

Step 5: Write stub interface routines for the client and the server.

Step 6: Modify the client program as necessary.

Step 7: Modify the server program as necessary.

Step 8: Compile and link the client using something like this:

```
gcc -g -o client client.c example_clnt.c -lnsl
```

Step 9: Compile and link the server using something like this:

```
gcc -g -o server server.c example_svc.c -lnsl
```

Step 10: Start the server on the remote machine, then invoke the client on the local machine.

11.7 JAVA PROGRAMMING

The growth of the Internet over the last few years has shown new promises for Java as a network programming tool. The language was designed with networking in mind, and for this reason most network programs in other languages such as C and C++ may be written very easily in the Java. One of the biggest advantages of using Java is that it is free but fully capable in terms of supporting network application developments.

The `java.net` package in the Java platform provides a `Socket` class that implements one side of a two-way connection between two Java programs. As in other languages supporting network programming, the `Socket` class sits on top of a platform-dependent implementation, hiding the details of any particular system from the Java program. By using the `java.net.Socket` class instead of relying on native code, Java programs can communicate over the network in a platform-independent fashion.

Additionally, `java.net` includes the `ServerSocket` class, which implements a socket that servers can use to listen for and accept connections to clients. If you are trying to connect to the Web, the `URL` class and related classes (`URLConnection`, `URLEncoder`) are probably more appropriate than the socket classes. In fact, URLs are a relatively high-level connection to the web and use sockets as part of the underlying implementation.

Through the classes in `java.net`, Java programs can use TCP or UDP to communicate over the Internet. The `URL`, `URLConnection`, `Socket`, and `ServerSocket` classes use TCP to communicate over the network. The `DatagramPacket`, `DatagramSocket`, and `MulticastSocket` classes are for use with UDP.

Connecting to a server from the client side is extremely easy with Java's built-in socket classes for communicating via HTTP. In Java, a socket is created by using the URL and port number of connection, very much as in a web connection. For example,

```
Socket s = new Socket("url", PortNumber);
```

Thus, the following code would open a socket to www.anywhere.com on port 80:

```
Socket s = new Socket("www.anywhere.com", 80);
```

Again, when selecting a port number, remember that port numbers between 0 and 1023 are reserved for privileged users (i.e., superuser or root). These port numbers are reserved for standard services, such as e-mail, FTP, and HTTP. When selecting a port number for your server, select the one that is greater than 1023.

Now, what happens if the application cannot reach the specified server? Just as in C or C++, it is essential to place the networking code within a try block that catches several errors such as UnknownHostException, MalformedURLException, and IOException. The foregoing code can be rewritten as follows:

```
Socket s;
try {
            s = new Socket("url", PortNumber);
```

```
}
catch (IOException e) {
          System.out.println(e);
}
```

A `try` block with `catch` is used in Java to handle the I/O exception. For programming a server, a socket may be opened as follows:

```
ServerSocket ss;
try {
   ss = new ServerSocket (PortNumber);
     }
     catch (IOException e) {
               System.out.println(e);
     }
```

For implementing a server, a socket object from the `ServerSocket` needs to be created to listen for and accept connections from clients.

```
Socket s = null;
try {
   serviceSocket = ss.accept();
     }
catch (IOException e) {
     System.out.println(e);
}
```

The class `DataInputStream` allows reading of lines of text and Java primitive data types in a portable way. It has methods such as `read`, `readChar`, `readInt`, `readDouble`, and `readLine`. The `DataInputStream` class may be used on the client side to create an input stream as follows:

```
DataInputStream input;
try {
     input = new DataInputStream(s.getInputStream());
}
catch (IOException e) {
     System.out.println(e);
}
```

`DataInputStream` may be used at the server side to receive input from the client:

```
DataInputStream input;
try {
     input = new DataInputStream(ss.getInputStream());
}
catch (IOException e) {
     System.out.println(e);
}
```

An output stream may be created on the client side to send information to the server socket by using the class `PrintStream` or `DataOutputStream` of `java.io`:

```
PrintStream output;
try {
      output = new PrintStream(s.getOutputStream());
}
catch (IOException e) {
      System.out.println(e);
}
```

The `PrintStream` class has methods for displaying textual representation of Java primitive data types. Its `Write` and `println` methods may be used. Also, the `DataOutputStream` method may be used:

```
DataOutputStream output;
try {
   output = new DataOutputStream(s.getOutputStream());
}
catch (IOException e) {
      System.out.println(e);
}
```

The `DataOutputStream` class allows writing of Java primitive data types; many of its methods write a single Java primitive type to the output stream. The method `writeBytes` may be used. On the server side, you can use `PrintStream` to send information to the client.

```
PrintStream output;
try {
      output = new PrintStream(ss.getOutputStream());
}
catch (IOException e) {
      System.out.println(e);
}
```

Always close the output and input streams before closing the socket. This is done at the client side by:

```
try {
                  output.close();
                  input.close();
      s.close();
}
catch (IOException e) {
      System.out.println(e);
}
```

```
D:\My Documents\Book\ch11stuff>java Client dads-computer
2099
Datagram Client sending on port:2099

D:\My Documents\Book\ch11stuff>
```

Figure 11.7 Client program run window.

On the server side it is:

```
try {
     output.close();
     input.close();
     serviceSocket.close();
     ss.close();
}
catch (IOException e) {
     System.out.println(e);
}
```

The following client–server Java program illustrates the use of the different classes we have discussed through a UDP connection. In this example, a server program accepts 10 numbers from a client program. The server listens on a particular port on the local host. The port, which will change each time the program is run, basically checks for the first available port and listens on that port number. Along with the program listings, Figures 11.7 (client program) and 11.8 (server program) show an example run of the program. Notice that the client and server programs are run in two different MS-DOS windows. Server is run first and then the client is run, to transfer the numbers to the server.

```
// Client.java
import java.io.*;          // Import some needed classes
import java.net.*;
```

```
D:\My Documents\Book\ch11stuff>
D:\My Documents\Book\ch11stuff>
D:\My Documents\Book\ch11stuff>java Server
Datagram Server listening on port:2099
Received from inet address: pm246-
25.dialip.mich.net/35.9.9.90 port number: 2102
0
1
2
3
4
5
6
7
8
9

D:\My Documents\Book\ch11stuff>
```

Figure 11.8 Server program run window.

```
import java.util.*;
import sun.net.*;

class Client {
    public static void main(String argv[]) {
     if (argv.length!=2) {
      System.out.println("java Client ⟨hostname⟩ ⟨port#⟩");
       return;
    }

    try {
      int port = Integer.parseInt(argv[1]);
      InetAddress in[] = InetAddress.getAllByName(argv[0]);
      DatagramSocket s = new DatagramSocket();
      byte buf[] = new byte[10];
      for (int i=0; i⟨buf.length; ++i)
            buf[i] = (byte)i;
      DatagramPacket dp = new
             DatagramPacket(buf,10,in[0],port);
      System.out.print("Datagram Client sending on port:");
      System.out.println(dp.getPort());
      s.send(dp);
      s.close();
    } catch (Exception e) {
    e.printStackTrace();
    }
   }
  }

//Server.java
import java.io.*;
import java.net.*;
import java.util.*;
import sun.net.*;

 // To run server from DOS Prompt as eg. X:\>java Server

class Server {
    public static void main(String argv[]) {
    try {
       DatagramSocket s = new DatagramSocket();
       System.out.print("Datagram Server listening on
              port:");
       System.out.println(s.getLocalPort());
       byte buf[] = new byte[10];
       DatagramPacket dp = new DatagramPacket(buf, 10);
       s.receive(dp);
       System.out.print("Received from inet address: "
              + dp.getAddress().toString());
```

```
            System.out.print(" port number: ");
            System.out.print(dp.getPort());
            for (int i=0; i<buf.length; ++i)
            System.out.println(buf[i]);
            s.close();
        } catch (Exception e) {
        e.printStackTrace();
        }
    }
}
```

11.8 CHAPTER SUMMARY

This chapter deals with the art and science of network programming. First we discussed the popular business models used to support the network programming of today and tomorrow, including the two- and three-tier architectures, which provide a setup for systems development without and with a reliable Internet integration, respectively. A three-tier system was shown to be a more reliable business model, with Internet/intranet developments being possible. In terms of network programming, a heavy burden is on programmers to develop client–server systems to support a vast array of applications involving conventional LANs with multimedia, TCP/IP, ATM, frame relays, mobile systems, and many other evolving technologies.

In this chapter we made an attempt to illustrate different programming techniques. We started with physical-level network programming, using the serial and parallel ports on computers. This approach, being inexpensive, provides an ideal experimental setup. Also, using this approach facilitates the setup of a communications network in a localized environment, since none of the TCP/IP details are involved. However, security may be a concern as the network grows.

NetBIOS is an answer for expanding networks to some extent inasmuch as it provides peer-to-peer connection. We have provided necessary details for a beginner to understand the concept and possibly start in this area of network programming. However, the bulk of the network programming that involves the Internet and wide-area networks is by using the sockets and TCP/IP concepts. Although we have not emphasized any particular client–server programming system that uses sockets, readers with background in C language are likely to gain most.

Sockets are supported by most of the languages that support network programming. Some examples are Visual Basic, Visual C/C++, and Java, in addition to traditional C and C++. There is a tremendous amount of literature available that shows how to do network programming with each language, and it is impossible to cover this subject in few subsections of a chapter. However, we have made an attempt to illustrate the basic concepts, especially using C, C++, and Java.

In addition, we introduced the use of RPC programming, which is yet another parallel approach of network programming. The use of this approach allows machines to communicate to one another on the network via procedure calls. This approach may be cumbersome, and it requires a good understanding of operating systems and systems programming, but yet some believed it to be the ultimate solution of network programming.

11.9 PROBLEMS

11.1 Describe the architecture and operations of a WWW client and server constructed by means of sockets.

11.2 In a network situation where multiple servers are supporting multiple services, how do incoming client messages get delivered to the correct server?

11.3 What type of socket connection (Unix domain or Internet domain) is used to send messages to a different machine in a network, and why? What is the disadvantage of using a stream connection with that type of socket connection on the same machine?

11.4 If `char buf[1024]` and `msg` are ASCII strings using C syntax, can we use a function call `write(socket, buf, sizeof(buf))` in either client or server? Explain.

11.5 Consider a stream socket program with `listen(socket, 3)` and long service time for each request. Does it makes sense to improve the service by forking a child process for each request? If yes, what are the consequences?

11.6 In a program, what is the consequence of setting the last parameter of `select()` call to Null with no data in and out?

11.7 What is the purpose of call `htons()`? Specifically, indicate from what type (`int/float, how many bytes?`) to what type (little/big endian?) conversion takes place.

11.8 An application on host A wants to send a string of 8192 bytes to an application on host B, and A and B are connected by TCP sockets. Host A executes the following statement:

```
n = write(sock, buffer, 8192);
```

The maximum segment size of the TCP connection has been negotiated to be 2048. However, the two machines are on an Ethernet, where the maximum frame size is 1500 bytes of data (plus header info).

(a) Describe the data that would be transmitted between each layer of the protocol stack on both machines for all packets.
(b) What difference would it make if the sender, instead of executing one write of size 8192, executed eight writes of size 1024 to send the same data?
(c) What difference would it make if the receiver, instead of executing one read of size 8192, executed eight reads of size 1024?

11.9 A server is accessed by clients over TCP connections overlaid on top of a T3 (45 Mbps) link. The T3 link has intermittent clock problems, as a result of which every second it loses a millisecond's worth of T3 frames. Consequently, all packets transmitted over the link during this time are lost. The average round-trip delay of TCP connections over this link is 0.25 second. What is the impact of the T3 problem on the performance of the TCP connections and the client–server interactions performance?

11.10 Write a client–server program using TCP socket connection in C or C++ in which client requests the server to return the sum of two integers entered as input. Use a port number to establish a connection between the client and server. First, run the server in the background mode to wait for a connection. Next, run the client, using the same port number as server to make a connection. The client must also accept two integers to be added by the server. A sample input/output is as follows:

```
$ Server 5555 &
[1] 1971
$ client 5555
Host Contacted
CLIENT: Please enter two integers, 0 to end: 25 5
SERVER: Request is 25 + 5
CLIENT: Result from server is 30
CLIENT: Please enter two integers: 0
[1]+ Exit 1          server 5555
$
```

11.11 Write a program in C or C++ that measures the throughputs and latencies of TCP and UDP connections by using a socket interface between a client and a server running on two Unix workstations. Perform the following measurements and plot/tabulate the results.

(a) The round-trip latency for TCP and UDP for message sizes of 1, 100, 200, 500, and 1000 bytes.
(b) The throughput of TCP and UDP for message sizes of 1, 2, 4, 10, 20, and 35 kilobytes. Plot the measured throughput as a function of message size.
(c) The throughput as a function of time by sending a 500-kilobyte file in a loop that sends each time a 1000 byte message using a TCP connection. Repeat the measurements for 1 and 2 MB files and message sizes of 2000 and 3000 bytes.

11.12 Explain what is being done in the statement

```
int port = Integer.parseInt(argv[1]);
```

11.13 Explain what is being done in the statement

```
DatagramPacket dp = new DatagramPacket(buf, 10);
```

11.14 If `s` and `dp` are the socket descriptor and datagram packet variables, which of the following would the code that sends data to a socket in Java typically be?

```
send.dp.()
send(s).and.stop
(dp).send()
s.send(dp)
```

11.15 Name the Java network class used by the client for binding a socket.

11.16 What are the differences between RPC and traditional procedure calls?

11.10 REFERENCES

BOOKS

AIX Version 4.3 Communications Programming Concepts. Armonk, NY: IBM, 1997.

Axelson, J., *Serial Port Complete Programming and Circuits for RS-232 and RS-485 Links and Networks*. Madison, WI: Lakeview Research, 1998.

Axelson, J., *Parallel Port Complete: Programming, Interfacing, and Using the PC's Parallel Printer Port*. Madison, WI: Lakeview Research, 1997.

Baker, R. H., *Networking the Enterprise: How to Build Client/Server Systems That Work*. New York: McGraw-Hill, 1994.

Berson, A., *Client–Server Architecture*. New York: McGraw-Hill, 1992.

Black, U. D., *TCP/IP and Related Protocols*. New York: McGraw-Hill, 1992.

Boar, B. H., *Implementing Client/Server Computing: A Strategic Perspective*. New York: McGraw-Hill.

Bonner, P., *Network Programming with Windows Sockets*. Englewood Cliffs, NJ: Prentice Hall, 1996.

Brain, M., *Win32 System Services*. Englewood Cliffs, NJ: Prentice Hall, 1995.

Campione, M., and K. Walrath, *The Java Tutorial: Object-Oriented Programming for the Internet*, 2nd ed. Reading, MA: Addison-Wesley, 1998.

Comer, D. E., *Internetworking with TCP/IP*, vol. I: *Principles, Protocols, and Architecture*. 3rd ed. Englewood Cliffs, NJ: Prentic Hall, 1995.

Comer, D. E., and D. L. Stevens, *Internetworking with TCP/IP*, vol. II: *Design, Implementation, and Internals*, 2nd ed. Englewood Cliffs, NJ: Prentice Hall, 1994.

Feit, S., *TCP/IP: Architecture, Protocols, and Implementation*. New York: McGraw-Hill, 1993.

Forouzan, B. A., *TCP/IP Protocol Suite*. New York: McGraw-Hill, 2000.

Grier, R., Z. Thomas, J. Shields, and Phounsavan, eds. *Visual Basic Programmer's Guide to Serial Communications*. Stanwood, WA: Mabry Publishing, 1997.

Grodzinsky, F. S., ed., *Networking and Data Communications Laboratory Manual*. Englewood Cliffs, NJ: Prentice Hall, 1999.

Harold, E. R., *Java Network Programming*. Sebastopol, CA: O'Reilly & Associates, 1997.

Hunt, C., *TCP/IP Network Administration*. Sebastopol, CA: O'Reilly & Associates, XXXX.

Inmon, W. H., *Developing client/server applications*. Wellesley, MA: QED Publishing Group, 1993.

Khanna, R., ed., *Distributed Computing: Implementation and Management Strategies*. Englewood Cliffs, NJ: Prentice Hall, 1994.

Leffler, S. M., M. K. McKusick, and J. Quartermen, *The Design and Implementation of the 4.3BSD UNIX Operating System*. Reading, MA: Addison-Wesley, 1989.

Linthicum, D., *David Linthicum's Guide to Client/Server and Intranet Development*. New York: Wiley, 1997.

Mahmoud, Q. H., *Distributed Programming with Java*. Greenwich, CT: Manning Publications, 1999.

Quinn, B., and D. Shute, *Windows Sockets Network Programming*. Reading, MA: Addison-Wesley, 1996.

Renaud, P. E., *Introduction to Client/Server Systems*. 2nd ed. New York: Wiley, 1996.

Robert, D. H., *Client/Server Programming with OS/2 2.0*. New York: Van Nostrand Reinhold, 1992.

Smith, P., *Client/Server Computing*. Indianapolis: SAMS Publishing, 1992.

Stallings, W., *Data and Computer Communications*, 5th ed. Englewood Cliffs, NJ: Prentice Hall, 1997.

Stevens, W. R., *UNIX Network Programming*. Englewood Cliffs, NJ: Prentice Hall, 1990.

Stevens, W. R., *Advanced Programming in the UNIX Environment*. Reading, MA: Addison-Wesley, 1992.

Stevens, W. R., *TCP/IP Illustrated*, vol. 1: *The Protocols*. Reading, MA: Addison-Wesley, 1994.

Stevens, W. R., *TCP/IP Illustrated*, vol. 3: *TCP for Transactions, HTTP, NNTP, and the UNIX Domain Protocols*. Reading, MA: Addison-Wesley, 1996.

Stevens, W. R., *UNIX Network Programming*, vol. 1; *Networking APIs: Sockets and XTI*, 2nd ed. Englewood Cliffs, NJ: Prentice Hall, 1998.

Stevens, W. R., *UNIX Network Programming*, vol. 2; *Interprocess Communications*, 2nd ed. Englewood Cliffs, NJ: Englewood Cliffs, NJ: Prentice Hall, 1999.

Travis, D. T., *Client/Server Computing*. New York: McGraw-Hill, 1993.

Wright, G. R., and Stevens, W. R., *TCP/IP Illustrated*, vol. 2: *The Implementation*. Reading, MA: Addison-Wesley, 1995.

ARTICLES

Adler, R. M., "Distributed Coordination Models for Client/Sever Computing," *Computer* 28, no. 4 (April 1995), 4–22.

Allman, M., V. Paxson, and W. R. Stevens, TCP Congestion Control. RFC 2581, 1999.

Dickman, A., "Two-Tier Versus Three-Tier Applications," *Informationweek* 553 (November 13, 1995), 74–80.

Edelstein, H., "Unraveling Client/Server Architecture," *DBMS* 34, no. 7 (May 1994).

Gallaugher, J., and S. Ramanathan, "Choosing a Client/Server Architecture. A Comparison of Two-Tier and Three-Tier Systems," *Information Systems Management Magazine* 13, no. 2 (spring 1996), 7–13.

Gilligan, R. E., S. Thomson, J. Bound, and W. R. Stevens, Basic Socket Interface Extensions for IPv6. RFC 2553, 1999.

Stevens, W. R., and J.-S. Pendry, "Portals in 4.4BSD," *Proceedings of the 1995 Winter USENIX Technical Conference*, New Orleans, 1995, pp. 1–10.

Stevens, W. R., and M. Thomas, Advanced Sockets API for IPv6. RFC 2292, 1999.

WORLD WIDE WEB SITES

NetBIOS
http://members.tripod.com/~Gavin_Winston/NETBIOS.HTM

NetBIOS
http://www.zdwebopedia.com/Networks/NetBIOS.html

NetBIOS
http://webopedia.internet.com/TERM/N/NetBIOS.html

BSD Sockets: A Quick And Dirty Primer. It also has other great Unix system programming information
http://www.cs.umn.edu/~bentlema/unix/

Sockets
http://www.sangoma.com/fguide.htm

Sockets
http://www.ecst.csuchico.edu/~beej/guide/net

Winsock
http://www.stardust.com

Winsock
http://sunsite.unc.edu/winsock
Winsock
http://www.sockets.com/mswsock.htm
The Unix Socket FAQ
http://www.ibrado.com/sock-faq/
Client–Server computing
http://pandonia.canberra.edu.au/ClientServer/socket.html
Schussel, G., "Client/Server Past, Present, and Future," [1995].
http://www.dciexpo.com/geos/
Serial communication
http://www.geocities.com/SiliconValley/Bay/8302/serial.htm
Serial Communication
http://www.ctv.es/pckits/tutorial.html

RFCs

RFCs may be obtained via ftp from one of the following primary repositories:

ftp://ftp.nis.nsf.net
ftp://ftp.nisc.jvnc.net
ftp://ftp.isi.edu
ftp://ftp.wuarchive.wustl.edu
ftp://ftp.src.doc.ic.ac.uk
ftp://ftp.ncren.net
ftp://ftp.sesqui.net
ftp://ftp.nic.it
ftp://ftp.imag.fr
http://www.normos.org

These are also available at ftp://ftp.cis.ohio-state.edu/pub/rfc

RFC-768, The User Datagram Protocol (UDP).
RFC-791, The Internet Protocol (IP).
RFC-793, The Transmission Control Protocol (TCP).
RFC-854, The Telnet Protocol.
RFC-951, The Bootstrap Protocol (BOOTP).
RFC-1350, The Trivial File Transfer Protocol (TFTP).
RFC-1790, An Agreement Between the Internet Society and Sun Microsystems, Inc., in the Matter of ONC RPC and XDR Protocols.
RFC-1831, RPC: Remote Procedure Call Protocol Specification Version 2.
RFC-1832, XDR: External Data Representation Standard.
RFC-1833, Binding Protocols for ONC RPC Version 2.
RFC-2292, Advanced Sockets API for IPv6.
RFC-2553, Basic Socket Interface Extensions for IPv6.
RFC-2581, TCP Congestion Control.

INDEX